Deep Learning and Signal Processing

Deep Learning and Signal Processing

Deep Learning and Signal Processing

Yecai Guo
Anhui Institute of Information Technology, China

Lixiang Ma
Anhui Institute of Information Technology, China

World Scientific

NEW JERSEY • LONDON • SINGAPORE • BEIJING • SHANGHAI • HONG KONG • TAIPEI • CHENNAI • TOKYO

Published by

World Scientific Publishing Co. Pte. Ltd.
5 Toh Tuck Link, Singapore 596224
USA office: 27 Warren Street, Suite 401-402, Hackensack, NJ 07601
UK office: 57 Shelton Street, Covent Garden, London WC2H 9HE

Library of Congress Cataloging-in-Publication Data
Names: Guo, Yecai author | Ma, Lixiang author
Title: Deep learning and signal processing /
Yecai Guo, Anhui Institute of Information Technology, China,
Lixiang Ma, Anhui Institute of Information Technology, China.
Description: New Jersey : World Scientific, [2026] | Includes bibliographical references and index.
Identifiers: LCCN 2025042208 | ISBN 9789819824175 hardcover |
ISBN 9789819824182 ebook for institutions | ISBN 9789819824199 ebook for individuals
Subjects: LCSH: Signal processing--Data processing | Deep learning (Machine learning)
Classification: LCC TK5102.9 .G859 2026
LC record available at https://lccn.loc.gov/2025042208

British Library Cataloguing-in-Publication Data
A catalogue record for this book is available from the British Library.

For any available supplementary material, please visit
https://www.worldscientific.com/worldscibooks/10.1142/14617#t=suppl

Desk Editors: Murali Appadurai/Veronica Lee

Typeset by Stallion Press
Email: enquiries@stallionpress.com

Preface

With the rapid development of deep learning technology, the latest deep learning models have far surpassed traditional machine learning algorithms and achieved remarkable success in fields such as image processing, biometric recognition, and intelligent healthcare. Therefore, more and more people are interested in deep learning, which has sparked a new wave in the field of artificial intelligence. As far as the field of signal processing is concerned, the coverage scope is quite extensive, and it is indeed difficult to use deep learning principles to solve specific problems in the field of signal processing. So, each chapter of this book is organized in a "foundation priority, progressive extension, and case introduction" manner, briefly describes the algorithm background, focuses on analyzing the algorithm principles in detail, and employs complete practical cases for illustrating the entry points and processes of applying deep learning networks to solve specific problems in the field of signal processing, in order to help readers personally practice.

This book is a collection of the author and his postgraduate team's insights and research results on the application of deep learning networks in the field of signal processing over many years. At the same time, it has absorbed some of the latest achievements published by other authors in domestic and foreign journals or in doctoral and master's theses. This book consists of ten chapters. In Chapter 1, an introduction to deep learning is provided; the main contents focus on analyzing the concepts, development, and supporting systems of deep learning. Chapter 2 is about artificial neural networks, which mainly discuss the principles of neural networks, training, prediction, and optimization algorithms, computational graphs, and BackPropagation (BP) algorithms, overfitting concepts and methods to prevent overfitting, etc. This chapter serves as the foundation for subsequent chapters, and the application of neural networks in the field of blind equalization is also studied. Chapter 3 is about Convolutional Neural Networks (CNNs), which analyzes the basic structure of conventional CNNs and studies extended models such as feature fusion CNNs and deep guided filtering networks, as well as examples of remote sensing image classification and motion blur removal. In Chapter 4, deep generative adversarial networks and their extended models, as well as the application of these networks and models, are studied. The extended models and their applications in image rain removal, stripe noise removal, and motion image deblurring are emerged. Chapter 5 discusses deep restricted Boltzmann machines, including Boltzmann machines, Restricted Boltzmann Machines (RBMs), sparse RBMs, competitive sparse RBMs, and classification RBMs. Examples of handwriting recognition are analyzed. Chapter 6 is about deep belief networks, which analyzes the extended models of Deep Belief Networks (DBMs) and their applications in early diagnosis of attention deficit hyperactivity disorder based on conventional DBMs. Chapter 7 is about the deep autoencoder. Based on the AutoEncoder (AE), its extended models (such as variational AE, stacked variational AE, deep convolutional variational AE, autoencoder echo state, deep canonical correlation AE, and dual adversarial AE) and their application models (such as mutual information sparse autoencoder soft measurement, deep autoencoding network fuzzy inference, feature clustering fast sparse autoencoder, and stacked denoising sparse autoencoder extreme learning machine) are analyzed and researched. Then, feature extraction and modulation recognition algorithms are given as examples. Chapter 8 is about recurrent neural networks. Starting from the analyses of the structure, principles, and computational architecture of Recurrent Neural Networks (RNNs), it discusses the problems of gradient vanishing and exploding, optimization algorithms (such as first-order descent algorithm and least squares algorithm) and their application examples in non-invasive load identification and image compression. Chapter 9 is about deep recursive cascaded convolutional neural networks. This chapter analyzes the structural principles and application methods of deep recursive neural networks, bilinear recursive neural networks, 3D convolutional neural networks, bidirectional recursive neural networks, 3D recurrent recursive convolutional neural networks, and so on. In the example of image rain removal algorithm and single-image super-resolution, the extended recursive convolutional neural network is used to enhance the functionality and practical application of the recursive convolutional neural network, achieving good results. Chapter 10 is about Long Short-Term Memory (LSTM) networks. Based on the analysis of the relationship between LSTM networks and RNNs, the structures and principles of dual-channel convolutional LSTM networks are analyzed. Taking non-cooperative underwater acoustic communication signal debugging pattern recognition as an example, the performance of the LSTM classifier was evaluated. Taking fingerprint outdoor positioning as an example, the accuracy of outdoor positioning has been improved by constructing a hybrid LSTM mechanism.

The authors of this book, Guo Yecai and Ma Lixiang, belong to Anhui Institute of Information Technology, Wuhu, Anhui, China. The purpose of publishing this book is to demonstrate to readers the thinking mode of deep learning and its application effectiveness in the field of signal processing. Through the analysis of application examples, the purpose is to build a bridge between the principles of deep learning and the field of signal processing, stimulate readers' creativity, and expand the theory and application of deep learning.

The following points need to be clarified: First of all, this book has referenced and cited other authors' related research results. Second, the publication of this book also received strong support from our postgraduate students. Third, this book was supported by Anhui Provincial University Key Research Projects (Grant 2025AHGXZK31026) and Anhui Province Discipline/Specialty Leader Cultivation Program (Grant DTR2024061). So, we would like to express our sincere gratitude to the authors cited in this book and postgraduate students! Once again, due to the limited level and experience of the authors and Project Units, the depth and breadth of the research in this book need to be further improved, and there may be some inappropriate aspects in the book. We also kindly ask readers to criticize and correct it!

The relevant network models in deep learning, as well as the algorithm structures, principles, core ideas, and practical cases of different network models, are analyzed and studied in this book.

The main content involves the structure, principles, and methods of deep learning networks and their development models, such as artificial neural networks, convolutional neural networks, deep generative adversarial networks, deep restricted Boltzmann machines, deep belief networks, deep autoencoders, recurrent recursive neural networks, deep recursive cascaded convolutional neural networks, and long short-term memory networks. Based on applying deep learning networks in fields such as channel adaptive equalization, image processing and motion blur removal, feature extraction and recognition, and disease and fault diagnosis, readers are provided with ideas and methods for using deep learning networks to solve specific problems.

About the Authors

Guo Yecai is Professor and doctoral advisor. He graduated from Anqing Normal University in 1986, majoring in physics. In 2003, he obtained a doctor's degree in underwater acoustic engineering from Northwest Polytechnical University. He is one of the winners of One Hundred Excellent Doctoral Dissertations in China, the academic and technical leader of Anhui Province in China, as well as currently the academic leader of Anhui Institute of Information Technology. Guo Yecai has presided over or undertaken more than 20 scientific research projects, such as special funds for National Excellent Doctoral Dissertation Authors, National Natural Science Foundation, as well as national and provincial teaching and research projects. He won more than 10 provincial science and technology achievement awards and teaching achievement awards and published a Nationally Planned Textbook and 5 provincial key textbooks. More than 50 patents have been authorized by the China National Intellectual Property Administration.

Ma Lixiang holds a PhD in Engineering and is both a Senior Engineer and Young Top-notch Talents under the High-end Talent Introduction and Cultivation Action Plan Project in Anhui Province, and currently serves as the Assistant to the President of Anhui Institute of Information Technology and the Executive Dean of the School of Electrical and Electronic Engineering at our university. He graduated from Shandong University with a bachelor's degree in 2008 and from the University of Chinese Academy of Sciences with a doctor's degree in 2013. His research interests include software-defined radio, radar signal processing, and the Internet of Things. Ma Lixiang has hosted provincial-level key research and development projects, participated in multiple provincial-level teaching and research projects, published more than 10 high-level papers, and won a first prize for teaching achievement awards in Anhui Province in China.

Contents

Chapter 1

Getting Started with Deep Learning

Summary

This chapter is divided into several sections. The first section defines the concept of "deep learning" and explores its developmental trajectory. The subsequent sections delve into the methods of acquiring knowledge in the field of deep learning, strategies for accelerating the learning process, and other fundamental aspects. The final section examines the hardware support systems, mainstream software frameworks, and application platform systems that facilitate the implementation of deep learning algorithms.

1.1 How Deep is Deep Learning?

1.1.1 Deep learning concepts

The concept of deep learning (DL) was first proposed by Hinton *et al.* [1] from the University of Toronto in 2006. Deep learning is a machine learning (ML) process that is based on sample data to obtain a deep network structure containing multiple layers through certain training methods [2]. Machine learning is typically predicated on the analysis of structured data employing conventional algorithms, including decision trees, deductive logic programming, clustering, and Bayesian networks. These algorithms facilitate decision-making and prediction processes concerning events in the real world. The incorporation of deep learning into machine learning has been instrumental in aligning it more closely with its original objective: artificial intelligence (AI), which is driven by the construction of models designed to emulate the neural connectivity structure of the human brain. This process entails the provision of hierarchical feature representations of data through the integration of low-level features to generate more abstract, high-level representations, attribute classes, or features. The utilization of artificial neural network during the extraction of knowledge from data facilitates the resolution and analysis of problems, thereby enabling the discovery of hierarchical representations that extend the capabilities of standard machine learning. These hierarchical representations have the capacity to solve more complex problems with greater precision and to potentially solve other problems with higher accuracy, requiring fewer observations and easier manual tuning. In traditional neural networks, the network weights are initialized randomly, which can result in the network converging to a local minimum. To address this challenge, Hinton pioneered the concept of deep learning by proposing the utilization of unsupervised pretraining methodologies to optimize the network's initial weights, subsequently followed by fine-tuning of these weights. The deep network structure obtained by deep learning contains a substantial number of individual elements (neurons). Each neuron is linked to numerous other neurons, and the strength of the connections (weights) between the neurons modifies and determines the function of the network during the learning process. The deep network structure obtained through deep learning conforms to the characteristics of neural networks, thus demonstrating that deep networks are indeed deep neural networks (DNNs). Deep neural networks consist of multiple single-layer nonlinear networks that are superimposed on each other [3,4]. Common single-layer networks are categorized into three types according to encoding and decoding: encoder part only, decoder part only, and both encoder and decoder parts. The encoder provides a bottom-up mapping from the input to the hidden feature space, and the decoder maps the hidden features to the input space with the goal of reconstructing the result as close as possible to the original input [5]. As illustrated in Figure 1.1, deep neural networks are classified into the following three categories:

(1) Feedforward deep networks (FFDNs): They consist of multiple encoding layers that are stacked on top of each other. Examples of

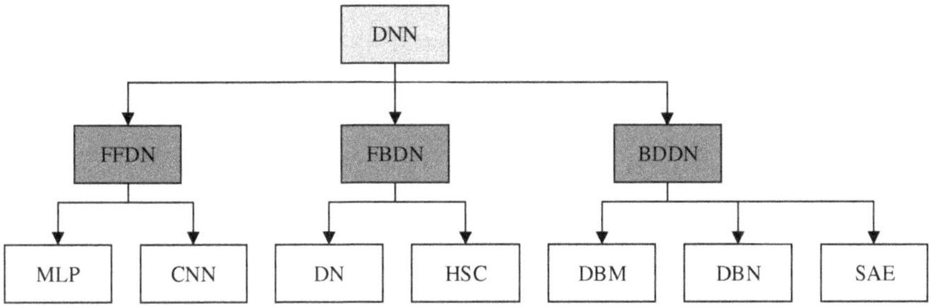

Figure 1.1. Deep neural network classification.

encoder layers include Multi-Layer Perceptrons (MLPs) [6,7], Convolutional Neural Networks (CNNs) [8,9], and other layers. According to the CNN, the sequence number is nine through ten.

(2) FeedBack Deep Networks (FBDNs): It is comprised of multiple decoder layers that are superimposed on each other. Examples of such networks include Deconvolutional Networks (DNs) [10] and hierarchical sparse coding (HSC) networks [11].

(3) Bi-Directional Deep Networks (BDDNs): It is composed of multiple encoder and decoder layers. Each layer may be a separate encoding or decoding process, or it may contain both encoding and decoding processes. Examples of such layers include those found in Deep Boltzmann Machines (DBMs) [10–13], Hierarchical Sparse Coding (HSC) [11], Deep Belief Networks (DBNs) [14], Stacked Auto-Encoders (SAEs) [15,16], and so on.

1.1.2 Deep learning development

1.1.2.1 *Evolution of deep learning development*

The advent of deep learning can be traced back to the evolution of neural networks, which reached a certain level of sophistication over time. The earliest neural network model can be traced back to the McCulloch–Pitts computational structure, which was proposed by McCulloch *et al.* in 1943 and is referred to as the MP model [17]. This model is capable of simulating the workings of human neurons. However, it requires the inconvenience of manually setting up the weights. In 1958, Rosenblatt proposed the perceptual machine model (perceptron) [18]. In comparison with the MP model, the perceptron model has the capacity to establish weights in a more automated and rational manner. However, in 1969, Minsky and Paper demonstrated that the perceptron model is capable of addressing only linearly differentiable problems. They further asserted the impracticality of multilayer neural network training and advanced the perspective that "research based on the perceptron will eventually lead to failure." In 1986, Geoffrey Hinton and David Rumelhart proposed the Backpropagation (BP) algorithm, which solved the problem of complex computation of two-layer neural networks and greatly reduced the amount of computation originally expected. The projected amount of computation was significantly reduced, thereby effectively refuting the viewpoint of Minsky *et al.* and marking the second climax of neural network research. The advent of massively parallel computers with shared memory in the late 1980s and early 1990s led to significant advancements in computational processing power, thereby facilitating the accelerated development of deep learning. In 1989, Yann LeCun and others proposed a Convolutional Neural Network (CNN), which is a deep neural network model containing convolutional layers. This was an earlier attempt at deep learning for image processing. In 2012, Hinton constructed a deep neural network and applied it to ImageNet, making qualitative improvements and breakthroughs [12]. In the same year, people became familiar with the Google Brain team. In 2013, the European Commission launched a supercomputer project that mimics the human brain, and Baidu announced the establishment of a deep learning organization. In 2014, the deep learning model known as "Top-5" achieved a first-place ranking in the ImageNet 2014 computer recognition competition. Concurrently, Tencent and Jingdong also established their own deep learning laboratories. From 2015 to the beginning of 2017, Google's AlphaGo, developed by the artificial intelligence team at DeepMind, a subsidiary of Google, achieved successive victories over professional human Go players. This "Go" garnered global attention and prompted human Go masters to engage in profound reflection. This phenomenon signifies a substantial indication of the development of deep learning as a methodical and expanding influence. The evolution of deep learning is illustrated in Figure 1.2.

The relationship between machine learning and deep learning can be conceptualized as a dynamic interplay of inclusion and being included, as illustrated in Figure 1.3.

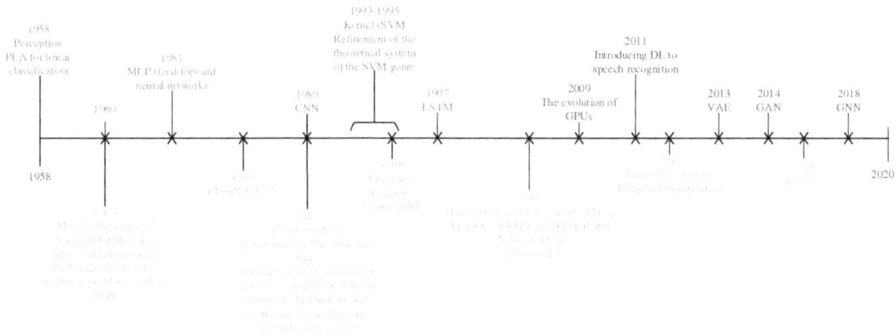

Figure 1.2. Deep learning development mainline.

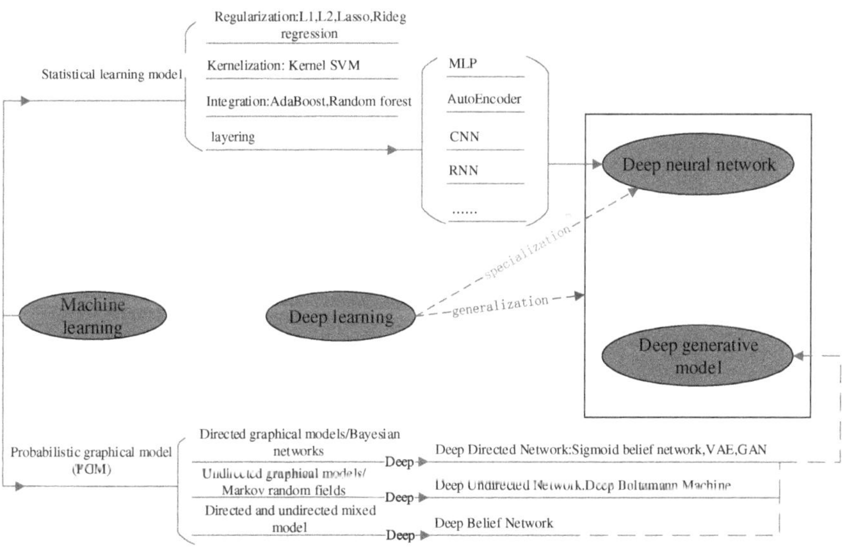

Figure 1.3. Relationship between machine learning and deep learning.

1.1.2.2 *Limitations and bottlenecks in deep learning*

Deep Neural Networks (DNNs) represent a sophisticated framework that has the potential to address a wide range of business challenges. Presently, deep learning exhibits certain limitations [19].

First, it should be noted that deep learning techniques are heuristic in nature. The capacity of deep learning to solve a given problem remains uncertain due to the absence of a mathematical theory that can demonstrate the existence of a "good enough" deep learning solution. The technique is heuristic, and its working is self-evident.

Second, the employment of deep learning techniques is characterized by an element of unpredictability. Deep learning is characterized by the presence of multiple hidden layers, which are often referred to as "black box models." These models have the potential to compromise compliance and present challenges to white box models.

Third, the systematization of deep learning is in its nascent stages. Presently, there is an absence of a universal deep learning network that caters to the diverse requirements of all industries or enterprises. Instead, each industry or enterprise must selectively integrate the available tools to develop its own customized solution, ensuring compatibility with subsequent software iterations.

Fourth, erroneous results can lead to unfavorable consequences. Presently, deep learning systems lack the capacity to attain 100% accuracy in problem-solving applications. The most salient risks and pitfalls associated with deep learning are analogous to those encountered in shallow machine learning.

The fifth issue pertains to the suboptimal learning rate of deep learning algorithms. The implementation of deep learning systems necessitates extensive training to ensure optimal performance.

Despite the successful implementation of deep learning in some domains such as image recognition and speech recognition, along with other fields, the emergence of prominent artificial intelligence enterprises, including YiTu, ShangTang, and Cambrian, has exposed the limitations of deep learning in various domains. These limitations persist in both industry and academia, hindering advancements in the field.

Initially, the data bottleneck must be addressed. The training of deep neural networks necessitates substantial datasets as training samples; the acquisition of sufficient labeled data is a prerequisite for the initiation of deep learning processes. Despite the recent initiatives by prominent entities such as Google to develop artificial data technology, the full scope of its impact remains to be thoroughly assessed.

Second, the cognitive bottleneck must be considered. This is determined by the characteristics of deep learning. Deep learning has been demonstrated to facilitate the execution of perceptual tasks with a higher degree of efficacy. Conversely, its capacity to support cognitive tasks is comparatively diminished. It has been observed that deep learning systems are not inherently equipped to develop the ability to understand, intuition, epiphany, and self-awareness. It has been hypothesized by scientists that the phenomenon under investigation may be attributable to the inherent limitations of human intellectual awareness. This hypothesis suggests that deep learning has, to a certain extent, surpassed human cognitive abilities and cognitive scopes.

1.2 How is Deep Learning Learned?

1.2.1 General approach to machine learning

Machine learning can be categorized into two primary groups, namely supervised learning and unsupervised learning, based on their respective methodologies. Specifically, supervised learning encompasses problems such as classification and regression, while unsupervised learning comprises problems such as clustering and correlation analysis. Deep learning is classified as a form of supervised learning. The general approach to supervised learning is illustrated in Figure 1.4 [20].

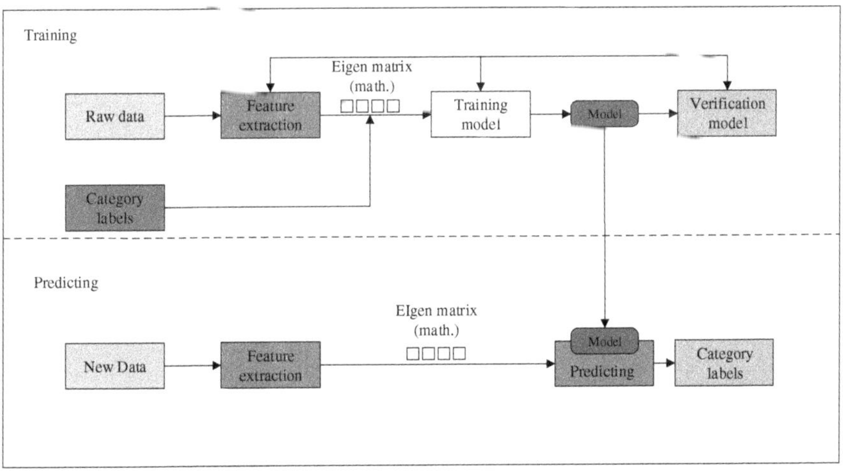

Figure 1.4. General approach to supervised learning.

1.2.2 General approach to deep learning

The advent of deep learning has led to the development of advanced algorithms that have significantly surpassed the predictive and classification capabilities of conventional machine learning algorithms for data. In contrast to conventional data extraction methods, deep learning does not require extracting features on its own, but automatically filters the data and extracts high-dimensional features. In the training phase of the model, a substantial number of samples are utilized, both labeled and unlabeled, for the purpose of training and acquiring knowledge, thereby constructing a deep neural network model. In the prediction phase, a more abstract feature representation of the data is extracted based on the learned model. As illustrated in Figure 1.5, the general approach to deep learning does not the feature extraction and a consequent reduction in the time required for engineers to complete their work, when compared with the general approach to supervised learning in traditional machine learning (Figure 1.4).

In these two phases, a substantial number of computational operations are necessary, and as the number of layers in the deep neural network model increases, the corresponding weight parameters grow by geometric multiples, thereby placing an increasingly high demand on the computational power of the hardware.

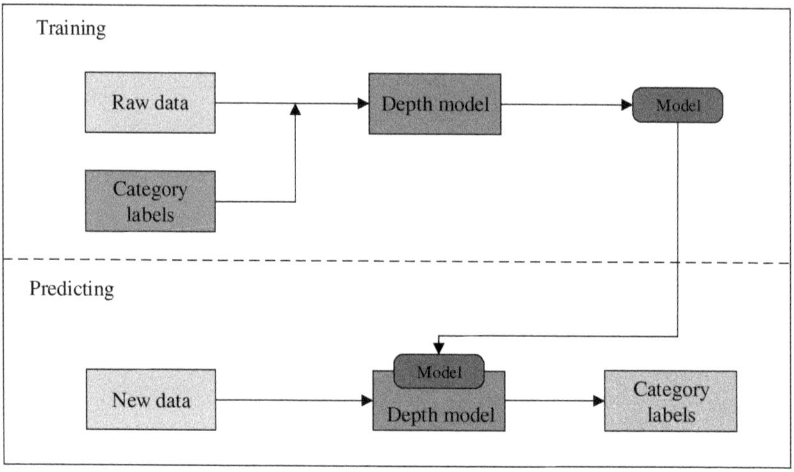

Figure 1.5. General approach to deep learning.

1.3 How Does Deep Learning Speed Up?

In order to expedite the process of deep learning, it is imperative to comprehensively understand the application system architecture of deep learning. This architecture, delineated in Figure 1.6, plays a pivotal role in facilitating the efficient execution of deep learning algorithms.

As illustrated in Figure 1.6, the application system is comprised of four distinct components: the hardware support system, the deep learning framework, the supporting technology system, and the application platform system. The application system is initiated with a substantial dataset, and a deep learning model is selected. Each model possesses certain internal parameters that must undergo adjustment to facilitate the model's learning of the data.

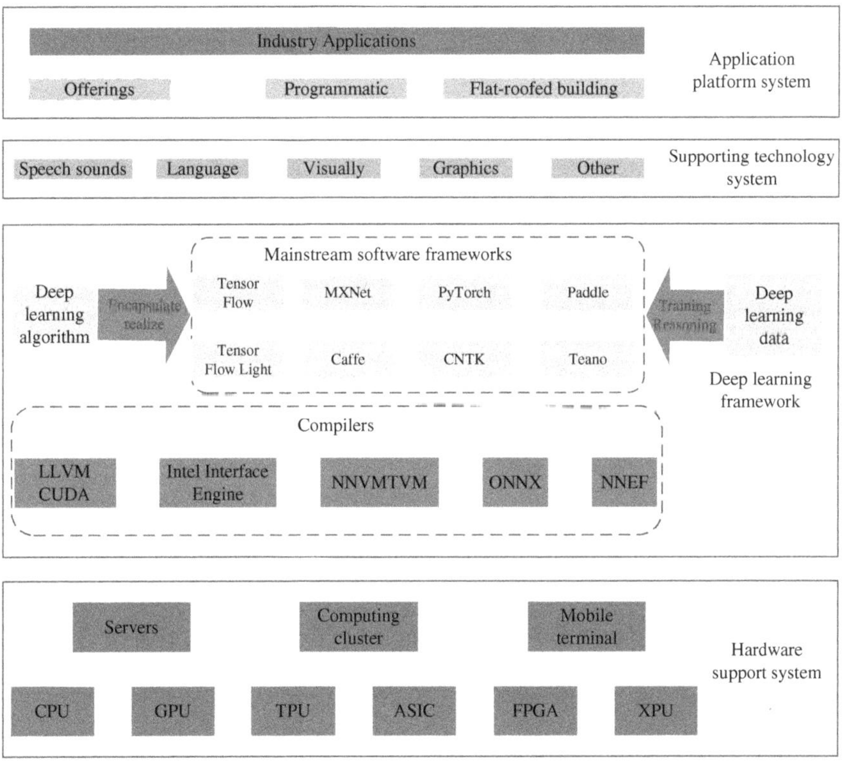

Figure 1.6. Application system architecture of deep learning.

This parameter tuning can be conceptualized as an optimization problem, wherein the tuning of these parameters is tantamount to optimizing specific constraint conditions. Baidu's Silicon Valley Artificial Intelligence Laboratory (SVAIL) has developed the Deep Bench benchmark for deep learning hardware, which focuses on measuring the hardware performance of the underlying computation rather than the performance of the learning model. This method aims to identify bottlenecks that slow down or make calculations inefficient. Therefore, the focus is on designing an architecture that performs the best basic operations for training deep neural network. So what are the basic operations? The current Deep Bench benchmark for deep learning proposes four basic operations:

(1) *Matrix multiplication*: Almost all deep learning models include this operation, which is computationally intensive.
(2) *Convolution*: It constitutes another prevalent operation that consumes the majority of floating-point operations per second (floating-point/second) in the model.
(3) *Recurrent layers*: The recurrent layers are defined as the feedback layer in the model, which is essentially a combination of the first two operations.
(4) *All reduce*: This is a sequence of operations that pass or analyze the learned parameters before optimization. This operation is particularly effective when performing synchronous optimization on deep learning networks distributed across hardware (as in the example of AlphaGo).

In addition to this, hardware accelerator for deep learning must be characterized by data-level and process-oriented parallelism, multi-threading, and high memory bandwidth. Furthermore, the hardware architecture must be low-power due to the extended training time required for data. Consequently, performance per watt emerges as a pivotal evaluation criterion for hardware architectures.

To address the challenges posed by the inefficiency and substantial energy consumption of CPUs in the context of large-scale data operations, two development pathways emerge. The first involves the extension of the conventional $Von \cdot Neumann$ architecture, which is primarily embodied by three categories of chips: Graphics Processing Units (GPUs), Field-Programmable Gate Arrays (FPGAs), and Application-Specific Integrated Circuits (ASICs). The second is chips designed with the structure of human brain neurons, which have been fully anthropomorphized as the goal, aiming to continuously approach the human brain in chip architecture. This type of chip is called a brain-like chip.

1.3.1 Accelerated chip based on Von · Neumann architecture

A computer based on the *Von · Neumann* architecture stores programs and the data used to process them in two separate areas in the same way, one for the instruction set and the other for the dataset. The computer is required to periodically communicate between these two areas, the Central Processing Unit (CPU) and memory, for each operation, thereby generating data traffic between these two components. The advent of deep learning algorithms has led to an escalating demand for enhancing chip computational capabilities, thereby accentuating the *Von · Neumann* bottleneck. In scenarios where the CPU attempts to process rudimentary instructions on voluminous datasets, the data traffic imposes a substantial burden on the system, significantly diminishing overall efficiency. This phenomenon results in the CPU's inactivity while the data are being processed. Moreover, traditional chips exhibit inefficiency. When the chip is functioning properly, the majority of the electrical energy is converted into heat. Consequently, a computer without a heat sink can rapidly melt itself due to the heat generated by its CPU. The utilization of energy by intelligent devices is a multifaceted issue, largely attributable to the intricacies inherent in the design of the chip. This has led to a situation where these devices frequently exhibit suboptimal energy efficiency, which is characterized by limited operational range. Despite the implementation of various improvement strategies, the challenges of achieving optimal thermal management and mitigating leakage remain substantial obstacles.

1.3.1.1 *GPU acceleration technology*

The GPU (graphics processing unit) is a hardware component that facilitates the acceleration of deep learning, data analytics, and engineering applications through the utilization of a substantial number of computing resources by multiple GPUs. These units collaborate with the CPU to optimize the processing of these applications. Following the release of the first GPU supporting CUDA (Compute Unified Device Architecture) in 2007 by NVIDIA, GPUs have undergone a marked increase in power and capability, enabling them to solve complex computational problems.

1.3.1.1.1 *Difference between GPU and graphics card*

Graphics cards, also known as display adapters, are classified into two primary categories: independent graphics cards and motherboard-integrated graphics cards. Among them, independent graphics cards consist of a GPU, video memory, and interface circuits. In contrast, integrated graphics cards lack independent video memory and instead utilize the memory resources of the motherboard.

A GPU, or graphics processor, is typically integrated into the graphics card. While the term "GPU" is often used to refer to the graphics card itself, it is important to understand that a GPU is, in essence, the "heart" of a graphics card. These two components are in a "parasitic and parasitized" relationship, meaning that one component relies on the other for its functionality. It is important to note that the GPU cannot function independently; rather, it requires the presence of ancillary circuits and interfaces to operate. At this juncture, the graphics card becomes the primary component responsible for processing graphical data.

1.3.1.1.2 *GPU vs CPU*

A comparison of GPUs and CPUs can be made by assessing the manner in which the two process tasks. As demonstrated in Figure 1.7, the CPU utilizes a limited number of core processing units to optimize serial sequential tasks, whereas the GPU employs a massively parallel architecture comprising thousands of smaller and more efficient processing units. These units are capable of processing multiple small parallel tasks with high efficiency, making them particularly well suited for deep learning, which necessitates highly efficient matrix operations and a substantial number of convolution operations. The primary factor contributing to the GPU's superior performance in matrix and convolution operations compared to the CPU is the GPU's high bandwidth, high cache, and numerous parallel units. The underlying reason for this disparity is the marked differences in

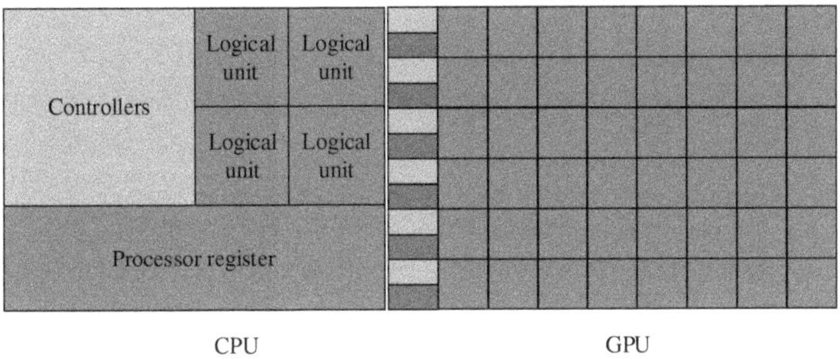

Figure 1.7. Internal structure of GPU and CPU.

the capabilities of the two types of processors. GPUs are characterized by their high bandwidth, substantial cache capacity, and a considerable number of parallel processing units. In contrast, CPUs are equipped with complex system instructions, enabling them to execute intricate task operations and scheduling processes. It is evident that these disparate features preclude the possibility of one processor type replacing the other in all contexts. In the context of concurrent execution of multiple tasks, the CPU is required to wait for bandwidth, while the GPU has the capacity to optimize it. Therefore, the CPU is adept at manipulating smaller memory blocks, while the graphics processing unit (GPU) is proficient in manipulating larger memory blocks.

1.3.1.1.3 *GPU type*

In the domain of deep learning accelerator GPUs, the predominant brands are currently AMD, NVIDIA, and Intel's Xeon Phi. Among them, NVIDIA's GPU has the highest prevalence, and its use has become remarkably straightforward. The development of deep learning networks on CUDA platforms has been significantly facilitated by NVIDIA's GPU, its computing acceleration standard library (cuDNN), and the resultant speed enhancement is up to fivefold compared to the absence of cuDNN on the same graphics card. In recent years, NVIDIA's GPU architectures have included Tesla, Fermi, Kepler, Maxwell, Pascal, Volta, Turing, and so on. In May 2017, NVIDIA announced that its new GPU architecture Volta could achieve the performance of the Pascal architecture increased by 4 times. The GV100 utilizes the first GPU to establish the Volta architecture, and the Tesla V100 adopts the first AI chip of the GV100 GPU. A comparison of the Tesla V100 and the Pascal architecture reveals that the former can improve the performance by 4–5 times during training and prediction phases of the deep neural network.

Graphics processing units (GPUs) have been shown to exceed CPUs in terms of floating-point computation and parallel processing capabilities.

Additionally, there has been an increasing trend of standard libraries for deep learning supporting GPU-based acceleration, including OpenCL, CUDA, and others. NVIDIA's GPU Cloud Platform NGC can provide frameworks such as Caffe, Caffe2, MXNet, CNTK, Theano, TensorFlow, Torch, and Software Development Kit (SDK), among others. This move will significantly promote the advancement of deep learning technology.

1.3.1.2 *FPGA*

Field-Programmable Gate Arrays (FPGAs) are also hardware supported by OpenCL. In comparison to GPUs, the FPGAs exhibit greater flexibility in terms of hardware configuration and frequently demonstrate superior performance in unit energy consumption when executing subroutines that are critical to deep learning (e.g., the computation of sliding windows). However, the configuration of the FPGAs necessitates hardware-specific knowledge and its difficulty between the General-Purpose Processors (GPPs) and Application-Specific Integrated Circuits (ASICs). The FPGAs offer the performance benefits of integrated circuits while also enabling the realization of sequential logic through the use of Flip-Flops (FFs) and combinational logic through the utilization of Look-Up Tables (LUTs). Contemporary FPGAs are equipped with hardened components, including full processor cores, communication cores, arithmetic cores, and Block Random-Access Memory (BRAM). These components are utilized to implement fundamental functions. Moreover, contemporary trends in FPGAs are gravitating toward a System on Chip (SoC) design approach, wherein the ARM coprocessor and the FPGA are typically situated on the same chip. The current FPGA market is dominated by Xilinx, which holds more than 85% of the market share. Furthermore, the FPGAs are rapidly replacing

ASICs and Application-Specific Standard Products (ASSPs) for fixed-function logic.

The FPGAs hold considerable promise for surpassing the acceleration capabilities of the GPP in the context of deep learning. GPP's execution at the software level is contingent on the conventional *von · Neumann* architecture, wherein instructions and data are stored in external memory and taken when required. A critical limitation arises from the communication between the processor and the memory, which significantly hinders the performance of the GPP, particularly with regard to the rate at which deep learning frequently requires access to stored information. In contrast, the programmable logic primitives of field-programmable gate arrays (FPGAs) do not rely on the *von Neumann* structure. These FPGAs can achieve data and path control in common logic functions and utilize distributed on-chip memories as well as deep exploitation of pipeline parallelism. This method naturally aligns with feedforward deep learning methods and supports partial dynamic reconfiguration, which means that when a part of the FPGA is reconfigured, it does not disrupt the ongoing calculations of other parts. This has an impact on large-scale deep learning modes and can be used for models that cannot be accommodated by a single FPGA. At the same time, intermediate results can be saved locally to reduce high global storage read costs. The FPGA architectures are tailored specifically for applications, and less emphasis is placed on adapting algorithms to a certain fixed computational structure when developing FPGA deep learning techniques. This method leaves more freedom to explore algorithmic-level optimizations, and the techniques that require many complex lower-level hardware control operations are difficult to implement in upper-level software languages; however, they can improve the execution efficiency of the FPGAs. This improvement in execution efficiency comes at the cost of requiring a significant amount of compilation (locality and loop) time.

The most prevalent languages utilized for the FPGAs are Verilog and VHDL, both of which are Hardware Description Languages (HDL). The primary distinction between these languages and conventional software languages lies in their functionality. Specifically, the HDL focuses exclusively on describing hardware, while software languages such as C language describe sequential instructions without the need to understand the execution details at the hardware level.

Effective description of hardware requires specialized knowledge in digital design and circuits. Although some lower-level implementation decisions can be left to automated synthesis tools, efficient design often cannot be achieved. Therefore, researchers tend to choose software design because it is already very mature and has a large number of abstract and convenient classifications to improve programmer efficiency. These trends have led to a greater preference for highly abstract design tools in the FPGA field.

The following timeline enumerates seminal advancements in the domain of FPGA deep learning research:

In 1987, the IEEE standard for VHDL was established.

In 1992, the GANGLION emerged as the inaugural FPGA neural network hardware implementation project (Cox *et al.*).

In 1994, Synopsys initiated the development of the first generation of the FPGA behavioral synthesis solutions.

In 1996, the VIP pioneered the first FPGA CNN implementation program (Cloutier *et al.*).

In 2005, the FPGA market value was close to 2 billion US dollars.

In 2006, the first processing power of 5 GOPS was realized on FPGAs using the BP algorithm.

In 2011, Altera introduced OpenCL to support the FPGAs, and large-scale research on FPGA-based CNN algorithms emerged (Farabet *et al.*).

In 2016, FPGA-based acceleration of data-center CNN algorithms emerged based on the Microsoft Catapult project (Ovtcharov *et al.*).

1.3.1.3 *ASIC*

Despite their high degree of parallel processing power, the GPUs are not designed for machine learning applications. The FPGAs, on the other hand, require users to program them. The design of these chips must prioritize three key characteristics: parallelization, low-power consumption, and high operational efficiency. Additionally, these chips should facilitate local on-the-fly computation, a capability in which Application Specific Integrated Circuits (ASICs) have demonstrated a clear advantage. However, the ASIC's extensive R&D cycle may hinder its ability to promptly adapt to market changes. Therefore, the SoC+IP model is more prevalent. The SoC has the capacity to integrate a multitude of disparate modules on the chip.

Each module on the SoC can be designated as an IP, which can be designed in-house or by another company and subsequently integrated into its own chip. In comparison with the ASIC, this model exhibits several advantages, including cost efficiency, shorter time-to-market, and adaptability to user requirements.

Additionally, there are companies in the market that specialize in ASIC development dedicated to machine learning. For instance, the Tensor Processor (full name: Tensor Processing Unit; abbreviation: TPU) developed by Google is a customized chip designed specifically for the deep learning language TensorFlow. Google's special preparation of this technology for TensorFlow obviates the need for any customizability, provided that it can adequately support all the instructions required by TensorFlow. Concurrently, the TPU operates TensorFlow with unparalleled efficiency, outperforming all other devices in this regard. Google developed the TPU for the most evident reason: to pursue extreme efficiency. As with the CPU/GPU, the TPU is programmable and can execute CISC instructions on a variety of large networks, including Convolutional Neural Networks (CNNs), Long and Short Term Memory (LSTM) networks, and so on. The internal structure of the TPU is shown in Figure 1.8. Moreover, the Wuji series, developed under the auspices of the Institute of Computing at the Chinese Academy of Sciences, has yielded notable

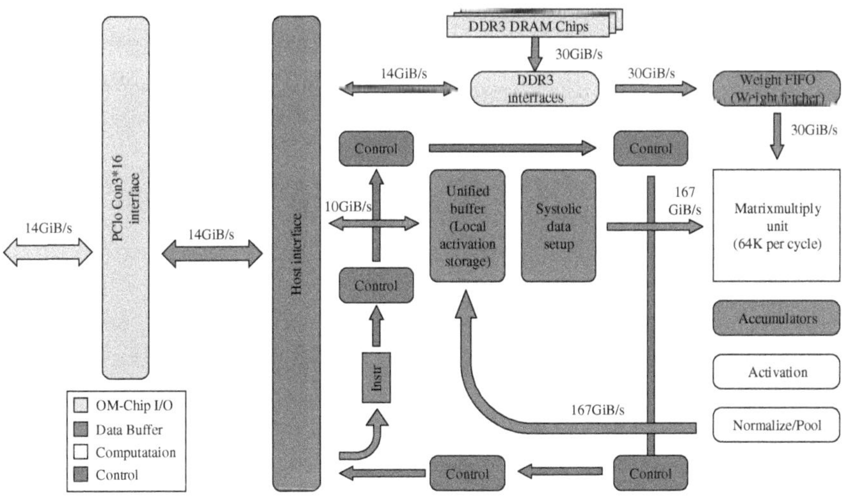

Figure 1.8. Internal structure of the TPU.

advancements. This includes the Cambrian 1, a pioneering processor architecture tailored for neural networks, the Cambrian 2, designed for large-scale neural networks, and the Cambrian 3, which caters to a diverse array of machine-learning algorithms. These achievements signify a significant milestone in the field of neural network research.

The ASICs have been demonstrated to exhibit superior performance in comparison to the GPUs and the FPGAs. However, it should be noted that their effectiveness is contingent upon the duration of the development cycle. The ASIC chips designed for machine learning necessitate greater financial investment and technological sophistication. Google had previously employed FPGAs for addressing the challenges posed by low cost, high efficiency, and performance. However, due to the significant difference between the performance of the FPGAs and that of the ASICs, the company eventually transitioned to utilizing custom ASICs.

1.3.2 Accelerated chip based on neuronal structure of human brain

The term "accelerated chips" refers to a type of semiconductor device that is based on the neuronal structure of the human brain. These chips are also referred to as "brain-like chips." The architecture of the brain-like chip emulates the structural characteristics of synaptic transmission within the human brain. Numerous processors are similar to neurons, and communication systems are similar to nerve fibers. The computation of each neuron is carried out locally. From an overall perspective, neurons work in a distributed manner, which means that the overall task is divided into some tasks, and each neuron is only responsible for a specific portion of the computation. The merits of this approach become evident when handling substantial volumes of data, and its power consumption is demonstrably lower than that of conventional chips.

The present state of research in the field of brain-like chips is predicated on the integration of microelectronics technology and novel neuromorphic devices. This integration is driven by the objective of surpassing the conventional computing architecture, thereby facilitating the convergence of storage and computation. The primary goals of this research include the substantial enhancement of computational performance, the optimization of integration, and the reduction of energy consumption.

A comparison of *von · Neumann*-based chips and brain-like chips reveals that the latter, despite their promising potential, are still in the research and development stage, or even the conceptual stage.

1.3.2.1 *Intel Pohoiki Beach chip system*

In September 2017, Intel unveiled its neuromimetic chip, dubbed "Loihi." Since then, new breakthroughs and progress have been continuously made, and the Intel Neuromorphic Research Community (INRC) was established to promote the development of neuromorphic computing. Intel also made an announcement regarding the new Neuromorphic Pohoiki Beach System-on-Chip. The system contains up to 64 Loihi chips with 132 billion transistors, a total area of 3,840 square millimeters, more than 8 million "neurons" (equivalent to the brains of some small rodents), and 8 billion "synapses." Intel Loihi chips are manufactured using a 14-nanometer process, with each integrating 2.1 billion transistors. These chips have a core area of 60 square millimeters and include three Quark × 86 CPU cores, 128 neuromimetic computational cores, 130,000 neurons, 130 million synapses, and programming toolchain support, including Python APIs. Contrary to conventional *von · Neumann* computational models of silicon chips, this chip emulates the neuromimetic computation of human brain principles. It is an asynchronous circuit that functions without a global clock signal. Rather, it utilizes an asynchronous pulsed neural network.

According to Intel, the system exhibits a capacity to execute artificial intelligence (AI) tasks at a rate that is one thousand times faster and ten thousand times more power-efficient than conventional central processing units (CPUs). The introduction of this neuromimetic system represents a significant advancement toward the goal of simulating the brain. Analogous to neurons in the human brain, Loihi possesses digital "axons" for transmitting electrical signals to neighboring neurons, "dendrites" for receiving signals, and "synapses" for establishing connections between the "axons" and "synapses". The "synapses" between the two are also used to establish connections. According to Intel, the systems based on such chips have been utilized for a variety of applications, including the simulation of tactile sensations in the skin, the control of prosthetic legs, the playing of foosball games, and so on.

The most recent 64-chip system has been disseminated among more than 60 INRC eco-partners. Recent assessments have demonstrated that when executing real-time deep learning benchmarks, the power consumption is approximately 109 times lower than that of traditional CPUs and 5 times lower than that of purpose-built IoT inference hardware. Notably, the real-time performance remains consistent even in scenarios where the network size is expanded by a factor of 50, resulting in a mere 30% increase in power consumption. In contrast, when utilizing IoT hardware, the power consumption escalates by more than fivefold, leading to a significant decline in real-time performance.

According to the plan, Intel was expected to manufacture 100 million neurons and 100 billion synapses by the end of 2019, and anticipated to possess 768 chips and 1.5 trillion transistors.

1.3.2.2 *IBM TrueNorth (SyNAPSE chip)*

In 2011, IBM pioneered the simulation of the human brain structure based on a silicon chip prototype that exhibited brain-like functions, marking a significant advancement in the field. However, the initial generation of True North demonstrated suboptimal performance. In 2014, the second generation of True North, which employs the Samsung 28 nm process, set a new benchmark by integrating 5.4 billion transistors into a single chip. This significant advancement in integrated circuit technology has been accompanied by a substantial reduction in power consumption, reaching a remarkable 20 mW/cm^2. This enhancement in power efficiency has led to a substantial improvement in the overall performance of the device. The True North chip, which contains neural network models, can be utilized as a real-time perceptual flow inference engine. It is characterized by its high accuracy, fast speed, and ultra-low power consumption.

The IBM TrueNorth (SyNAPSE chip) is composed of 4,096 cores, each of which has been designed to concisely mimic the neural structure of the human brain. Each core is equipped with 256 neurons (processors), 256 axons (memory), and 64,000 synapses (communication between neurons and axons). The interconnection of disparate chips in arrays is a viable option.

IBM has posited that the integration of 48 TrueNorth chipsets, each capable of constructing a network comprising 48 million neurons, would yield an intelligence level analogous to that of conventional rats.

Subsequent to 2014, there are reports that IBM has developed a new type of supercomputer driven by 64 "TrueNorth" brain-like chips to further reduce functionality, conduct real-time analysis of large deep neural networks, and apply them to high-speed aerial target recognition. If the power consumption of the system can reach the level of the human brain; theoretically, it can be further expanded on the basis of the 64-chip prototype, so as to be able to simultaneously handle any number of real-time recognition tasks.

1.3.2.3 *Qualcomm Zeroth chip*

In 2013, Qualcomm unveiled a Zeroth chip. This chip achieves pre-programming of behavior and outcomes through learning similar to the neurotransmitter dopamine (also known as "positive reinforcement"). In order to enable devices equipped with this chip to learn on their own and receive feedback from the surrounding environment at any time, Qualcomm has developed a software tool. The machine car loaded with this chip by Qualcomm has completed tasks such as pathfinding and obstacle avoidance using algorithms inspired by the human brain.

1.3.2.4 *DeepSouth chip from West well technology*

Shanghai Westwell Technology has unveiled two self-developed products: the 10 billion neuron human brain simulator (Westwell Brain) and the 50 million brain-like neuron chips (DeepSouths). DeepSouth is a commercially available chip with more than 5 billion "neural synapses" that can simulate up to 50 million "neurons." It has the ability to "self-learn and self-improve in real time", can complete calculations directly on the chip without the need for a network, and consumes only one-tenth to one-hundredth of the power consumed by traditional chips under the same task. The power consumption for the same task is only a few tenths to a few hundredths of that of traditional chips.

1.3.2.5 "*Darwinian*" *brain-like chip*

In 2015, the brain-like chip developed by Zhejiang University and Hangzhou University of Electronic Science and Technology represented a significant milestone in China's advancements in neural network technology. This pioneering silicon-based chip, measuring

25 square millimeters, boasts an impressive 5 million transistors and 2,048 bionic neurons crafted from silicon. Remarkably, it is capable of supporting over 4 million neural synapses and 15 distinct synaptic delays, a testament to the chip's sophisticated design and potential for future applications in neuroscience and engineering. This chip can receive and accumulate stimuli from the outside world, generate pulses (electrical signals) for information processing and transmission, and recognize the 10 digits from 1 to 10 handwritten by different people. This chip can control the movement direction of a basketball on a computer screen after receiving human brainwaves. After becoming familiar with and learning the operator's brainwaves, they will respond similarly when receiving the same stimuli in the future.

1.3.2.6 *AI-CTX chip*

AI-CTX is a diminutive, domestic brain-like chip. It has been demonstrated that each neuron of the chip exhibits analogous electrical characteristics and dynamic parameters, as well as rudimentary arithmetic and storage functions, akin to those observed in human brain neurons. The chip employs a distinctive wiring method that facilitates communication between chips and circumvents the conventional constraints imposed by physical limitations. This augmentation of the chip group's original network improves its functionality and capabilities. The chip demonstrates efficacy in the processing of data with time-related parameters, including temperature, air pressure, human body signals, and Internet of Things (loT) data. However, its functionality is not optimized for the processing of static hard disk data.

1.4 Mainstream Deep Learning Frameworks

A deep learning framework is a set of independent architectures, unified style templates, and reusable solutions developed specifically for the field of deep learning. It generally has the characteristics of high cohesion, strict specifications, scalability, maintainability, and high usability. With the increasing popularity of deep learning, more and more deep learning frameworks are being developed. The contemporary mainstream deep learning frameworks are illustrated in Table 1.1.

Table 1.1. Mainstream deep learning frameworks.

Frameworks	Development languages	Suitable models	Vantage	Drawbacks
Caffe 1.0	C++/CUDA	CNN	Build combined feedforward networks and image processing, finetune existing networks; stereotypical modeling, no need to write any code.	Not suitable for RNNs; too frequent and annoying operations for large CNNs; poor scalability, not refined enough; slow updates.
TensorFlow	C++/CUDA/ Python	CNN/RNN/RL	Computational graphs abstracted for easy understanding; faster compilation time than Theano; visualization with TensorBoard; support for data parallelism and model parallelism.	Slow, large memory footprint; no commercial support; not many models have been predefined; not easily instrumented; prone to errors in large software projects.
Torch	Lua/C/CUDA	CNN/RNN	Lots of modular components that are easy to combine; easy to write definition layers; lots of models for predefined types.	To learn Lua and use Lua as the main language; plug-and-play with relatively little code; no commercial support; low-quality documentation.
Theano	Python C++/CUDA	CNN/RNN	Python+NumPy implementation with simple interface; computational graph abstraction, easy to understand; RNN works well with computational graphs; there are many advanced package derivations.	Low raw Theano level; long compilation time for large models; underdeveloped support for predefined models; only single GPU support.

(Continued)

Table 1.1. (*Continued*)

Frameworks	Development languages	Suitable models	Vantage	Drawbacks
MXNet	C++/CUDA	CNN	Build combined feedforward networks and image processing, finetune existing networks; stereotypical models, no need to write any code; more academic users to dock models; support GPU, CPU distributed computing.	Not suitable for RNN; cluster O&M cost is higher than DL4J.
CNTK	Python C++ Brain Script	RNN/CNN	Highly customizable, allowing users to choose their own parameters, algorithms, and networks, strong scalability; much better than TensorFlow and Theano on Babi RNN and MNIST RNN tests but worse than TensorFlow on CNN tests.	Implemented in a new language—Network Description Language (NDL)—that lacks visualization.

Table 1.2. The specific content arrangement of every chapter.

Chapters and their titles	Contents		
	Basic network model	Extended network model	Application cases
Chapter 2: Artificial Neural Networks	**Basic model:** MP model; single-layer neural network. **Basic algorithm:** Stochastic Gradient Descent (SGD) algorithm. **Basic graphs:** Addition and multiplication nodes.	**Extended model:** Multi-layer neural network; artificial neural networks. **Extended algorithms:** Variants of SGD algorithm, such as MSGD algorithm, NGA algorithm, AdaGard algorithm, Adam algorithm, BP, and BPTT algorithms. **Extended computation graphs:** Computation graphs of Affine/Softmax layers, Softmax-with-loss layer, cross-entropy error, backpropagation cross-entropy error.	Momentum blind equalization algorithm on deed forward neural network.
Chapter 3	**Basic network:** Convolutional Neural Networks (CNNs). **Basic operations:** Common convolution and pooling operations.	**Extended networks:** LeNet-5, VGGNet, ResNets, Dense-Net, AlexNet, and inception; feature-fusion CNNs and depth-guided filtering networks. **Extended operations:** Depthwise separable convolution; dilated convolution; 3D convolution; grouping convolution; transposition convolution; flat convolution. Random pooling; bilinear pooling; unpooling; global average/max pooling; spatial pyramid pooling; multi-scale overlapping sliding pooling.	1. Remote Sensing Image Classification based on deep CNN. 2. Motion blur removal method based on deep convolutional neural Network

(*Continued*)

Table 1.2. *(Continued)*

Chapters and their titles	Contents		
	Basic network model	Extended network model	Application cases
Chapter 4	**Basic network:** Generative Adversarial Network (GAN).	**Extended networks:** Conditional GAN; wavelet GAN; multi-scale GAN.	Motion-blurred image restoration based on deep residual GAN
Chapter 5	**Basic network:** Boltzmann Machine (BM).	**Extended networks:** Restricted Boltzmann Machines (RBM), sparse RBM, competitive RBM, and classification RBM.	1. Gait feature extraction and recognition using restricted Boltzmann machine. 2. Speech DOA estimation based on RBM data compression.
Chapter 6	**Basic network:** Restricted Boltzmann Machines (RBM); Deep Belief Networks (DBMs).	**Extended networks:** Sparse DBNs, Gamma DBNs, adaptive DBNs, full-parameter dynamic learning DBNs, KPCA DBNs, and optimized DBNs.	Diagnosis of attention deficit hyperactivity disorder by deep belief network based on greedy approach.
Chapter 7	**Basic network:** AutoEncoder (AE).	**Extended networks:** Sparse AE; variational AE, stacked variational AE, deep convolutional variational AE, AE-based echo state network, deep classical correlation sparse AE, and dual adversarial AE.	1. Soft measurement model based on sparse autoencoding of mutual information. 2. Modulation recognition algorithm based on improved LDA and autoencoder.

Chapter 8	**Basic network:** Recurrent Neural Networks (RNNs). **Basic algorithms:** The first-order Stochastic Gradient Descent (SGD) method; Recursive Least Squares (RLS) algorithm.	**Extended networks:** RLS optimized RNNs algorithm (RLS-RNNs); RNNs based on SGD optimization; improvement of RLS-RNN. **Extended algorithms:** BPTT algorithm.	1. A non-invasive load identification method based on correlated RNN model. 2. An image compression method based on DTC-WT and RNN encoder.
Chapter 9	**Basic network:** Recursive Convolutional Neural Networks (RCNNs); Two-dimensional (2D) Convolutional Neural Networks (CNNs).	**Extended networks:** Deep RCNN; Bilinear Re-current Neural Network (BR-NN); Bi-directional RNN (Bi-RNN); 3D-RCNNs; particle swarm optimization algorithm-based BRNN.	1. Image rain removal method based on attention mechanism and gated recurrent unit. 2. Single-image super-resolution method based on cascaded recursive residual CNN.
Chapter 10	**Basic network:** Gate units; Long Short-Term Memory Network (LSTM). **Basic algorithms:** Backpropagation algorithm.	**Extended networks:** dual convolutional LSTM network. **Extended algorithms:** BPTT algorithm.	1. Non-cooperative underwater acoustic communication modulation signal recognition. 2. Fingerprint outdoor localization method using hybrid LSTM mechanism.

1.5 Main Contents of the Book

The content and architecture of this book are organized into 10 chapters. Except for Chapter 1, each chapter includes three parts: basic network model, extended network model, and their application cases. The specific content arrangement of every chapter is shown in Table 1.2.

Chapter 2

Artificial Neural Networks

Summary

The discussion commences with an examination of biological and artificial neurons, followed by an analysis of the neural network structure, principle, training, and prediction process. The subsequent analysis delves into the optimization algorithm, with a focus on its evolution and the computational diagrams of neural networks and activation functions. The following topics are analyzed in this study: the introduction of the regularization penalty term, the parameter paradigm penalty, L1/L2 parameter regularization, and the neural network BP algorithm. The three-layer neural network is used as an example in the analysis of the latter. The concept of overfitting and underfitting is also discussed, as are the following methods of preventing overfitting: reducing feature variables, weight regularization, cross-validation, dropout regularization, and Bayes rule regularization. This chapter provides a foundation for the subsequent chapters.

2.1 Introduction

Neural networks represent a significant machine learning technique and serve as the foundation for the most prevalent form of deep learning. The acquisition of knowledge in the domain of neural networks is conducive to not only the mastery of a potent machine learning methodology but also the enhancement of comprehension regarding deep learning techniques.

2.2 Neural Network Evolution

The material is presented in a straightforward, step-by-step manner.

Neural network is a machine learning technique that emulates the networks of the human brain with the objective of facilitating artificial intelligence-like learning. The neural network of the human brain is a highly complex organization. The adult brain contains approximately 100 billion neurons, as demonstrated in Figure 2.1.

The following investigation explores the manner in which neural networks employed in the context of machine learning attain this simulation, to yield the results that are surprisingly effective.

A classical neural network is composed of a three-layer network, including an input layer, an output layer, and a middle layer, which is also referred to as a hidden layer (see Figure 2.2).

As illustrated in Figure 2.2, the input layer contains three input units, the hidden layer has four units, and the output layer includes two units. In the design of a neural network, the number of nodes in the input and output layers is frequently predetermined, while the middle layer can be specified with a high degree of flexibility. The topology and arrows in Figure 2.2 signify the data flow during the prediction process, which deviates to a certain extent from the data flow during training. The key point in Figure 2.2 is not the circles (representing the "neurons") but rather the lines (representing the connections between "neurons"). Each connection is associated with a distinct weight, denoted as "weight," which requires training.

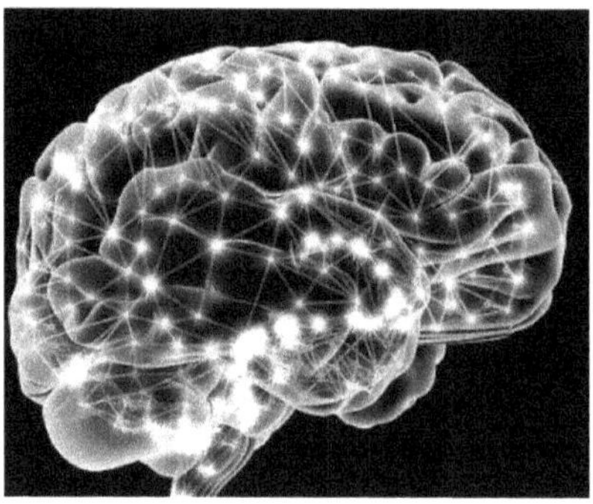

Figure 2.1. Schematic representation of the human brain.

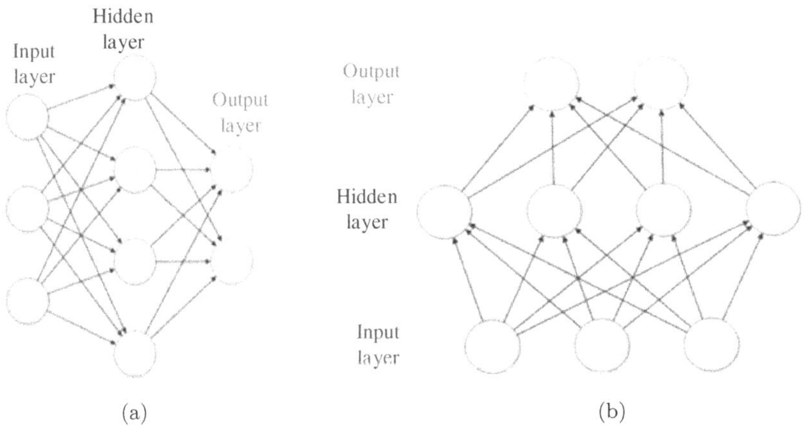

Figure 2.2. Neural network structure. (a) Left to right. (b) Bottom to top.

An artificial neural network is composed of many neuron models. These neurons are distributed in a parallel manner; each of them possesses a solitary output and has the capacity to establish connections with other neurons. There are many output connection methods in an artificial neural network; each of them corresponds to a specific connection weight coefficient. In essence, an artificial neural network can be defined as a directed graph that possesses the following characteristics:

(1) For each node, there is a state variable x_i.
(2) From node i to node j, there is a connection weight factor w_{ji}.
(3) For each node j, there is a threshold value or bias b_j.
(4) For each node j, a transformation function $f_j(x_i, w_{ji}, b_j)$ may be defined and $i \neq j$.

2.2.1 Neurons

2.2.1.1 *Biological neurons*

A neuron usually has multiple dendrites, mainly used to receive incoming information. And there is only one axon, with many axon endings at the tail end that can transmit information to multiple other neurons. The axon endings connect with dendrites of other neurons to transmit signals. The location of this connection is called a "synapse" in biology, as shown in Figure 2.3.

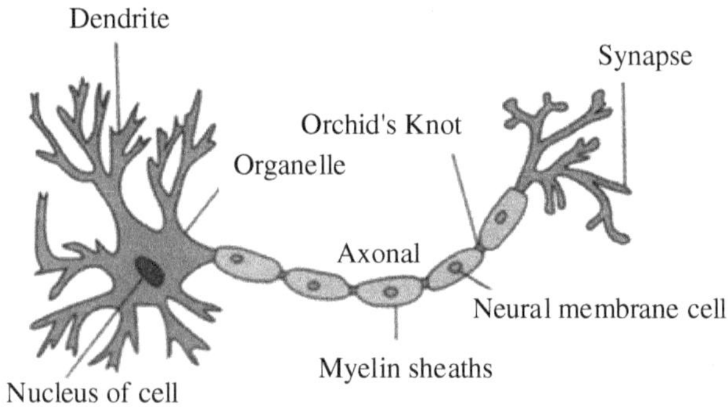

Figure 2.3. Biological neurons.

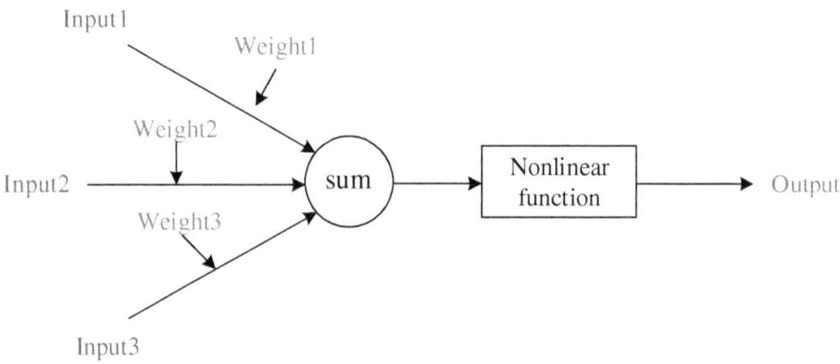

Figure 2.4. Neuron model.
Note: The lines of arrows in the middle, called "connections", have a weight on each connection. The training algorithm for a neural network is to optimize the values of the weights so that the whole network makes the best predictions.

2.2.1.2 *Artificial neuron*

In 1943, psychologist McCulloch and mathematician Pitts referred to the structure of biological neurons and published an abstract neuron model, abbreviated as the MP model. The MP model includes input, output, and computation functions. The input can be likened to the dendrites of a neuron, while the output can be likened to the axons of a neuron, and the computation function can be likened to the nucleus of a cell. Figure 2.4 illustrates a neuron with three inputs, one output, and two computational functions.

If we use $\boldsymbol{x}$ and $\boldsymbol{w}$ to denote the input and the weights, respectively, a directed arrow representing a connection can be interpreted in this way: At the beginning, the size of the transmitted signal is still $\boldsymbol{x}$. At the end, there is a weighting parameter $\boldsymbol{w}$ in the middle of the end, and after this weighting, the signal becomes $\boldsymbol{x} \cdot \boldsymbol{w}$, and so at the end of the connection, the size of the signal becomes $\boldsymbol{x} \cdot \boldsymbol{w}$, as shown in Figure 2.5.

In some models, directed arrows may signify an invariant transfer of values. In the neuron model, each directed arrow signifies a weighted transfer of values.

In the event that all variables within a neuron are represented symbolically, the output of the neuron is computed, as demonstrated in Figure 2.6.

In general, the output of a neuron can be represented as

$$y(\boldsymbol{w}, \boldsymbol{b}) = f\left(\sum_{i=1}^{N} w_i x_i + \boldsymbol{b}\right), \tag{2.2.1}$$

where $\boldsymbol{b}$ is the bias (threshold) of the neuronal unit. $\boldsymbol{w}$ is the connection weight vector, w_i is the ith connection weight coefficient in $\boldsymbol{w}$.

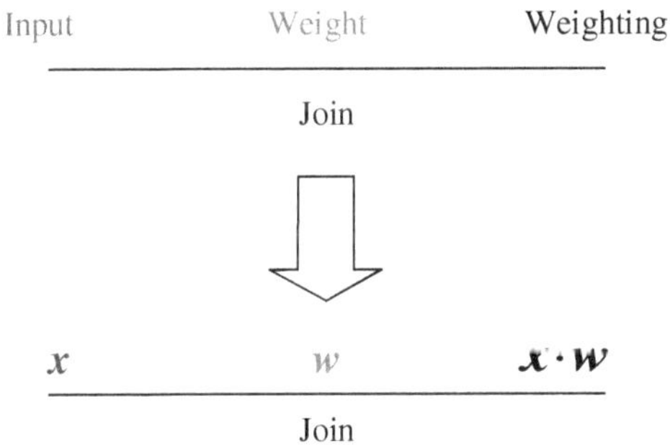

Figure 2.5. A connection.

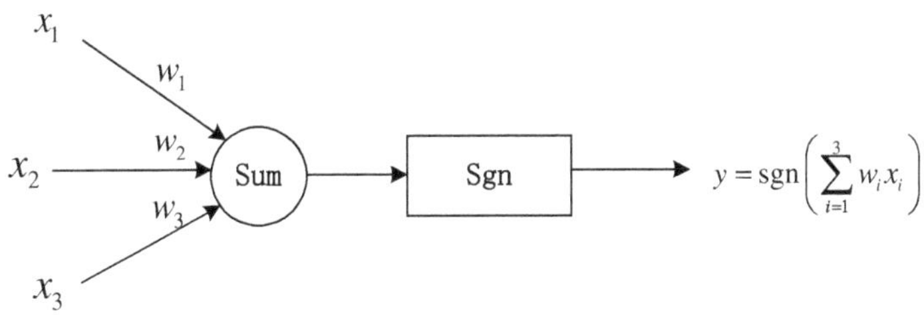

Figure 2.6. Neuron computation.

N is used to denote the number of input signals. y denotes the neuron output. $f(\cdot)$ is the output transformation function, which is sometimes referred to as the activation or activation function. This function is often represented using a binary function of 0 and 1 or an S-shaped function. As illustrated in Figure 2.6, the activation function is denoted as sgn.

As is evident, y is the value of a function f weighted and superimposed linearly on the inputs and weights. In the MP model, the function f is the sign function, which takes the sign of the function. The outputs of the function are 1 when the input is greater than zero. Conversely, its outputs are zero when the input is less than or equal to zero.

At present, several extensions to the artificial neuron model have emerged. First, the sum function is integrated with the sign function into a circle to represent the internal computation of the neuron. Second, the input x and output y are written to the upper left corner of the connection line. Finally, a neuron can generate multiple directed arrows representing the output, but their values are the same, as demonstrated in Figure 2.7.

Neurons can be seen as computing and storage units. Calculation is the calculation function of neurons on their inputs, while storage is the temporary storage of calculation results of neurons and transmitted to the next layer of the neural network.

In the context of a network comprising "neurons," the term "unit" is frequently employed to denote a "neuron" within the network. Concurrently, since the neural network is represented as a directed graph, the term "node" (node) is utilized to denote the same concept in some cases.

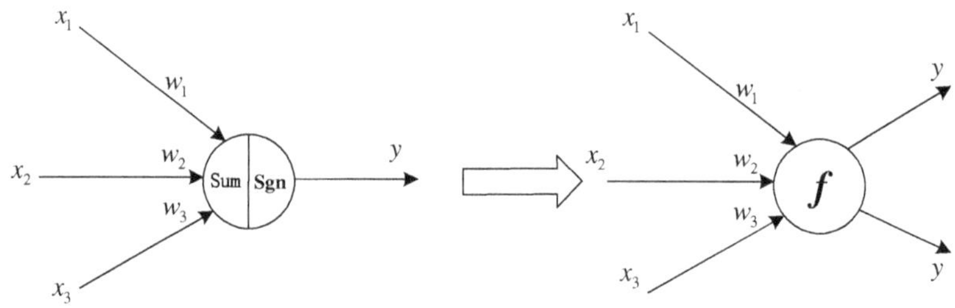

Figure 2.7. Neuron extension.

It is imperative to comprehend the intricacies of the neuron model. There is a data called a sample. The sample has four attributes, three of which are known and one is unknown. It is now necessary to predict the unknown attribute based on the three known attributes.

The specific way to do this is to use the neuron's formula for the calculation. The three known attribute values are recorded as x_1, x_2, x_3, and the unknown attribute y can be calculated using the formula.

Here, the known attributes are called features and the unknown attributes are called targets. Suppose that there is a linear relationship between the features and the target and the weights w_1, w_2, and w_3 have been obtained and are used to represent this relationship, then it is possible to predict the target of a new sample by neuronal modeling.

The MP model is the foundation for building the neural network. However, its weight values are pre-established and cannot be acquired through learning. In contrast, the *Hebb Learning Rule* considers that the strength of the synapses (also known as connections) between nerve cells in the human brain can undergo modification. Consequently, the machine can be configured to acquire knowledge through the adjustment of weight values, thereby establishing the foundation for subsequent learning algorithms.

2.2.2 Single-layer neural network

In 1958, F. Rosenblatt proposed a neural network consisting of two layers of neurons (see Figure 2.8). This pioneering artificial neural network was dubbed the "perceptron" or the "perceptual machine," and it was the first to demonstrate the capacity for learning at that time.

The neuron node must be added at the "input" position of the original MP model. It must be marked as an "input unit," and the weights w_1, w_2, and w_3 must be written in the middle of the "connection line." The perceptron model comprises two layers: the input layer and the output layer. The "input unit" in the input layer is responsible for the transmission of data without computation, while the "output unit" in the output layer performs computation on the input of the previous layer, and the layer that must be computed is referred to as the "computational layer," as well as a network

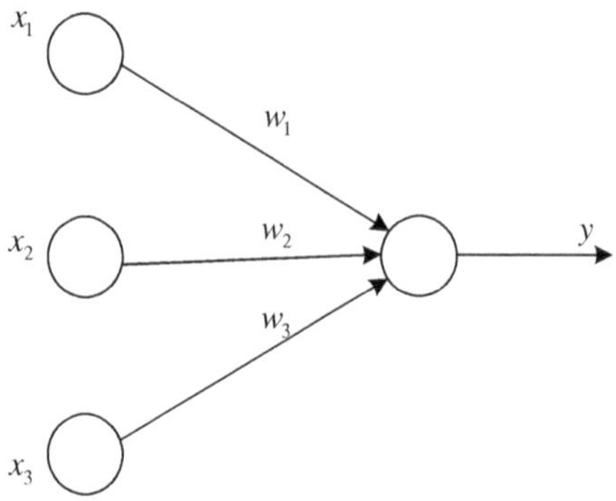

Figure 2.8. Single-layer neural network.

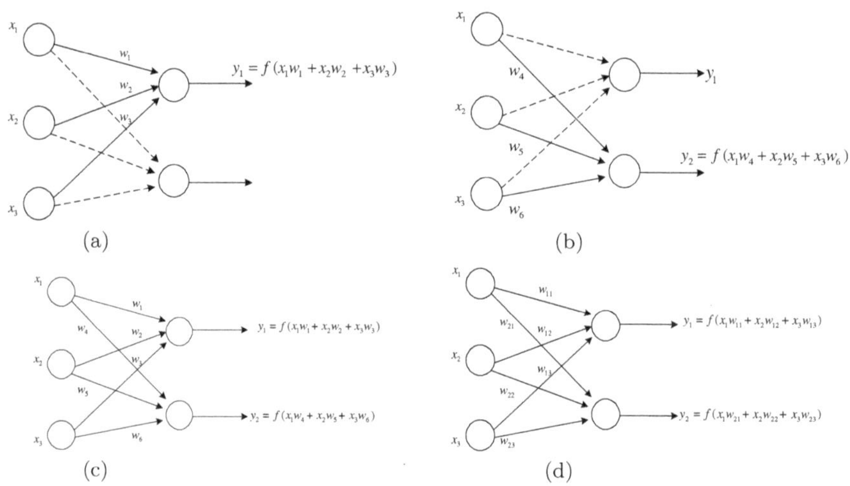

Figure 2.9. Single-layer neural network. (a) Output y_1. (b) Output y_2. (c) Output y_1 and y_2. (d) Output y_1 and y_2.

comprising a single computational layer is designated as a "single-layer neural network." If the predicted target is no longer a value but rather a vector, it is possible to add an additional "output unit" to the output layer. At this point, a single-layer neural network with two output units is illustrated in Figure 2.9.

In Figure 2.9(c), w_4, w_5, and w_6 are added later and do not reflect the relationship with w_1, w_2, and w_3. As illustrated in Figure 2.9(d), the elements w_{11} (corresponding to w_1), w_{12} (corresponding to w_2), and w_{13} (corresponding to w_3) can be regarded as three components of a weight vector. Similarly, the elements w_{21} (corresponding to w_4), w_{22} (corresponding to w_5), and w_{23} (corresponding to w_6) can be regarded as three components of another weight vector. The outputs y_1 and y_2 are also represented as vectors. Consequently, the output equation can be reformulated as

$$y_j(w_j, b_j) = f\left(\sum_{i=1}^{N} w_{ji}x_i + b_j\right), \tag{2.2.2}$$

where b_j is the bias (threshold) of the neuron unit, w_{ji} is the connection weight coefficient (w_{ji} takes a positive value for the activation state and w_{ji} takes a negative value for the suppressed state), N is the number of input signals, and y_j is the jth neuron of the output layer.

In contrast to the neuron model, the weights of the perceptron are determined through a process of training, which is analogous to the training of a logistic regression model for linear classification tasks. The classification effect can be visualized with decision demarcation, which is to draw a straight line in a two-dimensional data plane, as illustrated in Figure 2.10. In the case of three data dimensions, it is necessary to delineate a two-dimensional plane. Conversely, when the number of data dimensions is N, it is necessary to delineate an N-1-dimensional hyperplane.

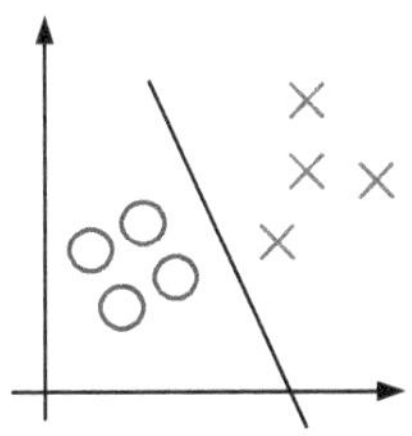

Figure 2.10. Single-layer neural network (decision demarcation).
Note: It is important to note that the perceptron is only capable of performing simple linear classification and is incapable of solving the XOR (hetero-or) problem.

2.2.3 Two-layer neural network

Minsky believed that single-layer neural networks could not solve the XOR problem, but when an additional computational layer was added, two-layer neural networks could not only solve the XOR problem but also have very good nonlinear classification performance. However, at that time, the computation of two-layer neural networks was a problem and there was no good solution.

In 1986, Rumelhart and Hinton proposed the Backpropagation (BP) algorithm, which solved the complex computational problem required for two-layer neural networks. A two-layer neural network consists of an input layer, an output layer, and a middle layer. At this point, both the middle layer and the output layer are considered as computational layers, as shown in Figure 2.11. Superscripts are used to distinguish variables between different levels.

The entire formula is expressed using matrix operations as

$$x^{(2)} = f(x^{(1)}w^{(1)}), \tag{2.2.3}$$

$$y = f(x^{(2)}w^{(2)}). \tag{2.2.4}$$

It can be seen that matrix representation is concise and not affected by an increase in the number of nodes (no matter how many nodes participate in the operation, there is only one variable at both ends of the multiplication).

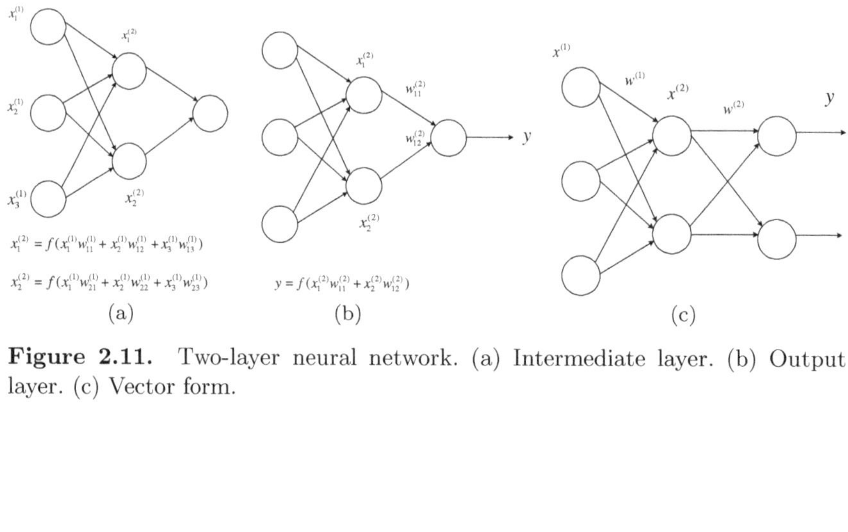

Figure 2.11. Two-layer neural network. (a) Intermediate layer. (b) Output layer. (c) Vector form.

2.2.4 Bias unit

In the preceding neural network configuration, there is an absence of any reference to bias nodes (bias units). It should be noted that these nodes are present by default, which exclusively perform storage functions and store the value invariably set to 1. In every layer of the neural network except for the output layer, a bias node is incorporated and connected to all nodes in the latter layer, and the bias parameter of the bias node is set to the vector $\boldsymbol{b}$, as illustrated in Figure 2.12.

In Figure 2.12, the bias node does not receive any inputs, as indicated by the absence of arrows pointing to it in the preceding layer. The presence of an explicit bias node in the structural diagram of certain neural networks is a distinguishing feature, while in others, this node is not present. In general, bias nodes are not explicitly drawn.

After taking the bias nodes into consideration, the matrix calculation formula for a neural network is defined as follows:

$$x^{(2)} = f(x^{(1)}w^{(1)} + b^{(1)}), \tag{2.2.5}$$

$$y = f(x^{(2)}w^{(2)} + b^{(2)}). \tag{2.2.6}$$

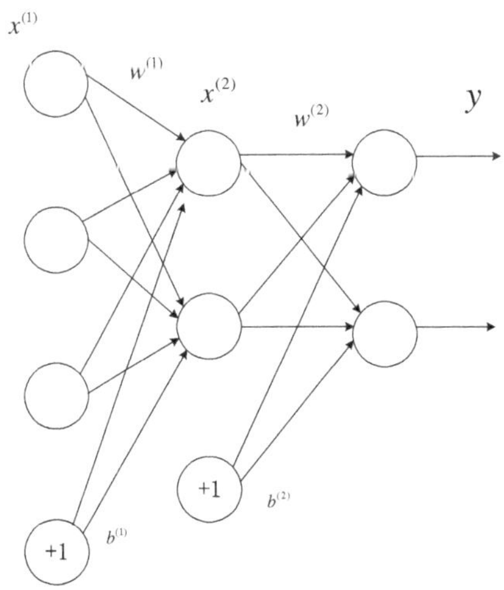

Figure 2.12. Two-layer neural network (considering bias nodes).

It is important to note that, in the context of a two-layer neural network, the sigmoid function is employed for underlying function f rather than the sigmoid function itself.

Indeed, neural networks can capture the true functional relationship between features and targets through the use of parameters and activation functions.

In contrast to single-layer neural networks, two-layer neural networks possess the capacity to infinitely approximate any continuous function. In the context of complex nonlinear classification tasks, two-layer neural networks, comprising one hidden layer, have demonstrated a capacity for effective classification. As illustrated in Figure 2.13, the red line and the blue line represent the data, while the red region and the blue region represent the region delineated by the neural network. The demarcation line between the two represents the decision demarcation.

As illustrated in Figure 2.13, the two-layer neural network demonstrates a remarkably smooth decision boundary and exhibits optimal classification performance. It is noteworthy that a single-layer network is capable of performing only linear classification tasks, and the final layer of a two-layer neural network is also a linear classification layer, thereby restricting its functionality to linear

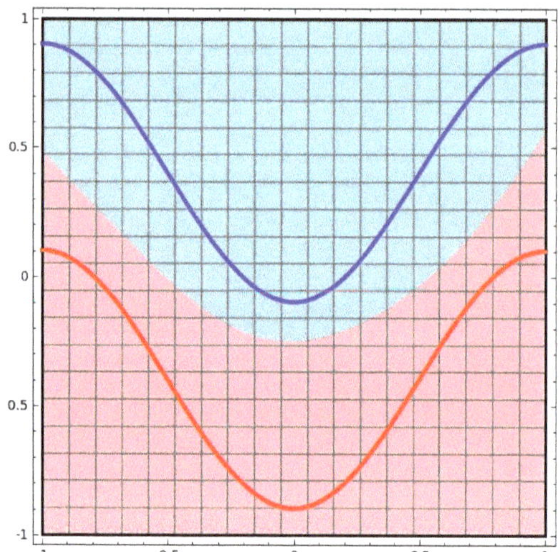

Figure 2.13. Two-layer neural network (decision demarcation).

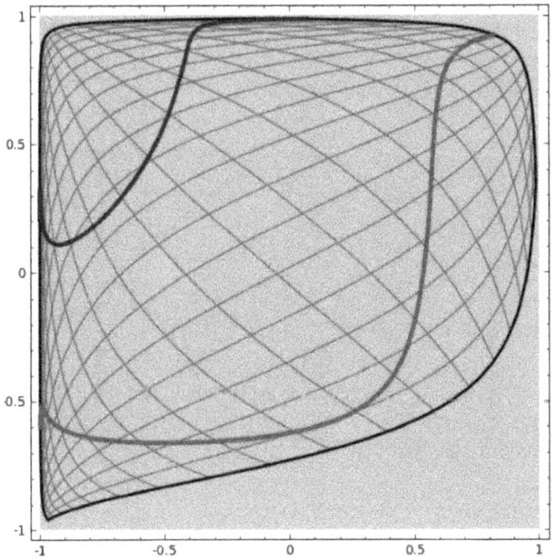

Figure 2.14. Two-layer neural network (spatial transformation).

classification tasks. However, it is still a subject of inquiry whether two linear classification tasks are combined to achieve a nonlinear classification task. The delineation of the decision output layer is illustrated in Figure 2.14.

As illustrated in Figure 2.14, the decision demarcation at the output layer maintains a linear trajectory. The key issue is that the data undergoes a spatial transformation from the input layer to the hidden layer. In a two-layer neural network, the hidden layer performs a spatial transformation on the original data so that it can be classified linearly. Subsequently, the decision demarcation in the output layer draws a linear classification demarcation to classify it. This indicates that the key to using two-layer neural networks for nonlinear classification is the hidden layer, whose parameter matrix functions to transform the original coordinate space of the data from linearly inseparable to linearly separable.

The two-layer neural network can simulate the real nonlinear functions in the data through a two-layer linear model. Therefore, the essence of multi-layer neural networks is complex function fitting.

When designing the number of hidden layer nodes, the number of input layer nodes needs to match the dimensionality of the features. The number of output layer nodes should match the dimension of

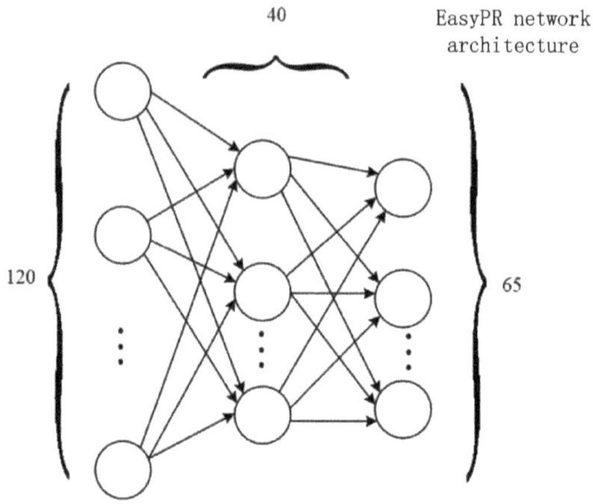

Figure 2.15. EasyPR character recognition network.

the target. The number of nodes of the middle layer is specified by the designer. The number of nodes will affect the overall performance of the model. How to determine the number of nodes in the middle layer? Usually, they are set by experience. A better method is to preset several optional values and observe the predictive performance of the entire model by switching between these values, selecting the value with the best performance as the final choice. This method is also known as Grid Searching (GS).

After understanding the two-layer neural network structure, one can comprehend similar networks. For example, the EasyPR character recognition network is shown in Figure 2.15 [21].

The EasyPR utilizes the images of characters to facilitate character text recognition. The input data is represented by a 120-dimensional vector, while the output as the text category to be predicted has a total of 65 categories. Experiments have shown that when the number of hidden layers is 40, the entire network performs better on the testing set. Therefore, the final structure of the network is chosen as 120, 40, and 65.

2.3 Neural Network Training and Prediction

The general methods of neural networks are divided into training and prediction stages. In the training phase, it is necessary to prepare the raw data and corresponding classification label data and obtain Model A through training. In the prediction phase, *Model A* is applied to new data to predict the category to which the new input data belongs.

2.3.1 Neural network training

To illustrate this point, the training of a two-layer neural network is considered.

In the perceptron model proposed by Rosenblatt, the parameters of the model can be trained. However, the use of the method is relatively simple and the common method in machine learning is not currently employed. This results in very limited scalability and applicability. The utilization of machine learning techniques for the training of neural networks has been initiated by researchers, commencing with two-layer neural networks.

The purpose of machine learning model training is to make the parameters as close as possible to the real model. The specific approach is to first assign random values to all parameters to predict the samples in the training data.

The predicted target for the sample is denoted as $\hat{\boldsymbol{y}}$, and the desired target is denoted as $\boldsymbol{y}$. The loss function is defined as

$$\text{Loss} = \frac{1}{2}(\boldsymbol{y} - \hat{\boldsymbol{y}})^2. \tag{2.3.1}$$

The objective is to minimize the loss sum across all training data.

If Equation (2.2.1) is substituted into $\hat{\boldsymbol{y}}$, then the loss function is defined as a function of parameter (parameter) $\boldsymbol{\theta} = \{\boldsymbol{w}, \boldsymbol{b}\}$. Now, a key issue is how to use Equation (2.3.1) to optimize parameter $\boldsymbol{\theta}$, with the objective of minimizing the value of the loss function. In general, the solution to this optimization problem is found through the application of a gradient descent method, which is employed for calculating the stochastic or instantaneous gradient of the parameters one at a time. It then allows the parameters to advance a distance in the opposite direction of the gradient. This process is repeated until the gradient cuts off near zero. At this point, all parameters have reached the state where the loss function reaches a minimum value.

In neural network models, due to their complex structure, the cost of calculating gradients each time is significant. Therefore, it is also necessary to use the backpropagation algorithm, which utilizes a neural network structure for computation. It does not calculate the gradients of all parameters simultaneously but instead performs backpropagation layer by layer from back to front. The backpropagation algorithm is based on the *Chain Rule*, which is a derivative rule in calculus used to find the derivative of a composite function. According to the *Chain Rule*, first, the gradient of the output layer, then the gradient of the second parameter matrix, and the gradient of the middle layer are calculated, the gradient of the first parameter matrix is sequentially calculated, and finally, the gradient of the input layer is calculated. After the calculation is completed, the gradients of the two parameter matrices are obtained.

The reason why machine learning problems are called learning problems rather than optimization problems is that they require the data to not only have a small error on the training set but also perform well on the testing set. Since the model ultimately needs to be deployed to real-world scenarios where training data have not been seen before. The enhancement of a model's predictive capabilities on a testing set is referred to as generalization, and a related method is known as regularization.

2.3.2 Neural network prediction

During the training phase, the weight vector $\boldsymbol{w}$ and bias b of the neural network are modified by the algorithm to minimize the loss function. When the loss function converges to a certain threshold or is equal to zero, the training is halted, and the weight vector and bias are obtained.

So far, all parameters in the artificial neural network structure (the number of nodes in the input layer, output layer, and hidden layer, weight vector $\boldsymbol{w}$, and bias b) are known. It is only necessary to input the vectorized data from the input layer of the artificial neural network, calculate it in the network along the direction of data flow, and transmit it to the output layer and output (one forward propagation) to complete one prediction and obtain the classification result.

2.4 Optimization Algorithm

Neural networks have five main hyperparameters: learning rate, weight initialization, the number of network layers, the number of neurons in a single layer (Units), and the regularization penalty term.

Optimization algorithm is exactly to minimize (or maximize) the loss function by adjusting the model parameters through improving the training method Loss(x).

There are some parameters inside the model that are used to calculate the degree of deviation between the expected value $\boldsymbol{y}$ and the predicted value $\hat{\boldsymbol{y}}$ of the target value in the testing set, based on the loss function Loss(x) formed on these parameters. For example, the weight vectors $\boldsymbol{w}$ and the b are such internal parameters, which are generally used to calculate the output values and play a major role in training the neural network model.

The internal parameters of the model play a pivotal role in the effective training of the model and the subsequent production of accurate results. This is also why various optimization strategies and algorithms are used to update and calculate network parameters that affect model training and model output, in order to approximate or reach optimal values.

Optimization algorithms can be divided into two categories:

(1) *First-order optimization algorithm*: This algorithm uses the gradient values of each parameter to minimize or maximize the loss function Loss($\boldsymbol{\theta}$). The most commonly used first-order optimization algorithm is the gradient descent method. For univariate functions, the derivatives are used for analysis. As for multivariate functions, the gradient is used instead of the derivative, and partial derivatives are used to compute the gradient. A major difference between the gradient and the derivative is that the gradient of the function forms a vector field.

(2) *Second-order optimization algorithm*: The second-order optimization algorithm employs second-order derivatives (also called the Hessian method) for minimizing or maximizing the loss function, which is not widely used due to its high computational cost.

The main neural network optimization algorithms are the gradient descent method, momentum algorithm, AdaGrad algorithm, RMSProp algorithm, AdaDelta algorithm, and Adam algorithm.

2.4.1 Gradient descent

The gradient descent method is a foundational technique in the training and optimization of intelligent systems.

The purpose of the gradient descent method is to regulate the variability by identifying the minimum difference square control, updating the model parameters, and ultimately converging the model.

In contemporary neural network models, the gradient descent method is predominantly employed for the purpose of updating model parameters to minimize the loss function. This process entails the adjustment and fine-tuning of model parameters in a single direction; this method has become the most prevalent optimization algorithm in the field of neural networks. The utilization of backpropagation algorithms facilitates the training of deep neural networks, which is a computational method that calculates and sums the product of the input signal and its corresponding weights in forward propagation. The activation function is a mathematical operation that converts the input signals into output signals.

Then, during the backpropagation process, the relevant errors are returned, the gradient of the error function relative to the parameters is calculated, and the model parameters are updated in the negative gradient direction.

The loss function of the nonlinear model is defined as the error function:

$$\begin{aligned}\text{Loss}(\boldsymbol{y}, \hat{\boldsymbol{y}}) &= \frac{1}{2N}\sum_{j=1}^{N}\left(y_j - f\left(\sum_{i=1}^{N} w_{ji}x_i + b_j\right)\right)^2 \\ &= \frac{1}{M}\sum_{j=1}^{M}\text{Loss}(y_j - \hat{y}_j), \end{aligned} \tag{2.4.1}$$

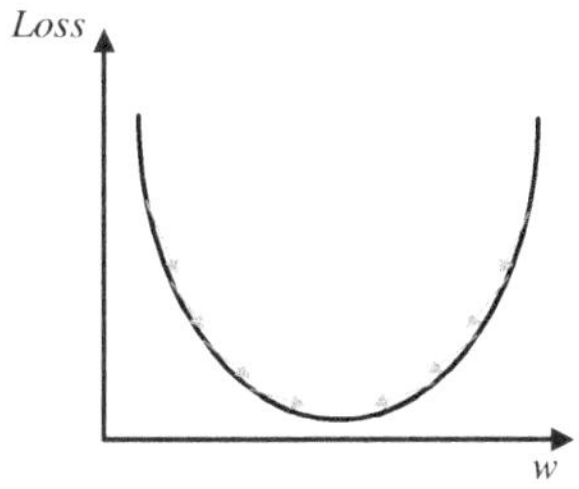

Figure 2.16. The direction of weight update opposite to the direction of gradient.

where $\text{Loss}(y_j - \hat{y}_j)$ is the loss function for the jth output and is defined as

$$\text{Loss}(y_j, \hat{y}_j) = \frac{1}{2}\left(y_j - f\left(\sum_{i=1}^{N} w_{ji}x_i + b_j\right)\right)^2. \tag{2.4.2}$$

Figure 2.16 shows the weight update process in the opposite direction to the gradient vector error, where the U-shaped curve is the gradient one.

Note that when the weight value $\boldsymbol{w}$ is too small or too large, there is a large error and the weights need to be updated and optimized to transform them to suitable values, so an attempt is made to find a local optimum in the direction opposite to the gradient.

The iterative formula for the gradient descent algorithm is written as

$$\boldsymbol{\theta}(k) = \boldsymbol{\theta}(k-1) - \eta g(k), \tag{2.4.3}$$

where $\boldsymbol{\theta} = \{\boldsymbol{w}, \boldsymbol{b}\}$ is the network parameter to be trained, η is the learning rate, and a constant, and $g(k)$ is the gradient, i.e.,

$$g(k) = \frac{\partial \text{Loss}(\boldsymbol{y}, \hat{\boldsymbol{y}})}{\partial \boldsymbol{\theta}(k)}. \tag{2.4.4}$$

The above is the most basic form of the gradient descent algorithm.

2.4.2 Variants of the gradient descent method

The aforementioned method involves the implementation of the gradient descent method to facilitate the adjustment of model parameters, including weights and biases. It is noteworthy that the updating of each parameter is executed by leveraging the entirety of the training sample dataset. This approach is formally referred to as the Batch Gradient Descent (BGD) method. In the BGD method, the entire dataset is involved in the gradient computation to obtain a standard gradient that facilitates the identification of the global optimal solution while requiring a minimal number of iterations. However, when dealing with voluminous datasets, the optimization process tends to be sluggish and challenging to manage and often results in memory exhaustion.

2.4.2.1 *Stochastic gradient descent method*

Stochastic gradient descent (SGD) method is the process of randomly sampling a sample from the training set each time to calculate the gradient of the loss function, and then the parameters are updated as follows:

$$\boldsymbol{\theta}(k) = \boldsymbol{\theta}(k-1) - \eta \frac{\partial \mathrm{Loss}(y_j, \hat{y}_j)}{\partial \boldsymbol{\theta}(k)}. \tag{2.4.5}$$

The problem with the SGD algorithm, however, is that it will eventually converge to a minimum due to frequent updates and fluctuations and will have an amount of overshoot due to frequent fluctuations.

Although it has been shown that when the learning rate is slowly reduced, the convergence pattern of the standard gradient descent method is the same as that of the SGD method.

2.4.2.2 *Mini-batch gradient descent method*

Mini-batch gradient descent method is to randomly sample m samples from the training set at a time to form a mini-batch to compute the loss function Loss and update the parameters. The Loss function is defined as

$$\mathrm{Loss}(\boldsymbol{w}, \boldsymbol{b}) = \frac{1}{2m} \sum_{j}^{j+m} \left(y_j - f\left(\sum_{i}^{i+m} w_{ji} x_i + b_j \right) \right)^2. \tag{2.4.6}$$

Then, it is updated according to Equation (2.4.5). The advantages of using the mini-batch gradient descent method are as follows:

(1) Fluctuations in updating parameters can be reduced, and better results and more stable convergence can be obtained.
(2) Common matrix optimization methods in the latest deep learning libraries can be used to make the gradient for mini-batches of data be efficiently calculated.
(3) Typically, the size of mini-batch samples ranges from 50 to 256 and can vary depending on the actual problem.
(4) Mini-batch gradient descent methods are usually chosen when training neural networks.

2.4.3 Optimization of gradient descent methods

2.4.3.1 *Momentum algorithm*

High variance oscillations in the SGD method make it difficult for the network to converge stably, so some researchers have used the momentum method to accelerate SGD training by optimizing the training in the relevant direction and weakening the oscillations in the irrelevant direction. In other words, this novel method adds β times the component of the updated vector from the previous time step to the current updated vector.

The momentum algorithm, also called the impulse algorithm, has iterative update formulas:

$$\boldsymbol{v}(k) = \beta\boldsymbol{v}(k-1) + \eta\boldsymbol{y}g(k), \tag{2.4.7}$$

$$\boldsymbol{\theta}(k) = \boldsymbol{\theta}(k-1) - \boldsymbol{v}(k), \tag{2.4.8}$$

where the momentum parameter $0 \leq \beta \leq 1$ and is usually set to 0.9, or some similar value. When $\beta = 0$, the momentum algorithm is equivalent to the mini-batch gradient descent method.

Momentum here is consistent with that in classical physics, as if a ball is thrown from a hill, its momentum is collected as it falls, and its velocity continuously increases. During the updating process of the parameter, it enables the network to converge more efficiently and stably and reduce oscillation processes. When its gradient points toward the actual direction of movement, the momentum term β increases. When the gradient is opposite to the actual direction of movement, β decreases. This approach means that the momentum term only updates the parameters of the relevant samples, so it reduces unnecessary parameter updates and results in faster and more stable convergence, as well as reducing oscillation processes.

In practical applications, it is necessary to smooth out some messy data, as shown in Figure 2.17(a).

Figure 2.17(b) shows the result of performing the smoothing process. The most common method is to slide a sliding window across the data points and calculate the average of the window to get a sliding average of the data. It is also possible to use an exponentially weighted average to do smoothing on the data.

If

$$\begin{cases} \boldsymbol{v}(0) = 0 \\ \boldsymbol{v}(k) = \beta\boldsymbol{v}(k-1) + (1-\beta)\boldsymbol{t}(k), \end{cases} \tag{2.4.9}$$

then the expansion is given by

$$\begin{aligned} \boldsymbol{v}(k) &= \beta\boldsymbol{v}(k-1) + (1-\beta)\boldsymbol{t}(k) = \beta^k\boldsymbol{v}(0) + \beta^{k-1}(1-\beta)\boldsymbol{t}(1) \\ &\quad + \beta^{k-2}(1-\beta)\boldsymbol{t}(2) + \cdots + \beta(1-\beta)\boldsymbol{t}(k-1) + (1-\beta)\boldsymbol{t}(k) \\ &= \beta^{k-1}(1-\beta)\boldsymbol{t}(1) + \beta^{k-2}(1-\beta)\boldsymbol{t}(2) + \cdots + \beta(1-\beta)\boldsymbol{t}(k-1) \\ &\quad + (1-\beta)\boldsymbol{t}(k). \end{aligned} \tag{2.4.10}$$

Equation (2.4.10) is called an exponentially weighted moving average. When k is relatively small, its recent data of an exponentially weighted moving average is too small, then it results in a relatively large estimation error, so it is modified as

$$\boldsymbol{v}(k) = \frac{\beta\boldsymbol{v}(k-1) + (1-\beta)\boldsymbol{t}(k)}{1-\beta^k}, \tag{2.4.11}$$

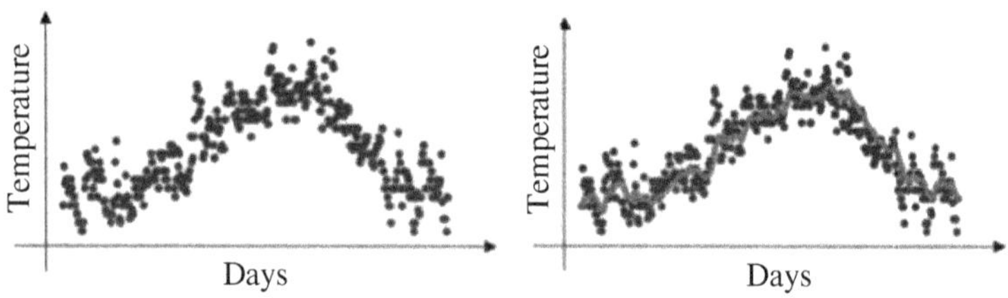

Figure 2.17. Smoothing process.

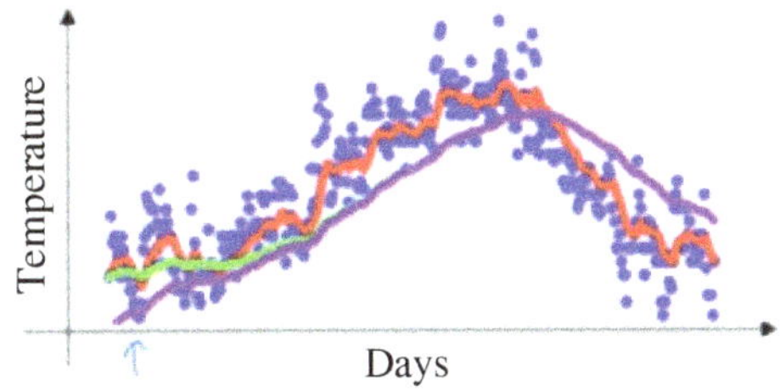

Figure 2.18. Correction results.

where $1 - \beta^k$ is the sum of all weights, which is equivalent to doing a normalization of the weights. In Figure 2.18, the purple line represents the result without correction, and after correction, it becomes the green curve. The two have significant differences between the first few data points, but they basically overlap later on.

2.4.3.2 *Nesterov gradient acceleration method*

There is a problem with the momentum algorithm:

If a ball rolling down a hill blindly slides down the slope, it is very inappropriate. A smarter ball should pay attention to where it is going, so the ball should slow down when tilting upwards again on an uphill slope.

In fact, when the ball reaches the lowest point on the curve, its momentum is quite high. Due to high momentum, it may completely miss the minimum value, so the ball does not know when to decelerate and continues to move upwards.

The method proposed by researcher Yurii Nesterov to solve the momentum problem is called the Nesterov Gradient Acceleration (NGA) method, which is a technique that endows the momentum term with the ability to predict, by using the momentum term $\beta \boldsymbol{v}(k-1)$ to modify the parameter $\boldsymbol{\theta}$. The approximation of the parameter at the next position is obtained by computing $\boldsymbol{w} - \beta \boldsymbol{v}(k-1)$, where the parameter is a rough concept. Thus, the gradient value of the current parameter $\boldsymbol{w}$ is not calculated, but the future is effectively foreseen by the approximate future position of the related parameter, i.e.,

$$\boldsymbol{v}(k) = \beta \boldsymbol{v}(k-1) + \eta \nabla_w \text{Loss}(\boldsymbol{w} - \beta \boldsymbol{v}(k-1)). \tag{2.4.12}$$

Then, the parameter $\boldsymbol{\theta}(k)$ is updated by

$$\boldsymbol{\theta}(k) = \boldsymbol{\theta}(k-1) - \boldsymbol{v}(k). \tag{2.4.13}$$

Now, by making network update fit to the slope of the error function and sequentially accelerating the SGD method, the corresponding parameters can also be adjusted and updated based on their importance to perform larger or smaller update amplitudes.

2.4.3.3 *AdaGrad algorithm*

The gradient descent method and momentum algorithms use a uniform learning rate, which is difficult to adapt to all dimensions. The AdaGrad algorithm is introduced to adjust the learning rate on each dimension according to the magnitude of the gradient value of the independent variable in each dimension. A mini-batch of stochastic gradient g is accumulated by element square to the variable $\boldsymbol{s}(k)$, i.e.,

$$\boldsymbol{s}(k) = \boldsymbol{s}(k-1) + g(k) \odot g(k). \tag{2.4.14}$$

The cumulative variable $\boldsymbol{s}(k)$ is used to control the learning rate η, at which point the parameters are updated with the following equation:

$$\boldsymbol{\theta}(k) = \boldsymbol{\theta}(k-1) - \frac{\eta}{\sqrt{s(k) + \varepsilon}} \odot g(k), \tag{2.4.15}$$

where η is the learning rate and ε is a constant with an added denominator to maintain numerical stability. Equation (2.4.15) shows that $\boldsymbol{s}$ is a smoothing of the square root of the gradient. When updating $\boldsymbol{\theta}$, the gradient is first divided by $\frac{1}{\sqrt{s(k)+\varepsilon}}$, which is equivalent to a normalization of the gradient. If the gradient oscillation is large in one direction, the step size should be reduced. If the oscillation is large, then s is also large in this direction, and the normalized gradient is small. If the gradient oscillation is small in one direction, its step size should be increased, and if the oscillation is small, s is also smaller in this direction and the normalized gradient is larger. Therefore, the step size on different dimensions can be adjusted by the AdaGrad algorithm to speed up the convergence.

However, when the learning rate drops faster early in the iteration and the current solution is still poor, the AdaGrad algorithm may

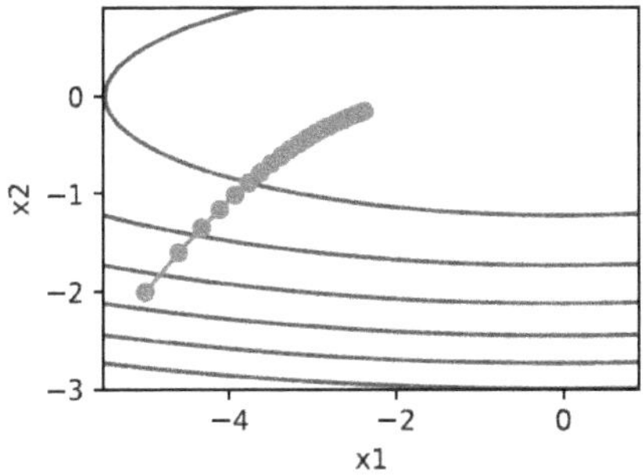

Figure 2.19. Iterative trajectory of the independent variable.

have a harder time finding a useful solution late in the iteration due to the small learning rate.

The objective function is given by

$$f(x) = 0.1x_1^2 + 2x_2^2, \tag{2.4.16}$$

when the AdaGrad algorithm iterates over the trajectory of the independent variables, as shown in Figure 2.19.

2.4.3.4 *RMSprop algorithm*

The exponentially weighted moving average in the momentum method is different from the gradient accumulation state variable in the AdaGrad algorithm in that it is the sum of all the mini-batch stochastic gradients by element squares up to time step k. The RMSProp algorithm performs an exponentially weighted moving average of these gradients by the element-squared accumulation state variable. Specifically, given the hyperparameters $0 \leq \gamma < 1$, the RMSProp algorithm calculates the following updating equation up to time step $k > 0$:

$$\boldsymbol{s}(k) = \gamma \boldsymbol{s}(k) - (1-\gamma) g(k) \odot g(k). \tag{2.4.17}$$

At this point, the adjusting equation of the parameter is still shaped as in Equation (2.4.15). Since the state variable $s(k)$ of the RMSProp algorithm is exponentially weighted moving averages of the squared terms, it can be viewed as weighted averages of the squared terms $g(k) \odot g(k)$ of mini-batches of stochastic gradients from the most recent $1/(1-\gamma)$ time steps. In this way, the learning rate of each element of the independent variable no longer decreases (or remains constant) all the time during the iterations. The RMSProp algorithm is shown as follows:

Algorithm 2.1. RMSProp Algorithm.

Parameters: Global learning rate η, Decay rate γ, Small constant ε, Usually set to 10^{-6} (for numerical stability when dividing by a small number)

Initialization: Initialize parameters $\boldsymbol{\theta}$, Initialize cumulative variables $s(k) = 0$

While not meeting the stopping criteria do

  A mini-batch of m samples $\{\boldsymbol{x}_1, \ldots, \boldsymbol{x}_m\}$ from the training set is taken, corresponding to the objective of $\boldsymbol{y}_i$

  Calculating the gradient: $g \leftarrow \frac{1}{m}\nabla_{\boldsymbol{\theta}} \sum_i \text{Loss}(f(\boldsymbol{x}_i; \boldsymbol{\theta}), \boldsymbol{y}_i)$

  Cumulative square gradient: $\boldsymbol{s} \leftarrow \gamma \boldsymbol{s} + (1 - \gamma) g \odot g$

  Calculate parameter's update: $\Delta\boldsymbol{\theta} = -\frac{\eta}{\sqrt{s+\varepsilon}} \odot g$

  Application updates: $\boldsymbol{\theta} \leftarrow \boldsymbol{\theta} + \Delta\boldsymbol{\theta}$

end While

For the objective function Equation (2.4.16), the trajectory of the independent variable iterations is observed by the RMSProp algorithm, as shown in Figure 2.20.

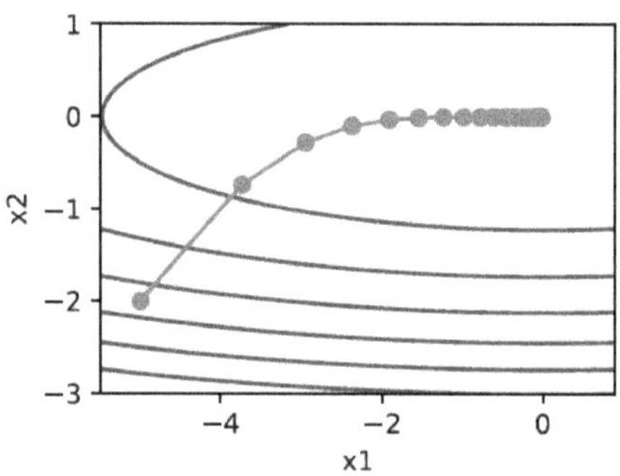

Figure 2.20. Iterative trajectory.

2.4.3.5 *AdaDelta algorithm*

The AdaDelta algorithm is also an improvement on the AdaGrad algorithm, based on a mini-batch stochastic gradient $g(k)$, an exponentially weighted moving average variable $\boldsymbol{s}(k)$ according to the element square is calculated by

$$\boldsymbol{s}(k) = \gamma \boldsymbol{s}(k) - (1 - \gamma) g(k) \odot g(k). \tag{2.4.18}$$

The AdaDelta algorithm maintains an additional state variable $\Delta \boldsymbol{u}(k)$ compared to the RMSProp algorithm, i.e.,

$$\Delta \boldsymbol{u}(k) = \gamma \Delta \boldsymbol{u}(k-1) - (1 - \gamma) g'(k) \odot g'(k). \tag{2.4.19}$$

Then, the independent variables are calculated by

$$g'(k) = \sqrt{\frac{\Delta u(k-1) + \varepsilon}{s(k) + \varepsilon}} \odot g(k), \tag{2.4.20}$$

$$\boldsymbol{u}(k) = \boldsymbol{u}(k-1) - g'(k). \tag{2.4.21}$$

Analyses of the above algorithms show that the AdaDelta algorithm differs from the RMSProp algorithm by using $\sqrt{\Delta u(k-1)}$ instead of η if the effect of ε is not considered.

2.4.3.6 *Adam algorithm*

The Adam algorithm is different from the traditional Stochastic Gradient Descent (SGD) method. The SGD method maintains a single learning rate for updating all weights, and the learning rate does not change during training. The Adam algorithm, on the other hand, designs independent adaptive learning rates for different parameters by computing first-order moment estimates and second-order moment estimates of the gradient. The Adam algorithm is a combination of the advantages of two stochastic gradient descent extensions, that is, the Adaptive Gradient (AdaGrad) algorithm maintains a learning rate for each parameter to improve performance on sparse gradients (i.e., natural language and computer vision problems).

The Root Mean Square Propagation (RMSProp) algorithm adaptively reserves a learning rate for each parameter based on the mean of the nearest magnitude of the weight gradient. This means that the algorithm has excellent performance on non-stationary and online problems.

The Adam algorithm not only calculates the adaptive parameter learning rate based on the first-order moment mean as the RMSProp algorithm does but also takes full advantage of the second-order moment mean of the gradient, i.e., the uncentered variance. Specifically, the Adam algorithm computes the exponential moving average of the gradient, and the hyperparameters β_1 and β_2 control the decay rate of these moving averages. The initial values β_1 and β_2 of the moving average are close to 1 (the recommended value) so that the bias of the moment estimate is close to 0. This bias is enhanced by computing the bias estimates first and then the bias-corrected estimates.

The Adam algorithm uses the momentum variable $\boldsymbol{v}(k)$ and the exponentially weighted moving average of the mini-batch stochastic gradient squared $\boldsymbol{s}(k)$ by element in the RMSProp algorithm and initializes each of them to zero at time step zero. Given the hyperparameters $0 \leq \beta_1 < 1$, the momentum variable $\boldsymbol{v}(k)$ at time step k, i.e., the exponentially weighted moving average of the mini-batch stochastic gradient $g(k)$, is updated by

$$\boldsymbol{v}(k) = \beta_1 \boldsymbol{v}(k-1) + (1-\beta_1) g(k). \tag{2.4.22}$$

As in the RMSProp algorithm, if the hyperparameters $0 \leq \beta_2 < 1$ are given, the mini-batch stochastic gradient $g(k) \odot g(k)$ is exponentially weighted by the term after squaring the elements to obtain

$$\boldsymbol{s}(k) = \beta_2 \boldsymbol{s}(k) + (1-\beta_2) g(k) \odot g(k).$$

Since the elements in $\boldsymbol{v}(0)$ and $\boldsymbol{s}(0)$ are initialized to zero at time step k, $\boldsymbol{v}(k)$ can be obtained by

$$\boldsymbol{v}(k) = (1-\beta_1) \sum_{i=1}^{k} \beta_1^{k-i} g(i).$$

After the weights of the mini-batch stochastic gradient at each past time step are summed, we yield

$$(1-\beta_1)\sum_{i=1}^{k}\beta_1^{k-i}=1-\beta_1^k.$$

Note that when k is small, the sum of the weights of the mini-batch SGD method at past time steps will be smaller. For example, $v(1) = 0.1g(1)$ when $\beta_1 = 0.9$.

In order to eliminate such effects, for any time step k, the weights of the mini batch stochastic gradients of the past time steps can be summed up to 1 via $\boldsymbol{v}(k)$ dividing by $1-\beta_1^k$, which is also called bias correction. In the Adam algorithm, the bias correction is applied to both the variables $\boldsymbol{v}(k)$ and $\boldsymbol{s}(k)$, i.e.,

$$\hat{\boldsymbol{v}}(k)=\frac{\boldsymbol{v}(k)}{1-\beta_1^k}, \tag{2.4.23}$$

$$\hat{\boldsymbol{s}}(k)=\frac{\boldsymbol{s}(k)}{1-\beta_2^k}. \tag{2.4.24}$$

Based on the above bias-corrected variables $\hat{\boldsymbol{v}}(k)$ and $\hat{\boldsymbol{s}}(k)$, the learning rate of each element in the model parameters is readjusted by element calculation:

$$g'(k)=\frac{\eta\hat{\boldsymbol{v}}(k)}{\sqrt{\hat{\boldsymbol{s}}(k)}+\varepsilon}. \tag{2.4.25}$$

Like the AdaGrad algorithm, the RMSProp algorithm, and the AdaDelta algorithm, each element of the objective function's independent variable has its own learning rate, respectively. Finally, the independent variables $\boldsymbol{\theta}(k)$ are updated by $g'(k)$ as follows:

$$\boldsymbol{\theta}(k)=\boldsymbol{\theta}(k-1)-g'(k). \tag{2.4.26}$$

The Adam algorithm is shown as Algorithm 2.2.

Algorithm 2.2. Adam's algorithm.

Parameter: Step size η (recommended default: 0.001)
The exponential decay rate of the moment estimate, β_1 and β_2 within $[0, 1)$ the interval. (Recommended defaults: 0.9 and 0.999, respectively)
Small constant ε for numerical stabilization (recommended default:)
Initialization: Parameters $\boldsymbol{\theta}$
First-order and second-order variables $\boldsymbol{s} = 0, \boldsymbol{v} = 0$;
Time step $k = 0$
While not meeting the stopping criteria do
A mini batch $\{\boldsymbol{x}_1, \ldots, \boldsymbol{x}_m\}$ containing m samples is taken from the training set, corresponding to the objective is $\boldsymbol{y}_i$.
Calculating the gradient: $g \leftarrow \frac{1}{m}\nabla_{\boldsymbol{\theta}} \sum_i L(f(\boldsymbol{x}_i; \boldsymbol{\theta}), \boldsymbol{y}_i)$
$k \leftarrow k + 1$
Updating biased first-order moment estimates: $\boldsymbol{s} \leftarrow \beta_1 \boldsymbol{s} + (1 - \beta_1) g$
Updating biased second-order moment estimates: $\boldsymbol{v} \leftarrow \beta_2 \boldsymbol{v} + (1 - \beta_2) g \odot g$
Correcting deviations from first-order moments: $\hat{\boldsymbol{s}} \leftarrow \frac{\boldsymbol{s}}{1-\beta_1^k}$
Correcting for deviations from second-order moments: $\hat{\boldsymbol{v}} \leftarrow \frac{\boldsymbol{v}}{1-\beta_2^k}$
Calculation update: $\Delta\boldsymbol{\theta} = -\eta \frac{\hat{\boldsymbol{s}}}{\sqrt{\hat{v}} + \varepsilon}$ (Apply operations element by element)
Application Updates: $\boldsymbol{\theta} \leftarrow \boldsymbol{\theta} + \Delta\boldsymbol{\theta}$
End while

2.5 Computation Charts

The core of a machine learning task is the definition of a model and the way in which the parameters of the model are solved. After abstracting these two aspects, a unique computational logic can be determined, and it is represented as a graph, called a computational graph, which is expressed as a directed acyclic graph and defines how the data flows and how the data is computed, and the interdependence between various computations.

As defined by the data structure, the graph consists of a vertex set $V(G)$ and an edge set $E(G)$, and is denoted as $G = (V, E)$, where $E(G)$ is a finite set of edges and edges are either unordered pairs of vertices (undirected graph) or ordered pairs (directed graph).

For the directed graphs, $E(G)$ is a finite set of directed edges (also called arcs (Arc)), and arcs are ordered pairs of vertices, denoted as $\langle v, w \rangle$, where v and w are the vertices. Among them, v is the end of the arc (the root of the arrow) and w is the head of the arc (at the arrow).

For an undirected graph, $E(G)$ is a finite set of edges, edges are unordered pairs of vertices, denoted as (v, w) or (w, v), and $(v, w) = (w, v)$.

2.5.1 Meaning of computational graph

A computational graph is a graphical representation of a computational process, a "language" for describing equations. This language is used to describe a function, and a neural network is just a function, so it needs a language for describing functions. In fact, there are many ways to define a graph; here, a node represents an operation (simple function), and an edge represents a variable, which can be a scalar, vector, or even tensor.

Both the computational graph and the backpropagation algorithm are important core concepts for deep learning to train neural networks.

2.5.1.1 *Forward propagation*

Forward propagation, also known as positive propagation, is the process of evaluating the value of a mathematical expression represented by a computational graph. Forward propagation means passing the value of a variable forward from the left side (input) to the right side where the output is located.

The multiplication operation of matrices in the forward propagation of a neural network is known as "affine transformation" in the field of geometry. Therefore, the implementation of the affine transformation processing is referred to as the "Affine layer".

2.5.1.2 *Backward propagation*

Backward propagation is the passing of values of variables from the right (outputs) to the left (inputs) with the aim of calculating the gradient of each input with respect to the final output, which are essential for training neural networks using the gradient descent method.

2.5.1.3 *Computational graph for propagation of additive/multiplicative nodes*

(1) Additive nodes of forward and backward propagation
Its computational graph is shown in Figure 2.21.

(2) Multiplication nodes of forward and backward propagation
Its computational graph is shown in Figure 2.22.

2.5.2 Computational graph for the AFFINE layer/SOFTMAX layer

2.5.2.1 *Computational graph for the AFFINE layer*

For example, the computational graph of forward propagation for equation $\boldsymbol{Y} = \boldsymbol{X} \cdot \boldsymbol{W} + \boldsymbol{B}$ is shown in Figure 2.23.

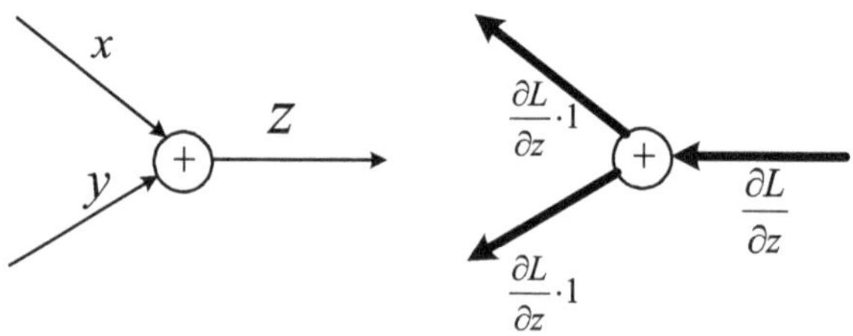

Figure 2.21. Addition nodes.
Note: The backpropagation of addition nodes only multiplies by 1, so the input value will flow unchanged to the next node.

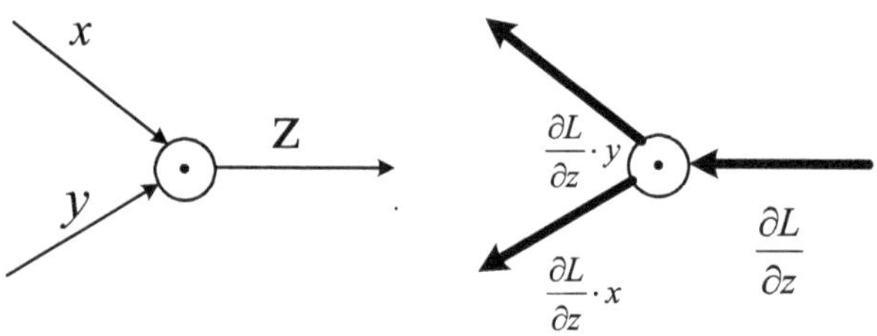

Figure 2.22. Multiplication nodes.
Note: The backpropagation of multiplication multiplies the upsteam value by the "flip value" of the input signals during forward propagation on and passes it downstream.

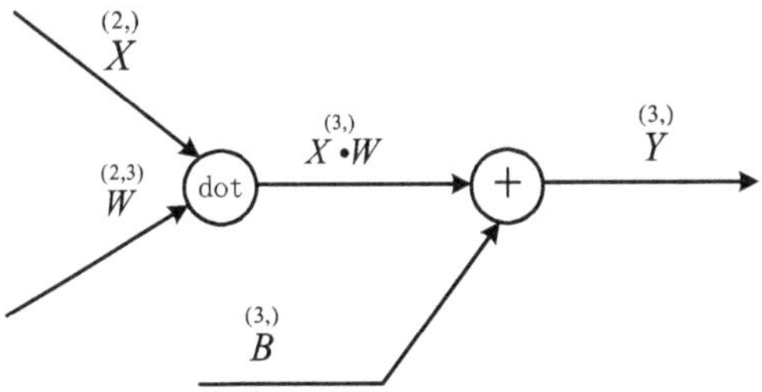

Figure 2.23. Computational graph.

In Figure 2.23, the variables are matrices, and the shape of each variable is labeled above that variable. $\boldsymbol{X}, \boldsymbol{W}$, and $\boldsymbol{B}$ are multi-dimensional arrays of shapes (2,), (2, 3), and (3,), respectively.

The backward propagation of the equation is written as $\boldsymbol{Y} = \boldsymbol{X} \cdot \boldsymbol{W} + \boldsymbol{B}$ and plotted as shown in Figure 2.24.

$$\frac{\partial L}{\partial \boldsymbol{X}} = \frac{\partial L}{\partial \boldsymbol{Y}} \cdot \boldsymbol{W}^{\mathrm{T}},$$

$$\frac{\partial L}{\partial \boldsymbol{W}} = \boldsymbol{X}^{\mathrm{T}} \cdot \frac{\partial L}{\partial \boldsymbol{Y}}.$$

2.5.2.2 *Computational graph of the Softmax-with-Loss layer*

The processing performed in neural networks has two phases: inference and learning. Inference in neural networks does not usually use the Softmax layer. The unregularized output of a neural network is sometimes called the "score". That is, when the inference of a neural network requires only one answer, it is only interested in the maximum value of the score. Therefore, a Softmax layer is not needed, whereas a Softmax layer is needed for the learning phase of the neural network.

After using the cross-entropy error as the loss function of the Softmax function, the backpropagation gives a "nice" result like $(\hat{y}_1 - y_1, \hat{y}_2 - y_2, \hat{y}_3 - y_3)$. In fact, such a "beautiful" result is not by chance; rather, the cross-entropy error function is deliberately designed to get such a result. In the regression problem, the output layer uses the "identity function" and the loss function uses the

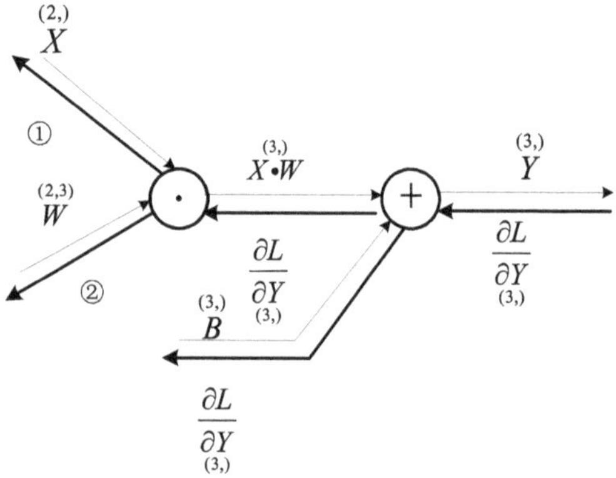

① $\underset{(2,)}{\frac{\partial L}{\partial X}} = \underset{(3,)}{\frac{\partial L}{\partial Y}} \cdot \underset{(3,2)}{W^T}$

② $\underset{(2,3)}{\frac{\partial L}{\partial W}} = \underset{(2,1)}{X^T} \cdot \underset{(1,3)}{\frac{\partial L}{\partial Y}}$

(a)

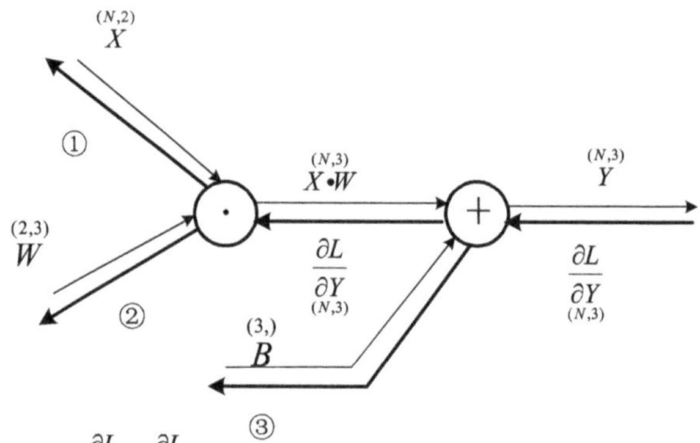

① $\underset{(N,2)}{\frac{\partial L}{\partial X}} = \underset{(N,3)}{\frac{\partial L}{\partial Y}} \cdot \underset{(3,2)}{W^T}$

② $\underset{(2,3)}{\frac{\partial L}{\partial W}} = \underset{(2,N)}{X^T} \cdot \underset{(N,3)}{\frac{\partial L}{\partial Y}}$

③ Sum in the direction of the first axis (0th axis) of $\underset{(3)}{\frac{\partial L}{\partial B}} = \underset{(N,3)}{\frac{\partial L}{\partial Y}}$

(b)

Figure 2.24. Backward propagation. (a) $\boldsymbol{X}, \boldsymbol{W}$, and $\boldsymbol{B}$ are multi-dimensional arrays of shapes (2,), (2, 3), and (3,), respectively. (b) $\boldsymbol{X}, \boldsymbol{W}$, and $\boldsymbol{B}$ are multi-dimensional arrays of shape (N, 2), (2, 3), (3,), respectively.

"sum-of-squares error" for the same reason. That is to say, the sum-of-squares error is used as the loss function of the identity function, and backpropagation can get "beautiful" results.

The Softmax function regularizes the input values before outputting them. The Softmax-with-Loss layer is represented in a computational graph, as shown in Figure 2.25.

In Figure 2.25, the Softmax and Cross-Entropy Error (CEE) layers are omitted from the computational graph.

2.5.2.2.1 *Forward propagation*

The Softmax function can be expressed as

$$y_i = \frac{\exp(a_i)}{\sum_{n=1}^{N} \exp(a_n)}. \tag{2.5.1}$$

The computational graph for the Softmax layer is shown in Figure 2.26. In Figure 2.26, the sum of the exponents (which corresponds to the denominator of Equation (2.5.1)) is abbreviated as S. The final output is denoted as (y_{p1}, y_{p2}, y_{p3}).

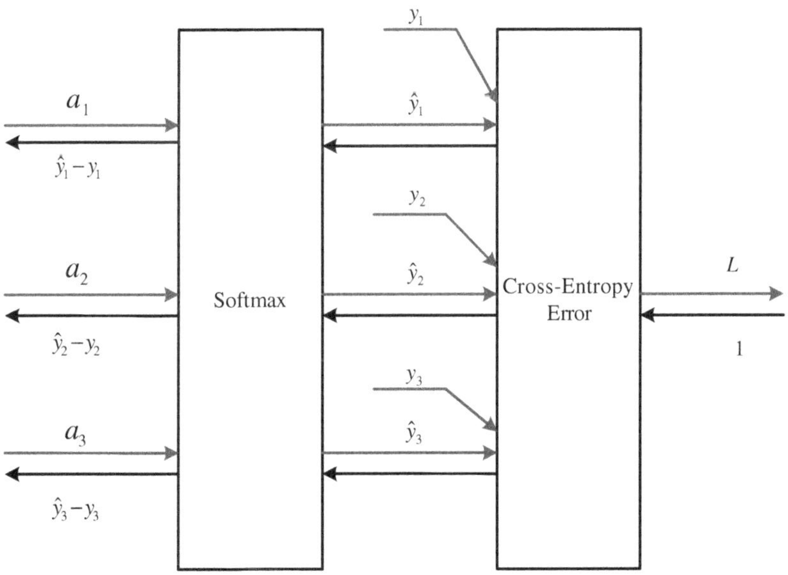

Figure 2.25. Computational diagram for the Softmax-with-Loss layer.

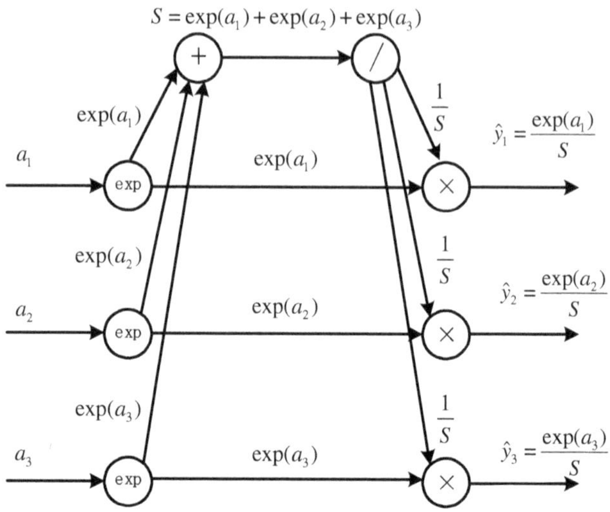

Figure 2.26. Calculated plot for Softmax layer (forward propagation only).

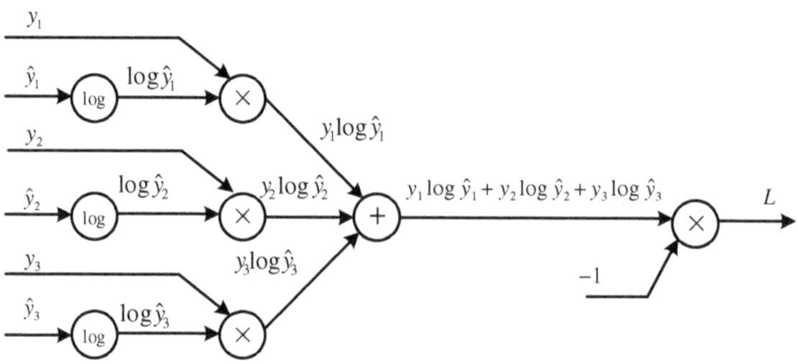

Figure 2.27. Computational graph of Cross-Entropy Error layer (forward propagation).

The cross-entropy error for the Cross-Entropy Error layer is defined as

$$L = -\sum_k y_k \log \hat{y}_k. \tag{2.5.2}$$

The computational graph for the CEE layer is shown in Figure 2.27.

2.5.2.2.2 *Backpropagation*

The computational graph of backpropagation for the CEE layer is shown in Figure 2.28.

Note the following points:

(1) The initial value of the backpropagation (the rightmost value in Figure 2.28) is 1 (because $\frac{\partial L}{\partial L} = 1$).
(2) The backpropagation of the "×" node flips the input value from the forward propagation, multiplies it by the derivative coming from upstream, and then passes it downstream.
(3) The "+" node passes the derivative coming from upstream to downstream unchanged.
(4) The formula followed by the backpropagation of the 'log' node is as follows:

$$y = \log x,$$
$$\frac{\partial y}{\partial x} = \frac{1}{x}.$$

The computational graph of backpropagation for Softmax-with-Loss is shown in Figure 2.29.

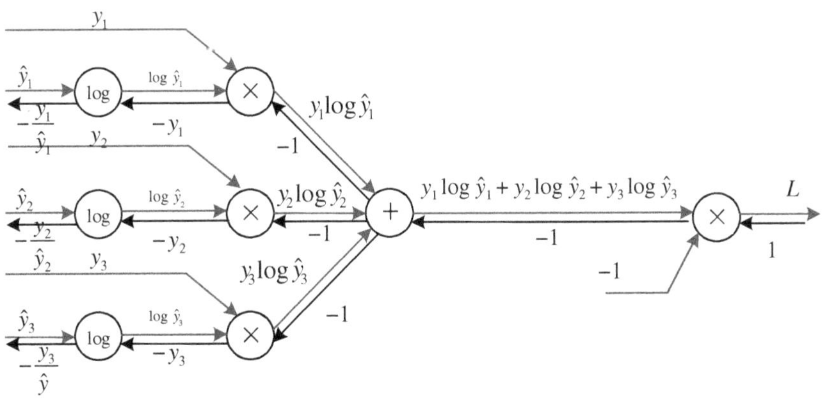

Figure 2.28. Backpropagation of cross-entropy error.

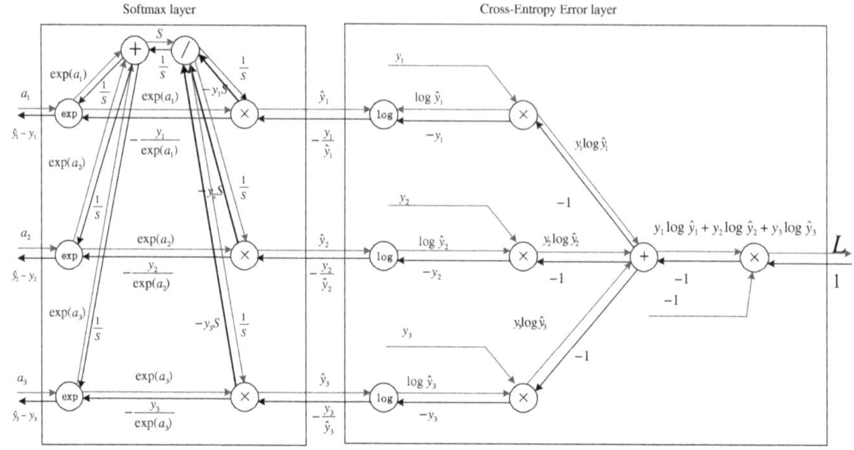

Figure 2.29. Computational functions for Softmax-with-Loss layers.

2.5.3 Computational graph of the activation function

2.5.3.1 *Sigmoid backpropagation*

2.5.3.1.1 *Functional equation*

Functional equation is given by

$$y = \frac{1}{1 + \exp(-x)}. \tag{2.5.3}$$

2.5.3.1.2 *Computational graph*

The computational graph of Equation (2.4.3) is shown in Figure 2.30.

The "/" node denotes $y = 1/x$, and its derivative is calculated by

$$\begin{aligned} \frac{\partial y}{\partial x} &= -\frac{1}{x^2} \\ &= -y^2. \end{aligned} \tag{2.5.4}$$

The node "exp" denotes $y = \exp(x)$, and its derivative is calculated by

$$\frac{\partial y}{\partial x} = \exp(x). \tag{2.5.5}$$

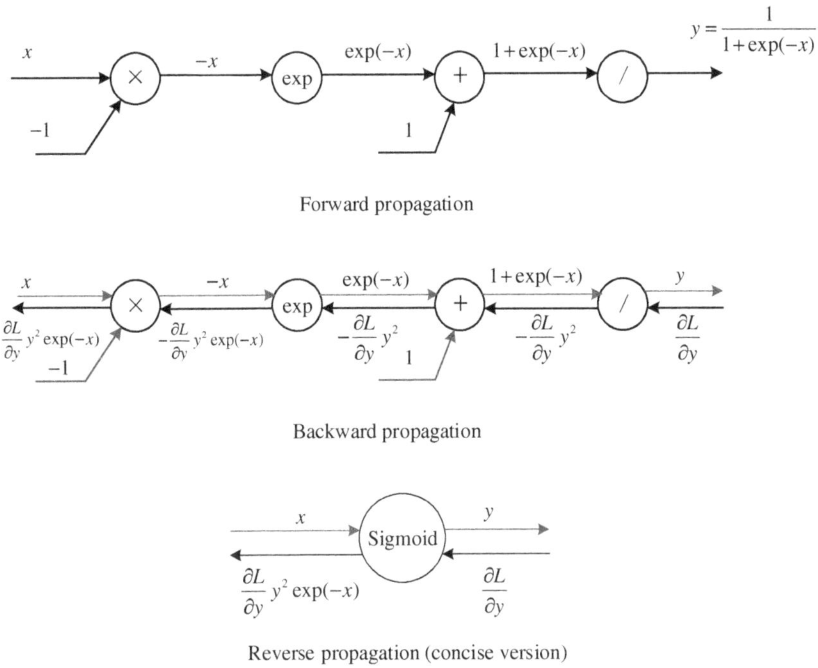

Figure 2.30. Computational graph.

2.5.3.1.3 *Simplified output*

The simplified output is represented as

$$\frac{\partial L}{\partial y} y^2 \exp(-x) = \frac{\partial L}{\partial y} y(1-y). \tag{2.5.6}$$

2.5.3.2 *ReLU backpropagation*

(1) The expression of the ReLU function is given by

$$f(x) = \max(0, x) \tag{2.5.7}$$

or

$$f(x) = \begin{cases} x & (x > 0) \\ 0 & (x \leq 0). \end{cases} \tag{2.5.8}$$

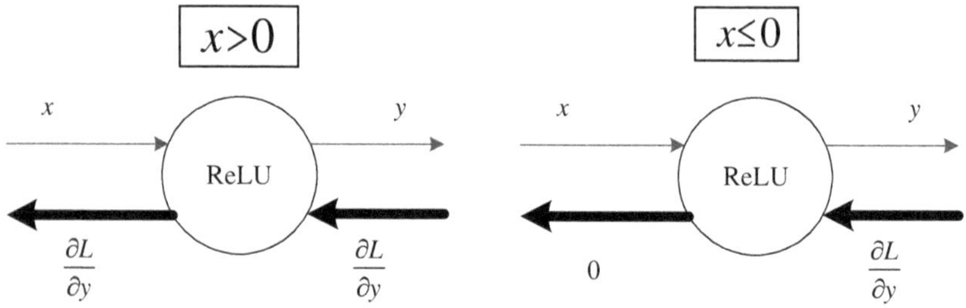

Figure 2.31. Calculated diagram of the ReLU layer.
Note: The ReLU layer is like a switch in a circuit.

(1) For forward propagation, the switch is set to ON when there is current passing through. The switch is set to OFF when there is no current passing through.
(2) For backpropagation, with the switch ON, the current will pass directly. With the switch OFF, no current will pass.

Through Equation (2.5.8), the derivative of $f(x)$ with respect to x is given by

$$\frac{\partial y}{\partial x} = \begin{cases} 1 & (x > 0) \\ 0 & (x \leq 0). \end{cases} \tag{2.5.9}$$

In Equation (2.5.9), if the input x in forward propagation is greater than 0, then backpropagation will steadfastly pass the upstream value to the downstream. Conversely, if the input x for forward propagation is less than or equal to 0, the signals passed downstream in backpropagation will stop here. The computational graph is shown in Figure 2.31.

2.6 Regularization Penalty Term

The learning ability of neural networks is too strong, the complexity is too high, the training time is too long, the activation function is inappropriate, or the amount of data is too small, which often leads to overfitting. Regularization methods are a general term for a class of methods that introduce additional information into the original model at this time to prevent overfitting and improve the model's generalization performance. In practical deep learning scenarios, it can be found that the best fitting model (in the sense of minimizing generalization error) is a properly regularized large model.

2.6.1 Parameter norm penalty

The capacity of the model can be reduced by adding a parameter norm penalty term $\Omega(\boldsymbol{w})$ to the objective function; this is a common class of regularization methods. The regularized loss function is defined as

$$\overline{\text{Loss}}(\boldsymbol{w}) = \text{Loss}(\boldsymbol{w}) + \alpha\Omega(\boldsymbol{w}), \tag{2.6.1}$$

where $\alpha \in [0, \infty)$ weighs the relative contribution of the parameter norm penalty term, and a larger α corresponds to more regularization.

Typically, in deep learning, only constraints are added to the weight $\boldsymbol{w}$ of the network, and no constraints are added to the bias terms. The main reason is that the bias term generally requires less data to fit accurately, is not regularized, and will not cause too much variance. In addition, the regularization of the bias parameters may cause significant underfitting.

2.6.2 L2 parameter regularization

The common L2 parameter regularization is used to bring the weights closer to the origin by adding a regular term $\Omega(\boldsymbol{w}) = \frac{1}{2}|\boldsymbol{w}|^2$ to the objective function. The L2 parameter regularization method is also called weight decay or Ridge regression.

The loss function after L2 parameter regularization is defined as

$$\overline{\text{Loss}}(\boldsymbol{w}) = \text{Loss}(\boldsymbol{w}) + \frac{1}{2}\alpha\boldsymbol{w}\boldsymbol{w}^{\mathrm{T}}. \tag{2.6.2}$$

The gradient corresponding to the loss function is calculated by

$$\nabla_{\boldsymbol{w}}\overline{\text{Loss}}(\boldsymbol{w}) = \nabla_{\boldsymbol{w}}\text{Loss}(\boldsymbol{w}) + \alpha\boldsymbol{w}. \tag{2.6.3}$$

The weights are updated using a single-step gradient descent method as

$$\boldsymbol{w}(k+1) = \boldsymbol{w}(k) - \eta(\alpha\boldsymbol{w}(k) + \nabla_{\boldsymbol{w}}\text{Loss}(\boldsymbol{w})) \tag{2.6.4}$$

or

$$\boldsymbol{w}(k+1) = (1 - \eta\alpha)\boldsymbol{w}(k) - \eta\nabla_{\boldsymbol{w}}\text{Loss}(\boldsymbol{w})). \tag{2.6.5}$$

Equation (2.6.5) shows that the addition of weight decay induces a modification of the learning rule, i.e., the weight vectors (multiplying the weight vectors by a constant factor) are shrunk before updating the general gradient at each step. Since η and α are both greater than 0, the weights of the model after regularization have smaller values after each update step relative to the model without regularization.

Assuming that it is a quadratic optimization problem (e.g., using a squared loss function), the model parameters can be further expressed as $\tilde{\boldsymbol{w}} = \frac{\lambda}{\lambda+\alpha}\boldsymbol{w}$, i.e., it is equivalent to adding a control factor to the original parameters, where λ is the eigenvalue of the parameter Hessian matrix. The following can be seen:

When $\lambda \gg \alpha$, the role of the penalty factor is relatively small.

When $\lambda < \alpha$, the corresponding parameter can be shrunk to zero.

In Figure 2.32, the solid line represents the contour of the unregularized objective function, and the dashed circle represents the contour of the L2 norm term. At the point $\tilde{\boldsymbol{w}}$, these two competing objectives reach equilibrium. In the horizontal direction, the point $\tilde{\boldsymbol{w}}$ moves horizontally from the point $\boldsymbol{w}^*$, the value of the objective function does not increase much, i.e., the objective function does not have a strong preference in this direction, and thus the regularization has a stronger effect in this direction, which is manifested by pulling $\boldsymbol{w}_1$ a longer distance toward the origin $\boldsymbol{w}^*$. On the other hand, in the direction of the vertical axis, the objective function is very sensitive to the movement away from the origin $\boldsymbol{w}^*$, i.e., the curvature of the

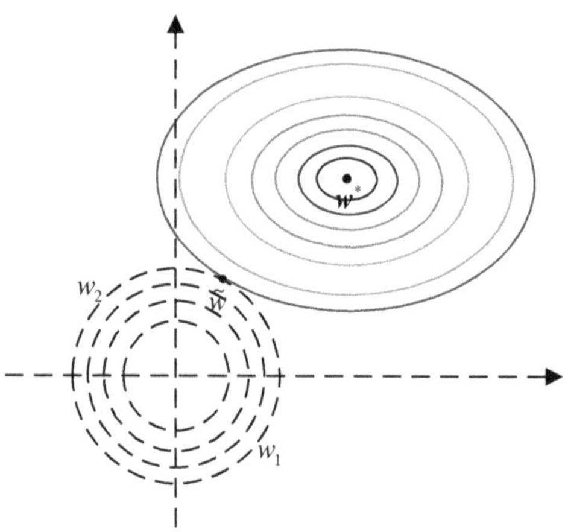

Figure 2.32. Contours of the objective function.

objective function is very high in this direction, so the regularization has less effect on $\boldsymbol{w}_2$.

The effect of smoothing the original function to some extent is reflected by adding the L2 norm penalty term to the original objective function and its gradient function.

For a class with a large number of stationary points (i.e., points with a gradient of zero), the L2 norm penalty term adding to the objective function means adding prior knowledge to distinguish the region where the derivative is originally zero (geometrically, it means that the region of the original platform tilts toward the zero point direction), which can help the optimization algorithm converge to at least one local optimal solution instead of staying at a saddle point.

By restricting the parameters $\boldsymbol{w}$ around the zero-point, the convergence procedure is accelerated and the difficulty of optimization has been reduced. We recall that for a class of common activation functions such as Sigmoid monotone bounded. According to the monotone boundedness theorem, for an arbitrarily small $\eta > 0$, a sufficiently large z_0 can be obtained to make $f'(z) = \eta(z < z_0)$. In other words, for this variable, a sufficiently large region of $(z, +\infty)$ can be found such that its derivative is close to zero. This means that the improvement of the variable based on the gradient descent method becomes extremely slow (recall the update of the backward propagation algorithm), and even the convergence speed is affected by the floating-point precision. In contrast, the above situation can be avoided by using the L2 norm to control the size of the variable around zero, which can accelerate the convergence of the optimization algorithm to a large extent.

2.6.3 L1 parameter regularization

L2 parameter regularization is the most common form of weight decay, and other methods can be used to limit the size of the model parameters. For example, L1 parameter regularization may be used.

L1 regularization of the model parameter $\boldsymbol{w}$ is defined as the sum of the absolute values of the weight vector $\boldsymbol{w}$ or the sum of the absolute values of each parameter., i.e.,

$$\Omega(\boldsymbol{w}) = ||\boldsymbol{w}||_1 = \sum_i |w_i|.$$

We will now discuss the impact of L1 regularization on a simple linear regression model. Like analyzing L2 regularization, bias parameters are not considered.

We are particularly interested in identifying the differences between L1 and L2 regularization. Similar to L2 weight decay, the strength of L1 weight decay can also be controlled by scaling the positive hyperparameter of the penalty term $\Omega(\boldsymbol{w})$. Therefore, the objective function of regularization is defined as

$$\overline{\text{Loss}}(\boldsymbol{w}) = \text{Loss}(\boldsymbol{w}) + \alpha|\boldsymbol{w}|_1. \tag{2.6.6}$$

The corresponding gradient is calculated by

$$\nabla_{\boldsymbol{w}}\overline{\text{Loss}}(\boldsymbol{w}) = \nabla_{\boldsymbol{w}}\text{Loss}(\boldsymbol{w}) + \alpha\,\text{sgn}(\boldsymbol{w}), \tag{2.6.7}$$

where $\text{sgn}(\boldsymbol{w})$ simply takes the plus and minus signs of the individual elements.

Obviously, the effect of L1 regularization is very different from that of L2. Specifically, the effect of regularization on the gradient is no longer to scale each element w_i in a linear manner. But instead, a constant with the same sign as $\text{sgn}(w_i)$ is added. After using this form of gradient, it may not be possible to obtain a direct arithmetic solution for the quadratic approximation of $\text{Loss}(\boldsymbol{w})$ (which can be achieved during L2 regularization).

Since $\alpha > 0$, the L1 parameter regularization, therefore, the weight vector updated at each step is closer to zero comparson the model without reqularization.

In the special case, for quadratic optimization problems, we assume that the corresponding Hessian matrix is diagonal, and the derived parameter recursion formula is given by

$$w_i = \text{sgn}(\boldsymbol{w}_i^*)\max\left(|\boldsymbol{w}_i^*| - \frac{\alpha}{\lambda_i}, 0\right). \tag{2.6.8}$$

Equation (2.6.8) shows that when $\boldsymbol{w}_i^* < \alpha/\lambda_i$, the corresponding parameter will be reduced to zero, which is also different from L2 regularization. Since L2 regularization does not directly reduce the parameter to zero but rather to a value very close to zero.

Overall, the difference between L1 regularization and L2 regularization is as follows:

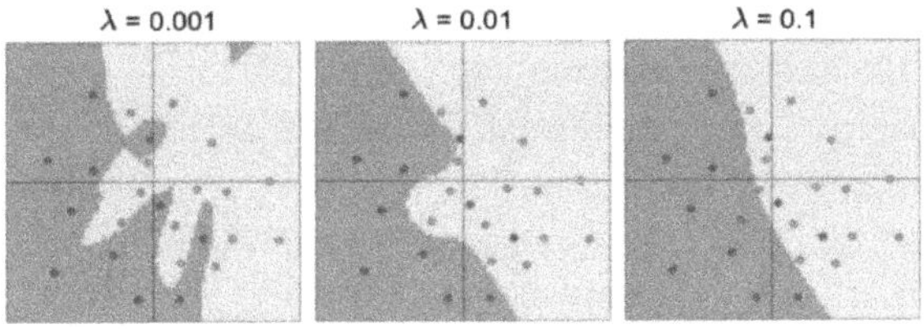

Figure 2.33. Penalty coefficient.

(1) Through the above analyses, L1 regularization can produce a more sparse model relative to L2 regularization, i.e., when L1 regularization can directly reduce the parameter to zero, the parameter $\boldsymbol{w}$ is relatively small and can directly be reduced to zero. Therefore, it can play an improvement role in feature selection, and the technique is also called LASSO.
(2) If analyzing from a probabilistic point of view, many of the norm constraints are equivalent to adding prior distributions to the parameters, where the L2 norm is equivalent to the parameters obeying a Gaussian prior distribution, and the L1 norm is equivalent to a Laplace distribution.

Figure 2.33 shows that the penalty coefficient $\lambda = 0.001$ is overfitting (due to insufficient penalization) and $\lambda = 0.1$ has a strong generalization ability.

2.7 BP Algorithm for Neural Network

2.7.1 BP algorithm idea

The multi-layer perceptron encounters a bottleneck in the problem of how to obtain the weights of the hidden layer. Since the weights of the hidden layer cannot be obtained directly, can the weights of the hidden layer be adjusted indirectly by first obtaining the output result and the error of the desired output through the output layer? The BP algorithm is designed using such an idea, and its basic idea is that the learning process consists of two processes: forward propagation of signals and backward propagation of errors.

When the propagation of signals is forward, the input samples are passed in from the input layer, processed layer by layer through each hidden layer, and then passed to the output layer. If the actual output of the output layer does not match the desired output, it is transferred to the backpropagation stage of the error.

In the backpropagation stage, the propagation of the output from the hidden layer to the input layer belongs to layer-by-layer backpropagation, and the error will be apportioned to all the units of each layer to obtain the error signals of each unit of each layer, which is used as a basis for correcting the weights of each unit.

2.7.2 Characterization of three elements of BP network

When analyzing an ANN, it is common to start with its three elements, i.e., network topology, activation function, and learning algorithm, as shown in Figure 2.34.

2.7.2.1 *Topology and activation function of BP network*

A Backpropagation (BP) network is actually a multi-layer perceptron, so its topology is the same as that of a multi-layer perceptron.

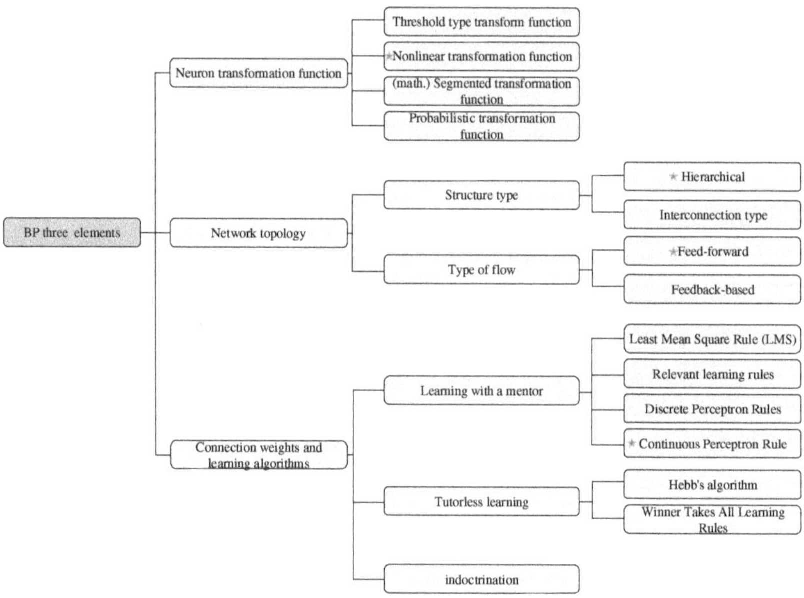

Figure 2.34. Three elements of BP network.

The single hidden-layer (three-layer) perceptron is most commonly used because it has been able to solve simple nonlinear problems. The topology of a three-layer perceptron is shown in Figure 2.35. A simplest three-layer BP configuration has a number of neurons (neurons) in each layer, which maintains connections to each neuron in the previous layer, and whose inputs are linear combinations of the outputs of each neuron in the previous layer. The output of each neuron is a function of its input, which is called the activation function and a nonlinear transformation function. It is characterized by the fact that the function itself and its derivatives are continuous, thus making it very convenient to handle.

In Figure 2.35, x_n denotes the input of the nth node of the input layer, $n = 1, 2, \ldots, N$, w_{ji} denotes the weight between the jth node of the hidden layer and the ith node of the input layer, b_j represents the threshold of the jth node of the hidden layer, $\phi(x)$ is the activation function of the hidden layer, w_{mj} indicates the weight between the mth node of the output layer and the jth node of the hidden layer, $j = 1, 2, \ldots, J, c_m$ stands for the threshold of the mth node of the output layer, $m = 1, 2, \ldots, J, \psi(x)$ is marked as the threshold of the mth node of the output layer, $m = 1, 2, \ldots, M$, and $\hat{z}_m$ is recorded as the output of the mth node of the output layer.

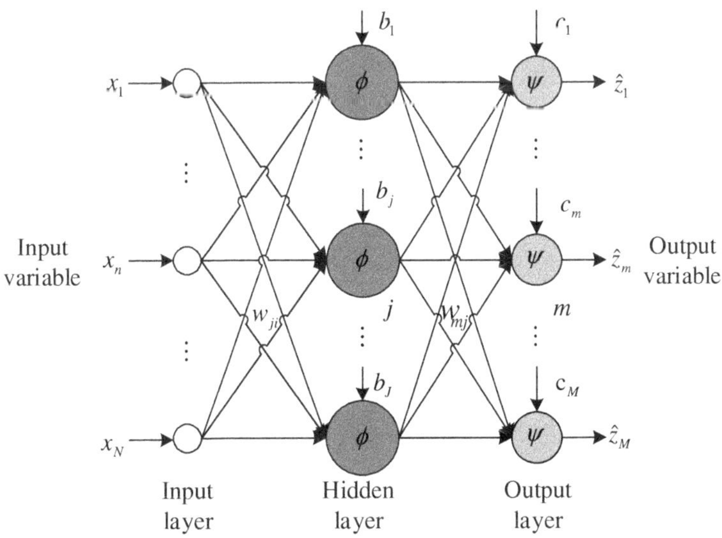

Figure 2.35. Three-layer neural network structure.

2.7.2.2 *BP algorithm*

2.7.2.2.1 *Forward propagation process of signals*

The input of the jth node of the hidden layer is written as

$$u_j = \sum_{i=1}^{N} w_{ji} x_i + b_j. \tag{2.7.1}$$

The output of the jth node of the hidden layer is given by

$$v_j = \phi(u_j) = \phi\left(\sum_{i=1}^{N} w_{ji} x_i + b_j\right). \tag{2.7.2}$$

The input to the mth node of the output layer is represented as

$$net_m = \sum_{j=1}^{J} w_{mj} v_j + c_m = \sum_{j=1}^{J} w_{mj} \phi\left(\sum_{i=1}^{N} w_{ji} x_i + b_j\right) + c_m. \tag{2.7.3}$$

The output of the mth node of the output layer is calculated by

$$\begin{aligned} \hat{z}_m = \psi(net_m) &= \psi\left(\sum_{j=1}^{J} w_{mj} v_j + c_m\right) \\ &= \psi\left(\sum_{j=1}^{J} w_{mj} \phi\left(\sum_{i=1}^{N} w_{ji} x_i + b_j\right) + c_m\right). \end{aligned} \tag{2.7.4}$$

2.7.2.2.2 *Backpropagation process of errors*

The backpropagation of error, that is, the output error of neurons in each layer, is first calculated layer by layer from the output layer, and then the weights and thresholds of each layer are adjusted according to the gradient descent method so that the final output $\hat{z}$ of the modified network can be close to the desired value z.

The quadratic error criterion function E_i for the ith sample is defined as

$$E_i = \frac{1}{2} \sum_{m=1}^{M} (z_{im} - \hat{z}_{im})^2. \tag{2.7.5}$$

The total error criterion function of the system for N training samples is defined as

$$E = \frac{1}{2}\sum_{i=1}^{N}\sum_{m=1}^{M}(z_{im} - \hat{z}_{im})^2. \tag{2.7.6}$$

According to the gradient descent method, the correction of the weights Δw_{mj} of the output layer, the correction of the thresholds Δc_m of the output layer, the correction of weights Δw_{ji} of the hidden layer, and the correction of the thresholds Δb_j of the hidden layer are corrected sequentially:

$$\begin{aligned} \Delta w_{mj} &= -\eta\frac{\partial E}{\partial w_{mj}}; \quad \Delta c_m = -\eta\frac{\partial E}{\partial c_m}; \\ \Delta w_{ji} &= -\eta\frac{\partial E}{\partial w_{ji}}; \quad \Delta b_j = -\eta\frac{\partial E}{\partial b_j}. \end{aligned} \tag{2.7.7}$$

The weight adjustment formula of the output layer is given by

$$\Delta w_{mj} = -\eta\frac{\partial E}{\partial w_{mj}} = -\eta\frac{\partial E}{\partial net_m}\frac{\partial net_m}{\partial w_{mj}} = -\eta\frac{\partial E}{\partial \hat{z}_m}\frac{\partial \hat{z}_m}{\partial net_m}\frac{\partial net_m}{\partial w_{mj}}. \tag{2.7.8}$$

The threshold adjustment formula of the output layer is given by

$$\Delta c_m = -\eta\frac{\partial E}{\partial c_m} = -\eta\frac{\partial E}{\partial net_m}\frac{\partial net_m}{\partial c_m} = -\eta\frac{\partial E}{\partial \hat{z}_m}\frac{\partial \hat{z}_m}{\partial net_m}\frac{\partial net_m}{\partial c_m}. \tag{2.7.9}$$

The weight adjustment formula of the hidden layer is given by

$$\Delta w_{ji} = -\eta\frac{\partial E}{\partial w_{ji}} = -\eta\frac{\partial E}{\partial u_j}\frac{\partial u_j}{\partial w_{ji}} = -\eta\frac{\partial E}{\partial v_j}\frac{\partial v_j}{\partial u_j}\frac{\partial u_j}{\partial w_{ji}}. \tag{2.7.10}$$

The threshold adjustment formula of the hidden layer is given by

$$\Delta b_j = -\eta\frac{\partial E}{\partial b_j} = -\eta\frac{\partial E}{\partial u_j}\frac{\partial u_j}{\partial b_j} = -\eta\frac{\partial E}{\partial v_j}\frac{\partial v_j}{\partial u_j}\frac{\partial u_j}{\partial b_j}, \tag{2.7.11}$$

again because

$$\frac{\partial E}{\partial \hat{z}_m} = -\sum_{i=1}^{N}\sum_{m=1}^{M}(z_{im} - \hat{z}_{im}), \tag{2.7.12}$$

$$\frac{\partial net_m}{\partial w_j} = v_j, \quad \frac{\partial net_m}{\partial c_m} = 1, \quad \frac{\partial u_j}{\partial w_{ji}} = x_i, \quad \frac{\partial u_j}{\partial b_j} = 1, \tag{2.7.13}$$

$$\frac{\partial E}{\partial v_j} = -\sum_{i=1}^{N}\sum_{m=1}^{M}(z_{im} - \hat{z}_{im})\psi'(net_m)w_{mj}, \tag{2.7.14}$$

$$\frac{\partial v_j}{\partial u_j} = \phi'(u_j), \tag{2.7.15}$$

$$\frac{\partial z_m}{\partial net_m} = \psi'(net_m). \tag{2.7.16}$$

So, in the end, we obtain

$$\Delta w_{mj} = \eta\sum_{i=1}^{N}\sum_{m=1}^{M}(z_{im} - \hat{z}_{im})\psi'(net_m)v_j, \tag{2.7.17}$$

$$\Delta c_m = \eta\sum_{i=1}^{N}\sum_{m=1}^{M}(z_{im} - \hat{z}_{im})\psi'(net_m), \tag{2.7.18}$$

$$\Delta w_{ji} = \eta\sum_{p=1}^{N}\sum_{k=1}^{K}(z_{pm} - \hat{z}_{pm})\psi'(net_m)w_{mj}\cdot\phi'(u_j)x_i, \tag{2.7.19}$$

$$\Delta b_j = \eta\sum_{i=1}^{N}\sum_{m=1}^{M}(z_{im} - \hat{z}_{im})\cdot\psi'(net_m)\cdot w_{mj}\cdot\phi'(u_j). \tag{2.7.20}$$

The flow chart of the BP algorithm is shown in Figure 2.36.

2.7.2.2.3 *Improvement of BP algorithm*

The essence of the BP algorithm is to solve the problem of the minimum value of the error function; due to its use of the steepest descent method in nonlinear programming, modifying the weights in the direction of the negative gradient of the error function, there are some problems such as low learning efficiency, slow convergence,

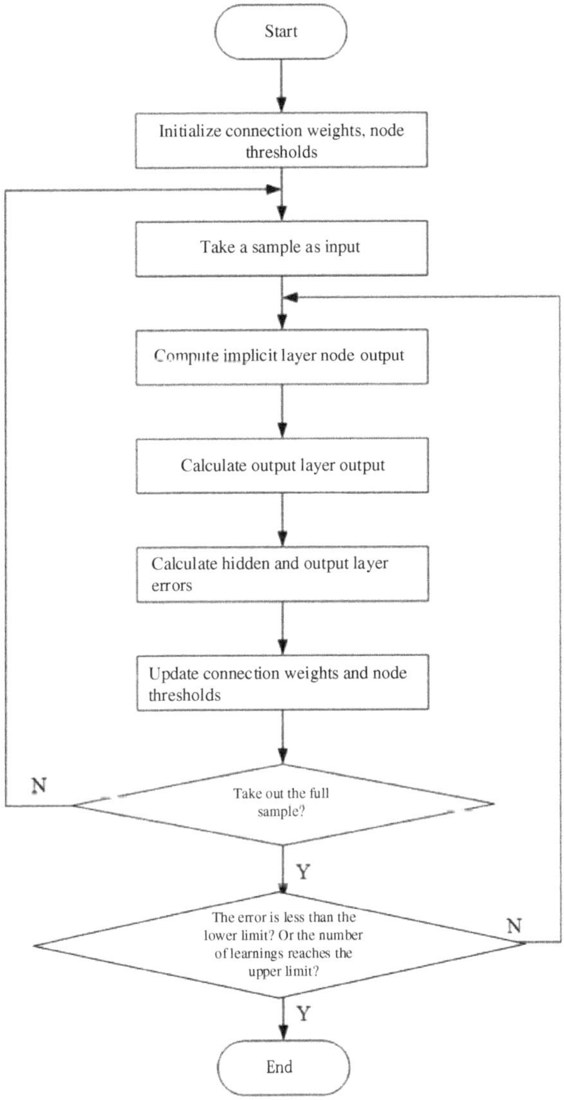

Figure 2.36. Flowchart of BP algorithm program.

and easy to fall into the local minimal state. For this purpose, the following improvements are proposed.

(1) *Additional momentum method*: The additional momentum method enables the network to consider not only the effect of the

error on the gradient but also the effect of the changing trendancy on the error surface when correcting the weights. Without the effect of additional momentum, the network may fall into shallow local minima. However, it is possible to slip past these minima using the effect of additional momentum.

This method adds a value proportional to the previous weight (or threshold) change to each weight (or threshold) change in backpropagation and generates new weight (or threshold) changes based on backpropagation.

The adjustment formulas for weights and thresholds with additional momentum factors are denoted as

$$\Delta w_{ji} \leftarrow (1 - m_c)\eta e_j y_i + m_c \Delta w_{ji}, \tag{2.7.21}$$

$$\Delta b_j \leftarrow (1 - m_c)\eta e_j + m_c \Delta b_j, \tag{2.7.22}$$

where $e_j = z_j - \hat{z}_j$ is the network error and m_c is the momentum factor, generally taken to be around 0.95.

The essence of the additional momentum method is to transfer the impact of the last weight (or threshold) change through a momentum factor. When the momentum factor is zero, the change in weight (or threshold) is only generated by the gradient descent method. When the momentum factor is set to 1, the new change of the weight (or threshold) is set as the change of the last weight (or threshold), while the change generated by the gradient method is ignored. In this way, when the momentum term is increased, it causes the adjustment of weights to change toward the average direction at the bottom of the error surface. When the network weights enter the flat area at the bottom of the error surface, the error e_j will become very small, and thus, $\Delta w_{ji}(k+1) = \Delta w_{ji}(k)$ to prevent the emergence of $\Delta w_{ji} = 0$, which helps make the network jump out from the local minima of the error surface.

According to the design principle of the additional momentum method, when the modified weights cause too much growth in the error, the new weights should be canceled and not adopted, and the momentum effect should be stopped to prevent the network from entering the larger error surface. When the new change rate of the error exceeds a predetermined maximum change rate of error for its old value, the calculated weight changes must also be canceled. Its maximum change rate of error can be greater than or equal to 1.

The typical value is set to 1.04. Therefore, when designing the training program for the additional momentum method, conditional judgment must be added to correctly use its weight correction formula.

The judgment conditions based on the momentum method are defined as

$$m_c = \begin{cases} 0 & E(k) > 1.04E(k-1) \\ 0.95 & E(k) < E(k-1) \\ m_c & \text{Other} \end{cases}, \tag{2.7.23}$$

where $E(k)$ is the sum of squares of the kth step errors.

(2) *Adaptive learning rate*: It is not easy to choose an appropriate learning rate for a particular problem. It is usually obtained empirically or experimentally, but even so, the learning rate that works well for the initial period at the beginning of training is not necessarily appropriate for later training. To solve this problem, automatic adjustment of the learning rate is implemented during training. Usually, the guidelines for adjusting the learning rate are as follows: Check whether the weights really reduce the error function; if they do, then the selected learning rate is too small and can be increased by an appropriate amount. If there is no reduction in the error function, overshooting will occur, then the learning rate should be reduced. The adjustment formula of adaptive learning rate is defined as

$$\eta(k+1) = \begin{cases} 1.05\eta(k) & E(k+1) < E(k) \\ 0.7\eta(k) & E(k+1) > 1.04E(k) \\ \eta(k) & \text{other} \end{cases}. \tag{2.7.24}$$

The selection range of the initial learning rate $\eta(0)$ can have great flexibility.

(3) Adjustment algorithm of moment-adaptive learning rate

The employment of the aforementioned momentum method enables the BP algorithm to identify the global optimal solution. Furthermore, the implementation of an adaptive learning rate within the BP algorithm has the capacity to reduce the duration of training. By combining these two methods to train a neural network, an adjustment algorithm of a momentum adaptive learning rate is obtained.

2.8 Overfitting and Underfitting

The terms "underfitting" and "overfitting" are used to describe situations in which a linear regression model does not accurately represent the data. "High bias" refers to a situation in which the model predicts values that are too high or too low compared to the actual values of the data. "High variance" refers to a situation in which the standard deviation of the model predictions is too high compared to the actual data. These concepts are illustrated in Figure 2.37. As illustrated in Figure 2.37(a), a straight line is employed for making the prediction model. Obviously, no matter how the starting point and slope can be adjusted, and straight line will not fit the given five training samples well, let alone new data. Figure 2.37(c) perfectly fits the training samples using high-order polynomials, which may result in significant errors when new data is provided. Figure 2.37(b) fits the training data perfectly without being too complicated and basically depicts the relationship between $x1$ and $x2$ in predicting clearly.

For logistic regression, the same problem also exists, as shown in Figure 2.38.

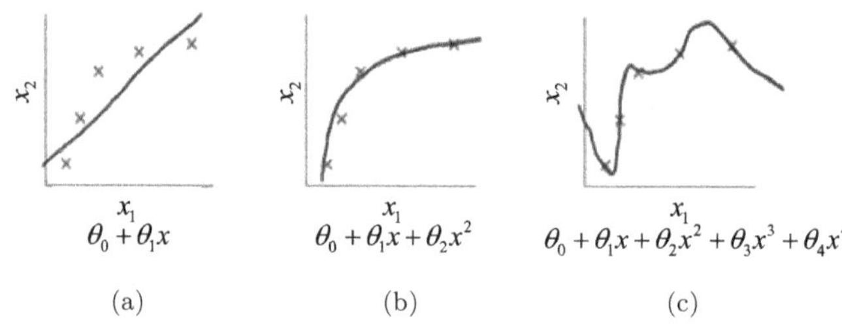

Figure 2.37. Fitting the training data. (a) Linear fitting. (b) Best fitting. (c) Overfitting.

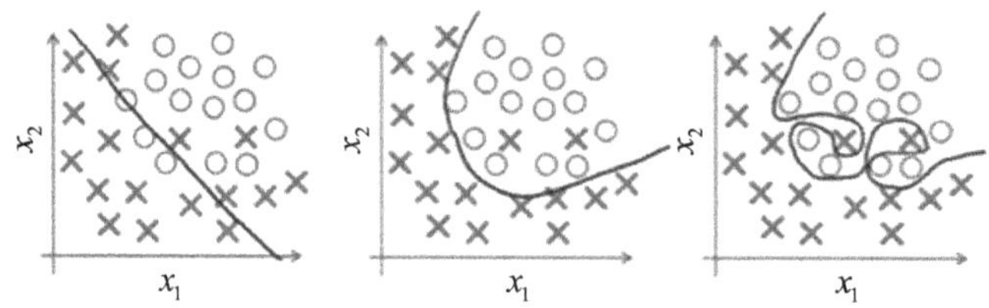

Figure 2.38. Logistic regression problem.

In machine learning, the process of learning the objective function from training data is called inductive learning. The performance of concepts learned from a machine learning model when encountering new data, such as prediction accuracy, is called generalization. Fitting refers to the degree of closeness to the objective function.

2.8.1 Basic concepts

2.8.1.1 *Overfitting*

An overfitted model performs well on the training set and poorly on the testing set. That is to say, the model performs very well in a known dataset, and much worse in a testing set with some new data. This is due to the fact that there are too many influencing factors to need to be considered and too many dimensions beyond the independent variable.

2.8.1.2 *Underfitting*

The model does not fit well enough, the model performs poorly on the training set, the data is not fully utilized, and the accuracy of the prediction is low.

2.8.1.3 *Bias*

First error = bias + variance. Bias is the error between the outputs of the model on the samples and the expected values, i.e., the accuracy of the model itself.

2.8.1.4 *Variance*

Variance is a measure of the difference between the outputs of the model and the expected values. The variance in statistics is the average of the squared values of the difference between each sample value of the model and the average of the values of all the samples, i.e., the stability of the model. Figure 2.39 gives the distance between the output values of the bias model and the expected output of the model (red centers), while variance refers to the distance between each output of the model and the expectation value.

Just like target shooting, low variance means that the distance between the aiming point and the bullseye (i.e., the red center) is

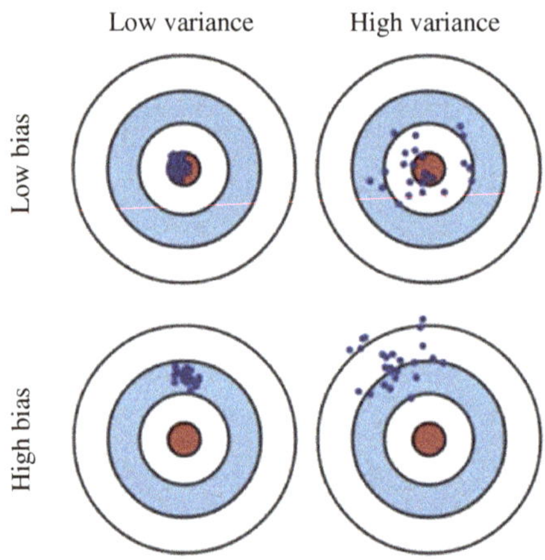

Figure 2.39. Distance between output values and output expectation (red center) in model.

close, while high variance means that the distance between the aiming point and the bullseye (i.e., the red center) is far. Low variance means that after aiming at a point, the position of the target in the bullet shot is relatively close to the position of the aimed point; high variance means that after aiming at a point, the position of the target in the bullet shot is relatively far from the position of the aimed point:

❶ With low bias and low variance, the predicted value is right on the bullseye (closest to the expected value) and more centralized (low variance) desired result.

❷ With low bias and high variance, the predicted values basically fall around the expected value but are scattered, at which point the variance is large, it indicates that the model is not stable enough.

❸ With high bias and low variance, the predicted value and the expected value have a large distance, but at this time, the value is very concentrated, and the variance is small. The stability of the model is better, but the prediction accuracy is not high, in the state of "predicting inaccuracies as always".

❹ With high bias and high variance, it is not the least expected results; the model is not only inaccurate but also unstable, and the difference between the predicted values is relatively large each time.

To prevent the model from learning incorrect or irrelevant patterns from the training data, the optimal solution is to get more data. Given enough data to train the model to as many exceptions as possible, it will keep correcting itself to get better results.

There are several methods to obtain more data, including the following:

① *Get more data from the data source.* This method is the easiest to think of, but in most cases, it is not easy to significantly increase the data itself. In addition, it is also not clear how much data to acquire to make the model perform better.
② *Estimate the parameters of the data distribution based on the current dataset and use that distribution to generate more data.* This method is generally not used because the process of estimating distribution parameters also introduces sampling error.
③ *Data augmentation*: The data are expanded by certain rules. For example, in object classification problems, the position, posture, scale, and overall image brightness of objects in the image will not affect the classification results. Therefore, the database can be expanded exponentially through methods such as image translation, flipping, scaling, and cutting.

2.8.2 Preventing overfitting by reducing feature variables

Overfitting is mainly caused by two reasons: too little data or too complex model. So, the overfitting problem can be prevented by using a model of appropriate complexity that fits the real rules well enough without fitting too many sampling errors:

(1) *Paremeter influence*: The number of network layers and neurons all can effect the fitting ability of the network.
(2) *Early stopping*: For each neuron, the performance of its activation function is different in different intervals, as shown in Figure 2.40.

When the network weights are small, the neuron's activation function works in the linear zone; at this time, the neuron's fitting ability is weak (similar to a linear neural network).

With the above consensus, it is possible to explain why limiting the training time (early stopping) is useful: because it is generally

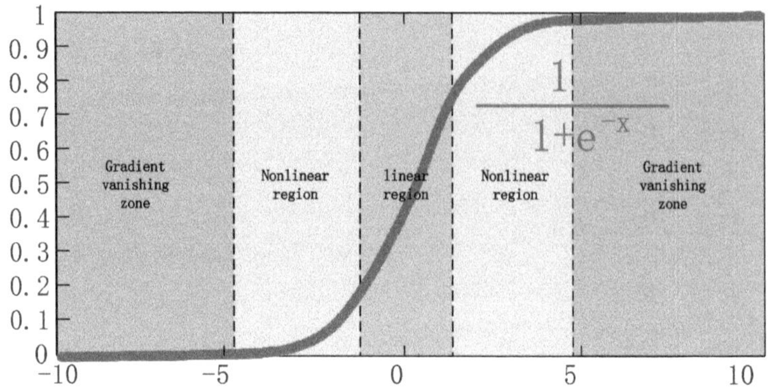

Figure 2.40. Performance of activation function in different zones.

initialized to smaller weights when initializing the network. The longer the training time, the larger some of the network weights may be, and if the training is stopped at the right time, it is possible to limit the network's capabilities to a certain range.

2.8.3 Preventing overfitting with weight regularization

A simple model with weight regularization (weight regularization) is a model with a smaller entropy of the distribution of parameter values. The weights of the model are forced to take only smaller values to limit the complexity of the model, whereas the distribution of weight values is more regular. The method is to add the cost associated with larger weight values to the network loss function.

For L1 regularization, the added cost is proportional to the absolute value of the weight coefficients (i.e., the L1 norm of the weights).

For L2 regularization, the added cost is proportional to the square of the weight coefficients (i.e., the L2 norm of the weights), also known as weight decay.

Regularization based on the L1 and L2 norms can be uniformly defined as L-P norms, i.e.,

$$L_p = \sqrt[p]{\sum_{p=1}^{N} x_l^p}, \tag{2.8.1}$$

where $x_l = (x_1, x_2, \ldots, x_N)$.

Figure 2.41. Different Euclidean distances.

The L-P norm is not a norm but a group of norms, which depends on the variation of P; the paradigms have different variations, and a classical one concerning the variation of P norms is shown in Figure 2.41.

Figure 2.41 shows the variation of the graph composed of points with a distance (norm) of 1 from the origin in three-dimensional space as p changes from infinity to zero. Taking the common L2 norm ($p = 2$) as an example, this norm is also known as Euclidean distance. The point with a Euclidean distance of 1 to the original point in space forms a sphere.

Weight decay (L2 regularization) can avoid overfitting of the model. The principle is as follows:

(1) Explained from the complexity of the model: Smaller weights w, in a sense, indicate lower network complexity and better fitting of data (this rule is also called Occam's razor).
(2) From a mathematical perspective, when overfitting, the coefficients of the fitting function are often very large, and the resulting fitting function fluctuates greatly. That is to say, the changes in function values are very drastic in certain small intervals. This means that the derivative (absolute value) of the function in certain small intervals is very large. Since the independent variable values can be large or small, only coefficients that are large enough can ensure a large derivative value. And regularization is achieved by constraining the norm of parameters to not be too large, which can reduce overfitting to a certain extent.

2.8.4 Preventing overfitting with cross-validation

Cross-validation is the process of slicing and dicing the sample data, combining them into different training and testing sets, using the training set to train the model, and using the testing set to evaluate how well the model predicts.

The meaning of cross is that a certain sample in a certain training set may become a sample in the testing set. There are three types of cross-validation:

2.8.4.1 *Simple cross-validation*

The sample data are divided into two parts (generally, 70% training set, 30% testing set), then the training set is used to train the model, and the testing set is utilized to validate the model and its parameters. It only needs to shuffle the original data and divide it into two groups, but there must be no crossover. Due to random grouping, the classification accuracy of the validation set has a great relationship with the grouping of the original data. Some data may have never been trained or tested, while others have been selected as training or testing data more than once.

2.8.4.2 *S-fold cross-validation*

S-fold cross-validation will randomly divide the sample into S parts, S-1 parts of them are randomly selected as the training set each time, and the remaining one as the testing set, as shown in Figure 2.42.

The selection method of k is as follows: When the amount of data is small, k can be set to larger. When the amount of data is large, k can be set to smaller. Here, the selection of k considers two extreme cases:

(1) Do not use cross-validation at all, i.e., $k = 1$, the model is easy to overfit, which can be understood as that the model has learnt all the features of the data, resulting in the model fitting the training data well, i.e., the bias is very small, but in fact, some of the features are not necessary to be learnt, and the result is a low bias and high variance.
(2) The leave-one-out method, i.e., $k = n$, where the variance of a single model evaluation gradually increases while the bias decreases as the value of k continues to rise (i.e., tends to overfit, since the full data is used again), and the computational effort increases dramatically.

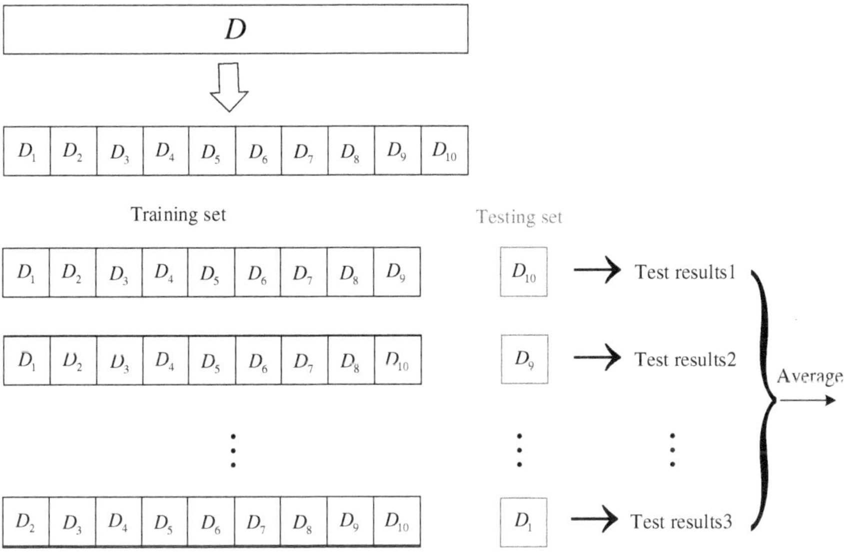

Figure 2.42. Principle of S-fold cross-validation.

In short, the use of some datasets will definitely increase bias and decrease variance compared to the entire dataset. That is to say, the selection of k is a trade-off between bias and variance.

2.8.4.3 *Leave-one-out cross-validation*

When S is equal to the number of samples N, so, for N samples, $N-1$ samples are chosen to train the data each time, and the remaining samples are used to verify that the model is good or bad.

2.8.4.4 *Self-help method*

If the sample is really too small, you can use the self-help method (bootstrapping) to perform validation. N samples are put back to form a training set (with duplicates), and the remaining samples are used as a testing set.

The purpose of cross-validation is as follows:

(1) The fundamental reason is that the data is limited. When the amount of data is not large enough, if all the data are used

to train the model, it is easy to lead to model overfitting, and the cross-validation must be used to assess the predictive performance of the model, especially the performance of the trained model on the new data, which can reduce overfitting to a certain extent.

(2) By cross-validation of data partitioning and integration of evaluation results, the variance in model selection can be effectively reduced.

2.8.5 Preventing overfitting with dropout regularization

Dropout regularization is a new approach to prevent the overfitting of the model. The idea of this method is that overfitting can be significantly reduced by ignoring half of the feature detectors in each training batch (making half of the hidden layer node values zero).

2.8.5.1 *Dropout method*

2.8.5.1.1 *Training phase*

(1) Dropout is to make the activation value of the hidden layer become zero in a certain proportion on the regular BP network structure, i.e., randomly disable a part of the hidden layer nodes according to a certain proportion. In the testing phase, on the basis that the nodes of the hidden layer are disabled, the input data can also be disabled in a certain proportion (20% for testing) to get a better result.

(2) A weight penalty term is not added, but an upper range for each weight is set. If the weights exceed this upper limit during the training process, the weights are set to the value of this upper limit.

This way of handling, no matter how large the weight update amount is, the weight will not be too large. In addition, the algorithm can be made to use a relatively large learning rate to speed up the learning process and allowed to search for better weights in a broader weight space without worrying about excessively large weights.

(3) Training process with dropout. To achieve the ensemble property, the neural network with dropout is shown in Figure 2.43. Its training and prediction then undergo some changes. Here, dropout can use different retention probabilities to determine neuron selection.

According to Figure 2.43, for the neural network without dropout, there are the following equations:

$$y_i^{(l+1)} = w_i^{(l+1)} x_i^{(l)} + b_i^{(l+1)}, \tag{2.8.2}$$

$$\ddot{z}_i^{(l+1)} = f(y_i^{(l+1)}). \tag{2.8.3}$$

For neural networks with dropout, there are the following equations:

$$r_i^{(l)} \sim \text{Bernoulli}(p), \tag{2.8.4}$$

$$\tilde{x}_i^{(l)} = r_i^{(l)} \otimes x_i^{(l)}, \tag{2.8.5}$$

$$y_i^{(l+1)} = w_i^{(l+1)} \tilde{x}_i^{(l)} + b_i^{(l+1)}, \tag{2.8.6}$$

$$\hat{z}_i^{(l+1)} = f(y_i^{(l+1)}). \tag{2.8.7}$$

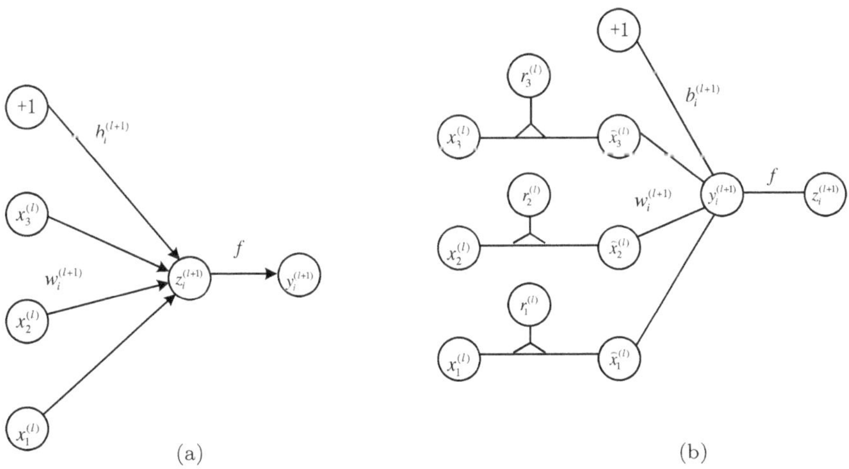

Figure 2.43. Network with dropout. (a) Standard network. (b) Dropout network.

Inevitably, a probabilistic process is added to each unit of the training network.

2.8.5.1.2 *Testing phase*

The output values of the nodes of the hidden layer are scaled down to $(1-v)$ times before the network is propagated forward to the output layer. For example, if the normal output of the hidden layer is y, it needs to be scaled down to $y(1-v)$. For prediction, the parameters of each cell are premultiplied by p, as shown in Figure 2.44.

In addition to this, there is another way in which the prediction phase remains unchanged while the training phase changes.

If the scaling factor of the inverted dropout is $\frac{1}{1-p}$, then

$$r_i^{(l)} \sim \text{Bernoulli}(p), \tag{2.8.8}$$

$$\tilde{x}_i^{(l)} = r_i^{(l)} \otimes x_i^{(l)}, \tag{2.8.9}$$

$$y_i^{(l+1)} = w_i^{(l+1)} \tilde{x}_i^{(l)} + b_i^{(l+1)}, \tag{2.8.10}$$

$$\hat{z}_i^{(l+1)} = \frac{1}{1-p} f(y_i^{(l+1)}). \tag{2.8.11}$$

2.8.5.1.3 *Dropout principle analyses*

Dropout can be regarded as a kind of model averaging. The so-called model averaging is to average the estimates or predictions from different models by certain weights. Model averaging is also called model combination, which generally includes combination estimation and combination prediction [22].

The dropout can reflect the characteristics of "different models"; its reason is to randomly choose to ignore nodes of the hidden layer, in each batch of the training process, due to randomly ignoring different nodes of each hidden layer, so each training of the network is different;

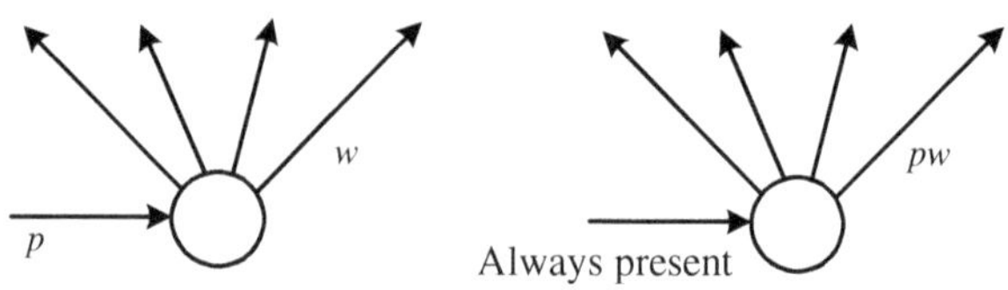

Figure 2.44. Scaling down the output values of the hidden layer nodes.

each training can be done alone with a "new" model. In addition, the hidden nodes appear randomly with a certain probability, so it cannot be guaranteed that every two hidden nodes appear at the same time every time. This way, the update of weights no longer depends on the joint action of the hidden nodes with a fixed relationship, which prevents the situation where some features are only effective in other specific features.

It can be seen that the dropout process is a very effective method of averaging neural network models by training a large number of different networks to average the prediction probabilities. Different models are trained on different training sets (each batch of training data is randomly selected), and finally, each model is "fused" with the same weights, similar to the boosting algorithm.

2.8.5.2 *Bayesian regularization*

2.8.5.2.1 *Bayesian theory*

Bayesian decision theory is an important part of subjective Bayesian inductive theory [23]. Bayesian decision-making is to estimate the partially unknown state with subjective probability under incomplete intelligence, then correct the probability of occurrence with the Bayesian formula, and finally make the optimal decision by using the expected value and the corrected probability.

Bayesian decision theory is a basic method in statistical modeling decision-making, and its basic idea is as follows:

(1) The class conditional probability density and prior probability are known.
(2) *A posteriori* probability is converted by using the Bayesian formula.
(3) Decision classification is done according to the size of a posteriori probability.

Let $S_1, S_2, \ldots, S_N$ be a division of the sample space S; if we denote $p(S_i)$ as the probability of occurrence of an event S_i, the conditional probability can be calculated by

$$p(S_j|x) = \frac{p(x|S_j)p(S_j)}{\sum_{i=1}^{N} p(x|S_i)p(S_i)}, \tag{2.8.12}$$

where $p(S_i) > 0$ and $i = 1, 2, \ldots, N$.

2.8.5.2.2 *Understanding regularization from a Bayesian perspective*

Given observational data S, the Bayesian approach estimates the parameter $\boldsymbol{w}$ by maximizing the posterior probability [23]. We obtain

$$\tilde{\boldsymbol{w}} = \arg\max_{w} p(\boldsymbol{w}|S) = \arg\max_{w} \frac{p(S|\boldsymbol{w})p(\boldsymbol{w})}{p(S)} = \arg\max_{w} p(S|\boldsymbol{w})p(\boldsymbol{w}), \tag{2.8.13}$$

where $p(S|\boldsymbol{w})$ is the likelihood function, i.e., the probability of occurrence of the observed data S when the parameter vector $\boldsymbol{w}$ is known, and $p(\boldsymbol{w})$ is the prior probability of the parameter vector $\boldsymbol{w}$.

For the likelihood function, there is

$$p(S|\boldsymbol{w}) = \prod_{n=1}^{N} p(S_n|\boldsymbol{w}). \tag{2.8.14}$$

Taking the logarithm of the posterior probability, we have

$$\tilde{\boldsymbol{w}} = \arg\max_{\boldsymbol{w}} p(S|\boldsymbol{w})p(\boldsymbol{w}) = \arg\max_{\boldsymbol{w}} \prod_{n=0}^{N} p(S_n|\boldsymbol{w})p(\boldsymbol{w}), \tag{2.8.15a}$$

$$\tilde{\boldsymbol{w}} = \arg\max_{\boldsymbol{w}} \sum_{n=0}^{N} \log p(S_n|\boldsymbol{w}) + \log p(\boldsymbol{w})$$

$$= \arg\min_{\boldsymbol{w}} \left[-\log \sum_{n=0}^{N} p(S_n|\boldsymbol{w}) - \log p(\boldsymbol{w}) \right]. \tag{2.8.15b}$$

While the probability distribution satisfies the normal distribution, i.e.,

$$p(w_i) = N(w_i|m_{w_i}, \sigma^2) = \frac{1}{\sqrt{2\pi\sigma^2}} e^{-\frac{(w_i - m_{w_i})^2}{2\sigma^2}}, \tag{2.8.16}$$

$$p(\boldsymbol{w}) = \prod_{i=1}^{N} p(w_i), \tag{2.8.17}$$

$$\begin{aligned}
\tilde{\boldsymbol{w}} &= \arg\min_{\boldsymbol{w}} \left[-\log \sum_{n=0}^{N} p(S_n|\boldsymbol{w}) - \log p(\boldsymbol{w}) \right] \\
&= \arg\min_{\boldsymbol{w}} \left[-\log \sum_{n=0}^{N} p(S_n|\boldsymbol{w}) - \sum_{m=0}^{M} \log p(w_m) \right] \\
&= \arg\min_{\boldsymbol{w}} \left[-\log \sum_{n=0}^{N} p(S_n|\boldsymbol{w}) + \sum_{m=0}^{M} \frac{1}{\sigma^2}(w_m - m_{w_m})^2 \right] \\
&= \arg\min_{\boldsymbol{w}} \left[-\log \sum_{n=0}^{n} p(S_n|\boldsymbol{w}) + \lambda \sum_{m=0}^{M} w_m^2 \right] \\
&\quad \left(m_{w_m} = 0, \sigma = \sqrt{\frac{1}{\lambda}} \right).
\end{aligned} \tag{2.8.18}$$

Compare the following equation:

$$\tilde{\boldsymbol{w}} = \arg\min_{w} \sum_{i} \mathrm{Loss}(y_i, f(x_i; \boldsymbol{w})) + \lambda\Omega(\boldsymbol{w}).$$

It can be seen that the likelihood function part corresponds to the loss function (empirical risk), while the prior probability part corresponds to the regularization term. The L2 regularization is equivalent to the prior probability distribution of the parameter $\boldsymbol{w}$ satisfying the normal distribution:

$$p(w_i) = N(w_i|m_{w_i}, b) = \frac{1}{2b} e^{-\frac{|w_i - m_{w_i}|}{b}}. \tag{2.8.19}$$

When the prior probability distribution satisfies the Laplace distribution, we get

$$\begin{aligned}
\tilde{\boldsymbol{w}} &= \arg\min_{\boldsymbol{w}} \left[-\log \sum_{n=0}^{N} p(S_n|\boldsymbol{w}) - \log p(\boldsymbol{w}) \right] \\
&= \arg\min_{\boldsymbol{w}} \left[-\log \sum_{n=0}^{N} p(S_n|\boldsymbol{w}) - \sum_{m=0}^{M} \log p(w_m) \right]
\end{aligned}$$

$$= \arg\min_{\boldsymbol{w}} \left[-\log \sum_{n=0}^{N} p(S_n|\boldsymbol{w}) + \sum_{m=0}^{M} \frac{1}{b} |w_m - m_{w_m}| \right]$$

$$= \arg\min_{\boldsymbol{w}} \left[-\log \sum_{n=0}^{N} p(S_n|\boldsymbol{w}) + \lambda \sum_{m=0}^{M} |w_m| \right] \quad \left(m_{w_m} = 0, b = \frac{1}{\lambda} \right). \tag{2.8.20}$$

The L1 norm is equivalent to the prior probability distribution of the parameter $\boldsymbol{w}$ satisfying the Laplace distribution.

Figure 2.45 shows the Laplace and Gaussian distribution curves. From Figure 2.45, we can know that the Laplace distribution is prominent near the value of zero, whereas the Gaussian distribution is gently distributed near the value of zero and sparsely distributed on both sides. Correspondingly, the L1 regularization tends to produce sparse models, and the L2 regularization penalizes heavily for parameters with high weights.

From a Bayesian perspective, the regularization term is equivalent to introducing the prior probability distribution of the parameters. The common L1/L2 regularization term is equivalent

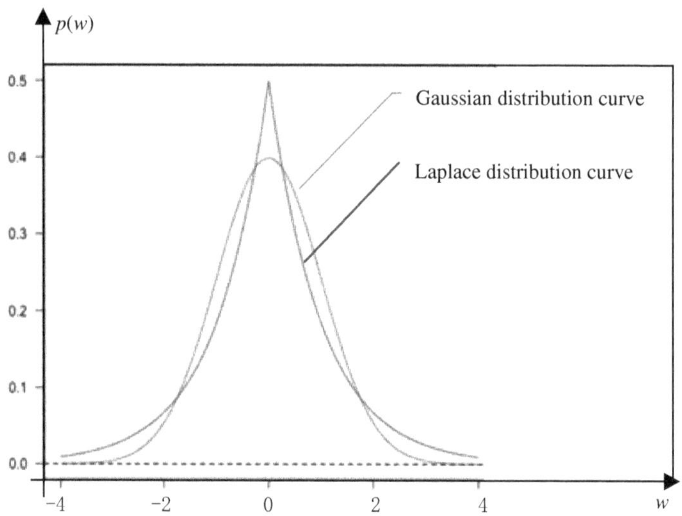

Figure 2.45. Laplace and Gaussian distribution curves.

to introducing the prior information that the parameters conform to the Laplace/Gaussian distribution, respectively.

2.9 Example 1: Momentum Blind Equalization Algorithm Based on Feedforward Neural Network

Feedforward neural networks are structured in a hierarchical form, where the signal goes from the input layer unit to its next layer, and any unit in the latter layer is connected to each unit in the previous layer, and its traditional training party is the backpropagation algorithm (BP algorithm). From a system perspective, feedforward neural networks represent the steady-state properties of nonlinear systems, and their good linear classification function has led to their widespread application in signal processing. In this section, the performance of the momentum blind equalization algorithm based on feedforward neural networks is investigated by adding a momentum term to the feedforward neural network blind equalization algorithm [24, 25].

2.9.1 Principle of the algorithm

The principle of feedforward neural network blind equalization algorithm is shown in Figure 2.46, where $\boldsymbol{a}(k)$ stands for the transmitted sequence, $\boldsymbol{c}(k)$ is the channel vector, $\boldsymbol{n}(k)$ denotes the noise vector (generally additive noise), $\boldsymbol{y}(k)$ represents the receiving sequence and is also the input sequence of the neural network, $z(k)$ marks the output sequence of the neural network, and $\hat{a}(k)$ is regarded as the output sequence of the decision device.

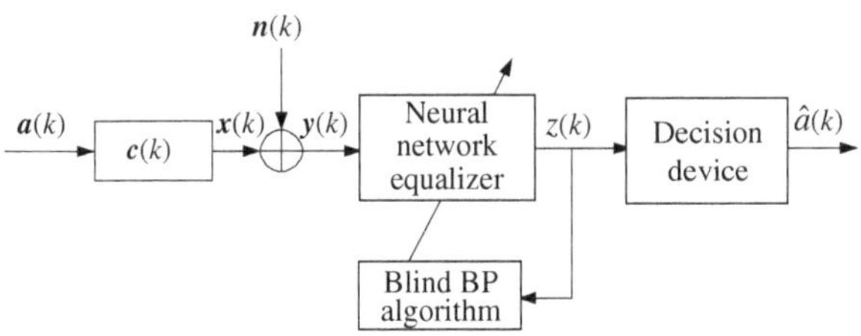

Figure 2.46. Principle of the blind equalization algorithm for neural networks.

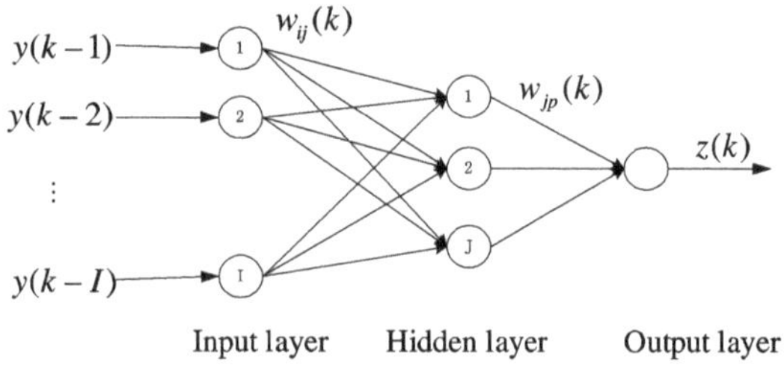

Figure 2.47. Structure of feedforward neural network.

Cybenc has proven that an arbitrary continuous function can be approximated with arbitrary accuracy using a feedforward neuron network containing a hidden layer, so a three-layer feedforward neural network (FNN) structure is used, as shown in Figure 2.47 [25].

In Figure 2.47, $w_{ji}(k)$ denotes the weights of the connection between the input layer and the hidden layer, N denotes the number of neurons in the input layer, and $i = 1, 2, \ldots, N, J$ denotes the number of neurons in the hidden layer, and $j = 1, 2, \ldots, J$. $w_{pj}(k)$ denotes the weights of the connection between the hidden layer and the output layer, p represents serial number of neuron in the output layer, the input of the input layer is marked as $\boldsymbol{y}(k) = \{y(k-1), y(k-2), \ldots, y(k-I)\}^{\mathrm{T}}$, $u_j^J(k)$ denotes the input to the jth hidden layer, $u_j^J(k)$ denotes the input to the jth hidden layer, and $j = 1, 2, \ldots, J$. The output of the jth hidden layer is denoted as $v_j^J(k)$, the pth output layer unit has an input of $u_p^P(k)$, and $p = 1, 2, \ldots, P$, and $\hat{z}(k)$ denotes the total output of neural networks. The activation function of the neuron is chosen as

$$F(x) = x + \alpha \sin(\pi x), \tag{2.9.1}$$

where $-\infty < x < \infty$, $0 < \alpha < 1$, and x is represented as $u_j^J(k)$ or $u_p^P(k)$, and the function $F(x)$ has good recognition ability for input signals.

According to Figure 2.45, it is obtained that

$$\boldsymbol{x}(k) = \mathbf{c}^T \boldsymbol{a}(k), \tag{2.9.2}$$

$$\boldsymbol{y}(k) = \boldsymbol{x}(k) + \boldsymbol{n}(k), \tag{2.9.3}$$

$$\boldsymbol{z}(k) = \boldsymbol{w}^T(k)\boldsymbol{y}(k), \tag{2.9.4}$$

$$\hat{a}(k) = \mathrm{dec}(z(k)). \tag{2.9.5}$$

The error function of Constant Modulus Algorithm (CMA) is defined as

$$e(k) = [|z(k)|^2 - R^2], \tag{2.9.6}$$

where

$$R^2 = E(|a(k)|^4)/E(|a(k)|^2), \tag{2.9.7}$$

where $a(k)$ is the modulus of the transmitted signals.

The cost function is defined as

$$J(k) = E[e^2(k)]. \tag{2.9.8}$$

The instantaneous value of the gradient of the cost function is calculated by

$$\begin{aligned}\frac{\partial J(k)}{\partial \boldsymbol{w}(k)} &= 2e(k)\frac{\partial e(k)}{\partial \boldsymbol{w}(k)} \\ &= 2e(k)\frac{\partial}{\partial \boldsymbol{w}(k)}\left[z(k)z^*(k) - R^2\right] \\ &= 2e(k)\frac{\partial}{\partial \boldsymbol{w}(k)}\left[z(k)z^*(k)\right] \\ &= 2e(k)\frac{\partial}{\partial \boldsymbol{w}(k)}\left[\boldsymbol{w}^H(n)\boldsymbol{y}(k)(\boldsymbol{y}^H(k)\boldsymbol{w}(k))\right] \\ &= 4e(k)\boldsymbol{y}(k)z^*(k). \end{aligned} \tag{2.9.9}$$

According to the steepest descent algorithm, the iterative formula for the weights can be obtained by

$$\boldsymbol{w}(k+1) = \boldsymbol{w}(k) - \mu\frac{\partial J(k)}{\partial \boldsymbol{w}(k)}, \tag{2.9.10}$$

where μ is the step size. The selection of μ is very important; the larger the value of μ, the faster the convergence of the algorithm, but the larger the steady state error. If μ is too large, it will also cause the algorithm to be unstable. The smaller the value of μ, although the instability can be avoided and the steady state error will not increase, the convergence speed will become slow. In order to improve the convergence speed in the case of a smaller value of μ, the simplest and most effective way is to add the "momentum term".

According to Figure 2.47, the state equations of the feedforward neural network are given by

$$u_j^J(k) = \sum_{i=1}^{I} \boldsymbol{w}_{ji}(k)\boldsymbol{y}(k-i), \tag{2.9.11}$$

$$v_j^J(k) = F\left(\sum_{i=1}^{I} \boldsymbol{w}_{ji}(k)\boldsymbol{y}(k-i)\right), \tag{2.9.12}$$

$$u_p^P(k) = \sum_{j=1}^{J} \boldsymbol{w}_{pj}(k)v_j^J(k), \tag{2.9.13}$$

$$z(k) = F\left(\sum_{j=1}^{J} \boldsymbol{w}_{pj}(k)v_j^J(k)\right). \tag{2.9.14}$$

With the introduction of the "momentum term", the iterative formula of $w_{ji}(k)$ in the blind equalization algorithm based on feedforward neural network is written as

$$\boldsymbol{w}_{ji}(k+1) = \boldsymbol{w}_{ji}(k) + \mu_{ji}\frac{\partial J(k)}{\partial \boldsymbol{w}_{ji}(k)} + \mu_{jim}[\boldsymbol{w}_{ji}(k) - \boldsymbol{w}_{ji}(k-1)], \tag{2.9.15}$$

where μ_{jim} is the momentum term, usually constant, and $0 < \mu_{jim} < 1$, μ_{ji} is the iteration step of the weight vector:

$$\frac{\partial J(k)}{\partial \boldsymbol{w}_{ij}(k)} = \frac{\partial J(k)}{\partial z(k)}\frac{\partial z(k)}{\partial u_p^P(k)}\frac{\partial u_p^P(k)}{\partial v_j^J(k)}\frac{\partial v_j^J(k)}{\partial u_j^J(k)}\frac{\partial u_j^J(k)}{\partial \boldsymbol{w}_{ij}(k)}, \tag{2.9.16}$$

where

$$\frac{\partial J(k)}{\partial z(k)} = 2|z(k)|(|z(k)|^2 - R^2), \tag{2.9.17a}$$

$$\frac{\partial z(k)}{\partial u_p^P(k)} = F'(u_p^P(k)), \tag{2.9.17b}$$

$$\frac{\partial u_p^P(k)}{\partial v_j^J(k)} = \boldsymbol{w}_{pj}(k), \tag{2.9.17c}$$

$$\frac{\partial v_j^J(k)}{\partial u_j^J(k)} = F'(u_j^J(k)), \tag{2.9.17d}$$

$$\frac{\partial v_j^J(k)}{\partial \boldsymbol{w}_{ij}(k)} = \boldsymbol{y}_i(k). \tag{2.9.17e}$$

The iterative process for the weight vectors of the output layer remains unchanged as

$$\boldsymbol{w}_{pj}(k+1) = \boldsymbol{w}_{pj}(k) - \mu_{pj}\frac{\partial J(k)}{\partial \boldsymbol{w}_{pj}(k)}, \tag{2.9.18}$$

where μ_{pj} is the step size, and

$$\frac{\partial J(k)}{\partial \boldsymbol{w}_{pj}(k)} = \frac{\partial J(k)}{\partial z(k)}\frac{\partial z(k)}{\partial u_p^P(k)}\frac{\partial u_p^P(k)}{\partial \boldsymbol{w}_{pj}(k)}, \tag{2.9.19}$$

where

$$\frac{\partial u_p^P(k)}{\partial \boldsymbol{w}_{pj}(k)} = v_j^J(k). \tag{2.9.20}$$

Equations (2.8.1)–(2.8.12) constitute the Momentum term-based Feedforward Neural Network (MFNN) blind equalization algorithm.

2.9.2 Simulation experiments and result analyses

To verify the performance of the proposed MFNN blind equalization algorithm, its performance of the MFNN algorithm was compared with that of the Feedforward Neural Network (FNN) constant blind equalization algorithm.

【Experiment 2.1】 Multi-path acoustic channel was given by $\boldsymbol{h} = [0.35\ \ 0\ \ 0\ \ 1]$, the transmit signal was 4QAM, the Signal-to-Noise Ratio (SNR) was 20dB, the length of the feedforward neural network equalizer was set to 16, and the center tap was initialized to 0, the step size $\mu_{MFNN} = 0.4 \times 10^{-3}$, $\mu_{FNN} = 0.2 \times 10^{-2}$, $\mu_{ijm} = 0.6$,

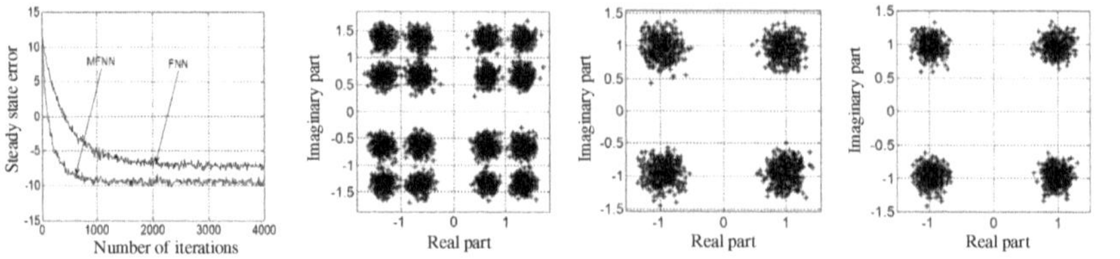

(a) Error curves (b) Input of Equalizer (c) Output of FNN (d) Output of MFNN

Figure 2.48. Simulation results.

and $\mu_{jp} = 0.001$. The results of 600 Monte Carlo simulations are shown in Figure 2.48.

Figure 2.48 shows that the steady-state error of the proposed MFNN is about 10dB, and it converges to a stable state after about 1000 iterations, while the steady-state error of the FNN is about 7dB, and it takes about 2100 iterations to reach a stable state. The output constellation diagram of the MFNN is more compact and clear. Therefore, the performance of the MFNN is better.

【Experiment 2.2】 A time-varying channel was constructed using a method of combining three channels. The transmitter transmitted a total of 15,000 signal samples. When the transmitter first transmitted the signal, which passed through the minimum-phase hydroacoustic channel $\boldsymbol{h}_1 = [1\ 0.5\ 0.25\ 0.125]$. At the moment of the $n_1 = 1000$th sampling point, the channel changes to the two-path hydroacoustic channel $\boldsymbol{h}_2 = [0.35\ 0\ 0\ 1]$, and at the moment of the $n_2 = 8000$th sampling point, the channel changes again to the mixed-phase hydroacoustic channel $\boldsymbol{h}_3 = [0.3122\ \ -0.1040\ 0.8908\ 0.3134]$. The zero-pole plots of the three channels are shown in Figure 2.49.

Figure 2.49 shows that as the transmitted signal is sampled at different times, and the zero points of the channel are spreading outward from the inside of the unit circle. Therefore, a time-varying channel is simulated by these three hydroacoustic channels, which can reflect a better representation of the time-varying characteristics of the channel.

The transmit signal was 8PSK, the SNR was set to 20dB, and the weight length of the equalizer was 16. The step size $\mu_{MFNN} = 0.00007$, $\mu_{FNN} = 0.00006$, $\mu_{ijm} = 0.8$, and $\mu_{jp} = 0.001$, the 4th tap coefficient was set to 1, and the rest was 0. The 600 times Monte Carlo true results are shown in Figure 2.50.

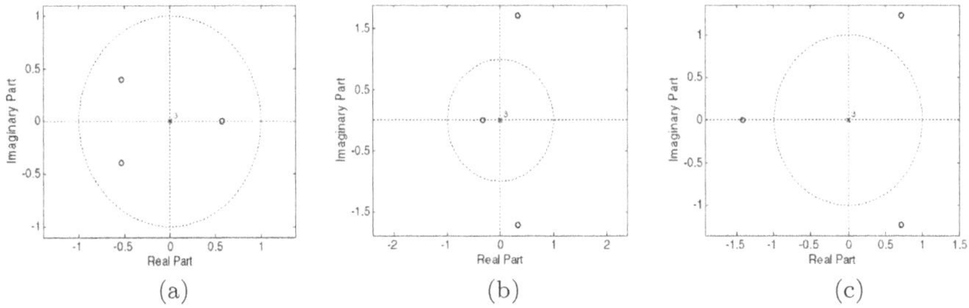

Figure 2.49. Zero-pole plots for three channels. (a) Zero-pole plot for channel (h1). (b) Zero-pole plot for channel (h2). (c) Zero-pole plot for channel (h3).

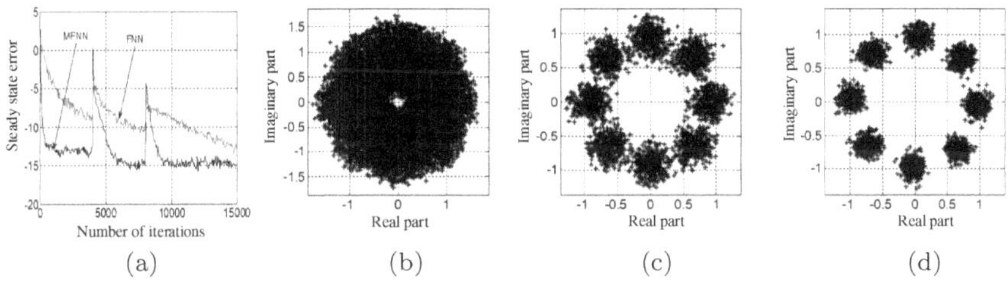

Figure 2.50. Simulation results. (a) Curve of error. (b) Input of equalizer. (c) Output of FNN. (d) Output of MFNN.

Figure 2.50 shows that in time-varying channels, when the channel changes, the proposed MFNN has a stronger restarting ability to track the time-varying channel quickly compared to the FNN. Moreover, the MFNN outputs a more tightly concentrated constellation diagram and a clearer eye opening. Therefore, the MFNN has more practical value.

Chapter 3

Convolutional Neural Networks

Summary

Starting from the basic structure of conventional Convolutional Neural Networks (CNNs), this chapter elaborates on the basic concepts of convolutional kernels, stride, padding, depth, convolution, and correlation through using illustrations and focuses on analyzing the functions of the input layer, convolution layer, activation layer, pooling layer, and output layer, as well as discusses the process of dimensional changes and variations of convolution and pooling operations. Several conventional CNN structures and principles are analyzed, including LeNet-5, VGGNet, ResNets, DenseNet, AlexNet, and Inception. Based on this, the structures and principles of feature fusion CNNs and deep guided filtering networks are studied. Taking the remote sensing image classification and motion blur removal method based on deep CNNs as examples, this chapter elaborates in detail on how to expand the application model of the CNNs, establish the bridge between the CNN principles and application fields, and create problem-solving methodologies.

3.1 Introduction

In the learning process of traditional neural network, the input of neural network is typically one-dimensional features, i.e., a one-dimensional vector. When dealing with inputs with multi-dimensional features, dimensionality reduction is often employed.

Although this compressed feature learning method can meet research objectives, it tends to overlook the spatial characteristics of sample features in classification and recognition tasks, so it results in lower recognition accuracy. Due to the introduction of Convolutional Neural Network (CNN), it has been applied in some fields such as image segmentation, image style transfer, image recognition, and speech recognition.

This chapter provides a detailed analysis of the structure, principles, methodologies, and applications of CNNs.

3.2 Structure of Convolutional Neural Networks

The reason why Convolutional Neural Networks (CNNs) excel in the field of images is that they can preserve the original spatial features of the digital images and reduce the amount of computational data.

A conventional CNN consists of an input layer, a convolutional layer, a pooling layer, a fully connected layer, an output layer [8, 26], and so on. Its thinking orientation and structural schematic are

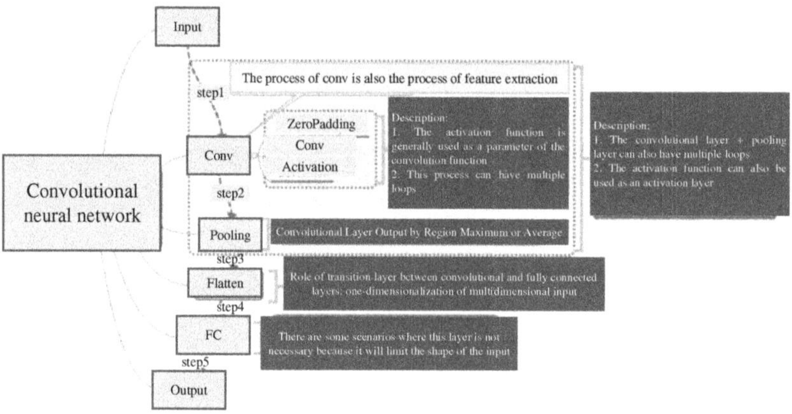

Figure 3.1. Thinking orientation of the CNN.

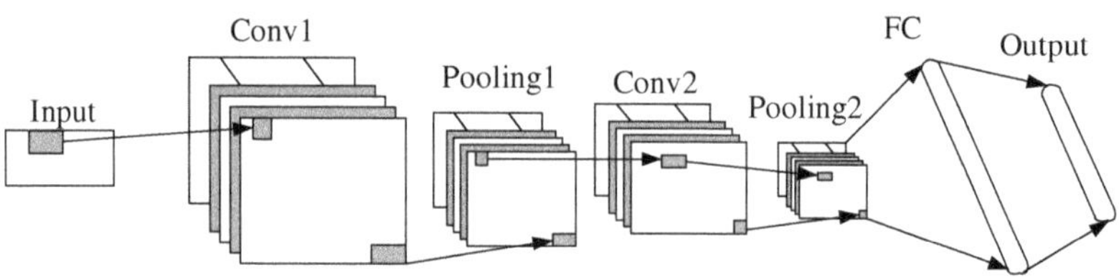

Figure 3.2. Schematic diagram of the CNN.

shown in Figures 3.1 and 3.2, respectively. The convolutional layer extracts local features from the samples, the pooling layer reduces the sample dimension output by the convolutional layer, and the fully connected layer is the basic structure of traditional neural networks used to output results.

3.2.1 Basic concepts

3.2.1.1 *Convolutional kernel*

The convolution kernel (also known as Filter) refers to the observation range. Unlike the human eye, the observation range of a computer is much smaller. Generally, 3×3, 5×5, and 7×7 matrices are used as convolution kernels, as shown in Figure 3.3. The size of the convolution kernel is generally determined by the size of the input images. The larger the input images, the larger the convolution kernel required.

3.2.1.2 *Strides*

The human eye can easily locate target content, whereas computers cannot. Computers must scan the entire image row by row to find the target, and the scanning interval is called the stride. To avoid missing feature values, the stride is typically set to 1 (though larger strides may be used for larger images), as shown in Figure 3.4.

3.2.1.3 *Padding*

3.2.1.3.1 *Only scan convolutional pixels*

Valid padding (padding = 'valid'): The convolutional pixels are only scanned, as shown in Figure 3.5.

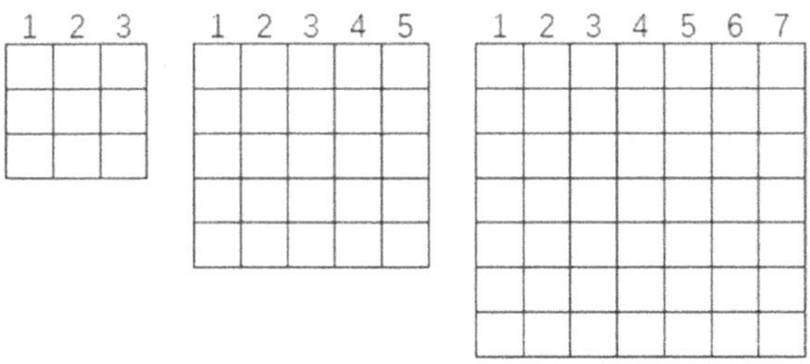

Figure 3.3. Convolutional kernel.

Note: Convolutional kernels are usually odd-numbered.

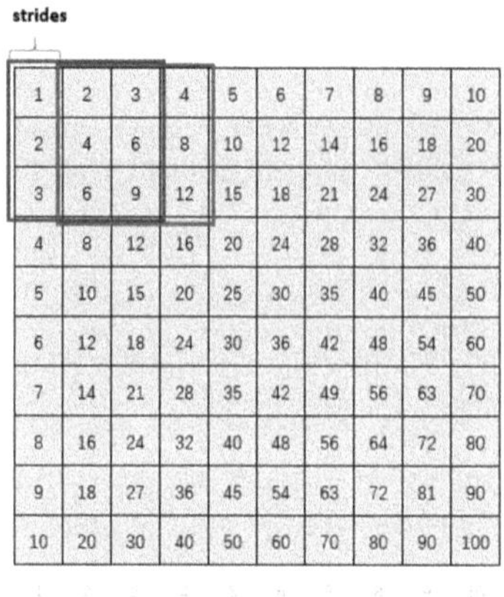

Figure 3.4. The meaning of a step size of 1.

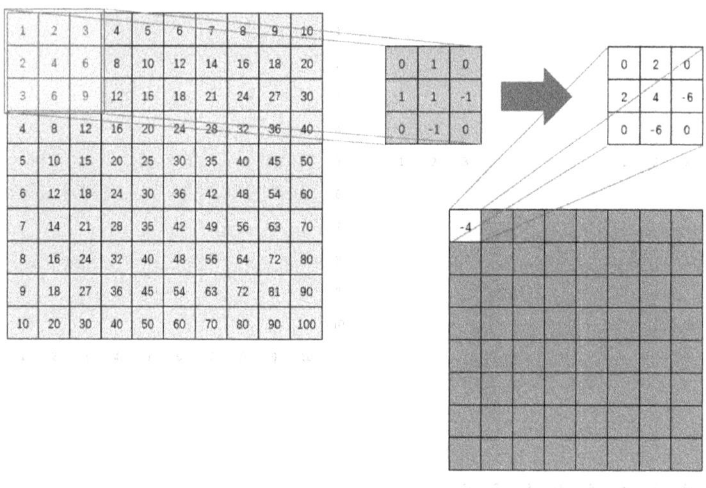

Figure 3.5. Only scanning convolvable pixels starting at position (2,2).

If a 3×3 convolution kernel for convolution is used, it is usually necessary to start from position (2,2) (where the subscript starts from 1) because if starting from (1,1), there is no data on the left or top of that point, and similarly, it ends with $(n - 1,\ n - 1)$. If the stride is set to 1, the result obtained after the convolution operation will be 2 pixels less in length and width than the original images.

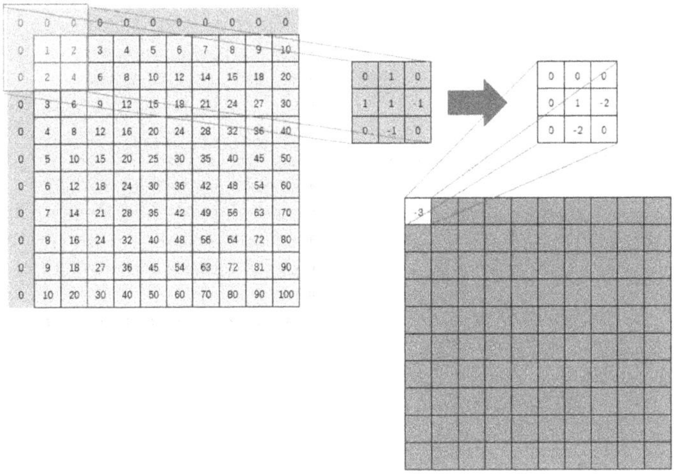

Figure 3.6. Scanning all pixels, starting from position (1,1).

3.2.1.3.2 *Scan all pixels with zero-padding at edges*

Same padding (padding = 'same'): All pixels with zero-padding at edges are scanned, as shown in Figure 3.6.

This operation starts from position (1,1) regardless of the convolutional kernel size, and any insufficient edges are filled with zeros. Therefore, when the stride is set to 1, the result obtained after the convolution operation is the same size as the original images.

3.2.1.4 *Depth*

A convolutional kernel is applied to an image to yield a result, whereas multiple convolutional kernels are applied to the same images to yield multiple results, and the number of these results is precisely called the depth.

Why do we need to perform multiple convolution operations on the same image?

First, in the fully connected layer, the weight from the neuron of the upper layer to a certain neuron in the current layer is denoted as $\boldsymbol{w}$, and the number of parameters $\boldsymbol{w}$ between the two layers is the product of the number of neurons in each layer. In contrast,

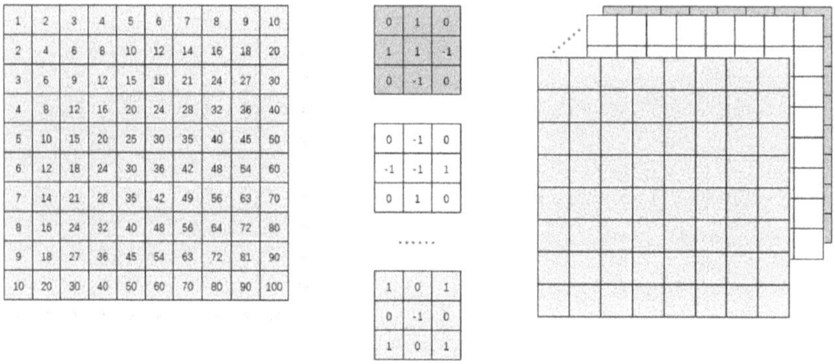

Figure 3.7. Number of parameters in fully connected layer.

when convolving an image with the same kernel, the convolutional kernel serves the same meaning as a '$\boldsymbol{w}$' in a fully connected layer. A convolution kernel corresponds to a parameter (or can also be as shown in Figure 3.7 understood as a convolution kernel corresponding to one or a group of neurons), and during training neural network, the values of each convolution kernel can be adjusted.

3.2.1.5 *Convolution and cross-correlation*

Convolutional neural networks are named after convolution operations, which is a widely used technique in signal processing, image processing, and other fields. In deep learning, cross-correlation operations are often used instead of convolution ones. While their effects are the same as the CNNs, there are some subtle differences between convolution and cross-correlation, which need to be carefully understood.

The mathematical definition of convolution is written as

$$f(t) \otimes g(t) = \int_{-\infty}^{\infty} f(\tau)g(t-\tau)d\tau. \tag{3.2.1}$$

Equation (3.2.1) indicates that convolution operation is the integration of two functions, one of which undergoes inversion and shift operations. Figure 3.8 visually demonstrates the convolution operation.

In Figure 3.8, the top-left images show the original functions $f(t)$ and $g(t)$, which are also called filters, where $g(t)$ is first flipped, as shown in the bottom-left two images, and then we will make the

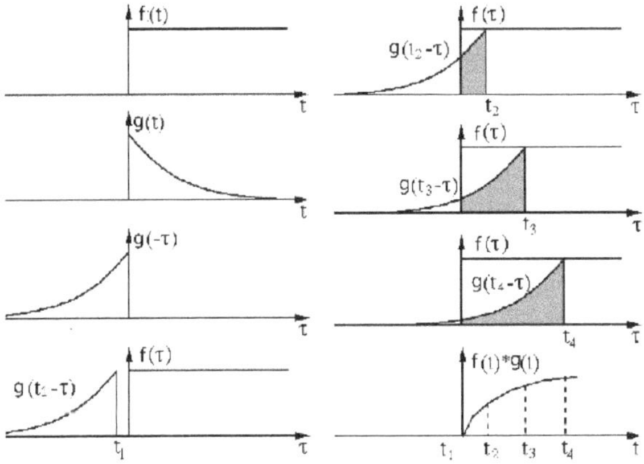

Figure 3.8. Inversion and shift operations.

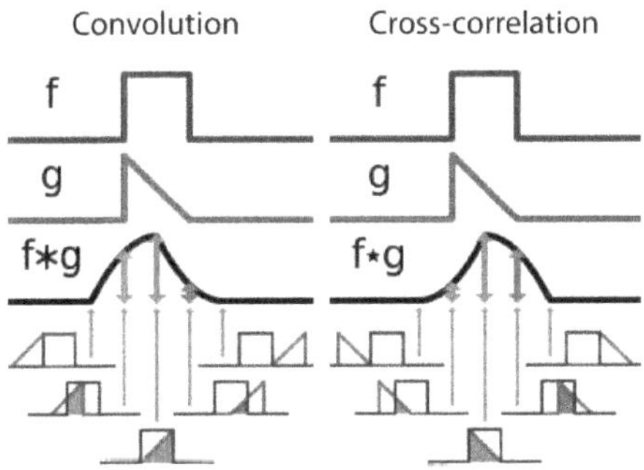

Figure 3.9. Cross-correlation vs. convolution.

flipped $g(t)$ slide along the forward horizontal axis, as depicted in the top-right two images. At each position, the area of the intersection range between $f(t)$ and the flipped $g(t)$ is calculated, and the intersection area of each specific point is the convolution value of that point. The convolution of the entire operation is the sum of all convolution values.

Cross-correlation operation, on the other hand, involves a sliding dot product or inner product of two functions without flipping the filter. The cross-correlation is the sum of the overlapping areas between the filter and the function. Figure 3.9 illustrates the difference between cross-correlation operation and convolution operation.

In deep learning, the filters in convolution operations are not flipped, so it means the operation is effectively cross-correlation.

However, since the filters in the CNN are randomly initialized, the distinction between cross-correlation operation and convolution operation is negligible in practice. The learned cross-correlation filters can be viewed as flipped versions of actual convolution filters.

3.2.2 Input layer

Like traditional neural networks, the input to the CNN model requires preprocessing. Common preprocessing methods for images in the input layer include mean subtraction, normalization, dimensionality reduction of Principal Component Analysis (PCA), and Support Vector Machine (SVM), as shown in Figure 3.10.

3.2.2.1 *Mean subtraction*

All dimensions of the input data are centralized to 0, all samples are summed and averaged, and then the mean sample is subtracted from all samples to remove the mean. This process is called mean subtraction.

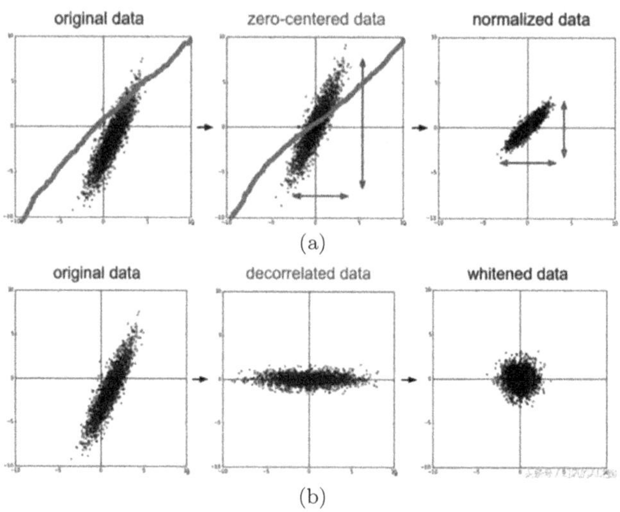

Figure 3.10. Input preprocessing. (a) Mean subtraction and normalization. (b) Decorrelation and whitening.

3.2.2.2 *Normalization*

The data amplitude is normalized to the same range, typically in [−1, 1], for each feature.

3.2.2.3 *PCA/Whitening*

It is very necessary to use the PCA for dimensionality reduction and to eliminate correlations between dimensions to ensure feature independence. Whitening is to make the amplitude on each feature axis be normalized.

3.2.3 Convolutional layer

3.2.3.1 *Standard convolution*

Consider a convolutional layer with a 3×3 kernel, 3 input channels, and 4 output channels. The operation involves convolving the input data with four kernels of size $3 \times 3 \times 3$ to obtain a single-channel output, which results from summing the convolutions of each kernel channel with the corresponding input channel. Thus, four kernels require the parameters about $(3 \times 3 \times 3) \times 4 = 108$.

3.2.3.2 *Local receptive fields*

In human vision, the human brain doesn't recognize an entire image at the same time but can perceive local features first, then integrates them at higher levels to obtain global information. This local perception process is illustrated in Figure 3.11.

The formulas for the convolutional layer are given by

$$\begin{aligned} F_1 &= w_{11} \otimes R + w_{12} \otimes G + w_{13} \otimes B + b_1 \\ F_2 &= w_{21} \otimes R + w_{22} \otimes G + w_{23} \otimes B + b_2 \\ F_3 &= w_{31} \otimes R + w_{32} \otimes G + w_{33} \otimes B + b_3 \\ F_4 &= w_{41} \otimes R + w_{42} \otimes G + w_{43} \otimes B + b_4 \end{aligned} \tag{3.2.2}$$

where F represents each feature map, w is the convolutional kernel, and $\otimes$ denotes the convolution operation.

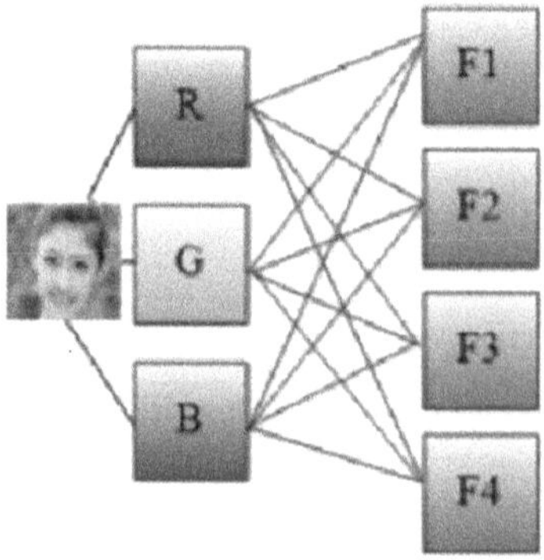

Figure 3.11. Local perception process.

What does convolution operation mean? For example, a $32\times32\times3$ image is subjected to calculatation process with a convolution kernel size of $5 \times 5 \times 3$, as shown in Figure 3.12.

For a three-channel convolution, for example, the process of convolving a $7 \times 7 \times 3$ image with a $3 \times 3 \times 3$ kernel with 2 padding is shown in Figure 3.13.

3.2.4 Activation layer

Activation function is used to map the nonlinear output of the convolutional layer, which can be expressed as

$$\boldsymbol{X}_n^{(l)} = f\left(\sum_n \boldsymbol{W}_{mn}^{(l)} \otimes \boldsymbol{X}_m^{(l-1)} + \boldsymbol{b}_n^{(l)}\right), \tag{3.2.3}$$

where $\boldsymbol{X}_n^{(l)}$ is the nth output feature map of the lth layer, $\boldsymbol{W}_{mn}^{(l)}$ is the weight between connecting the nth feature map of the lth layer to the mth feature map of the $(l-1)$th layer, $\otimes$ is the convolution operation, $\boldsymbol{X}_m^{(l-1)}$ is the mth output feature map of the $(l-1)$th layer, and $\boldsymbol{b}_n^{(l)}$ is the bias for the nth feature map of the lth layer. Additionally, $f(\cdot)$ denotes the activation function for the current convolutional layer. Common nonlinear activation functions include Sigmoid, Tanh, ReLU, and LReLU.

The activation function is often applied after weighting the input data. In the case of not using an activation function (which is equivalent to the excitation function being f $(x) = x$), the output of each

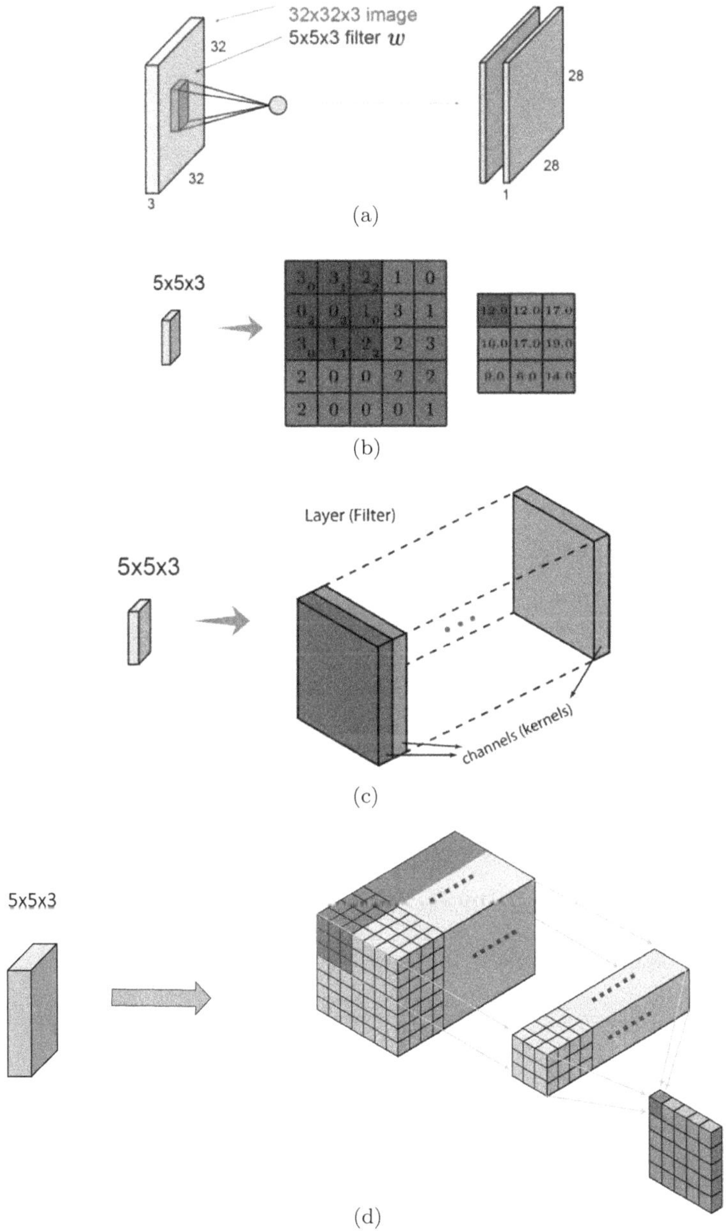

Figure 3.12. Convolution operation. (a) Convolution conditions and results. (b) Convolution calculation. (c) Convolutional kernel. (d) Layer-by-layer convolution.

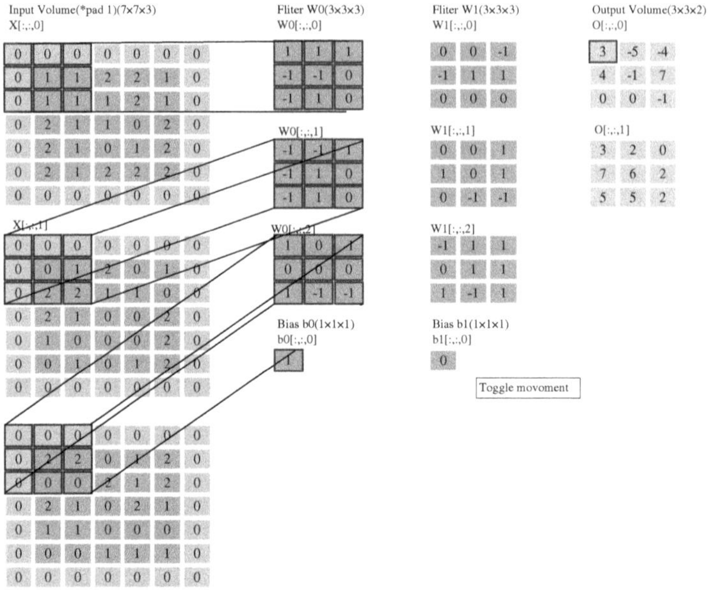

(a)

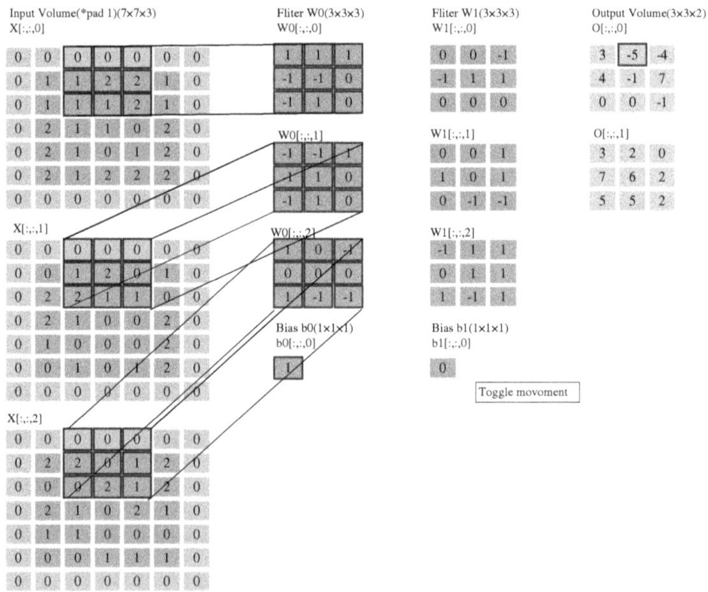

(b)

Figure 3.13. Three-channel convolution process. (a) Top-left 3×3 region convolved with Kernel. (b) After shifting right by 2 steps. (c) After further shifting right by two steps.

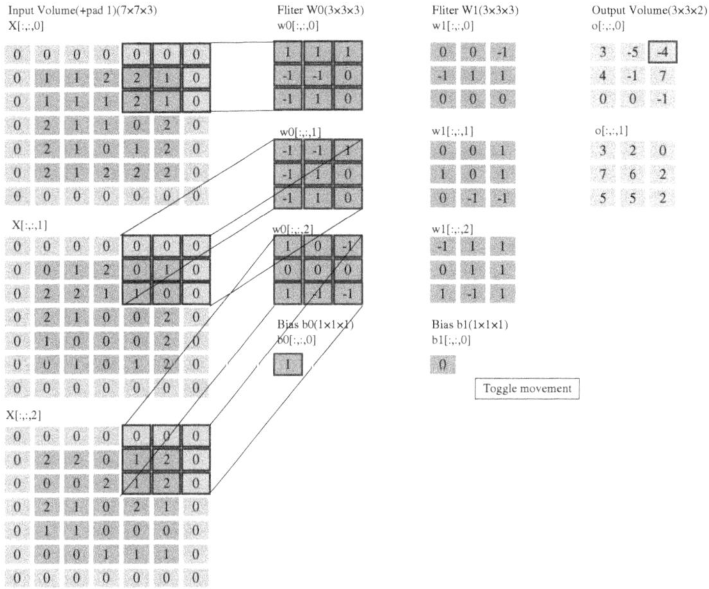

(c)

Figure 3.13. (*Continued*)

layer is a linear function of the input of the previous layer. Therefore, it is easy to conclude that no matter how many neural network layers there are, the output is a linear combination of inputs, and the effect is the same as without hidden layers. This is the most primitive perceptron. After nonlinear factors are introduced through activation functions in the network, it can enhance the ability of deep neural networks to fit arbitrary functions and is conducive to expressing complex features of convolutional layers:

Sigmoid:

$$\text{Sigmoid}(x) = \frac{1}{1+e^{-x}}, \tag{3.2.4}$$

Tanh:

$$\text{Tanh}(x) = \frac{e^{x}-e^{-x}}{e^{x}+e^{-x}}, \tag{3.2.5}$$

ReLU:

$$\text{ReLU}(x) = \begin{cases} x & \text{if } x \geq 0 \\ 0 & \text{if } x \leq 0 \end{cases}, \tag{3.2.6}$$

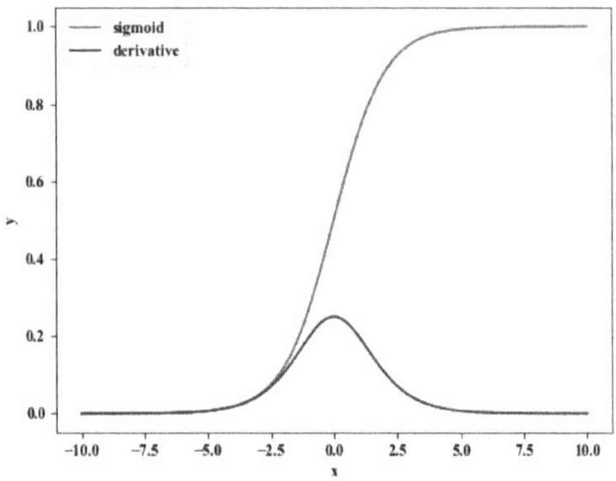

Figure 3.14. Sigmoid function and its derivative.

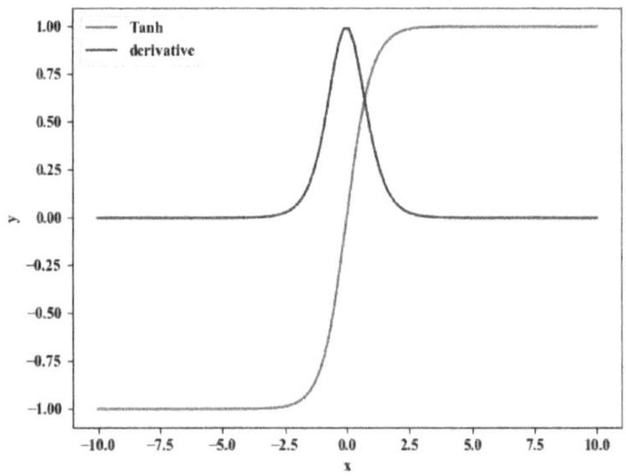

Figure 3.15. Tanh function and its derivative.

LReLU:

$$\mathrm{LReLU}(x) = \begin{cases} x & \text{if } x \geq 0 \\ \alpha x & \text{if } x \leq 0 \end{cases}, \tag{3.2.7}$$

where α is a small non-zero constant and represents the non-zero slope. The curves of the above functions are shown in Figures 3.14–3.17. Sigmoid and Tanh, being doubly saturated activation functions, focus on signals in the central region, which is beneficial for feature space mapping. However, in the backpropagation, the increasing layers can lead to gradient vanishing (i.e., gradients approach zero). In contrast, although ReLU function is not differentiable everywhere, it facilitates gradient descent and backpropagation in the positive region, and neurons can be activated only when inputs exceed a

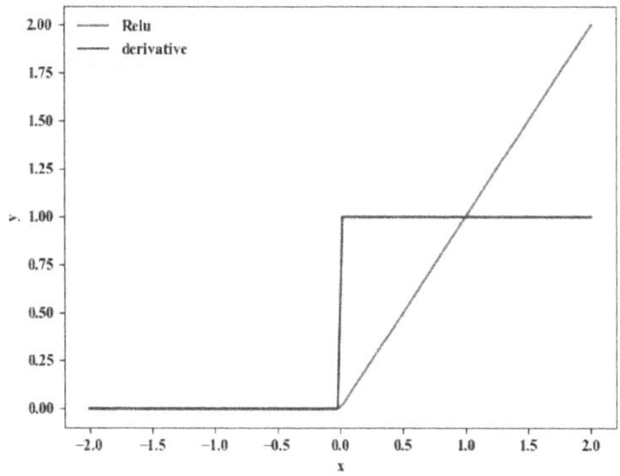

Figure 3.16. ReLU function and its derivative.

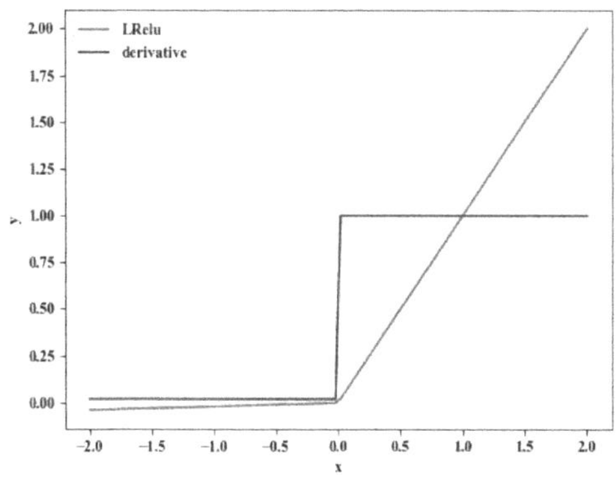

Figure 3.17. Leaky ReLU function and its derivative.

threshold. The ReLU function makes all negative values in the feature map set to zero during training, so the dynamic sparsity may be introduced. Due to its slope of 1 and single-sided saturation, gradients can effectively propagate during backpropagation. Moreover, the ReLU function avoids complex exponential operations, so the training speed of deep neural networks can be improved. To address "dead neurons" (neurons that is never activated due to negative inputs), the Leaky ReLU (LReLU) function introduces a small leak (non-zero slope) in the negative region. Since its derivative is non-zero everywhere, the LReLU function can reduce dead neurons and improve gradient-based learning.

For activation layers, the ReLU function is preferred due to its fast convergence, but it may sometimes underperform. If the ReLU loses efficacy, the LReLU function or Maxout may be considered. Tanh functions work well in text and audio processing.

3.2.5 Pooling layer

The pooling layer, also called subsampling or downsampling, can be used to reduce feature dimensionality, compress data and parameters, mitigate overfitting, and improve model robustness.

3.2.5.1 *Functions of the pooling layer*

Pooling is often applied after convolutional layers to reduce feature dimensions, decrease parameters, and prevent overfitting. Its key functions include the following:

3.2.5.1.1 *Increasing the receptive field*

The so-called receptive field refers to the area size where one pixel corresponds to the original image. If there is no pooling, a 3×3 convolution with a stride of 1 will output a receptive field of 3×3 for one pixel. Another 3×3 convolution with stride=1 is added again to result in a receptive field of 5×5. What would happen if we add a 3×3 pooling in the middle of each convolution now? It is obvious that the receptive field rapidly increases, which is one of the major functions of pooling. The increase in the receptive field is necessary for the improvement of model ability, as the saying goes, "Mount Taishan can't be seen if one leaf blocks the vision".

3.2.5.1.2 *Translation invariance*

In the pooling layer, the pooling function uses the overall statistical characteristics of adjacent outputs at a certain location to replace the network's output at that location. I hope that no matter what kind of pooling function is used, when a small amount of translation is applied to the input, most of the outputs of the pooling function will not change, which is called the invariance of translation. Since pooling constantly abstracts the features of a region without caring

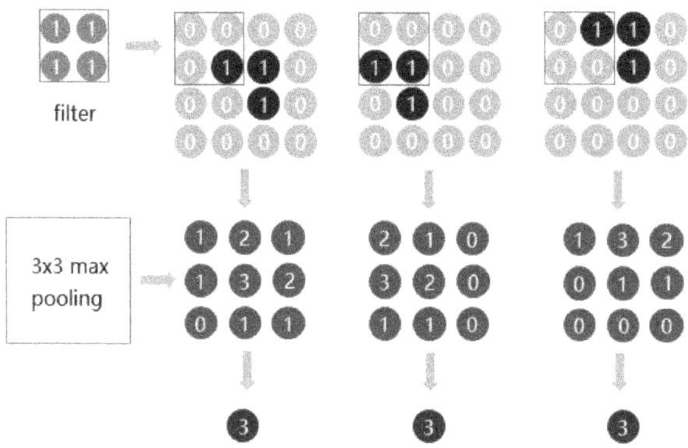

Figure 3.18. Translation invariance.

about its position, it increases translation invariance to some extent. The translation invariance of pooling is shown in Figure 3.18.

The top-right corner shows three images with different horizontal folding positions, which are processed with the convolution kernels in the top left corner and then subjected to a 3×3 size pooling operation to obtain the same recognition result.

3.2.5.1.3 *Reducing optimization difficulty and parameters*

Convolution with a stride greater than 1 can be used instead of pooling, but each feature channel in pooling is downsampled separately. Compared with convolution-based downsampling, it does not require parameters and is easier to optimize. Global pooling can greatly reduce the number of model parameters and optimization workload.

3.2.5.2 *Common pooling operations*

The most common pooling operations include average pooling and max pooling. Average pooling computes the mean of a region, while max pooling selects the maximum value of the one.

3.2.5.2.1 *Max pooling*

Max pooling can be overlapping or non-overlapping. For example, when *stride=kernel size*, max pooling is non-overlapping. When *stride* < *kernel size*, max pooling is overlapping. Overlapping pooling can improve accuracy and mitigate overfitting.

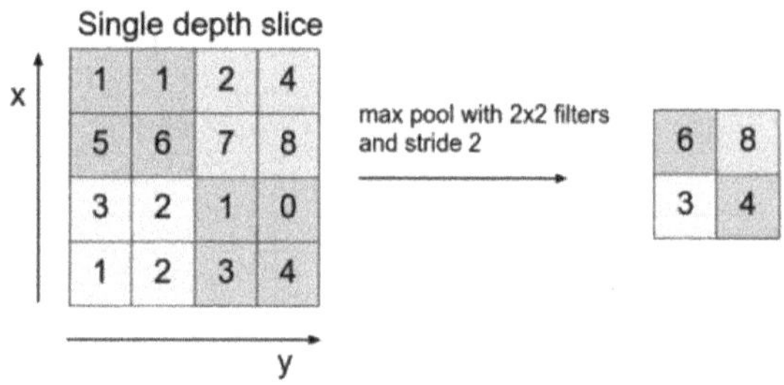

Figure 3.19. Max pooling.

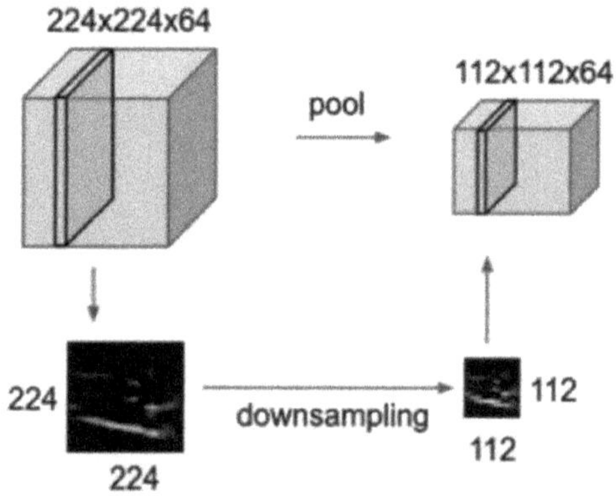

Figure 3.20. Pooling multiple feature maps.

In max pooling, the entire image is divided into several equally sized small pooling blocks without overlapping. The largest number within each block is only taken; other nodes are discarded, and the original planar structure is retained to obtain the output, as shown in Figure 3.19.

Correspondingly, for multiple feature maps, the original 64 images with a size of 224×224 are transformed into 64 images with a size of 112×112 after max pooling, as shown in Figure 3.20, thus achieving the purpose of downsampling.

3.2.5.2.2 *Average pooling*

Figure 3.21 shows a 4×4 feature map neighborhood where values are scanned using a 2×2 filter with a stride of 2, and the average

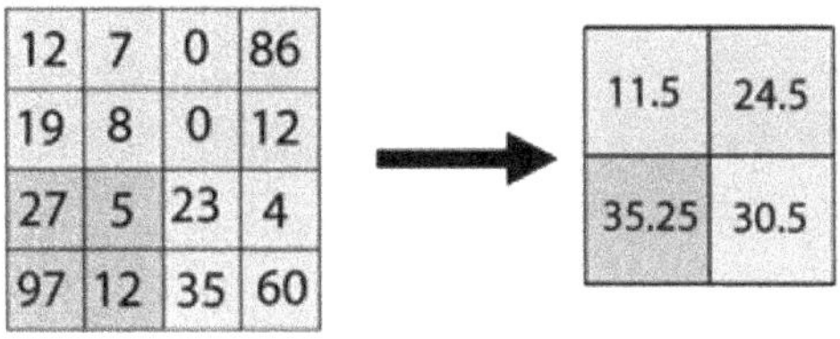

Figure 3.21. Average pooling.

value is calculated and outputted to the next layer. This is called average pooling.

3.2.6 Fully connected layer

After an image passes through several convolutions, activations, and pooling, it reaches the output layer. The CNN model will send a learned high-quality feature map to the fully connected layer. In fact, before the fully connected layer, if the number of neurons is too large and the learning ability is strong, overfitting may occur. Therefore, dropout operation can be introduced to solve this problem by randomly deleting some neurons in the neural network. Local Response Normalization (LRN) method and data augmentation can also be performed to increase robustness.

After reaching the fully connected layer, it can be understood as a simple multi-class neural network (such as a BP neural network), which obtains the final output through the Softmax function. The entire model has been trained.

All neurons between the two layers have connections of weights, and the fully connected layer is usually located at the tail of the CNN. That is to say, it is connected in the same way as traditional neural network neurons, as shown in Figure 3.22.

3.2.7 Dimensional transformation process

In the CNNs, feature extraction is completed by convolving the previous layer with a convolutional kernel, and pooling is performed in the next layer. In this section, the dimensional changing process of convolutional and pooling layers is mainly analyzed. When using all zeros padding (if the step size is 1, it can avoid the size change of the node matrix after passing through the convolutional layer), the dimension

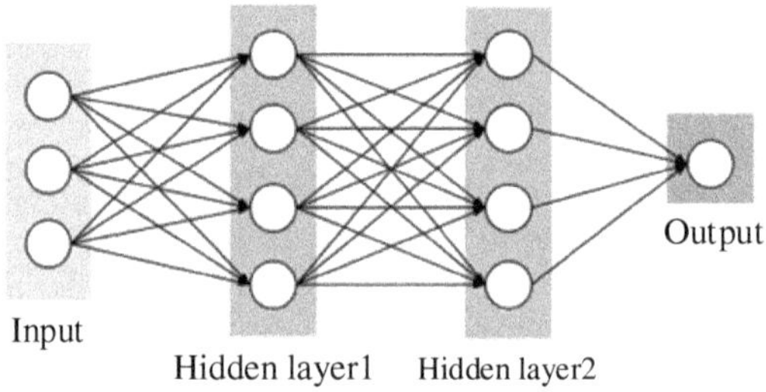

Figure 3.22. Output layer.

calculation formulas of outputs in the convolutional/pooling layer are given by

$$\text{out}_{\text{length}} = \lceil \text{in}_{\text{length}}/\text{stride}_{\text{length}} \rceil, \tag{3.2.12a}$$

$$\text{out}_{\text{width}} = \lceil \text{in}_{\text{width}}/\text{stride}_{\text{width}} \rceil, \tag{3.2.12b}$$

where $\text{out}_{\text{length}}$ represents the length of the output matrix of the convolutional layer, which is equal to the length of the matrix of the input layer divided by the upward integer of the step size in the length direction. $\text{out}_{\text{width}}$ represents the width of the output matrix of the convolutional layer, which is equal to the width of the matrix of the input layer divided by the upward integer of the step size in the width direction.

If all zeros aren't used for padding, the output dimension calculation formula for convolutional/pooling layers is written as

$$\text{out}_{\text{length}} = \lceil (\text{in}_{\text{length}} - \text{filter}_{\text{length}} + 1)/\text{stride}_{\text{length}} \rceil, \tag{3.2.13a}$$

$$\text{out}_{\text{width}} = \lceil (\text{in}_{\text{width}} - \text{filter}_{\text{width}} + 1)/\text{stride}_{\text{width}} \rceil, \tag{3.2.13b}$$

where $\text{filter}_{\text{length}}$ and $\text{filter}_{\text{width}}$ are the kernel's height and width.

If the dimension of the matrix of the input layer is $32 \times 32 \times l$, where l represents a grayscale image, the size of the convolution kernel in the first layer is 5×5, the depth is 6 (i.e., there are 6 convolution kernels), without using all zeros padding, and the step size is 1, then the output dimension of the convolution layer is 28 (i.e., $32 - 5 + 1 = 28$), and the parameters of the convolution layer are 156 (i.e., $5 \times 5 \times 1 \times 6 + 6 = 156$). Here, it can be found that the number of parameters in the convolution layer is independent of the size of the image, only related to the size, depth, and depth of node matrix in the current layer. This allows the CNN to be extended to image data of any size. The output dimension of the convolution layer, which is the input dimension of the next layer (usually the pooling layer), is 4704 (i.e., $28 \times 28 \times 6 = 4704$) nodes. The pooling kernel size of the pooling layer is 2×2, with a stride of 2. The output dimension of the pooling layer is 13.5 (i.e., $(28-2+1)/2 = 13.5$), and the result of rounding up is 14. The pooling layer does not affect the depth of the node matrix, and the output dimension of the pooling layer is $14 \times 14 \times 6$.

3.3 Training Process of Convolutional Neural Networks

The training process of the CNNs, including deep neural networks derived from them, consists of two parts: forward propagation and backward propagation.

Assuming that the network has L layers, where the number of convolution kernels in the convolutional layer is K, the size of the convolution kernels is F, the size of the padding is P, and the convolution step size is S. The steps of forward propagation are as follows:

Step 1: Fill the samples

Dimension is expanded to the given length according to the padding size P.

Step 2: Initialize the weights $\boldsymbol{w}$ and biases $\boldsymbol{b}$

The weights $\boldsymbol{w}$ and biases $\boldsymbol{b}$ of the network are initialized into appropriate vectors.

Step 3: Propagate forward

If the first layer is the input layer, then forward propagation starts from layer 2 with $l = 2$ and ends at layer $L - 2$.

(a) If the l-th layer is a convolutional layer, then the output of that layer is defined as

$$a^{(l)} = f(z^{(l)}) = f(\boldsymbol{a}^{(l-1)} \otimes \boldsymbol{w}^{(l)} + b^{(l)}). \tag{3.3.1}$$

(b) If the l-th layer is a pooling layer, then the output of that layer is written as

$$a^{(l)} = \text{Pool}(a^{(l-1)}). \tag{3.3.2}$$

(c) If the l-th layer is a fully connected layer, then the output of that layer is defined as

$$a^{(l)} = \sigma(z^{(l)}) = \sigma(\boldsymbol{w}^{(l)}\boldsymbol{a}^{(l-1)} + b^{(l)}). \tag{3.3.3}$$

Step 4: Select activation function

If the activation function of the last layer in the network is selected as the Softmax function, the output of the Lth layer can be written as

$$a^{(L)} = \text{Softmax}(\text{z}^{(L)})\text{=Softmax}(\boldsymbol{w}^{(L)}\boldsymbol{a}^{(L-1)} + b^{(L)}). \tag{3.3.4}$$

The above process is that of forward propagation in the CNN. In order to prevent overfitting, dropout coefficients are set during network configuration, whose function is to randomly inactivate neurons in the network, i.e., to force neurons to zero.

In the backpropagation phase, the loss function of the network is selected as the Categorical Cross-entropy Loss (CCL) function, which describes the distance between the output probability of the network and the actual output probability. That is, the smaller the cross-entropy, the closer the probability distributions of the two. Assume that the loss function is recorded as Loss($\boldsymbol{w}$,$\boldsymbol{b}$) and the learning rate is η. The update formulas for the weights and biases of the lth layer in the backpropagation phase are obtained by

$$\boldsymbol{w}^{(l)}(k+1) = \boldsymbol{w}^{(l)}(k) - \eta\frac{\partial}{\partial \boldsymbol{w}^{(l)}}\text{Loss}(\boldsymbol{W}, \boldsymbol{b}), \tag{3.3.5}$$

$$\boldsymbol{b}^{(l)}(k+1) = \boldsymbol{b}^{(l)}(k) - \eta\frac{\partial}{\partial \boldsymbol{b}^{(l)}}\text{Loss}(\boldsymbol{W}, \boldsymbol{b}). \tag{3.3.6}$$

The network iteratively optimizes weights and biases to minimize the loss function until convergence or a stopping criterion is met.

3.4 Variations of Convolution Operations

3.4.1 Depthwise separable convolution

For the sake of comparison, the conventional convolution process will still be presented here, as shown in Figure 3.23.

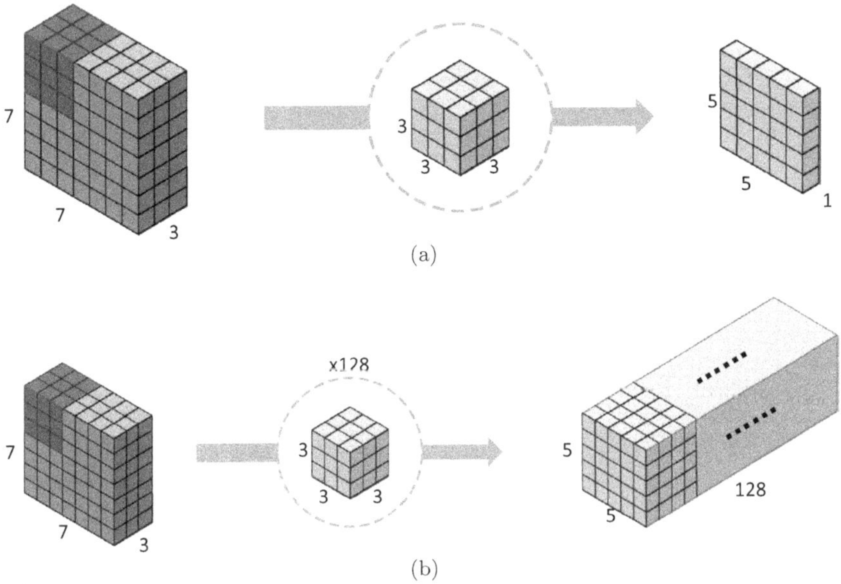

Figure 3.23. Standard convolution. (a) Standard 2D convolution for one output layer with one filter. (b) Standard 2D convolution for 128 output layers with 128 filters.

3.4.1.1 *Depthwise convolution*

One convolution kernel in DepthWise Convolution (DWC) [26] is responsible for one channel, and one channel is only convolved by one convolution kernel. In the previous conventional convolution process, each convolution kernel simultaneously operates on each channel of the input images. Now, instead of using a single filter with a size of $3 \times 3 \times 3$ in 2D convolution, three kernels are used separately. The size of each filter is $3 \times 3 \times 1$. Each kernel convolves with one channel of the input layer (only one channel, not all channels!). Each of these convolutions can provide a map of size $5 \times 5 \times 1$. Then, these mapping maps are stacked together to create a $5 \times 5 \times 3$ image. After performing this operation, an output with a size of $5 \times 5 \times 3$ is obtained, as shown in Figure 3.24. The number of feature maps after performing depthwise convolution is the same as the number of channels in the input layer, i.e., it is impossible to expand the feature maps. Moreover, this operation independently convolves each channel of the input layer, without effectively utilizing the feature information

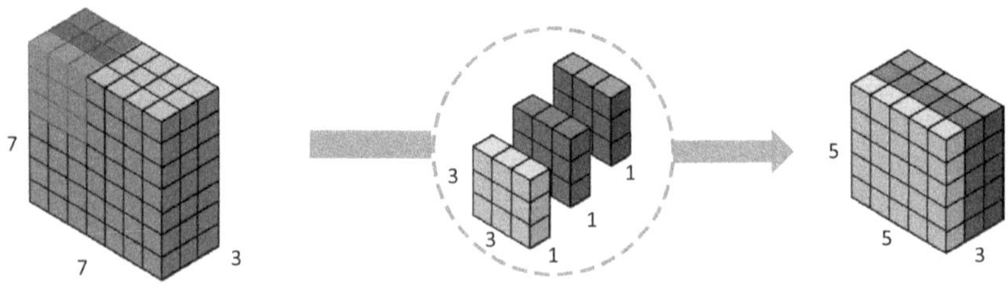

Figure 3.24. Depthwise convolution.

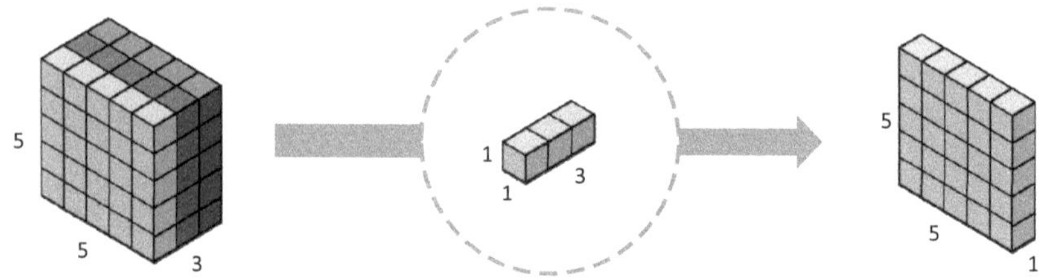

Figure 3.25. Mapping of $5 \times 5 \times 1$ feature map.

of different channels at the same spatial position. Therefore, it is necessary to convolve these feature maps layer by layer to combine them and generate new feature maps.

3.4.1.2 *Pointwise convolution*

PointWise Convolution (PWC) operates similarly to standard convolution, but its kernel size is $1 \times 1 \times C$, where C is the number of channels from the previous layer. This convolution operation performs a weighted combination of the previous feature maps along the depth direction to generate new feature maps. The number of output feature maps equals the number of convolution kernels.

To increase the depth of the network, a 1×1 convolution with a kernel size of $1 \times 1 \times 3$ is applied. By convolving the $5 \times 5 \times 3$ input image with each $1 \times 1 \times 3$ kernel, a $5 \times 5 \times 1$ feature map is produced, as illustrated in Figure 3.25.

Thus, after applying 128 1×1 convolutions, a layer of size $5 \times 5 \times 128$ is obtained, as shown in Figure 3.26.

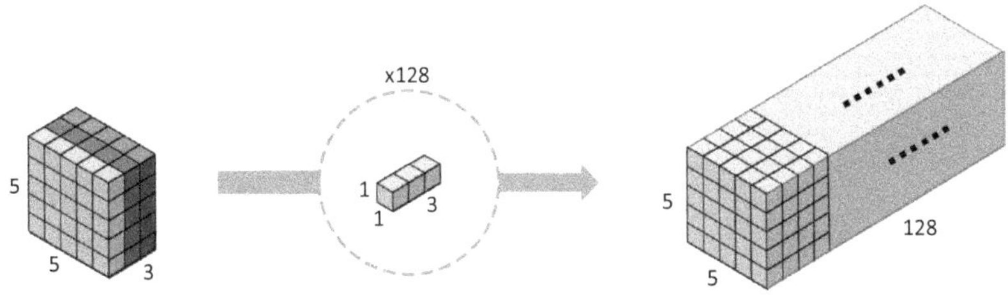

Figure 3.26. Mapping of 5 × 5 × 128 feature map.

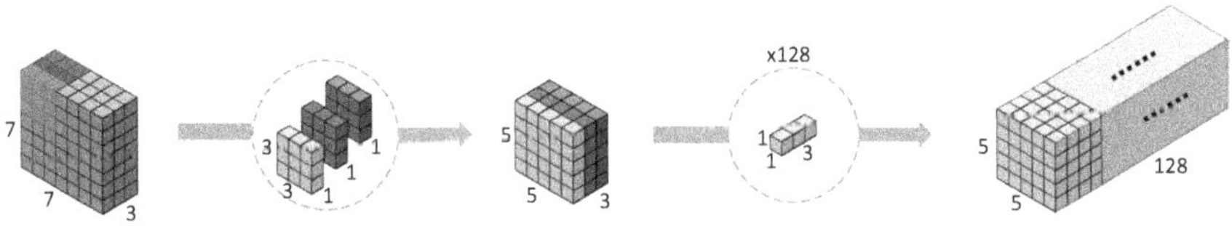

Figure 3.27. Depthwise separable convolution.

3.4.1.3 *Depthwise separable convolution*

Through the above two steps, depthwise separable convolution will also transform the input layer of size $7 \times 7 \times 3$ into an output layer of size $5 \times 5 \times 128$. The entire process of depthwise separable convolution is shown in Figure 3.27.

Compared to 2D convolution, depthwise separable convolution requires significantly fewer operations.

For an input image of size H × W × C, if the 2D convolution is performed using Nc kernels of size $h \times h \times C$ (with a stride of one, padding of zero, and where h is even), the total number of multiplications needed to transform the input layer of size H × W × C into the output layer of size $(H - h + 1) \times (W - h + 1) \times$ Nc is calculated by

$$N_{1Eol} = Nc \times h \times h \times C \times (H - h + 1) \times (W - h + 1). \quad (3.4.1)$$

On the other hand, for the same transformation, the number of multiplications in performing depth-wise separable convolution is calculated by

$$\begin{aligned} N_{2Eol} &= C \times h \times h \times 1 \times (H - h + 1) \times (W - h + 1) \\ &\quad + Nc \times 1 \times 1 \times C \times (H - h + 1) \times (W - h + 1) \\ &= (h \times h + Nc) \times C \times (H - h + 1) \times (W - h + 1). \quad (3.4.2) \end{aligned}$$

The ratio of multiplication times required for depthwise separable convolution and 2D convolution is defined as

$$\frac{N_{1Eol}}{N_{2Eol}} = \frac{1}{N_c} + \frac{1}{h^2}. \tag{3.4.3}$$

When $N_c \gg h$, Equation (3.3.3) can be approximated as $1/h^2$. Based on this, if a 3×3 filter is used, 2D convolution requires 9 times more multiplications than depthwise separable convolution. For a 5×5 filter, 2D convolution requires 25 times more multiplications than its depthwise separable counterpart.

Depthwise separable convolution will be used to reduce the number of parameters in the convolution. As a result, for smaller models, if depthwise separable convolution is used instead of 2D convolution, the CNN model's ability may significantly decrease. The CNN model obtained in this way may be suboptimal. Of course, if depthwise separable convolution is used properly, it can help you achieve efficiency improvements without compromising the performance of your model.

3.4.2 Dilated convolution

Dilated convolution, also known as expansion convolution, is simply a process of inserting some spaces (zeros) between the elements of the convolution kernel to expand its effective size [26]. In this way, the receptive field of the images can be increased without pooling loss information, and the size of the kernel is the same as that of a conventional convolution kernel, with no change in the number of parameters.

Assuming that a variable *rate* is used to measure the expansion coefficient of dilated convolutions, the relationship between the actual convolution kernel size (K) after inserting some dilations and the original convolution kernel size (k) is expressed as

$$K \leftarrow K + (k - 1)(rate - 1), \tag{3.4.4}$$

where k is the original convolution kernel size, *rate* is the convolution expansion rate, and K is the actual convolution kernel size after expansion. In addition, the convolution method of dilated convolution is the same as conventional convolution. The receptive field of the convolutional kernels varies with different expansion rates. For

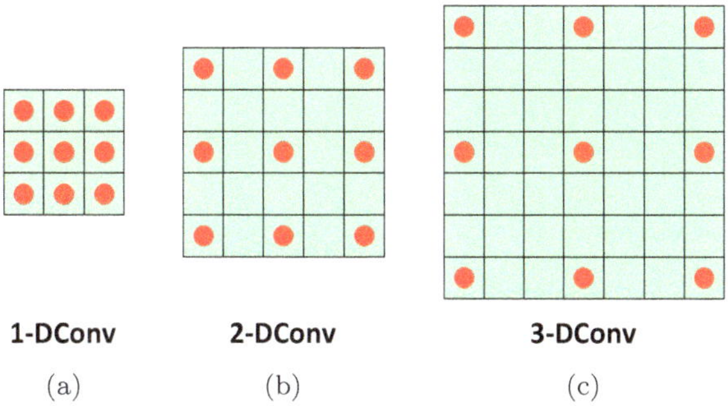

Figure 3.28. Relationship between dilation rate and receptive field in dilated convolution. (a) Rate = 1. (b) Rate = 2. (c) Rate = 3.

example, the receptive field of the convolution kernel at $rate = 1, 2, 4$ is shown in Figure 3.28.

In Figure 3.28, the convolution kernel does not have red dot positions marked by zeros, and the marked position with a red dot is the same as that of a conventional convolution kernel. The 3×3 red dots indicate that the output image is 3×3 pixels after convolution. Although the outputs of all three dilated convolutions are of the same size, there are significant differences in the receptive fields observed by the model.

The receptive field size of the lth convolutional or pooling layer in the network is given by

$$r^{(l)} = r^{(l-1)} + \left((k-1) \cdot \prod_{i-1}^{l-1} s^{(i)} \right), \tag{3.4.5}$$

where k represents the kernel size of the convolutional or pooling layer at the lth layer, $r^{(l-1)}$denotes the receptive field size of the previous layer, and $s^{(i)}$ is the stride of the ith convolutional or pooling layer.

If the size of the initial receptive field is set to 1, then we have the following:

For a 3×3 convolution (stride=1): $r = 1 + (3 - 1) = 3$, the receptive field size is 3×3.

For a 2×2 pooling (stride=2): $r = 3 + (2 - 1) = 4$, the receptive field size is 4×4.

For a 3×3 convolution (stride=3): $r = 4 + (3 - 1) \times 2 \times 1 = 8$, the receptive field size is 8×8.

For a 3×3 convolution (stride=2): $r = 8 + (3-1) \times 3 \times 2 \times 1 = 20$, the receptive field size is 20×20.

The method for calculating the receptive field of dilated convolution is the same as above. The so-called dilated convolution can be understood as expanding the size of the convolution kernel. In the following, the changes in the receptive field of dilated convolution (the kernel size is 3×3, stride=1, and the subsequent convolution process is based on the previous one) are introduced:

1-dilated conv: A convolution with rate=1 is essentially a standard 3×3 convolution. Therefore, $r = 1+(3-1) = 3$ ($r = 2^{\log_2 1+2} - 1 = 3$); the receptive field is 3×3.

2-dilated conv: A convolution with rate=2 can be interpreted as expanding the kernel to 5×5. Therefore, $r = 3 + (5-1) \times 1 = 7$ ($r = 2^{\log_2 2+2} - 1 = 7$); the receptive field size becomes 7×7.

4-dilated conv: A convolution with rate=4 can be interpreted as expanding the kernel to 9×9. Thus, $r = 7 + (9-1) \times 1 \times 1 = 15$ ($r = 2^{\log_2 4+2} - 1 = 15$); the receptive field size is equal to 15×15.

It can be seen that the receptive field will change exponentially via superimposing convolution with the above process. The receptive field size is calculated based on the stacking order. If three 2-dilated convolutions with the same size of 3×3 are used independently, the receptive field can be calculated using the receptive field formula of the standard convolution (e.g., 2-dilated is equivalent to a 5×5 convolution) by the following:

First layer of 2-dilated 3×3 conv: $r = 1 + (5-1) = 5$.
Second layer of 2-dilated 3×3 conv: $r = 5 + (5-1) \times 1 = 9$.
Third layer of 2-dilated 3×3 conv: $r = 9 + (5-1) \times 1 \times 1 = 13$.

3.4.3 3D convolution

3D convolution is built on the basis of 2D convolution and is a generalization of 2D convolution [27]. Figure 3.29 shows a 3D convolution with a filter depth smaller than the input layer depth (kernel size < channel size). Consequently, 3D filters can be moved in all three directions (height, width, and channel of the image). At each position, element-wise multiplication and addition provide a numerical

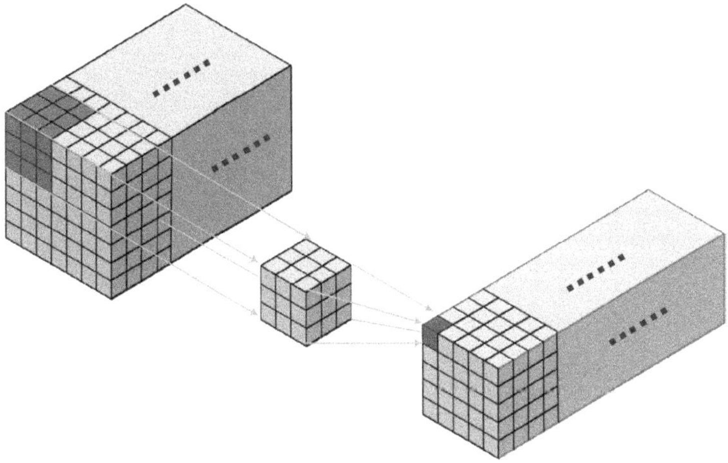

Figure 3.29. 3D convolution.

value. Since the filter slides over a 3D space, its output values are also arranged in 3D space. That is to say, the output is a 3D data.

Similar to 2D convolution, which encodes the spatial relationships of objects in the 2D domain, 3D convolution can describe the spatial relationships of objects in 3D space. For certain applications, such as 3D segmentation/reconstruction in biomedical imaging, such 3D relationships are important. For example, in the CT and MRI, targets such as blood vessels can meander in 3D space.

3.4.4 Grouping convolution

3.4.4.1 *Principle of grouping convolution*

Grouping convolution (GC) first appeared in the AlexNet [28], where conventional convolution operations perform overall convolution calculations on the input images, as shown in Figure 3.23. A convolutional layer with two filter banks is shown in Figure 3.30.

In Figure 3.30, the input data is divided into two groups (number of groups $= g$).

Note: *This grouping is only based on depth (C_{in}), where certain channels are grouped together, and the specific number is determined by C_{in}. Due to changes in the output data, the corresponding convolution kernels also need to make the same changes. If the input*

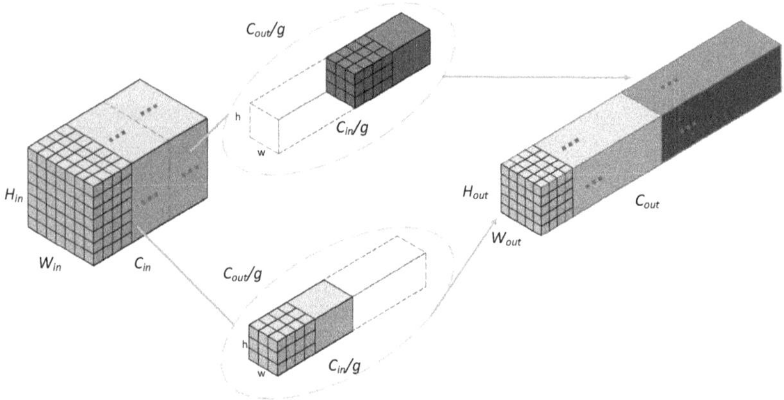

Figure 3.30. Schematic of grouped convolution.
Note: Grouping convolution divided into 2-filter banks.

data is divided into g groups, the depth of the convolution kernels in each group is C_{in}/g, while the size of the convolution kernels remains unchanged. At this point, the number of convolution kernels in each group is C_{out}/g, instead of the original C_{out}. Then, the convolution kernels of each group are used to convolve with their corresponding input data within the group in order to obtain the output data, and then they are combined in a concatenated manner. The final channel of the output data still is C_{out}. That is to say, after the number of groups g is determined, the parallel operation of g identical convolution processes will be performed. In each process (each group), the input data is $H_{\text{in}} \times W_{\text{in}} \times C_{\text{in}}$, the size of the convolution kernel is $h_1 \times w_1 \times C_{\text{in}}$, there are a total of C_{out}, and the output data is $H_{\text{out}} \times W_{\text{out}} \times C_{\text{out}}$.

3.4.4.1.1 *Analyses of parameter quantity*

The size of the input feature map is $H \times W \times C$, and the number of channels in the output feature map is C_{out}. If the input data is divided into g groups, the number of input feature maps in each group is C_{in}/g, and the number of output channels in each group is C_{out}/g (because the number of output channels $C_{\text{out}}$is predefined, the average number of output feature maps distributed to each group should be C_{out}/g).

3.4.4.1.2 *Quantity of convolutional kernel parameters*

That is to say, each group needs to have C_{out}/g convolution kernels, so there are a total of convolution kernels C_{out} (the total number of convolution kernels is the same as conventional convolution, but due to sparse connections, the number of parameters for each convolution kernel is reduced to $C/g \times K \times K$, and the total number of parameters for each convolution kernel is $C_{\text{out}} \times C/g \times K \times K$).

3.4.4.1.3 *Analyses of computational complexity*

Assuming that the size of the feature map of the convolutional layer remains unchanged, the computational complexity of the convolutional layer is calculated by

$$N_{con} = C_{\text{out}} \times H \times W \times \frac{C_{\text{in}}}{g} \times (2 \times K^2 - 1). \qquad (3.4.6)$$

Equation (3.2.6) shows that grouped convolution reduces both parameters and computational complexity to $1/g$ of the original ungrouped convolution. Comparison between conventional convolution and grouped convolution is shown in Figure 3.31.

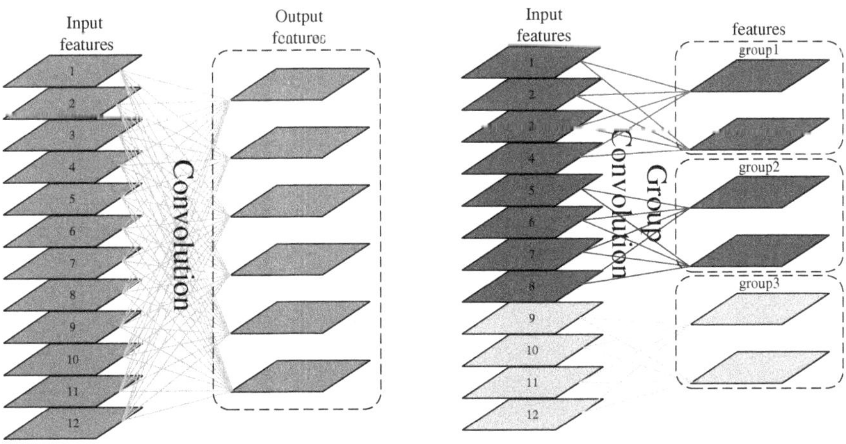

Figure 3.31. Comparison standard convolution with grouped convolution.

3.4.4.2 *Advantages of grouping convolution*

The grouping convolution greatly reduced parameters. For example, when the input channels are 256, the output channels are also 256, the kernel size is 3×3, and the parameters for ungrouped convolution are $256 \times 3 \times 3 \times 256$. When implementing the grouping convolution, if the number of groups is 8, the input and output channels of each group are both 32, and the parameters are $8 \times 32 \times 3 \times 3 \times 32$, which is one eighth of the original one. And the feature maps outputted by each group in the grouping convolution should be combined in a connected manner.

Alex believes that grouping convolution can increase the diagonal correlation between filters and reduce training parameters; it makes it less prone to overfitting, similar to the regularization effect.

3.4.5 Transposition convolutions

Transposition convolutions, also known as deconvolution.

Note: Here, the deconvolution is not a mathematical deconvolution, nor is it a fractional stride convolution. The reason it is called transposition convolution is because it transposes the convolution kernel in conventional convolution operations, and then the output of the convolution is taken as the input of the transposition convolution, and the output of the transposition convolution is the input of the convolution.

The calculation process of convolution and its transposition convolution is exactly the opposite, as shown in Figure 3.32.

The size of the convolution kernel for conventional convolution is 3×3, with a stride of 2 and padding of 1. The convolution kernel outputs element 1 at the red box position and element *2* at the green

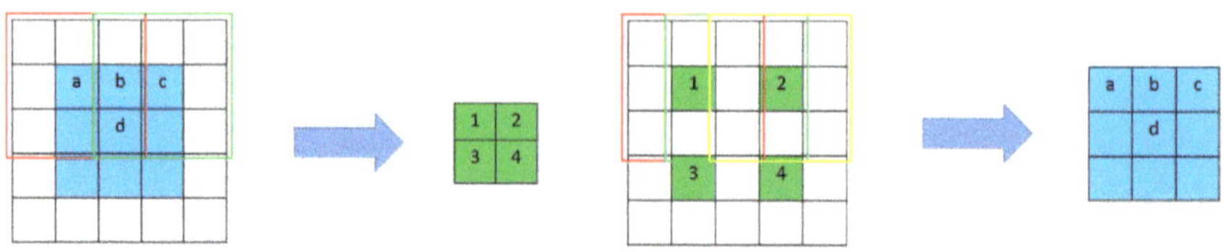

(a) (b)

Figure 3.32. Contrasts of two convolutions. (a) Standard convolution. (b) Transposed convolution.

box position in Figure 3.32. It can be observed that input element a is only related to an output element, i.e., element 1, while input element b is related to both output elements 1 and 2. Similarly, c is only related to element 2, while d is related to all four elements 1, 2, 3, and 4. So, when performing transposition convolution, this connection relationship should still be kept unchanged.

Transposition convolution (deconvolution) requires taking the green feature map in Figure 3.32(a) as input and the blue feature map as output, while ensuring that the connection relationship remains unchanged. That is to say, a is only related to 1, b is related to elements 1 and 2, and so on. How can we achieve this effect? You can first use 0 to interpolate the green feature map. The number of interpolations is to make the interval between adjacent green elements the stride of the original convolution. At the same time, the edges also need to be padded with *zeros* equal to the number of interpolations, as shown in Figure 3.32(b).

At this point, the sliding step size of the convolution kernel is not 2 but 1, and the step size is reflected in the process of interpolating and adding *zeros*.

In the CNNs, transposition convolution is primarily used for upsampling feature maps. For example, to double a feature map's spatial dimensions, a transposition convolution with stride=2 needs to be used.

To better understand transposition convolution, let w represent the kernel, *Large* the input image, and *Small* the output image. After convolution (implemented as matrix multiplication), the large images are downsampled into small images. The convolution implementation of this matrix multiplication is $w \times$ *Large*=*Small*, as shown in Figure 3.33. The input is flattened into a 16×1 matrix and the convolution kernel is transformed into a (4×16) sparse matrix. Then, matrix multiplication between the sparse matrix and the flattened input is done. Afterwards, the obtained matrix (4×1) is converted into a 2×2 output.

The *Large* input image (4×4) can be transformed into the *Small* output image (2×2) through matrix multiplication.

Now, if both sides of the equation in Figure 3.32 are multiplied by the transposition matrix T and the property that "a matrix multiplied by its transposition produces an identity matrix" is applied, we obtain $T \times$ *Small*=*Large*, as shown in Figure 3.34.

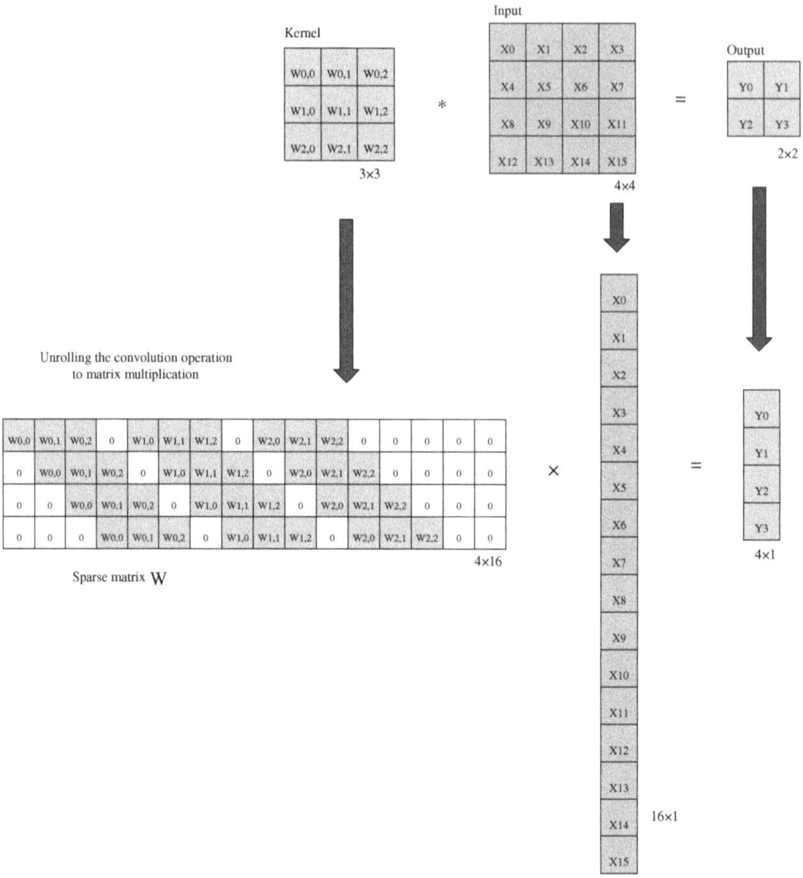

Figure 3.33. Matrix implementation of convolution.

3.4.6 Flat convolution

Flat convolution is between local convolution and traditional convolution, and its similarity with local convolution lies in the fact that adjacent units have different parameters. The difference is that there will be t different convolution kernels used in a loop, which means that convolution kernels separated by interval t will share parameters. The principle of flat convolution is shown in Figure 3.35.

In Figure 3.35, S represents the convolution kernel and x denotes the feature value. While adjacent convolution kernels have their own parameters, the parameters are reused every interval t kernels (here, $t = 2$).

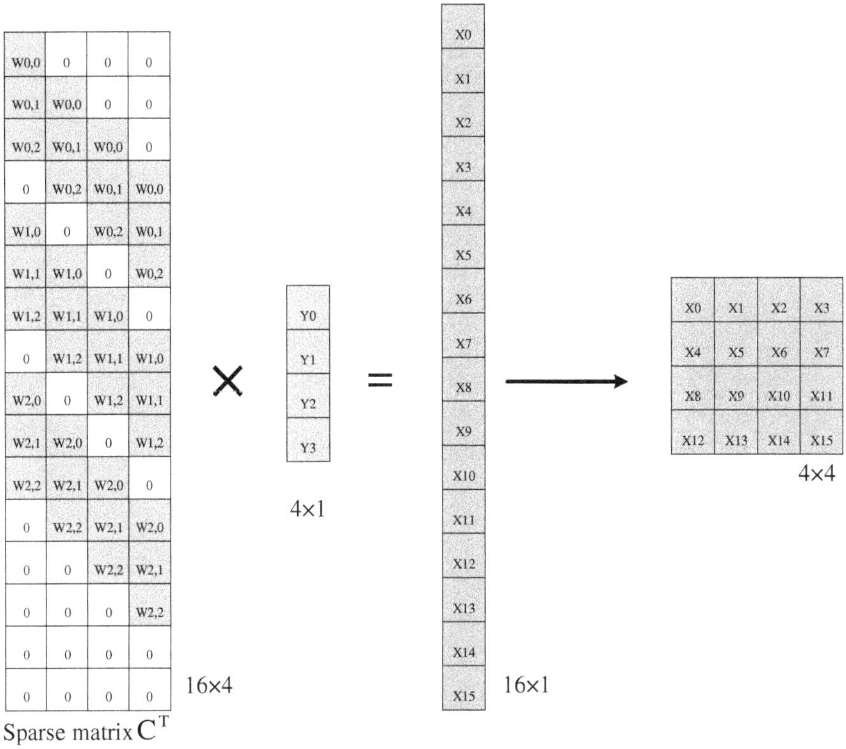

Figure 3.34. Matrix implementation of transposed convolution.

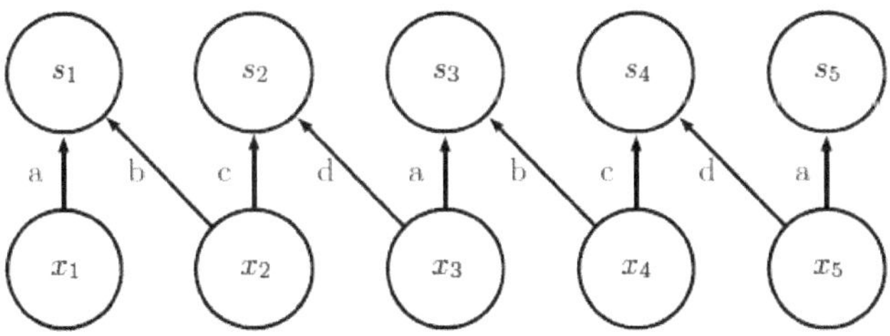

Figure 3.35. Flat convolution.

3.4.7 Key ideas of convolutional operations

Convolutional operations primarily rely on three key ideas to improve machine learning systems: sparse interactions, parameter sharing, and equivariant representations.

3.4.7.1 *Sparse interactions*

Sparse interactions (also called sparse connectivity or sparse weights) in the CNNs are achieved by making the kernel size much smaller than the input size. This differs from the matrix multiplication in fully connected layers, as the kernel only processes a limited number of inputs, thereby reducing the number of parameters.

For example, a 3×3 kernel only receives input from 9 pixels. This block size is called the receptive field or Field Of View (FOV), and it indicates that the kernel primarily learns local correlations.

The CNNs can process time-series data because they utilize the local correlation of data, which are inherent in speeches and texts.

The convolution operation in the CNN is mainly used to obtain local features of images or text. In computer vision, this operation is called filtering, which aims to obtain the output of the local domain. The conventional convolution operation is essentially a weighted average, and its linear operation is the simplest operation to obtain local features. The use of BP algorithm is also particularly direct.

3.4.7.2 *Parameter sharing*

Parameter sharing refers to using the same parameters in multiple functions of the CNN model. In the CNNs, a convolutional kernel is typically applied to every position of the input, and this parameter sharing ensures that only one set of parameters needs to be learned rather than learning a separate parameter for each position of the input.

Simply put, when using convolution operations, a convolution kernel is used to traverse all positions of the feature map from left to right and from top to bottom according to the stride.

3.4.7.3 *Equivariant representations*

Parameter sharing enables neural network layers to have equivariant properties toward translation. The so-called equivariance means that if an input changes, the output also changes in the same way. If the input is slightly shifted, the result of the convolution operation is the same. However, convolution is not naturally equivariant to some other transformations. For example, for image scaling or rotation transformations, other mechanisms are needed to handle these transformations.

3.5 Variants of Pooling Operations

In addition to max pooling, there are also some other pooling operations.

3.5.1 Random pooling

According to relevant theories, the error in feature extraction mainly comes from two aspects:

(1) The variance increase of estimated values is affected by limited neighborhood size.
(2) The parameter error of the convolutional layer causes a shift in the estimated mean.

Generally speaking, average pooling can reduce the first type of error and preserve more background information of the images. Max pooling can reduce the second type of error and preserve more texture information.

Random pooling falls between the two operations mentioned above, whose process is to assign probabilities to pixels based on their numerical values and then sampling is carried out according to these probabilities. On average, it is similar to average pooling. In a local sense, it follows the principle of maximum pooling.

3.5.1.1 *Principle of random pooling*

Random pooling can be seen as normalizing feature map values within a pooling window. In the random pooling process, the sampling and selection based on the probability of the normalized feature map are randomly implemented. That is, the probability of selecting elements with larger values is also higher, unlike maximum pooling, which always takes only the maximum value element.

The random pooling process is shown in Figure 3.36. Figure 3.36 indicates that the larger the feature area, the higher the probability of its selection. For example, the one in the bottom left corner should

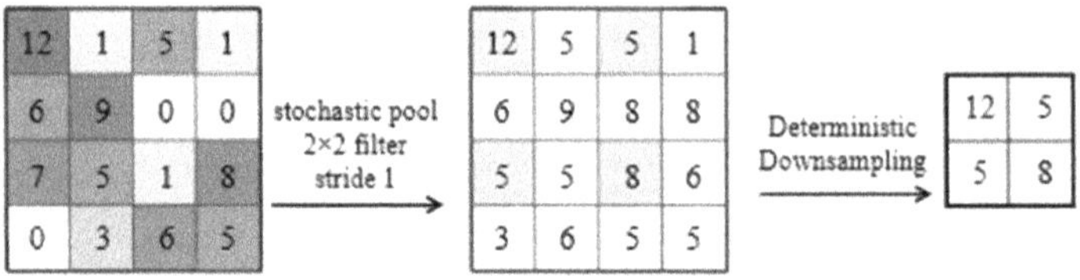

Figure 3.36. Stochastic pooling.

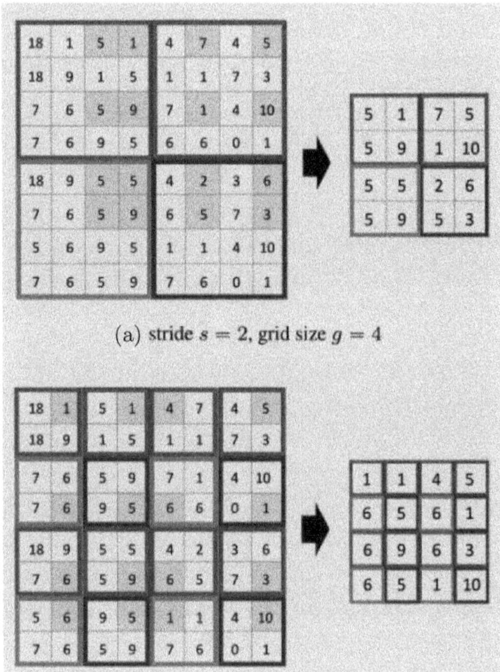

(a) stride $s = 2$, grid size $g = 4$

(b) stride $s = 2$, grid size $g = 2$

Figure 3.37. Changing grid size g.

have been selected as 7, but due to the introduction of probability, 5 also has a certain chance of being selected.

The level of distortion/stochasticity is controlled by adjusting the network size g, as shown in Figure 3.37.

As the network size increases, the training error also grows, corresponding to randomness being greater. The testing error first decreases (stronger regularization) and then rises (when training

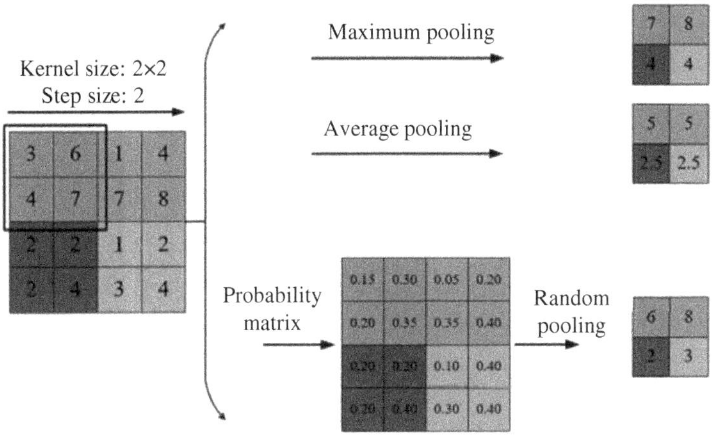

Figure 3.38. Comparison of random pooling, average pooling, and max pooling.

error becomes too high). A comparison between random pooling and max/average pooling is shown in Figure 3.38.

3.5.1.2 *Calculation of random pooling*

The computational steps for random pooling are as follows:

Step 1: Obtain probability matrix

The elements in the grid are divided by their sum to obtain a probability matrix.

Step 2: Select grid cell

Randomly select a grid cell according to the probabilities.

Step 3: Calculate the pooled value

The pooled value is the value of the selected grid cell.

When using random pooling (i.e., the test process), the inference process is also very simple; the weighted average of the matrix region needs only to be calculated.

When taking the derivative of backpropagation, it is only necessary to retain the value of the selected node's position recorded in the forward propagation, and all other values are set to zero, which is very similar to max pooling's backpropagation.

3.5.2 Bilinear pooling

Bilinear pooling is primarily used in fine-grained classification networks for feature fusion [29]. Given features x and y extracted from the same sample, bilinear pooling fuses them via an outer product to improve classification performance.

3.5.2.1 *Concept of bilinear pooling*

The main idea of bilinear pooling is that the processing methods for two different image features are different. Usually, different features of an image are concatenated, summed, or maximally pooled. There are two main pathways for human visual processing: the ventral stream for object recognition and the dorsal stream for object location discovery. Based on this principle, it is hoped that the two features can respectively represent the position of the image and the recognition of the shape, and the classification effect of fine-grained images can be improved on the basis of jointly exerting the functions of the two different features.

3.5.2.2 *Implementation of bilinear pooling*

When the features x and y originate from separate feature extractors, this pooling is termed Multi-modal Bilinear Pooling (MBP). If the features x and y are identical ($x = y$), it is referred to as Homogeneous Bilinear Pooling (HBP) or second-order pooling. The principle of bilinear pooling is illustrated in Figure 3.39, while gradient computation in a bilinear CNN model is demonstrated in Figure 3.40.

3.5.3 Unpooling

Unpooling is an upsampling operation, as shown in Figure 3.41.

The upsampling process is described as follows:

Step 1: Record information of positions

During pooling (typically max pooling), the positions of maximum values are recorded.

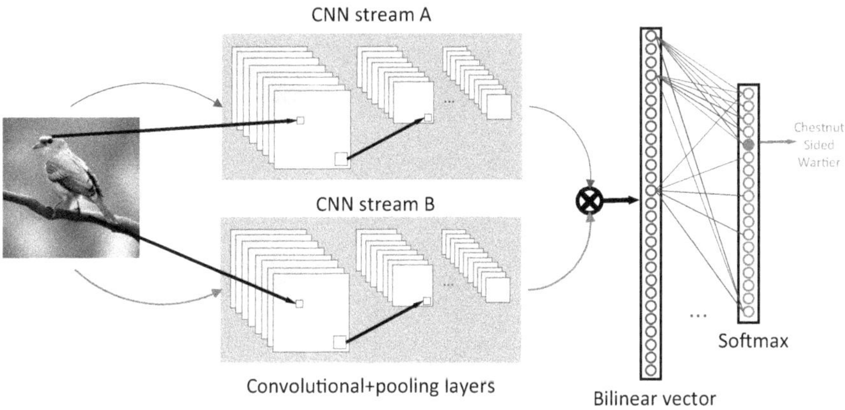

Figure 3.39. Bilinear pooling.

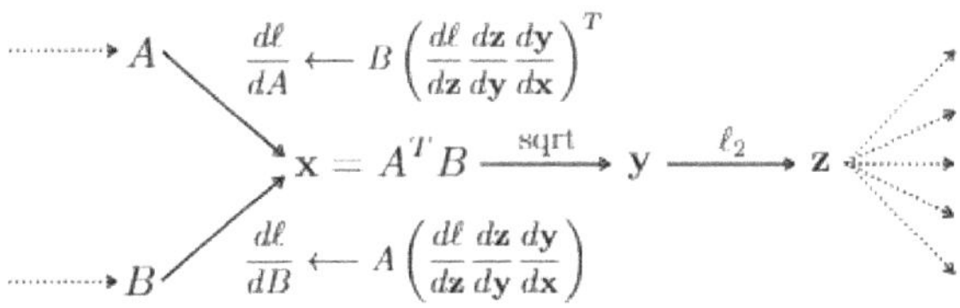

Figure 3.40. Gradient computation in bilinear CNN model.

Step 2: Calculate network layers

Intermediate computations are performed through several network layers.

Step 3: Reconstruct the feature map

In the upsampling phase, the saved max locations from Step 1 are used to reconstruct the feature map of the next layer.

Unpooling operation is not entirely the inverse operation of pooling, and the feature mapping after pooling needs to undergo several operations before the non-pooling operation can be performed, and non-maximum positions are filled with zero. However, this does not fully restore the information.

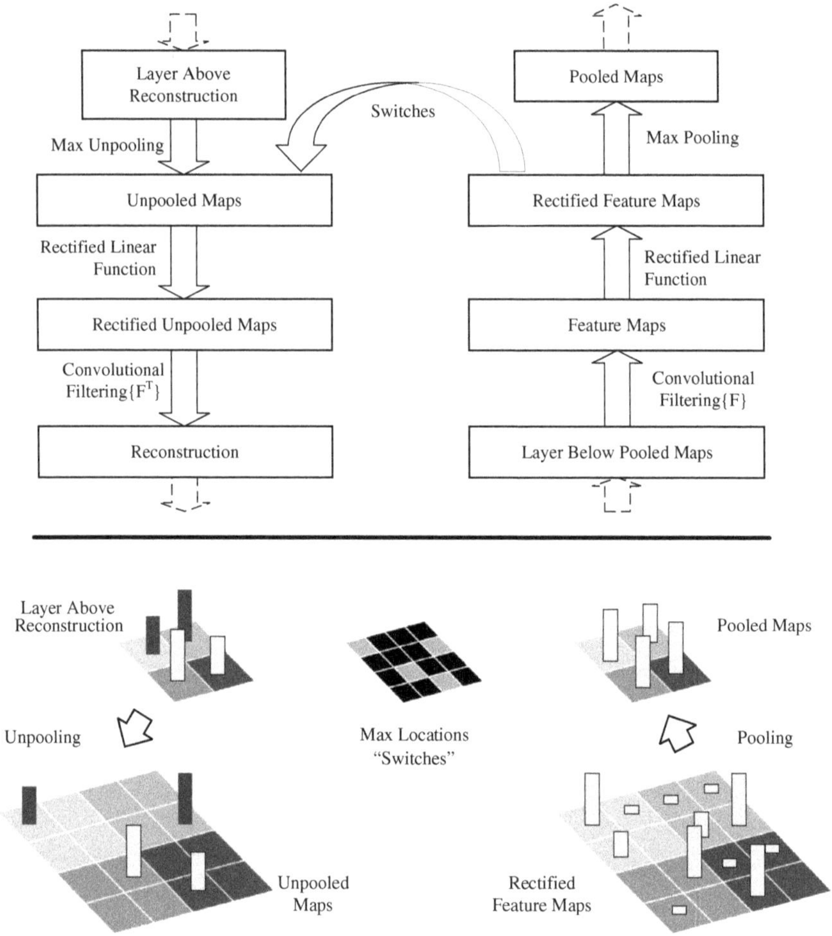

Figure 3.41. Unpooling.

3.5.4 Global average/Max pooling

3.5.4.1 *Global average pooling*

3.5.4.1.1 *Principle of global average pooling*

Global average pooling is generally used to replace fully connected layers. In classification networks, fully connected layers have become almost standard. In the last few layers, feature maps are reshaped into vectors, which are then multiplied to reduce their dimensionality.

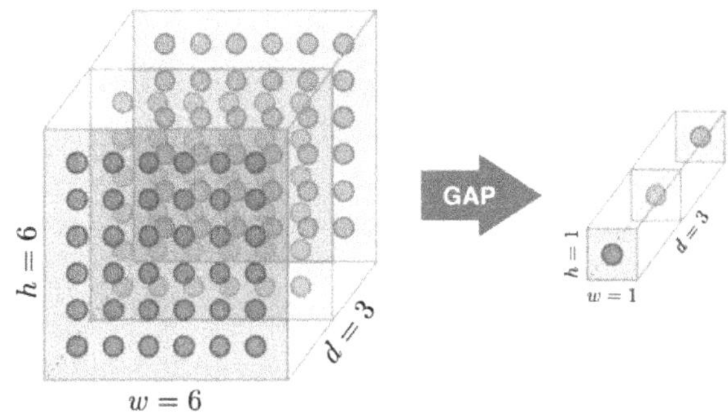

Figure 3.42. Global average pooling yielding 1×1 feature maps.

These vectors are then input into the Softmax layer to obtain the corresponding scores for each category. Too many fully connected layers not only increase network parameters but also lead to overfitting. In response to overfitting, fully connected layers are generally paired with dropout operations. And global average pooling directly pools the entire feature map (whose number is equal to the number of categories), and then average pooling results are input into the softmax layer to obtain the corresponding score for each category. In backpropagation differentiation, its parameter updates are similar to average pooling.

For predicting K classes, the last convolutional layer in the feature extraction stage generates K feature maps. Global average pooling then produces K feature maps of size 1×1, as shown in Figure 3.42.

These 1×1 feature maps are fed into the fully connected layer. Since the globally pooled values correspond to single pixels, the final fully connected operation becomes effectively a weighted addition operation. This structure is more intuitive and has better generalization performance compared to direct full connectivity, as shown in Figure 3.43.

3.5.4.1.2 *Advantages*

Global average pooling significantly reduces the number of network parameters (for classification networks, fully connected layers account for a large proportion) while naturally mitigating overfitting. It also assigns class-specific meaning to each channel of the output

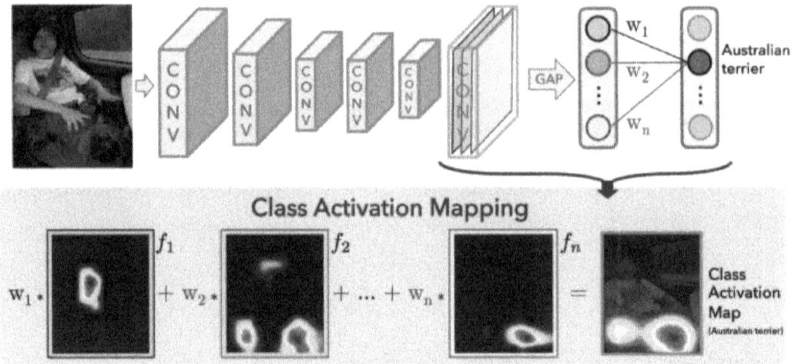

Figure 3.43. Weighted summation operation.

feature maps and eliminates the black-box nature of fully connected operations. Specific advantages are as follows:

(1) Global average pooling offers a more straightforward convolutional structure for establishing relationships between feature maps and class comparison with fully connected layers.
(2) Global average pooling layer requires no parameters to prevent overfitting in this layer.
(3) Spatial information is summed by global average pooling, and robustness to spatial variations in the input is enhanced.
(4) It computes the mean of each channel's feature map in the final convolutional layer.

3.5.4.2 *Global maximum pooling*

3.5.4.2.1 *1D maximum pooling*

As the name suggests, max pooling selects the maximum value within a pooling window. Figure 3.44 shows that compared to 1D max pooling, 1D global max pooling has three fewer parameters, which effectively describe one thing: the window size during pooling.

Figure 3.44 shows that when performing 1D max pooling on a $1 \times 1 \times 5$ network with a window size of 2, there are two issues as follows:

(1) What is the stride for sliding the window? (usually defaults to pool_2.)

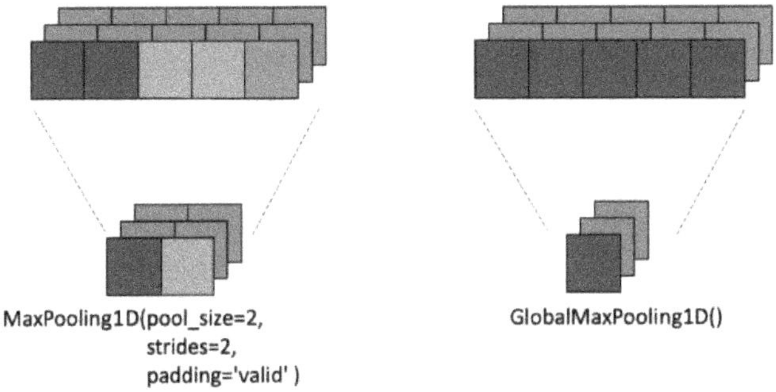

Figure 3.44. 1D maximum pooling. [81]

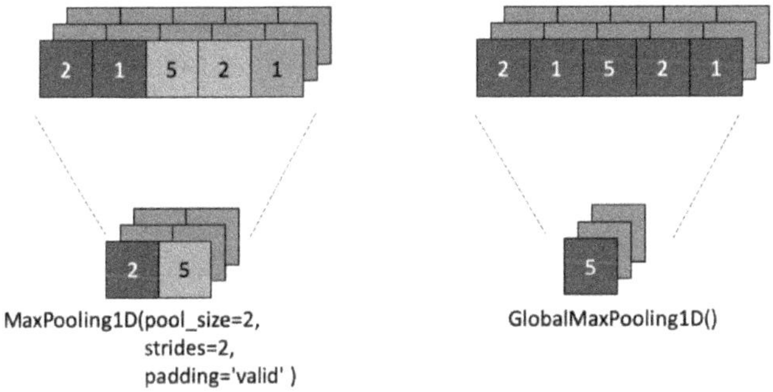

Figure 3.45. 1D max pooling. [81]

(2) If the stride is selected as 2 in Figure 3.44, the last row still needs to be taken (select 'valid' in Figure 3.44, which means no padding will be done, and abandon the last line).

For 1D global max pooling, the pool size is fixed, i.e., the dimension of the longest domain ("global" max pooling). So, since there is no sliding space, there is no need to consider the step size. Therefore, filling does not exist.

Correspondingly, the dimensions of the two pooled results are also different, as shown in Figure 3.45.

3.5.4.2.2 *2D maximum pooling*

Similar to 1D global max pooling, 2D global max pooling performs max pooling over the entire spatial dimensions, as shown in Figure 3.46.

In NLP, global max pooling is more commonly used, while in computer vision, max pooling (non-global) is more prevalent.

3.5.5 Spatial pyramid pooling

Spatial Pyramid Pooling (SPP) in deep convolutional networks (SPP Net) aims to address a transitional layer between the convolutional layer and the fully connected layer, as shown in Figure 3.47. In the CNN, convolution and pooling do not require knowing the image input size, while fully connected layers clearly require image results because the size of the connection weight matrix $\boldsymbol{w}$ in the fully connected layer is fixed after training. For example, from convolution to fully connected layers, the sizes of input and output are 50 and 30 neurons, respectively, so the weight matrix is (50, 30). Therefore, spatial pyramid pooling is proposed to address the transition from convolutional layers to fully connected layers.

For simplicity, a two-layer network, whose input is an image of arbitrary size with 256 channels and whose output is a 256-dimensional vector of 21 elements, is considered. The input image is divided using different scales, as illustrated in Figure 3.48.

In Figure 3.47, an input image is divided into 21 blocks of 16+4+1 using 3 different sizes of scales. A feature is extracted from each of

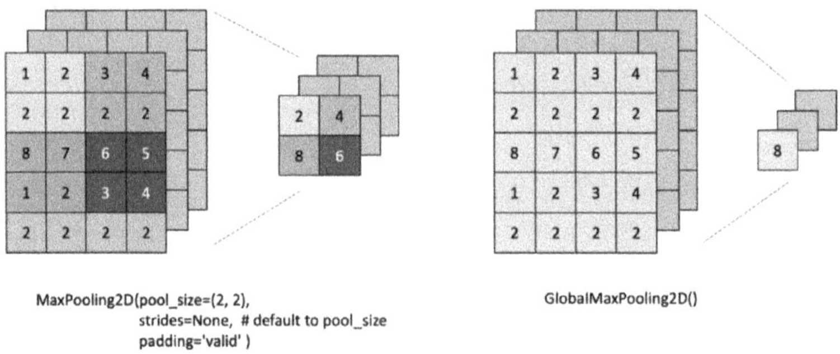

Figure 3.46. 2D max pooling.

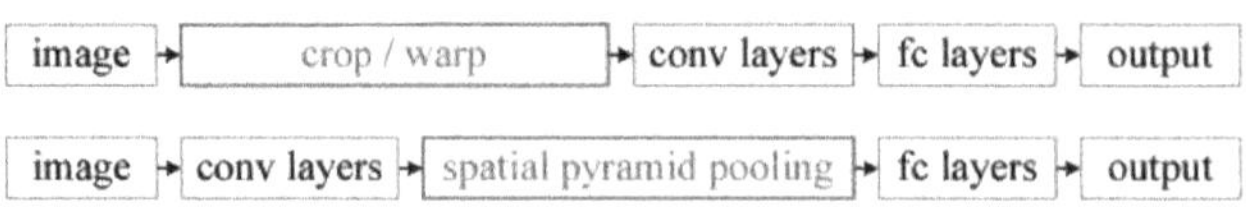

Figure 3.47. Position of the spatial pyramid pooling layer.

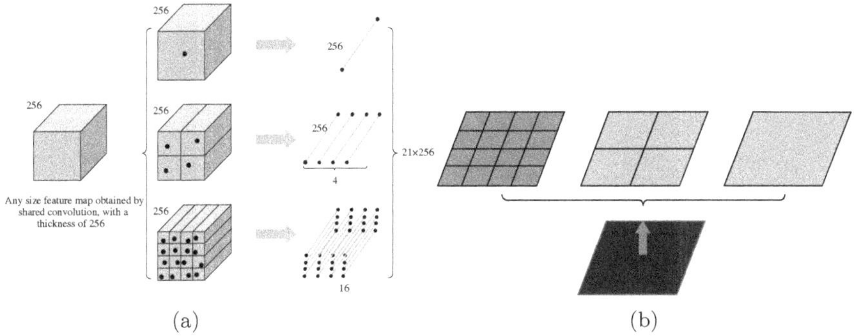

Figure 3.48. Image scale partitioning. (a) 3D view (256 channels). (b) 2D view (1 channel).

these 21 blocks, resulting in the extraction of a 21-dimensional feature vector.

The first type of scale: A complete image is divided into 16 blocks, each with a size of $(w/4, h/4)$. Outputs are exactly 16 vectors, whose dimensions are 1×256.

The second type of scale: An entire image is divided into 4 blocks, with each block with a size of $(w/2, h/2)$. Outputs are exactly 4 vectors, whose dimensions are 1×256.

The third type of scale: An entire image is taken as a block, with a size of (w, h), and it is outputted as a 1x256 vector.

The results pooled by the three partitioning methods are concatenated to obtain $(1 + 4 + 16) \times 256 = 21 \times 256$ features.

The process of maximizing the spatial pyramid pooling is actually to calculate the maximum value of each block (with a thickness of 256) from these 21 image blocks, in order to obtain an output neuron. Finally, an image of any size (with a thickness of 256) is converted into a 21-dimensional feature of a fixed size with a thickness of 256 (other dimensions of output can also be designed to increase the number of pyramid layers or change the size of the grid).

In the division of the above-mentioned three different scales, each scale is called a layer of the pyramid, and the size of each image block is called the window size. If a certain layer of the pyramid needs to output $n \times n$ features, then pooling with a window size of $(W/n, H/n)$ is required.

When there are many layers of the network, an image of any size is input to the network, and convolution and pooling can be

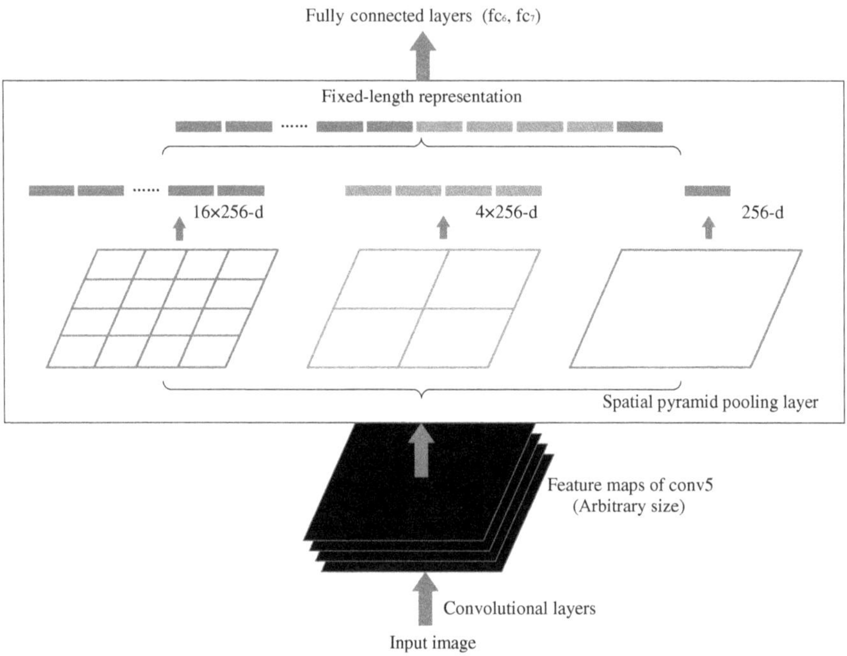

Figure 3.49. Spatial pyramid pooling process.

continuously performed until the last few layers of the network, i.e., when it is about to be connected to the fully connected layer, pyramid pooling is used to convert any size of feature map into a fixed size feature vector. This is called spatial pyramid pooling (multi-scale feature extraction extracts fixed-size feature vectors), as shown in Figure 3.49.

In deep networks, when an arbitrarily sized image is input, convolutions and pooling can proceed on it until the final layers (before fully connected layers). At this stage, spatial pyramid pooling is applied to convert variable-sized feature maps into fixed-size feature vectors (multi-scale feature extraction), as shown in Figure 3.49.

The convolutional layers include convolution, activation functions, and pooling operations. Suppose that the final layer outputs N_f feature maps, the spatial pyramid pooling layer consists of three groups of max pooling operations with dynamic parameters (kernel size and stride determined by $a \times b$). None of these pooling operations use padding, and overlapping is allowed.

The first group (far right) generates a 1×1 pooling map for each feature map. In other words, it adopts max pooling with a kernel size of $a \times b$ and a stride of $[a, b]$.

The second group (middle) generates a 2×2 pooling map for each feature map. This corresponds to max pooling with a kernel size of $\lceil a/2 \rceil \times \lceil b/2 \rceil$ and a stride of $\lfloor a/2 \rfloor, \lfloor b/2 \rfloor$.

The third group (far left) generates a 4×4 pooling map for each feature map. This is achieved using max pooling with a kernel size of $\lceil a/4 \rceil \times \lceil b/4 \rceil$ and a stride of $\lfloor a/4 \rfloor, \lfloor b/4 \rfloor$.

Finally, the pooling maps from all three groups are flattened and concatenated to produce a fixed-size output of dimension $(1 + 4 + 16) \times N_f$.

3.5.6 Multi-scale overlapping sliding pooling

Conventional pooling uses parameters with a pooling kernel of 2×2 and a stride of 2, whose main function is to reduce the size of the feature map and achieve acceleration. However, this single scale pooling parameter in the high-level of convolutional networks does not consider the saliency of features at different scales and the relationship between adjacent feature elements is used to result in a decrease in accuracy. To address this issue, the Scalable Overlapping Slide Pooling (SOSP) method is used to pool from coarse to fine granularity. The SOSP method mainly pools the input feature map through pooling kernels of several scales. In Figure 3.50, the first scale is used to pool the input feature map based on a kernel of size (n, m), a padding of $(n/2, m/2)$, and a stride of size (1,1) to generate a coarse-grained feature map. In the second scale, a kernel size of $(n, m/2)$, a padding of $(n/2, m/4)$, and a stride of (1,1) are used to pool the input feature map to obtain a medium granularity feature map. In the third scale, the kernels with size $(n/2, m)$, padding with size $(n/4, m/2)$, and stride size (1,1) are used to pool the input feature map to obtain a medium granularity feature map. The fourth scale uses a kernel with a size of $(n/2, m/2)$, a padding size of $(n/4, m/4)$, and a stride size of (1,1) to pool the input feature map and obtain a fine-grained feature map. Similarly, by using scales 5 and 6, etc., fine-grained feature maps can be obtained. Due to the step size of pooling being (1,1) and the filling size being half the size of the pooling kernel, the feature maps obtained by multi-scale overlapping pooling have the

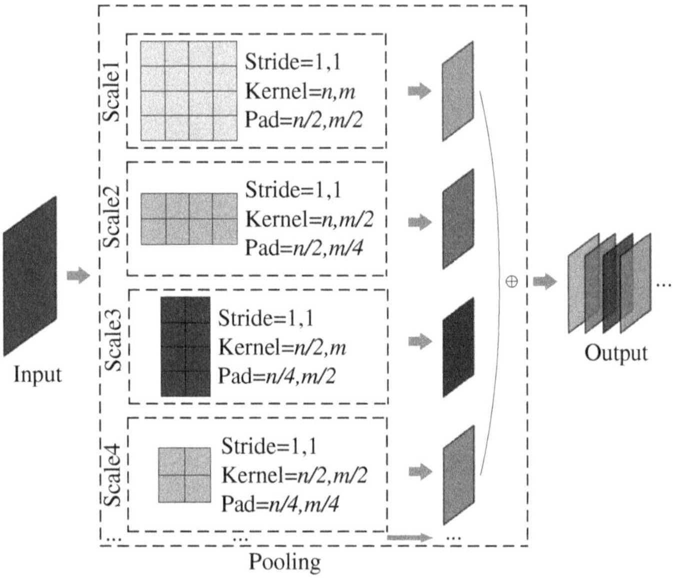

Figure 3.50. Multi-scale overlapping sliding pooling.

same size. Finally, several scale feature maps are merged together by channel and outputted.

Figure 3.50 illustrates this process.

3.6 Common CNN Architectures

3.6.1 LeNet-5 model

The LeNet-5 model, proposed by Professor Yann LeCun in 1998, was the first successful CNN applied to digit recognition and achieved 99.2% accuracy on the MNIST dataset. The LeNet-5 architecture is shown in Figure 3.51 [30].

The LeNet-5 architecture is based on the view that (especially) the features of an image are distributed across the entire image, and a convolution operation with learnable parameters is an effective way to extract similar features at multiple locations with a small number of parameters.

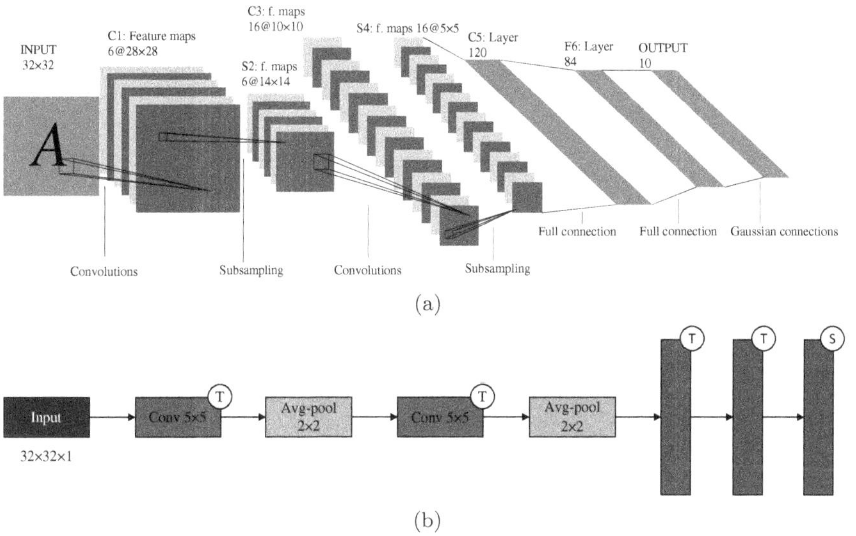

Figure 3.51. LeNet-5 model. (a) LeNet-5 structure. (b) Schematic framework.

The characteristics of LeNet-5 are as follows:

(1) CNNs use three layers in series: Convolution, pooling, and nonlinearity.
(2) Convolution operation is used to extract spatial features.
(3) Mapping is used to subsample the spatial mean.
(4) There is a nonlinearity with hyperbolic tangent (Tanh) or S-shaped (Sigmoid) form.
(5) Multi-layer neural network is used as the final classifier.
(6) Sparse connection matrix between layers is used to avoid high computational costs.

LeNet-5 model is one of the simplest architectures, which has 2 convolutional layers and 3 fully connected layers (so "5" is the sum of the number of convolutional layers and fully connected layers in the neural network). The commonly known average pooling layer is called a subsampling layer, which has trainable weights (different from the current design of the CNNs). This architecture has approximately 60000 parameters.

LeNet-5 model has become a standard "template": stacking convolutional and pooling layers, and ending with one or more fully connected layers.

3.6.2 VGGNet model

VGGNet model was developed by the Visual Geometry Group at Oxford University and researchers from Google DeepMind, whose primary contribution was demonstrating the importance of network depth (a key factor in the performance of the algorithm). Building upon this architecture, subsequent models such as ResNet (152-1000 layers), GoogLeNet (22 layers), and VGGNet (19 layers) emerged, which incorporate many techniques such as new optimization algorithms and multi-model fusion. To this day, the VGGNet model is still widely used for image feature extraction. The VGG16 architecture is shown in Figure 3.52.

In Figure 3.52, the input is a 224×224 RGB image. During preprocessing, the mean values of the three channels are calculated and subtracted from each pixel (resulting in fewer iterations and faster convergence).

The image is processed through a series of convolutional layers using very small 3×3 kernels (with some layers employing 1×1 kernels).

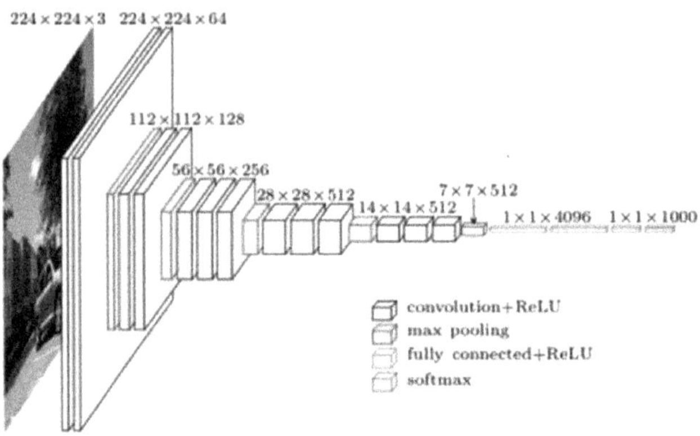

Figure 3.52. VGG16 architecture.

Convolutional layers use a stride of 1 pixel, with 3×3 convolutions padded by 1 pixel. Five max-pooling layers (window size 2×2, stride 2) are interspersed between convolutional layers.

After the convolutional layer, there are three fully connected (FC) layers. Each of the first two fully connected layers has 4096 channels, while the third fully connected layer has 1000 channels for classification. The configuration of the fully connected layer is the same for all networks.

The fully connected layer is a Softmax layer and used for classification.

All hidden layers (in the middle of each convolution layer) use ReLU as the activation function. VGGNet does not use local response normalization (LRN), which does not improve performance on the ILSVRC dataset but leads to more memory consumption and computation time.

3.6.3 ResNet model

For deep neural networks, VGGNets demonstrate that increasing network depth is an effective way to improve accuracy. However, due to the gradient vanishing problem, network depth cannot be infinitely increased [31]. The gradient dispersion problem is caused by the continuous accumulation of errors during the backpropagation process, which results in almost zero gradient values in the first few layers and makes convergence impossible. The test results indicate that for deep networks with more than 20 layers, the convergence performance deteriorates as the number of layers increases. A network of 50 layers has twice the error rate as a network of 20 layers. This phenomenon is called the degradation problem of deep networks.

The degradation problem indicates that not all systems can be easily optimized. Can the depth of the internet be further increased? Residual Network (ResNets) is precisely able to avoid gradient vanishing and is easier to optimize.

Neural networks actually map a spatial dimension vector $\boldsymbol{x}$ to another spatial dimension through a nonlinear transformation $H(\boldsymbol{x})$. But it is very difficult to optimize $H(\boldsymbol{x})$, so we turn to finding the residual $F(\boldsymbol{x})$ of $F(\boldsymbol{x}) = H(\boldsymbol{x}) - x$. Assuming that solving

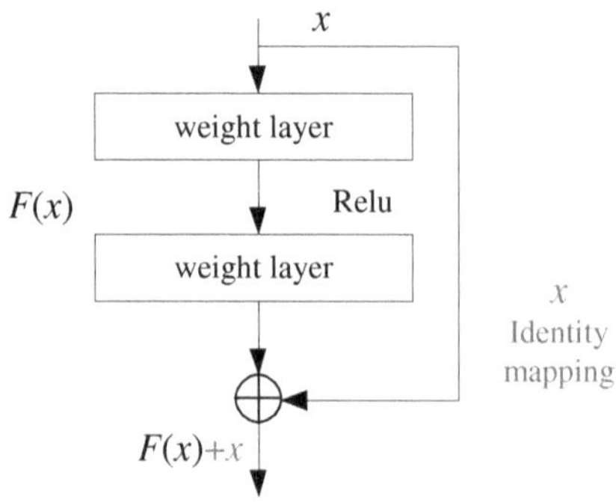

Figure 3.53. Basic structure of residual network.

$F(x) = H(\boldsymbol{x}) - \boldsymbol{x}$ is simpler than finding $H(\boldsymbol{x})$, the final output $\boldsymbol{x}$ can be achieved by adding $F(\boldsymbol{x}) + \boldsymbol{x}$, as shown in Figure 3.53. Figure 3.54 shows a residual network and a smoothing network. Thirty-four layers denote that there are 34 layers with trainable parameters, and the pooling layer doesn't contain trainable parameters. The only difference between the residual network shown in Figure 3.54(a) and the flat network shown in Figure 3.54(b) is the shortcut connection. These two networks double the number of filters when the feature mapping is halved to ensure consistent computational complexity for each layer.

ResNet uses identity mapping and does not have parameters on shortcut connections, so the computation complexities of the flat network and those of the residual network in Figure 3.54 are the same, which is equal to 3.6×10^9 FLOPs.

The introduction of residual networks enables neural networks to be as deep as possible without problems of the decreased accuracy and is widely used in computer vision tasks.

3.6.4 DenseNet model

DenseNets are a type of deep neural network with dense connectivity. In the network, each convolutional layer takes additional input from the previous layer and transmits its output features to all subsequent convolutional layers. In other words, the input of a certain layer is the set of outputs from all previous layers, and the output of that layer is a part of the input for all subsequent layers. The advantage of

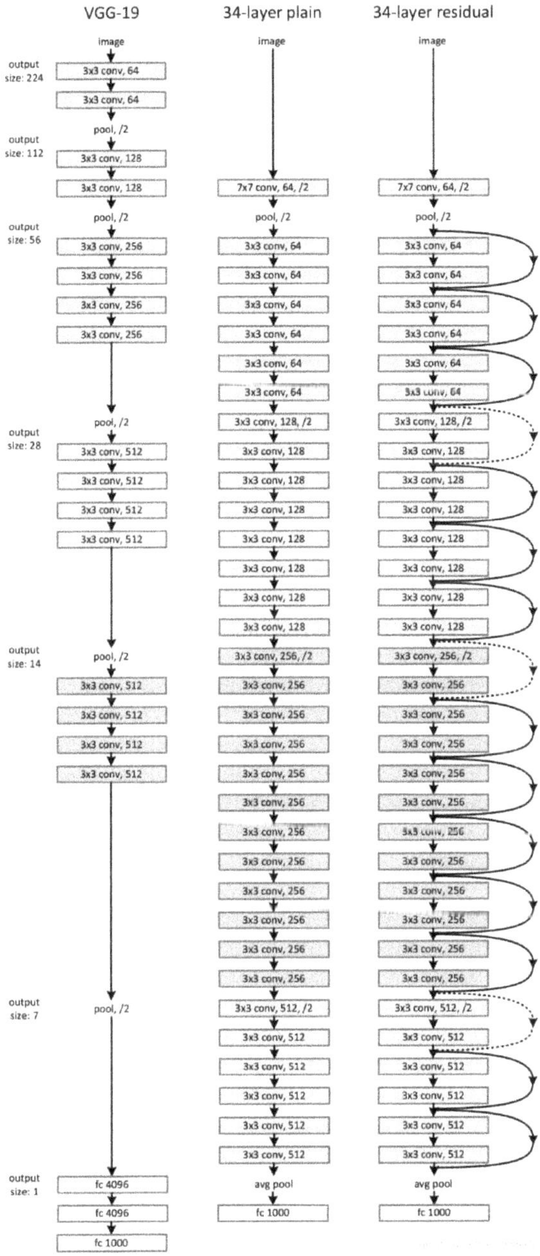

(a) (b) (c)

Figure 3.54. Residual network and plain network [31]. (a) VGG-19. (b) 34-layer plain network. (c) 34-layer residual network.

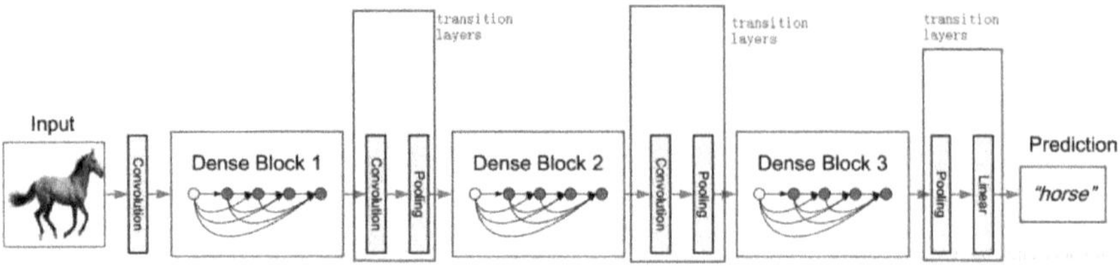

Figure 3.55. Framework of DenseNets.

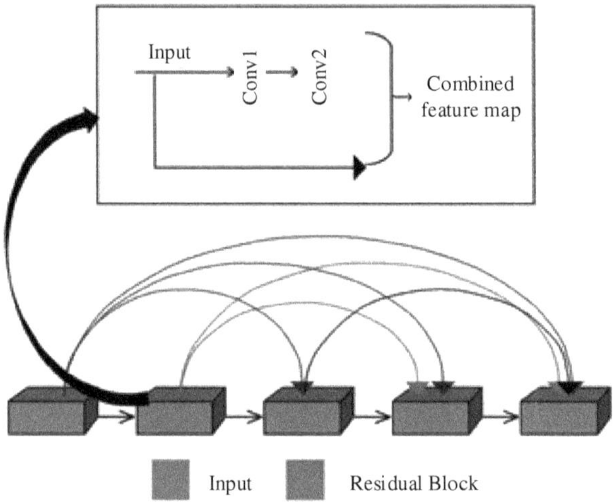

Figure 3.56. Structure of dense block.

this dense connection structure is that each layer enhances the propagation of features and significantly solves the problems of gradient explosion and vanishing. A dense network (DenseNet) structure is shown in Figure 3.55.

In Figure 3.55, there are three dense blocks. This structure divides DenseNet into multiple dense blocks, to hope that the size of the feature maps within each dense block is uniform, so that there will be no size problem in the connections. The structure of a dense block is shown in Figure 3.56.

3.6.5 AlexNet model

In 2012, Alex Krizhevsky, a student of Hinton, proposed the AlexNet model, which belongs to a deep CNN model and can be considered

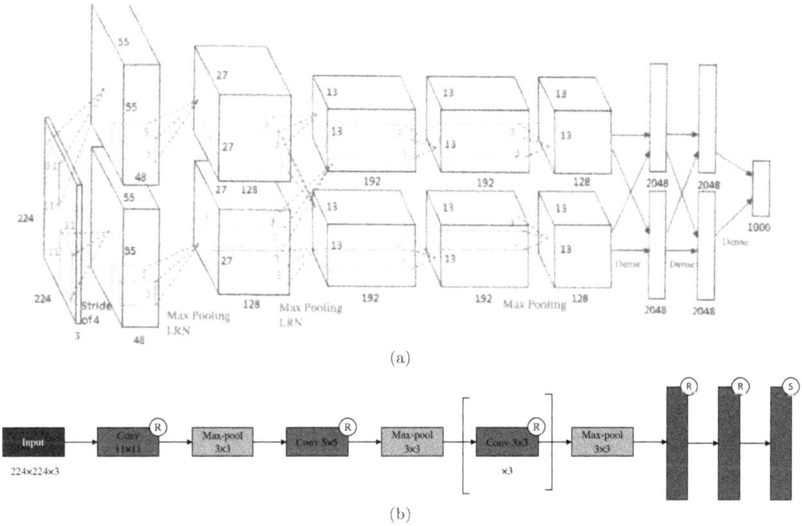

Figure 3.57. AlexNet architecture. (a) AlexNet Structure. (b) AlexNet schematic framework.

as a deeper and wider version of the LeNet model. AlexNet model contains several relatively new technological points and is also the first successful application of Tricks such as the ReLU, Dropout, and the LRN in the CNN. At the same time, AlexNet model also uses GPU for computation acceleration, and its author has broadened their CUDA code sources for training CNNs on GPU. AlexNet model contains 630 million connections, 60 million parameters, and 650000 neurons. The architecture comprises five convolutional layers; among them, three convolutional layers are followed by max-pooling layers. In addition, there are three fully connected layers in the architecture, as shown in Figure 3.57.

3.6.5.1 *Architecture of AlexNet model*

AlexNet model takes $224 \times 224 \times 3$ images as input and outputs conditional probabilities for 1000 classes. The network follows a pyramid structure:

(1) *First convolutional layer*: Use two $11 \times 11 \times 3 \times 48$ kernels with stride $S = 4$ and zero-padding $P = 3$ and output two $55 \times 55 \times 48$ feature map groups.

(2) *First pooling layer*: Apply 3×3 max-pooling with stride S =2 and output two $27 \times 27 \times 48$ feature map groups.
(3) *Second convolutional layer*: Use two $5 \times 5 \times 48 \times 128$ kernels with stride $S = 1$ and padding $P = 2$ and output two $27 \times 27 \times 128$ feature map groups.
(4) *Second pooling layer*: Apply 3×3 max-pooling with stride $S = 2$ and output two $13 \times 13 \times 128$ feature map groups.
(5) *Third convolutional layer*: Merge two paths using a $3 \times 3 \times 256 \times 384$ kernel with stride $S = 1$ and padding $P = 1$ and output two $13 \times 13 \times 192$ feature map groups.
(6) *Fourth convolutional layer*: Use two $3 \times 3 \times 192 \times 192$ kernels with stride $S = 1$ and padding $P = 1$ and output two $13 \times 13 \times 192$ feature map groups.
(7) *Fifth convolutional layer*: Use two $3 \times 3 \times 192 \times 128$ kernels with stride $S = 1$ and padding $P = 1$ and output two $13 \times 13 \times 128$ feature map groups.
(8) *Third pooling layer*: Apply 3×3 max-pooling with stride S = 2 and output two $6 \times 6 \times 128$ feature map groups.
(9) *Three fully connected layers*: Number of neurons are 4096, 4096, and 1000, respectively.

Local Response Normalization (LRN) is applied after the first two pooling layers to enhance generalization.

3.6.5.2 *The new technological features in AlexNet model*

AlexNet model further developed the ideas of LeNet and applied the basic principles of the CNN to deeper and wider networks. The main new technologies used by the AlexNet model are as follows:

(1) ReLU is successfully used as the activation function for the CNN and verified its effectiveness over Sigmoid in deeper networks, and the gradient dispersion problem of Sigmoid in deeper network is successfully solved. The emergence of the AlexNet model further develops the ReLU activation function.
(2) During training, the Dropout method is used to randomly ignore a portion of neurons, thereby the overfitting of the model is avoided. In the AlexNet model, Dropout is used in the last few fully connected layers to make Dropout more practical.

(3) In the CNN, the use of overlapping max pooling avoids the blurring effect of average pooling, and the stride is smaller than the size of the pooling kernel, so there are overlap and coverage between the outputs of the pooling layers and feature richness is improved.
(4) A Local Response Normalization (LRN) layer is proposed to create a competitive mechanism for the activity of local neurons, make the values with larger responses relatively larger, and suppress other neurons with smaller feedback, thereby, the generalization ability of the model is enhanced.
(5) The CUDA is used to accelerate the training of deep convolutional networks, and the powerful parallel computing capabilities of GPUs are leveraged to handle a large amount of matrix operations during training neural network. The AlexNet model is distributed on two GPUs and half of the neuron parameters are stored in the video memory of each GPU. Since communication between GPUs is convenient and they can access each other's memory without the need for host memory, it is very efficient to use multiple GPUs simultaneously. Meanwhile, AlexNet's design allows communication between GPUs to only occur at certain layers of the network; this design controls the performance loss of communication.
(6) The expanded data greatly reduces overfitting and improves generalization ability. At the same time, the Principal Component Analysis (PCA) method is applied to process the RGB data of the images, and a Gaussian perturbation with a standard deviation of 0.1 is applied to the principal components to add some noise. This Trick can further reduce the error rate by 1%.

3.6.6 Inception model

3.6.6.1 *Inception V1 module structure*

In convolutional networks, how to select appropriate kernel sizes is very crucial. In Inception networks, a convolutional layer contains multiple convolution operations of different sizes, called Inception modules. Inception networks are composed of multiple Inception modules and a small number of pooling layers stacked together.

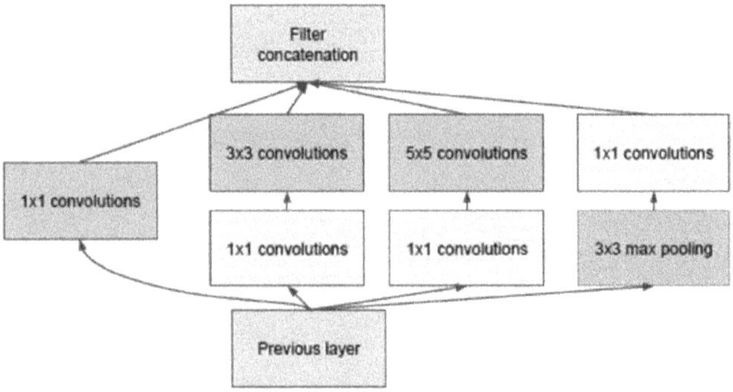

Figure 3.58. Inception V1 module.

The Inception module simultaneously uses convolution kernels of different sizes, such as 1×1, 3×3, and 5×5, and concatenates (stacks) the obtained feature maps in depth as the output feature map.

The Inception V1 module structure shown in Figure 3.58 has four branches:

The first branch performs a 1×1 convolution on the input, which is actually an important structure proposed in NIN (Network in Network). The 1×1 convolution is an excellent structure that can organize information across channels, improve network expression ability, and also increase and decrease the dimensionality of output channels. The four branches of the Inception module all use 1×1 convolutions for low-cost (much less computationally intensive than 3×3) cross-channel feature transformations.

The second branch first uses 1×1 convolution and then connects to a 3×3 convolution, which is equivalent to performing two feature transformations.

The third branch starts with a 1×1 convolution and then connects to a 5×5 convolution.

The fourth branch is to use 1×1 convolution directly after 3×3 max pooling.

Thus, it can be seen that some branches only use 1×1 convolution, while others employ 1×1 convolution once again when using convolutions of other sizes. This is because 1×1 convolution has a cost-effectiveness ratio and can add one layer of feature transformation and nonlinearity with a small amount of computation. The four branches of the Inception module are merged through an aggregation operation at the end (aggregated in terms of the number of output channels). The Inception module including three different sizes of convolutions and one maximum pooling can improve the network's adaptability to different scales. The Inception module can efficiently expand the depth and width of the network, improve accuracy, and prevent overfitting.

In the Inception network, the earliest version of Inception V1 is the well-known GoogLeNet model, which consists of 9 Inception V1 modules, 5 pooling layers, and other convolutional and fully connected layers, totaling 22 layers, as shown in Figure 3.59.

To solve the problem of gradient vanishing, the GoogLeNet model introduces two auxiliary classifiers in the middle layer of the network to enhance supervised information. In Inception networks, a representative improved version is Inception V3 network, which replaces large convolution kernels with multiple layers of small convolution kernels and reduces computational and parameter complexity while maintaining the same receptive field. Specifically, it includes the following: (1) Use a two-layer 3×3 convolution instead of the 5×5 convolution in Inception V1 and (2) use continuous $K \times 1$ and $1 \times K$ instead of $K \times K$ convolution. In addition, in the Inception V3 network, the optimization methods such as label smoothing and batch normalization were also introduced to train.

3.6.6.2 *Inception V2*

In February 2015, Batch-normalized Inception (BN) was introduced as Inception V2.

Batch normalization can calculate the mean and standard deviation of all feature maps on the output of a layer and normalize their responses using these values. This is equivalent to data whitening; thus, all neural maps have responses within the same range and zero mean. When there is no need to learn *offset* from input data in the next layer, this contributes to training and also focuses on how to best combine these features.

In Inception V2, two 3×3 convolutions are used instead of the 5×5 large convolution (to reduce parameters and alleviate overfitting), as shown in Figure 3.60.

The BN is a highly effective regularization method that can accelerate the training speed of large convolutional networks by many

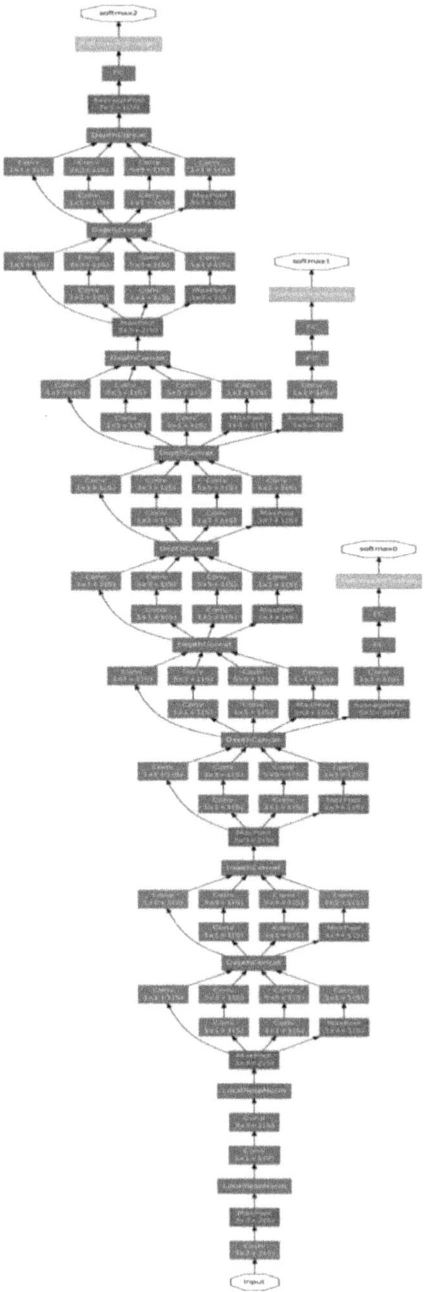

Figure 3.59. GoogLeNet network architecture.

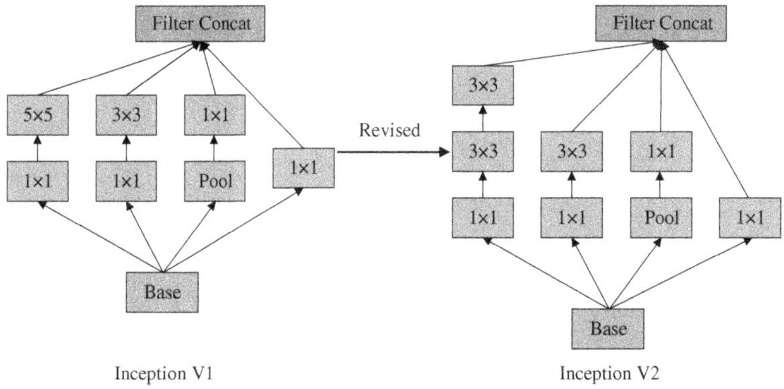

Figure 3.60. Inception V2 module.

times, while significantly improving the classification accuracy after convergence. When the BN method is applied to a certain layer of a neural network, it normalizes the interior of each mini-batch data to a normal distribution of $N(0,1)$ to reduce the Internal Covariance Shift (changes in the distribution of internal neurons).

In training traditional deep neural networks, the input distributions at each layer all are varying; it is difficult to result in training and only a small learning rate is required to solve this problem. In order to effectively solve the problem, the BN method is used for each layer, and the learning rate can be increased many times. Since the number of iterations, which is required to achieve the previous accuracy, is only 1/14, the training time is greatly reduced. After achieving the previous accuracy, training can continue and ultimately achieve a top-5 error rate of 4.8%, far exceeding the performance of the Inception V1 model, and this performance is better than the human eye level. Since the BN method also plays an important role in regularization to some extent, Dropout can be reduced or eliminated, so the network structure can be simplified.

Of course, the BN method only introduced to the Inception V2 model cannot make its performance have a significant improvement. Thereby, it is necessary to appropriately adjust the Inception V1 model. The specific adjustment plan is as follows: (1) The learning rate is enhanced and the decay rate of learning is accelerated to adapt to BN normalized data. (2) Dropout is removed and L2 regularization (as BN has already played a role in regularization) is alleviated.

(3) The LRN is removed. (4) The training samples are more thoroughly shuffled, and the optical distortion of the data is reduced during data augmentation (because BN training is faster and each sample is trained fewer times, more realistic samples are more helpful for training). After implementing these measures, the training speed of the Inception V2 model has improved 14 times under the condition of achieving the same accuracy as Inception V1, and the Inception V2 model has a higher upper limit accuracy during convergence.

3.6.6.3 *Inception V3*

One of the most important improvements of Inception V3 lies in the factorization of convolutions, which decomposes 7×7 convolution into two one-dimensional convolutions (1×7, 7×1), decomposes 3×3 into (1×3, 3×1) (as shown in Figure 3.61) to accelerate computation speed (redundant computing cap can be used to deepen the network), and further increases network depth and nonlinearity by splitting one convolution into two convolutions. Additionally, the input of the network has changed from 224×224 to 299×299, so a more finely module of $35 \times 35/17 \times 17/8 \times 8$ is designed.

Inception V3 modulus has optimized the structure of the Inception module, which now has three different structures: 35×35, 17×17, and 8×8, as shown in Figure 3.62. These Inception modules only appear at the back of the network, while the front is still a traditional convolutional layer. And Inception V3 modulus uses branches

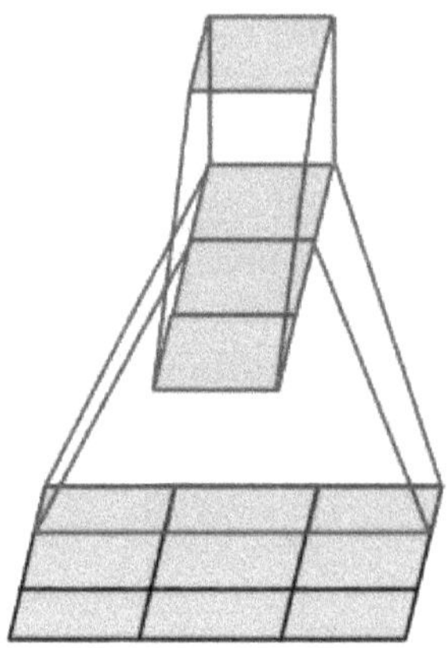

Figure 3.61. Decomposing a 3×3 convolution into 1×3 and 3×1 convolutions.

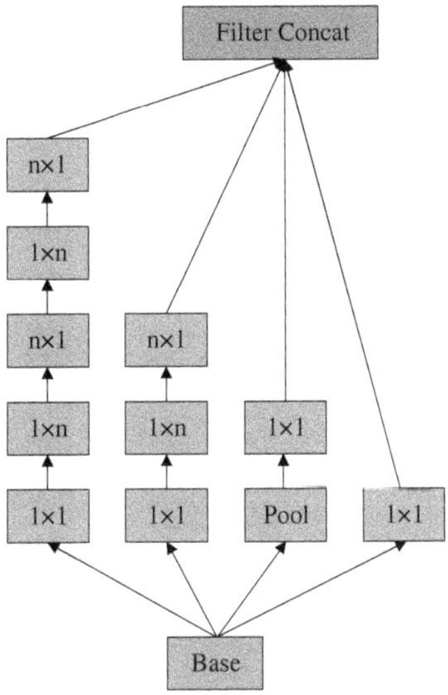

Figure 3.62. Three Inception modules in Inception V3 model.

in not only the Inception module but also the branches (in an 8×8 structure), which can be said to be Network in Network (NIN).

While performing Inception calculations, the Inception module can also reduce data size by providing pooling. This is basically similar to running a convolution in parallel with a simple pooling layer, as shown in Figure 3.63.

Inception also utilizes a pooling layer and Softmax as its final classifier. Inception V3 modulus is shown in Table 3.1.

3.6.6.4 *Inception V4 model*

Inception V4 model is illustrated in Figure 3.64.

3.6.7 Xception model

Xception model was proposed by Google as another improvement over Inception V3 modulus; the original convolution operations in Inception V3 modulus are primarily replaced with depthwise separable convolution, and the CNN based on entire channel-independent convolutional layers is introduced. The Xception architecture is shown in Figure 3.65. Its core ideas are as follows:

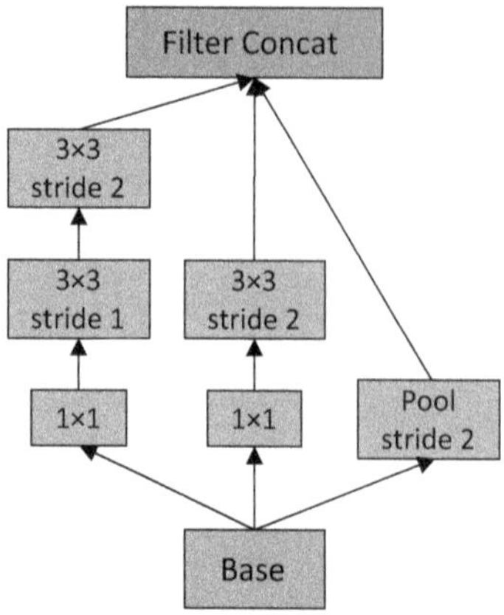

Figure 3.63. Pooling layer.

Table 3.1. Inception V3 network architecture.

Style	Kernel size/	Step
Conv	3×3 / 2	$299 \times 299 \times 3$
Conv	3×3 / 1	$149 \times 149 \times 32$
Conv	3×3 / 1	$147 \times 147 \times 32$
Pooling	3×3 / 2	$147 \times 147 \times 64$
Conv	3×3 / 1	$73 \times 73 \times 64$
Conv	3×3 / 2	$71 \times 71 \times 80$
Conv	3×3 / 1	$35 \times 35 \times 192$
Inception module group	The three Inception module	$35 \times 35 \times 288$
Inception module group	The Five Inception module	$17 \times 17 \times 768$
Inception module group	The three Inception module	$8 \times 8 \times 1280$
Pooling	8×8	$8 \times 8 \times 2048$
Linear	Logits	$1 \times 1 \times 2048$
Softmax	Categorized output	$1 \times 1 \times 1000$

First, cross-channel (or cross-feature map) correlations are captured by using 1×1 convolutions.

Second, spatial correlations within each channel are captured by using standard 3×3 or 5×5 convolutions.

This idea that is applied to the extreme means performing a 1×1 convolution on each channel, then a 3×3 convolution on each output is performed. This is the same as replacing the initial module with channel-independent convolutions.

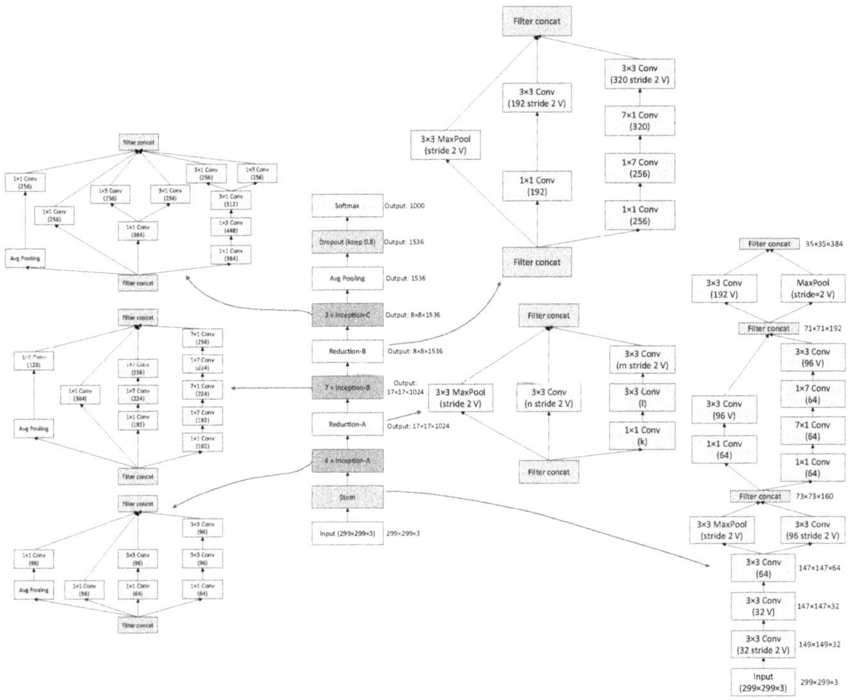

Figure 3.64. Inception V4 model.

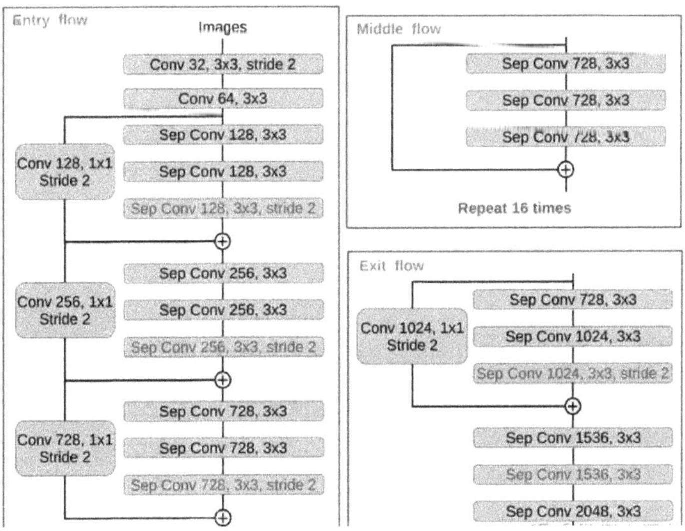

Figure 3.65. Xception architecture.

3.7 Extended CNN Architectures

3.7.1 Feature fusion CNN architecture

By combining feature fusion techniques with convolutional neural networks, a feature fusion convolutional neural network was proposed [31]. This proposed network employs Deep Linear Discriminant Analysis (DLDA) for optimizing sample features, and the optimized statistical features are input into a convolutional neural network to automatically learn the importance of sample features for better classification.

The use of LDA's linear discriminant equation as the objective function of neural networks has always been a hot topic of research [32]. The purpose is to integrate nonlinear neural networks into classical LDA methods. With the continuous updating of network weights, the variance of input samples of the same class will become smaller, and there will be better separation of features between similar and dissimilar samples. Assume that N samples $\{x_1, x_2, \ldots, x_N\}$ belong to the cth category and $c \in \{1, 2, \ldots, C\}$; whereas, the dimension of each sample is denoted as $D_n(n = 1, 2, \ldots, N\}$. The purpose of the LDA is to find the optimal projection matrix $\boldsymbol{w}$ to maximize the following object function, i.e.,

$$\arg\max_{\boldsymbol{W}} \frac{|\boldsymbol{w}\boldsymbol{S}_b\boldsymbol{w}^T|}{|\boldsymbol{w}\boldsymbol{S}_w\boldsymbol{w}^T|}, \tag{3.7.1}$$

where $\boldsymbol{S}_b$ is the interclass scattering matrix and $\boldsymbol{S}_w$ is the intraclass scattering matrix. They are defined as

$$\boldsymbol{S}_b = \boldsymbol{S}_t - \boldsymbol{S}_w, \tag{3.7.2}$$

where

$$\boldsymbol{S}_t = \frac{1}{N-1}(\boldsymbol{X} - \boldsymbol{m})(\boldsymbol{X} - \boldsymbol{m})^T, \tag{3.7.3}$$

$$\boldsymbol{S}_c = \frac{1}{N_c - 1}(\boldsymbol{X}_c - \boldsymbol{m}_c)(\boldsymbol{X}_c - \boldsymbol{m}_c)^T, \tag{3.7.4}$$

$$\boldsymbol{S}_w = \frac{1}{C}\sum\nolimits_c \boldsymbol{S}_c, \tag{3.7.5}$$

where $\boldsymbol{S}_t$ is the total scattering matrix and $\boldsymbol{X}$ and $\boldsymbol{X}_c$ represent the sample data and class-C sample data, respectively. $\boldsymbol{m}$ and $\boldsymbol{m}_c$ denote the mean of all samples and that of class-C sample. N is the number of total samples. We optimize the maximum value of the objective function; essentially, it's about finding the maximum eigenvalue $\boldsymbol{\lambda}$ in equation $\boldsymbol{S}_b\boldsymbol{\beta} = \boldsymbol{\lambda}\boldsymbol{S}_w\boldsymbol{\beta}$.

In order to overcome the problem of large eigenvalues being valued and small eigenvalues being ignored in optimization objectives, J. H. Friedman proposed to regularize the intraclass scattering matrix by adding multiples of the identity matrix, which is expressed as

$$\boldsymbol{S}_b\boldsymbol{\beta}_i = \lambda_i(\boldsymbol{S}_w + \alpha\boldsymbol{I})\boldsymbol{\beta}_i, \tag{3.7.6}$$

where $\lambda_i(i = 1, \ldots, C-1)$ represents the computed eigenvalues and $\boldsymbol{\beta}_i(i = 1, \ldots, C-1)$ denotes the corresponding eigenvectors.

The LDA-based neural network optimization algorithm continuously maximizes the eigenvalues during training. The optimization problem is described as

$$\arg\max_{\Theta} \frac{1}{C-1} \sum_{i=1}^{C-1} \lambda_i. \tag{3.7.7}$$

The problem with Equation (3.7.7) is that it can result in better separation of classes that are originally farther apart, while it is difficult to separate adjacent class samples. To solve this problem, a threshold is set to focus on J feature values to emphasize the contribution of small feature values. The specific calculation problem is described as

$$\arg\max \frac{1}{J} \sum_{i=1}^{J} \lambda_i \quad \{\lambda_1, \ldots, \lambda_J\} = \{\lambda_j | \lambda_j < \min\{\lambda_1, \ldots, \lambda_{C-1}\} + \varepsilon\}. \tag{3.7.8}$$

The operational steps for optimizing sample statistical features using the DLDA are illustrated in Figure 3.66.

The feature optimization network based on the DLDA consists of three fully connected layers, with the first two using Sigmoid activation and the last using linear activation. The number of neurons in the third fully connected layer can match the signal sampling points through zero-padding. If there are few sampling points in the

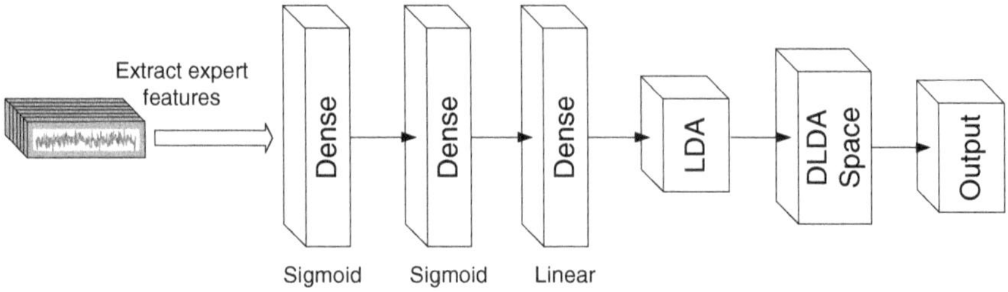

Figure 3.66. Feature optimization network based on the DLDA.

samples, the number of neurons in the third layer aligns with the lengths of the sample signals. Through continuous training, this network enhances the separability of input sample statistical features while allowing the controlled output feature dimensions for convenient feature fusion.

The feature optimization network based on the DLDA consists of three fully connected layers, with the activation function of the first two layers being the Sigmoid function and that of the last layer being a linear function. The number of neurons in the fully connected layer of the third layer can be kept consistent with the number of signal sampling points through zero-padding operation. If sample sampling points in the samples are fewer, the number of neurons in the third layer should be kept consistent with the length of the sample signal. By continuously training the network, the statistical features of the input samples will exhibit better separability, and the feature dimensions of the output can be controlled; it will provide convenience for feature fusion.

3.7.2 Deep guided filtering network

In response to the problem of limited texture detail information in original low dynamic range images, which makes it difficult to restore image details while restoring brightness and contrast, in this section, the use of deep fast guided filtering algorithm is studied to decompose the image into a low-frequency base layer and a high-frequency detail layer, and then to use the CNNs to learn corresponding features, so that the guided filtering network can focus on color conversion and high-frequency detail restoration of the images, and finally fuse high and low frequency features to restore high-quality images.

3.7.2.1 *Differentiable fast guided filtering layer*

Guided filter is a type of edge-preserving filter that is used to calculate the filtering output via referencing the content information of the guided image (which can be the input image itself or other specific task images) and uses optimization methods to maintain the similarity of image edge gradients. Guided filter has been applied to various computer vision tasks, such as rain removal, defogging, and enhancement. Fast guided filtering algorithm is an improved version of traditional guided filtering algorithm, which uses joint sampling methods to reduce computational complexity and improve application performance while maintaining effectiveness [33, 34].

Given a full-resolution image I_h and its corresponding low-resolution image I_l and low-resolution output O_l, a fast guided filter is used to generate a high-resolution output O_h. The optimization process of the guided filtering algorithm driven by local linear models is given by

$$O_l^i = a_l^k I_l^i + b_l^k, \quad \forall i \in W_k, \tag{3.7.9}$$

where i represents the index of image pixels and k is the index of a local square window W with radius r. Both a_l and b_l are calculated by minimizing the reconstruction error between I_l and O_l, while a_h and b_h are obtained through bilinear upsampling to a_l and b_l, respectively. The high-resolution output Q is then derived by using linear transformation, i.e.,

$$O_h = a_h \odot I_h + b_h, \tag{3.7.10}$$

where $\odot$ denotes element-wise multiplication. Typically, the low-resolution output O_l is generated by a transformation function $f(x)$. The guided filter is now formulated as a fully differentiable module and combined with $f(x)$ to perform end-to-end training. The computational graph of the differentiable guided filter layer is shown in Figure 3.67. The specific calculation process is shown in Table 3.2.

In Figure 3.67, a_l and b_l are calculated using mean filter f_m and the local linear model of given inputs I_l and O_l, then bilinear upsampling f_{mp} is applied to generate a_h and b_h as inputs to the linear transformation layer for further generating O_h. The radius r of f_m is set to 1 by default. ε is the regularization parameter controlling smoothness, with a default value of $1e^{-8}$. $F(I)$ is regarded as the

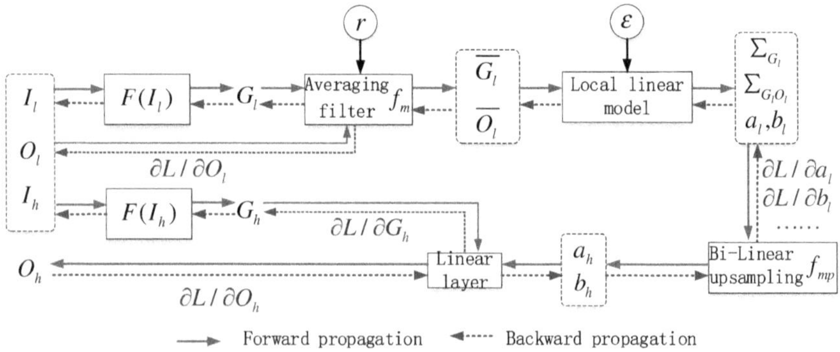

Figure 3.67. Computational graph of differentiable guided filtering layer.

Table 3.2. Guided filter layer computation process.

Input: Low-resolution image I_l, high-resolution image I_h
Low-resolution output O_l, radius r and regularization term ε
Output: High-resolution output O_h

1 $G_l = F(I_l)$, $G_h = F(I_h)$
2 $\overline{G_l} = f_m(G_l, r)$, $\overline{O_l} = f_m(O_l, r)$
$\overline{G_l^2} = f_m(G_l \odot G_l, r)$, $\overline{G_l O_l} = f_m(G_l \odot O_l, r)$
3 $\sum_{G_l} = \overline{G_l^2} - \overline{G_l} \odot \overline{G_l}$, $\sum_{G_l O_l} = \overline{G_l O_l} - \overline{G_l} \odot \overline{O_l}$
4 $a_l = \sum_{G_l O_l} / (\sum_{G_l} + \varepsilon)$, $b_l = \overline{O_l} - a_l \odot \overline{G_l}$
5 $a_h = f_{mp}(a_l)$, $b_h = f_{mp}(b_l)$
6 $O_h = a_h \odot G_h + b_h$

transformation function and consists of two 1×1 convolutional layers with batch normalization and ReLU activation function; among them, the first convolution has 16 output channels while the second one has 3 output channels. $F(I)$ acts on I_h and I_l to generate three-channel guidance maps G_h and G_l, to enable the guided filtering layer to adapt to image enhancement tasks.

The gradient propagation equations for the differentiable guided filtering layer are shown in Table 3.3. By modeling each operation as a differentiable function, the gradients of O_h can be propagated in an opposite direction through the computational graph to O_l, I_l, and I_h to further optimize the transformation function $f(x)$.

Table 3.3. Backward gradient propagation equations for guided filter layer.

Input: Low-resolution image I_l, High-resolution image I_h
Low-resolution output O_l, Derivative of high-resolution output ∂O_h
Output: Gradients of all inputs
1 $\partial b_l = \partial O_h \cdot \nabla_{b_l} f_{mp}$, $\partial a_l = \partial O_h \odot G_h \cdot \nabla_{a_l} f_{mp} - \partial b_l \odot \overline{G_l}$
2 $\partial \sum_{G_l O_l} = \partial a_l / (\sum_{G_l} + \varepsilon)$
$\partial \sum_{G_l} = -\partial a_l \odot \partial \sum_{G_l O_l} / (\sum_{G_l} + \varepsilon)^2$
3 $\partial \overline{O_l} = \partial b_l - \partial \sum_{G_l O_l} \odot \overline{G_l}$
$\partial O_l = \partial \sum_{G_l O_l} \cdot \nabla_{G_l \otimes O_l} f_m \odot G_l + \partial \overline{O_l} \cdot \nabla O_l f_m$
4 $\partial \overline{G_l} = -\partial b_l \odot a_l - \partial \sum_{G_l O_l} \odot \overline{O_l} - 2\partial \sum_{G_l} \odot \overline{G_l}$
$\partial G_l = \partial \sum_{G_l O_l} \cdot \nabla_{G_l * O_l} f_m \odot O_l + 2\partial \sum_{G_l} \cdot \nabla_{G_l * G_l} f_m \odot G_l + \partial \overline{G_l} \cdot \nabla_{G_l} f_m$
5 $\partial I_l = \partial G_l \cdot \nabla_{I_l} F$, $\partial I_h = \partial O_h \odot a_h \cdot \nabla_{I_h} F$

3.7.2.2 *Structure of deep guided filtering network*

The differentiable fast guided filtering layer is integrated into the CNN and trained in a coarse-to-fine manner. This framework generates edge-preserving outputs with low computational cost. The overall architecture of the deep guided filtering algorithm is shown in Figure 3.68. Given an original input image I_h, a low-resolution image I_l can be obtained through downsampling. The transformation function $f(x)$ of the low-resolution CNN $LC(I_l)$ learns the mapping from I_l to O_l, and the inputs I_h, I_l and O_l are then fed into the guided filtering layer to generate the high-resolution output O_h. The low-resolution CNN consists of multiple convolutional layers and modulates image signals through element-wise multiplication to enhance spatial variability. Residual structures are employed for stabilizing training and the residual module in the network is utilized to supplement the attention module.

3.8 Example 2: Remote Sensing Image Classification Based on Deep CNN

In this section, the CNN model is applied to remote sensing image recognition and classification, and a neural network for farmland

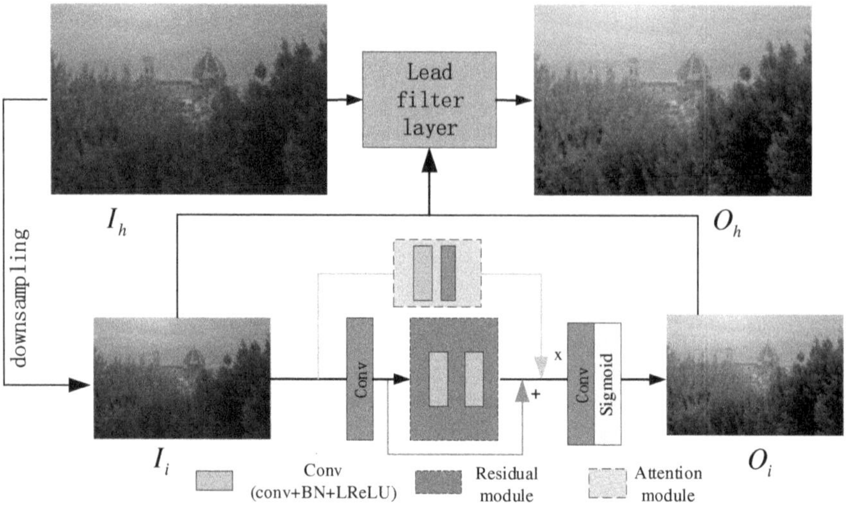

Figure 3.68. Deep guided filter layer network framework.

identification is designed. The network based on the classic AlexNet [32] is improved and adjusted. The number of convolutional layers and sizes of the filter are modified to improve recognition accuracy. Through experiments, it has been proven that if the number of fully connected layers is reduced when there are fewer classification categories, its network parameters will be decreased and the training speed will be improved. While the Batch Normalization (BN) algorithm is added to the network for the reduction of layers, the convergence speed and classification accuracy [35] of the network can be further enhanced.

3.8.1 Basic CNN framework for remote sensing image recognition

The improved CNN for remote sensing image recognition is shown in Figure 3.69. The recognition image of the CNN framework is a fused hyperspectral remote sensing image of land and resources, which contains multiple different land categories. So, on the premise of minimizing the size of remote sensing images, uniform cropping should be applied to the remote sensing images. The CNN needs to perform multiple convolutions on the input images, and when the CNNs extract image features, they require the size of the image to be no

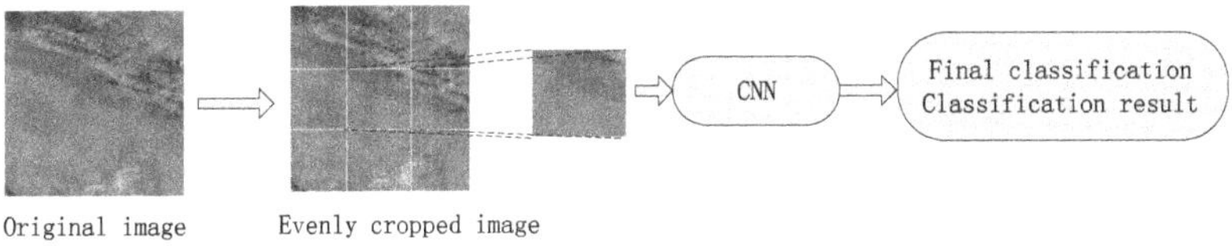

Figure 3.69. CNN-based remote sensing image farmland recognition process.

smaller than the local receptive field of the network, and it is essential for the image to have an appropriate size. Based on the above two aspects and experimental comparison, remote sensing images of 41×41 for classification are selected.

3.8.2 Remote sensing image classification based on modified AlexNet model

3.8.2.1 *Network architecture*

The improved CNN model for remote sensing image recognition is shown in Figure 3.70. Similar to the classic AlexNet model, it consists of an input layer, convolutional layers, pooling layers, fully connected layers, and an output layer; among them, convolutional and pooling layers are interconnected through interaction. The network input is uniformly cropped remote sensing images.

In deep networks, the first convolutional layer uses 96 kernels with a size of 13×13 and a stride of 4 pixels (which is the distance between the centers of the receptive field of adjacent neurons in the same kernel map) to filter the input images. By using 13×13 filters in the first layer, more features can be extracted from remote sensing images. The output images of the first convolutional layer are fed into 96 max pooling layers with a size of 3×3 filters and then are made to pass through a BN layer. The second convolutional layer is connected to the output of the first max pooling layer, with a convolution kernel size of 5×5 and a total of 256. The size of the convolution kernels in the third and fourth layers is 3×3, with 384 and 256 kernels, respectively. The entire network convolutional layer only has a max pooling layer between the third and fourth layers. Due to the different image sizes processed by the AlexNet model, if the 5th convolutional layer is retained, it will increase the network's data volume, increase computational costs, and may also cause a decrease

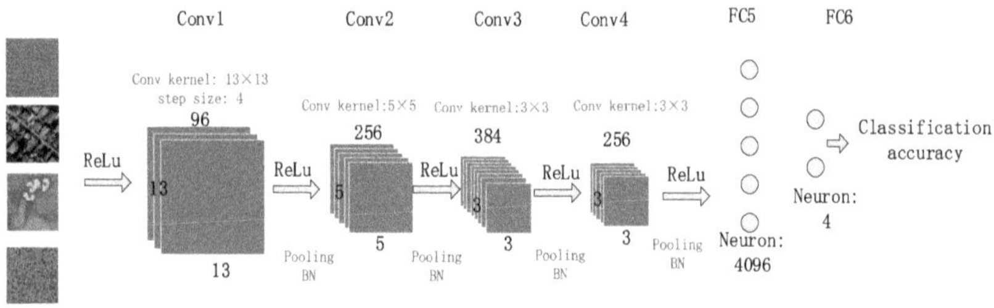

Figure 3.70. Improved CNN architecture for remote sensing image recognition.

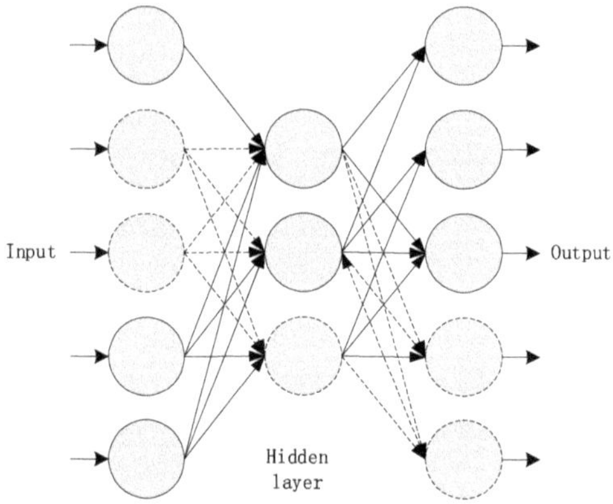

Figure 3.71. Dropout schematic.

in recognition accuracy due to excessive convolution of the images. Therefore, based on considering training efficiency and accuracy, a set of 384 with a kernel size of 3×3 convolutional layers was removed during network training.

After passing through the final convolutional layer, the generated $256 \times 1 \times 1$ feature maps are fed into the fully connected layer. Due to the fact that there are only four categories of classification targets in this section, there are fewer recognition tasks compared to the ImageNet model. Two fully connected layers are used in the fully connected layer, where the number of neurons in each layer is 4096 and 4, respectively.

AlexNet model employs Dropout [36] to avoid overfitting of data in the network. Fifty percent of the neurons are randomly discarded in the fully connected layer during each training, as shown in the Dropout schematic in Figure 3.71. The dashed line in Figure 3.71

Table 3.4. Convolutional kernel parameters for different networks.

Network model	Conv1	Conv2	Conv3	Conv4	Conv5	Conv6	Fc7	Fc8	Fc9
Network 1	13×13	3×3	3×3	3×3	3×3	3×3	4096	4096	4
AlexNet	11×11	5×5	3×3	3×3	3×3	–	4096	4096	4
Network 2	13×13	3×3	3×3	3×3	–	–	4096	4096	4
Network 3	13×13	3×3	3×3	3×3	3×3	–	4096	4	–
This paper's algorithm	13×13	5×5	3×3	3×3	–	–	4096	4	–

represents randomly discarded neurons. This approach is also preserved in the improved network, which uses fewer neurons and fully connected layers while also incorporating a dropout rate different from the AlexNet model.

A group of 5×5 filters was replaced with two groups of 3×3 filters in the VGG model [36] and this approach has achieved good results. In the selection of network structure, the different layers of networks are compared and convolution kernels of different sizes are used. The parameters of different convolution kernels in the network are shown in Table 3.4.

Rectified Linear Units (ReLU) are employed as activation functions.

3.8.2.2 *Training and learning methods*

The first layer's feature extraction and representation of images is expressed as

$$\boldsymbol{y}_1 = \max(0, \boldsymbol{w}_1 \otimes \boldsymbol{x} + \boldsymbol{b}_1), \tag{3.8.1}$$

where $\boldsymbol{x}$ represents the input images, $\boldsymbol{w}_1$ and $\boldsymbol{b}_1$ denote the local receptive field and bias, respectively, and $\otimes$ represents the convolution operation. $\boldsymbol{w}_1$ consists of n_1 filters of size $f_1 \times f_1$, where f_1 is the local receptive field. That is, the image is convolved n_1 times with $\boldsymbol{w}_1$, and convolution kernel is of $c \times f_1 \times f_1$. In this layer, n_1 feature maps are outputted. $\boldsymbol{b}_1$ is an n_1-dimensional vector, where each element corresponds to a local receptive field. The ReLU is employed as an activation function, i.e., $\max(0, x)$. For this layer, $n_1 = 96$, and the filter size is set to $f_1 = 13$.

The first layer extracts n_1-dimensional features from the images, while the second layer maps the n_1 feature maps from the previous layer to an n_2-dimensional feature vector. The structure from the second layer to the fourth layer is similar to that of the first layer, i.e.,

$$\begin{cases} \boldsymbol{y}_2 = \max(0, \boldsymbol{w}_2 \otimes \boldsymbol{y}_1 + \boldsymbol{b}_1) \\ \boldsymbol{y}_3 = \max(0, \boldsymbol{w}_3 \otimes \boldsymbol{y}_2 + \boldsymbol{b}_2) \\ \boldsymbol{y}_4 = \max(0, \boldsymbol{w}_4 \otimes \boldsymbol{y}_3 + \boldsymbol{b}_3) \end{cases}, \tag{3.8.2}$$

where $\boldsymbol{w}_2$ contains n_2 filters. For the second layer, $n_2 = 256$, and the filter size is reduced to 5×5. Similarly, the third and fourth layers have $n_3 = 384$ and $n_4 = 256$ filters, respectively, with a fixed size of 3×3. The ReLU is used as the activation function.

Like classical classification networks, the proposed algorithm alternates between convolutional and pooling layers. Pooling layers downsample the convolved output vectors to reduce parameters. In this section, max-pooling is chosen.

The output layer is fully connected to the previous layer, and its generated feature vector is fed into a logistic regression layer for classification. Weights in the network are updated based on the back-propagation algorithm.

Different from classic AlexNet and VGG model, in this section, there are two fully connected layers. While maintaining classification accuracy, the proposed model can reduce approximately 5×10^7 of data volume. The second fully connected layer (i.e., the sixth Layer) serves as the output layer, which contains m neurons corresponding to q categories of remote sensing land. Four types of remote sensing images are classified by the proposed algorithm, and at this time, $q = 4$. The probability of the network output is recorded as $p = [p_1, p_2, p_3, p_4]$, where p_j is defined as

$$p_j = \frac{\exp(\boldsymbol{y}_6^j)}{\sum\limits_{i=1}^{4} \exp(\boldsymbol{y}_6^i)}, \tag{3.8.3}$$

where $\boldsymbol{y}_6$ is the input to the Softmax function, j represents the current class number being calculated, $j = 1, 2, 3, 4$, and p_j denotes probability of the jth class true output.

3.8.3 Simulation experiments and results analyses

3.8.3.1 *Dataset*

The accessible remote sensing image datasets partially come from the UC Merced Land Use dataset [37]. There are 10 classes in the UC Merced Land Use dataset; among them, farmland, buildings, green lands, and deserts are selected. Sample images from the UC Merced Land Use dataset are shown in Figure 3.72.

Due to dataset scarcity, partially fused remote sensing images from Chapters 3 and 4 of Ref. [35] were selected as supplementary data. The classification of remote sensing images by neural networks was to classify RGB images, so fused MS images from Chapter 3 of Ref. [35] were preprocessed before use. After removing the NIR channel from the processed MS, cropping and segmentation were performed. This greatly increased the amount of data, improved the fitting ability of the network, and avoided overfitting of experimental results. The website of the United States Geological Survey (USGS) and the fused remote sensing images selected from chapters three and four of Ref. [35] were divided into four types: farmland, buildings, deserts, and vegetation (4500 images per type), with a total of 18,000

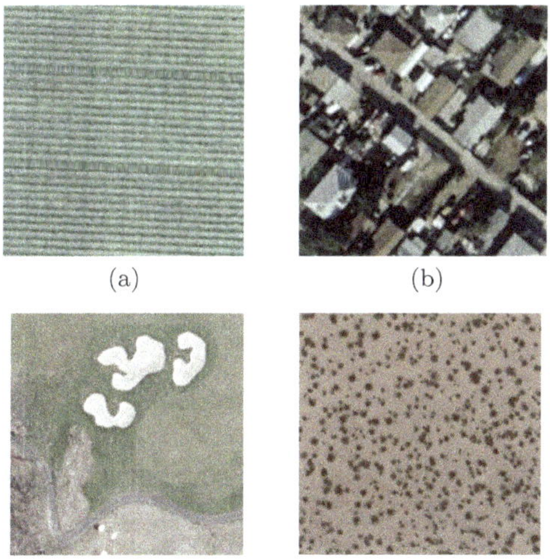

(a) (b) (c) (d)

Figure 3.72. UC Merced Land Use dataset.

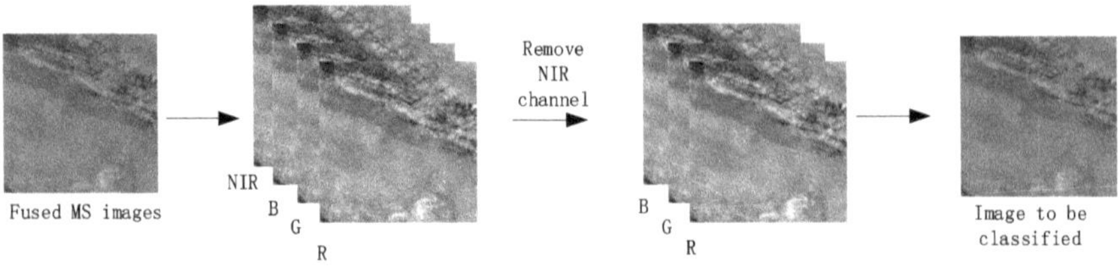

Figure 3.73. Classification-ready images extracted from fused data.

Table 3.5. Recognition accuracy for different image sizes.

Image size	35 × 35	37 × 37	39 × 39	41 × 41	43 × 43
Recognition accuracy	94.75%	85.72%	95.62%	97.75%	95.89%

images. Three thousand seven hundred and fifty images of each class were selected as the training set for the training model, and the remaining 750 images were used as the testing set. The obtained images from the fused image are classified and shown in Figure 3.73.

Multiple experiments were conducted on images of different sizes in the dataset. The results of models generated from 10,000 training iterations were compared and shown in Table 3.5.

Table 3.7 demonstrates that the model has higher recognition accuracy in recognizing images of size 41 × 41. In combination with the second section of this chapter, when determining the dataset size, it is most appropriate for us to choose images of size 41 × 41.

Compared to classic ImageNet tasks, this proposed algorithm employs more training samples for each category. Subsequent experiments have confirmed that larger datasets fully exploit the algorithm's advantages.

3.8.3.2 *Model validation*

The improved model, based on AlexNet, was evaluated using recognition accuracy, data size of model, and training convergence iterations as performance metrics. A series of experiments were conducted on the overall network architecture, filter sizes, and training hyperparameters.

3.8.3.2.1 *Model based on BN*

Table 3.6 shows that the original model and the BN-integrated model achieved the recognition accuracies of 92.1% and 87.3%, the convergence iterations were 149,000 and 35,000, and the model data sizes were 83M and 102M, respectively. Thus, after the BN is directly added to the AlexNet model, it accelerated convergence speed, but it reduced accuracy by 4.8% and increased data volume by 22.8%.

3.8.3.2.2 *Reduced-layer model*

Due to differences in remote sensing image sizes, the CNN architecture was modified accordingly. The classic AlexNet model processes 256×256 images, whereas this network handles 41×41 images. For smaller images, a 4-layer model was adopted, which used 2 fully connected layers. It significantly improved accuracy and reduced training iterations by integrating BN into the first four convolutional layers. Performance comparisons of different models with/without BN structure are shown in Table 3.7.

The improved model reduces the network's depth from 8 to 6 layers, decreases convergence iterations from 149×10^4 to 0.45×10^4, and significantly shortens training time. Recognition accuracy

Table 3.6. Performance comparison of AlexNet models before and after introducing the BN.

Performance metrics		Number of model convergence	Network data Volume	Recognition accuracy
Network model	Original model	14.9×10^5	7.33×10^8	92.14%
	Plus BN model	3.5×10^5	10.3×10^8	87.39%

Table 3.7. Performance comparison of different model types.

Performance metrics		Number of model convergence	Network data volume	Recognition accuracy
Network model	Original model	149×10^4	7.33×108	92.14%
	Plus BN model	3.5×10^4	10.3×10^8	87.39%
	Revised model	0.45×10^4	8.35×10^8	98.15%

improves from 92.14% to 98.15%, while the model data size increases from 7.33×10^8 to 8.35×10^8 after BN integration. The comparison of the performance between different models demonstrates that overall performance is enhanced in the modified model.

3.8.3.2.3 *Change in convolution kernel size*

We have conducted experiments on the influence of the size of the first-layer convolution kernel on recognition accuracy in improving the network model. Due to the size of the recognized image being 41×41, the convolution kernel size of the first convolutional layer is changed while ensuring that the subsequent network structure remains unchanged. Four sizes of convolution kernels, namely 10×10, 11×11, 12×12, and 13×13, were selected for the experiment. In the case of similar network structures, simply changing the size of the convolution kernel for experimentation resulted in the same amount of network data. The influence of convolution kernels of different sizes on the recognition accuracy of the network is shown in Figure 3.74.

Figure 3.74 shows that the classic AlexNet model with an 11×11 kernel exhibits greater fluctuation. The four networks with different kernel sizes achieve similar accuracy after convergence, but after

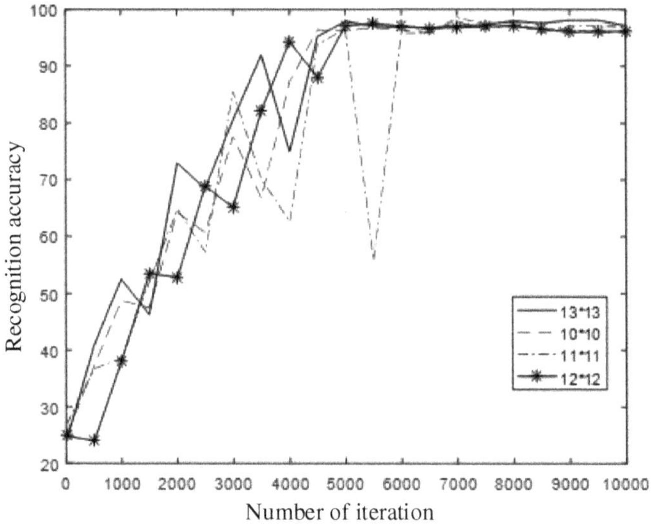

Figure 3.74. Impact of different kernel sizes on model recognition accuracy.

Table 3.8. Performance comparison of networks with different depths.

Performance metrics		Number of model convergence	Network data volume	Recognition accuracy
Network model	Network 1	1.9×10^4	9.82×10^8	90.33%
	AlexNet	149×10^4	7.33×10^8	92.14%
	Network 2	3.6×10^4	8.74×10^8	94.70%
	Network 3	1.2×10^4	7.78×10^8	98.21%
	This paper's algorithm	0.45×10^4	8.35×10^8	98.15%

7,500 iterations, the 13×13 kernel network significantly outperforms the others. When maintaining the overall network structure and selecting the first-layer filter of the network, a 13×13 kernel is optimal.

3.8.3.2.4 *Comparison of different networks*

As noted by Simonyan [36], a large kernel is replaced with multiple smaller ones to deepen the network and to yield better classification results. Performance comparisons of various networks are shown in Table 3.8.

Table 3.8 indicates that a single 5×5 filter replaced with two 3×3 filters can increase the depth of the model and improve the recognition accuracy of networks in the case of reducing the data volume of the model. However, as the number of network layers increases, the number of network iterations required for training the network also increases, and the increase in recognition accuracy is not significant. Therefore, in this section, we choose a network with 4 convolutional layers, 2 fully connected layers, and a larger convolutional kernel for the selection of network layers.

3.8.3.3 *Hyperparameter settings*

Proper hyperparameters can accelerate convergence and enhance accuracy. For the model (4 conv + 2 FC layers) with a 13×13 kernel, each configuration was tested 10 times to enhance the robustness of the model. Table 3.9 shows accuracy under varying iterations with a learning rate of 0.001.

Table 3.9. Recognition accuracy across iterations (Learning rate = 0.001).

Number of iterations	Test 1	Test 2	Test 3	Test 4	Test 5	Test 6	Test 7	Test 8	Test 9	Test 10
2000	63.15	61.17	62.89	65.74	64.89	63.22	54.74	75.63	74.56	67.38
4000	85.87	78.34	80.75	52.39	73.38	81.37	82.03	91.07	80.72	84.58
6000	93.81	92.88	95.36	98.11	49.72	95.36	97.28	97.41	95.95	95.07
8000	93.17	95.35	97.27	98.52	85.48	95.73	98.70	97.70	95.82	98.45
10000	95.74	97.69	98.17	99.23	89.33	95.67	99.32	98.58	98.29	98.37

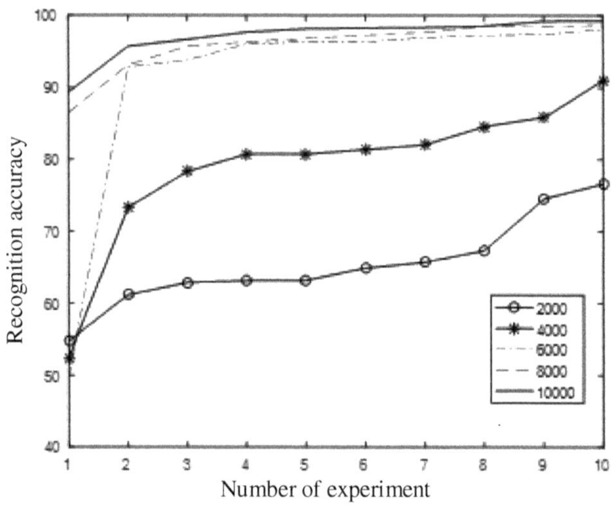

Figure 3.75. The effect of the number of iterations on the recognition accuracy when the learning rate is 0.001.

For comparative analysis, the data of each experiment in Table 3.11 were arranged in descending order, and the effect of the number of iterations on the recognition accuracy, when the learning rate is 0.001, is shown in Figure 3.75. Figure 3.75 shows that when the learning rate is 0.001, the recognition accuracies of models under condition of the 2000 and 4000 iterations are lower, the robustness of the model under 6000 iterations is poorer, the recognition accuracies of the models in the case of 8000 and 10,000 iterations are closer, and the recognition accuracy of the model after the 10,000 iterations is slightly higher than that of the model after its 8,000 iterations.

Table 3.10. Recognition accuracy for different learning rates at 10,000 iterations.

Learning rate	Test 1	Test 2	Test 3	Test 4	Test 5	Test 6	Test 7	Test 8	Test 9	Test 10
0.001	93.11	95.32	97.24	98.57	85.48	95.74	98.77	97.79	95.84	98.42
0.003	98.85	97.94	98.62	98.13	97.69	98.57	97.66	98.24	97.52	98.64
0.005	97.68	98.87	97.35	95.80	98.66	97.27	91.73	98.24	93.15	95.96
0.01	81.31	77.74	75.43	77.45	25.71	47.98	75.52	55.77	88.24	31.40
0.015	70.89	53.98	25.52	40.74	75.23	41.95	69.67	39.46	70.82	64.22

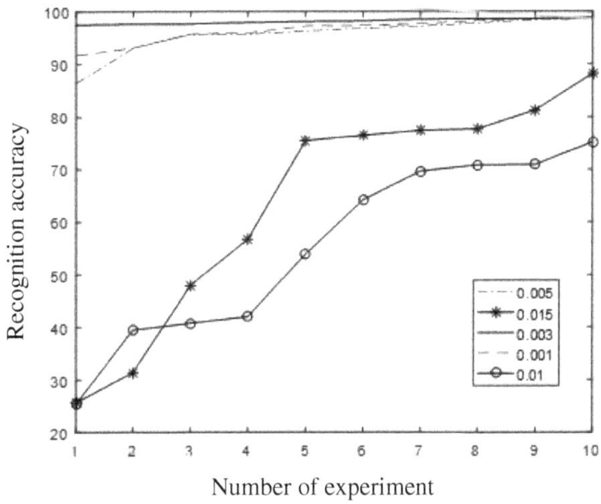

Figure 3.76. Accuracy of different learning rates at 10,000 iterations.

In addition, 10 experiments were conducted with 10,000 iterations of the model as a reference object at learning rates of 0.001, 0.003, 0.005, 0.01, and 0.015, respectively. The recognition accuracies are shown in Table 3.10.

For comparative analysis, the experimental data in Table 3.12 were arranged in order from smallest to largest, and the recognition of accuracies of different learning rates at 10,000 iterations is shown in Figure 3.76. Figure 3.76 shows that learning rates of 0.01 and 0.015 result in poor robustness of the model and lower recognition accuracy due to cost function oscillations. Recognition accuracies are closer at learning rates of 0.001, 0.003, and 0.005; the model with the learning rate of 0.003 has the best robustness.

3.8.3.4 Effect of Dropout value on model recognition accuracy

Dropout [37] is a technique used by the classical AlexNet model to prevent overfitting. Reference [38] achieved better results by keeping only 30% of the number of neurons in the last fully connected layer, while neural networks incorporating the BN can appropriately increase the number of neurons in the fully connected layer. Combining the above two points, the recognition accuracies of different dropout models were verified, as shown in Figure 3.77.

Figure 3.77 shows that the models with dropout values of 0.2, 0.4, and 0.8 have lower recognition accuracy than the models with other dropout values. Dropout values of 0 and 0.6 have a significant advantage in recognition accuracy before 3500 iterations and after 3500 iterations, and the recognition accuracy of the model with a dropout value of zero starts to oscillate after convergence and is lower than that of the model with the dropout values of 0.5 and 0.6, and the recognition accuracy of the model is highest when the dropout value is 0.6. Therefore, the dropout method not only avoids overfitting of the neural network but also effectively enhances the robustness of the model, which is consistent with the conclusion described in Ref. [37]. In summary, it can be seen that the recognition of the model with BN together and the dropout method is better.

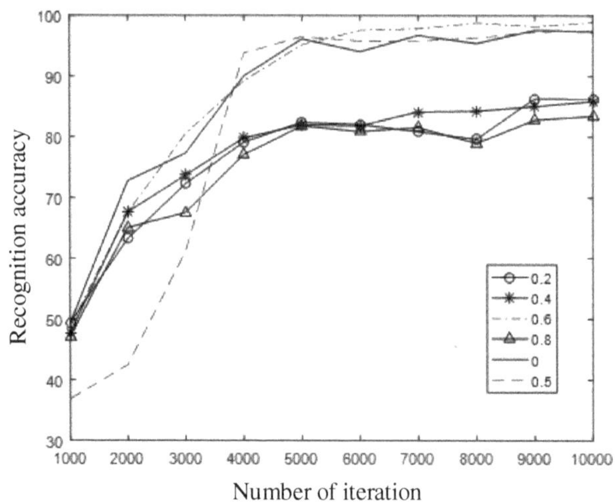

Figure 3.77. Recognition accuracy of the model with different Dropout.

In summary, in the deep CNN-based small block remote sensing image classification method, the network has used larger 13×13 filters as well as fewer network layers. While reducing the amount of data, the feature learning ability of the CNN model is fully utilized to greatly expand the dataset, avoid the phenomenon of overfitting, and improve the recognition accuracy, as well as enhance the robustness of the CNN model. Compared with the classical AlexNet model, the improved AlexNet model has higher recognition accuracy and faster convergence speed.

3.9 Example 3: Motion Blur Removal Method Based on Deep Convolutional Neural Network

Regarding the existing image deblurring methods that are difficult to deal with large blurring areas, difficult to balance the recovery effect and processing speed, and difficult to make full use of image feature information, in this section, a motion blur removal method based on deep CNNs is proposed [39]. Based on the multi-scale image in the image Gaussian pyramid, the autoencoding-based network model was constructed; in terms of expanding the image receptive field, the dilated convolution module is used to extract the multi-scale feature information of the image, and the residual module is utilized to broaden the depth of the network in order to solve the problem of the loss of the image details during the training process, as well as the recursive module is employed for strengthening the edge features to finally complete the end-to-end deblurring task of motion blurred images. Experimental results on the GOPRO dataset and real motion blur image sets have shown that the proposed method in this section has fewer parameters, runs faster, and is able to handle motion blur images of different sizes while guaranteeing the deblurring effect.

3.9.1 Multi-scale strategies in neural networks

3.9.1.1 *Multi-scale strategies in conventional algorithms*

In previous algorithms, a coarse-to-fine processing approach is commonly adopted when dealing with large fuzzy kernels. At the coarsest level, a smaller fuzzy kernel K is initialized, and then a gradient latent image ∇P is generated by running an iterative inference scheme while keeping the fixed fuzzy kernel K. Then, the fuzzy kernel and gradient latent image are inferred on the image pyramid of each layer, where the initial value of each layer is the upsampling value of the previous layer's fuzzy kernel K and gradient latent image ∇P. Finally, the fuzzy kernel K_L corresponding to the full resolution image is derived, and the restored image can be obtained through a non-blind deblurring method.

In addition, since the size of the blur encountered in the image may greatly vary from a few pixels to hundreds of pixels, the blur processing method is different. If a very large blur kernel is used for initialization, small blurred regions will be difficult to process. On the contrary, if the initialization blur kernel is too small, large blurred regions will be cropped. Therefore, in the setting of the initial blur kernel, the size of the initial blur kernel can be determined by observing the blur artifacts.

In the processing method from coarse to fine, on the one hand, because the extraction of image edge features is set by human beings, there will be a certain deviation and selection differences between different methods; on the other hand, due to the selection of the image pyramid of the whole image as an iterative object, even if it is carried out on the smaller size of the image, it will lead to spend longer and longer time on the estimation process of the fuzzy kernel due to the amount of the parameters, which affects the overall fuzzy kernel estimation correctly. Therefore, it is more convenient to use the image pixel values directly as reference values, as opposed to extracting the gradient image as reference values. In addition, processing for better results can be obtained by using an image block during processing. Figure 3.78 shows the estimation process of the multi-scale fuzzy kernel.

3.9.1.2 *Multi-scale image information extraction*

In the CNN, based on the above traditional algorithm from coarse to fine processing, an image Gaussian pyramid framework is constructed in the image preprocessing stage, the fuzzy image and the clear image pairs of different resolution versions are regarded as inputs to the

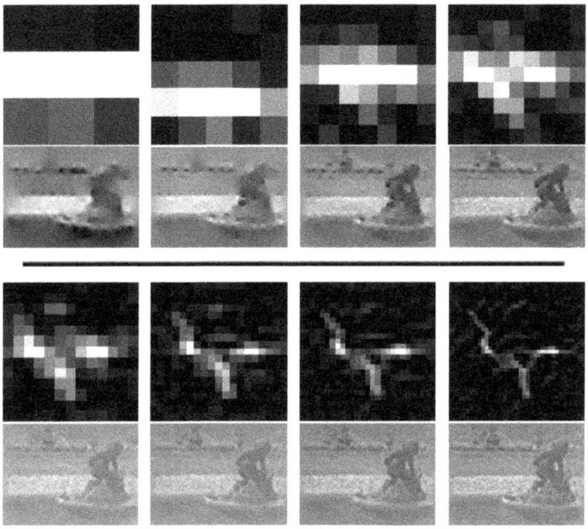

Figure 3.78. Estimation process of multi-scale fuzzy kernel.

neural network iteratively in turn, and then the obtained image block in the preprocessing stage is randomly cropped to obtain multi-scale image information.

To demonstrate the effectiveness of taking multi-scale image information, Figure 3.79 shows the relationship between images, image blocks, and blur kernels at high/low resolutions. Among them, X_S and B_S are the low-resolution images obtained by downsampling the clear image X and the blurred image B, respectively, P'_X, P'_B and P_X, P_B are the image blocks at the same position cropped from the corresponding images, and the corresponding fuzzy kernels k and k_S are measured according to the method of Ref. [40]. It can be seen in Figure 3.79 that the multi-scale image information extraction method is based on the traditional coarse-to-fine processing, on the one hand, because the blur in the low-resolution image is shrunk and the edge localization error is smaller, so the latent image and fuzzy kernel are processed and predicted more accurately at low resolution. On the other hand, the details of the clear image are reproduced much better than the fuzzy image at low resolution, and the downsampling parts of the clear images can be used as *a priori* parts of the images [41].

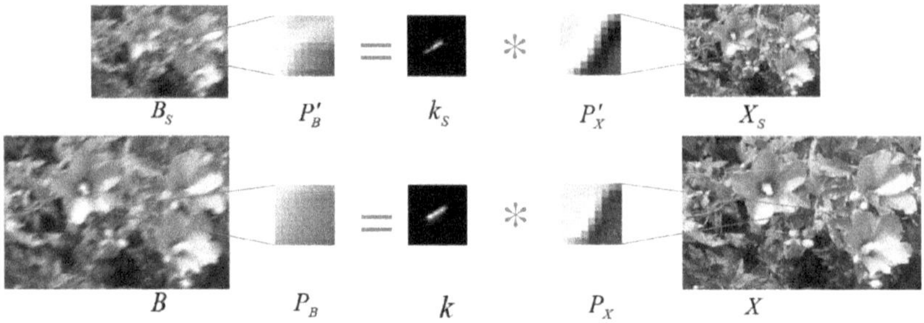

Figure 3.79. Image, image block, and blur kernel relationship at high/low resolution.

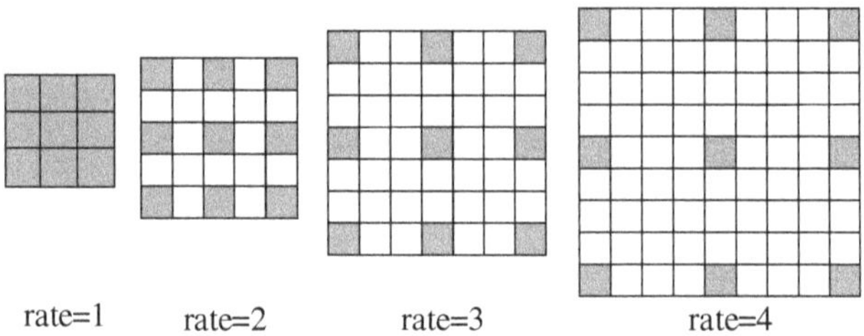

Figure 3.80. Dilated convolution with the expansion rate is 1, 2, 3, and 4, respectively, for a kernel size of 3×3.

3.9.2 Motion blur removal based on deep convolutional neural network

3.9.2.1 *Multi-scale feature information extraction*

Dilated convolution increases the receptive field of an image without pooling loss information by filling in holes between convolution kernels, i.e., filling in "zeros", and has the same size as an ordinary convolutional kernel with the same number of parameters [42]. The receptive field of dilated convolution increases exponentially with the increase in the expansion rate (dilate rate), and Figure 3.80 shows the schematic diagram of dilated convolution with different receptive fields.

A deeper network obviously improves the performance of the network, and the vertical expansion of the network while guaranteeing a certain network depth can also improve the performance of the network [43, 44]. Although the structure of the Dense convolutional

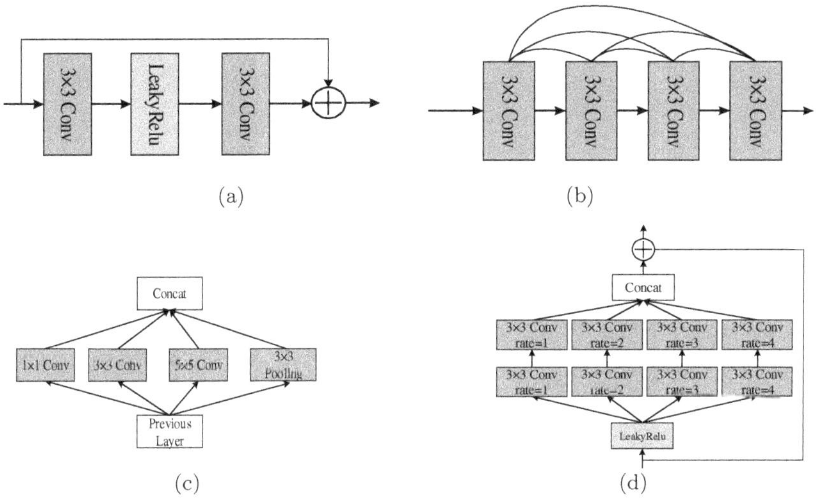

Figure 3.81. Schematic diagram of the network structure of ResNet, DenseNet, Inception, and DCB. (a) Resblock. (b) DenseBlock. (c) Inception. (d) DCB.

Network (DenseNet) is expanded in the stacking of feature information, the information is overly redundant and convergence speed is too slow, and it can only extract single-scale feature information. For visual tasks that require multi-scale, such as super-resolution and deblurring, multi-scale feature information is particularly important [45]. In addition to multi-scale processing in the pre-processing stage, different sizes of convolution kernels can also bring different scales of features, and the strategy of separating-transforming-converging in the Inception model fits with the method in this section. Figure 3.81 gives the schematic diagrams of the residual structure, DenseNet structure, Inception structure, and DCB structure, respectively.

At the level of the vertical expansion of the network, the convolution is decomposed by referring to Inception's feature fusion method into the connection of Depthwise Convolution (DCV) and Pointwise Convolution (PC) when assuming that the channel links and spatial links are not correlated in Xception model [46], and the ResNext structure also reduces the number of parameters and improves the accuracy by aggregating multiple topology structures with the same structure [47]. In addition, the fusion features obtained by concatenating in cascade form in the fusion strategy contain more effective spatial information representations than the aggregated features obtained by feature summing, which can enhance the edge details more effectively. At the same time, in the underlying image restoration tasks, residual learning allows information to jump directly to the output layer by establishing residual connections between different levels of convolutional layers, allowing the network to learn certain feature maps of the input and mitigating the gradient vanishing problem in deep neural networks. Combining the above points, the DCB fuses the feature layers in each branch by utilizing dilated convolutions with different expansion rates, while ensuring the same resolution, and residual connections are finally established to complete the learning of feature layers of different scales, with a larger sensory field and a smaller number of parameters.

3.9.2.2 *Structure design of convolutional neural network-based motion image deblurring*

In the image deblurring task, it is necessary to maintain a large receptive field to deal with larger blurred regions, and the receptive field is reflected as the size of the region where the pixels on the feature map outputted by the convolutional neural network map on the original image at each time. There are usually three ways to increase the receptive field: increasing the pooling layer, enlarging the convolutional kernel size, and increasing the number of convolutional layers; at the same time, it will also face many problems, such as loss of information, too many parameters, and gradient disappearance. In order to solve the above problems, the following neural network model is constructed.

Figure 3.82 shows the overall framework of the deep convolutional neural network proposed in this chapter, and the overall structure of the encoder–decoder (Eocoder–Decoder) is used, and the inputs are in the form of a cascade of fuzzy and clear images of different resolutions. In the encoder stage, a 5×5 convolutional layer is used to extract the information of the image feature layer and a 3×3 dilated convolution module is used to obtain the different expression information of multiple feature layers, and two convolutional layers with a step size of 2 are used as downsampling to increase the sensory field of the image and to learn the high-dimensional expression features of the information. In the transition stage, 10 weight-shared residual blocks (Residual Block, ResBlock) are used as the information

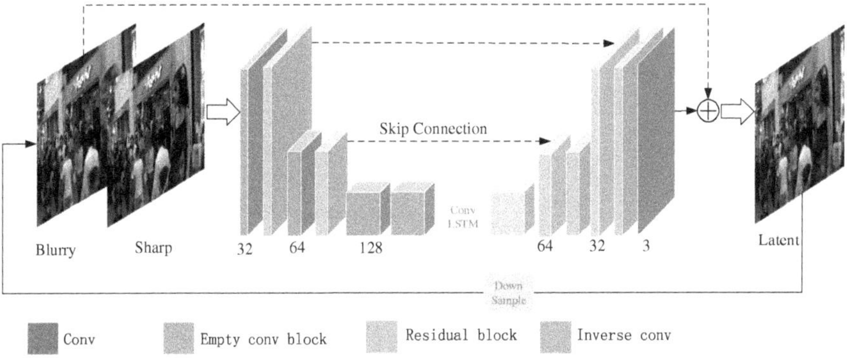

Figure 3.82. Deep convolutional neural network structure based on autoencoder.

transition framework between encoding and decoding. In the decoding part, two layers of deconvolution with a step size of 2 and the same dilated convolution module are used correspondingly to gradually increase the resolution of the feature layer, and finally, the latent image is obtained by summing the initial inputs with the outputs of the 5×5 convolutional layers. A Leaky Rectified Linear Unit (LReLU) activation function with a slope of 0.2 in the negative half of the region is connected after each convolutional and deconvolutional layer. Skip connection is added between the bottom information and the top information to compensate for the loss of information during downsampling. The transition framework adds a Convolutional Long Short-Term Memory (CLSTM) module as a recursive module to extract features from the underlying information and deepen the edge feature information of the restored images.

3.9.2.3 *Weight-sharing residual network*

The ResBlock has demonstrated good superiority in the propagation from low-level information to high-level information, as shown in Figure 3.83. The traditional ResBlock structure, after its Batch Normalization (BN) layer and the output layer are removed, is called the LReLU layer. On the one hand, when the distribution of training data and testing data is too different, the addition of the BN layer often can produce artifacts. On the other hand, the LReLU after the output layer plays a role in linear transformation and can only retain all complete information in low dimensions. After adding

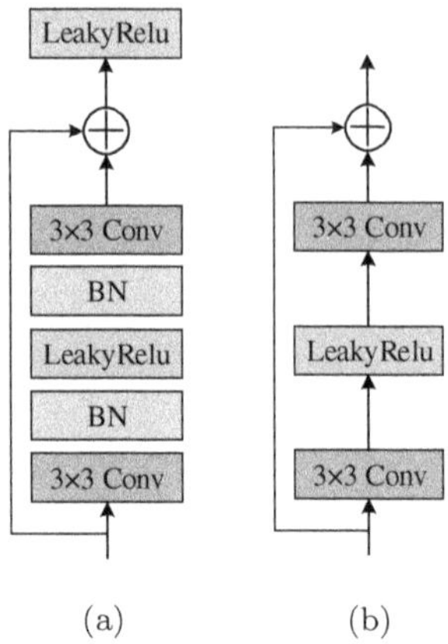

Figure 3.83. Residual block and the proposed residual block. (a) Traditional residual block. (b) MS-CNN [9].

it, the accuracy of some feature layers will be reduced. Figure 3.82 shows the traditional residual block and the residual block used in this section.

The traditional ResNet structure increases the number of parameters and deepens the depth of the network by stacking independent residual blocks with multiple parameters. However, the number of network's parameters mainly comes from the residual blocks, so the weight sharing is adopted in the ResNet structure, and the depth of the network is deepened by repeatedly looping a residual block without increasing the number of parameters, and the parameters of each residual block are shared so that the growth of the number of parameters is reduced under the premise of the number of parameters increases exponentially and the loss of the accuracy of the recovery results is small. The parameter sharing of each residual block reduces the growth of the number of parameters while the number of parameters increases exponentially, and the accuracy loss of the recovery results is small. Corresponding to the recursion of the scale, the recursive layer of the CLSTM in the encoding stage can play a certain role in removing image blurring by combining the feature information of each scale independently. Figure 3.84 shows the traditional ResNet and the ResNet used in this section.

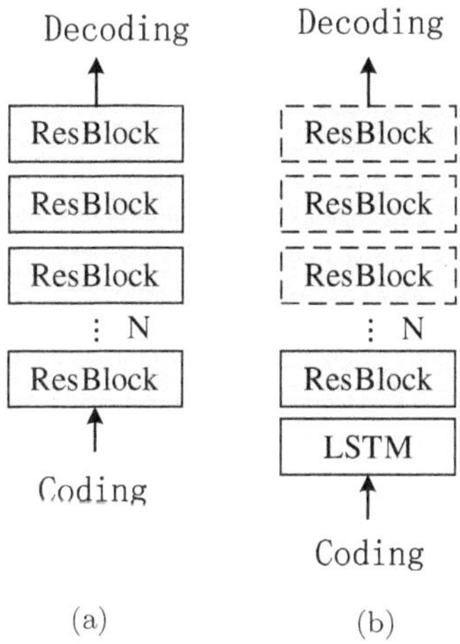

Figure 3.84. ResNet and the proposed network. (a) ResNet. (b) Recursive ResNet.

3.9.2.4 *Training process and loss function*

Given a series of blurred images $\boldsymbol{B}^{(i)}$ and their corresponding clear images $\boldsymbol{X}^{(i)}$, where i is the serial number of the corresponding image and $\boldsymbol{\theta}$ is the optimized parameter of the network, the output of the network at scale S is defined as

$$\boldsymbol{I}_s^{(i)} = \text{Net}_{ED}(\boldsymbol{B}_s^{(i)};\ \boldsymbol{\theta}). \tag{3.9.1}$$

The training objective is to make the blurred picture similar to the real picture at each scale. This process can be described as

$$\min \text{Loss}_{db}(\boldsymbol{I}_s^{(i)},\ \boldsymbol{X}_s^{(i)}), \tag{3.9.2}$$

where Loss_{db} is the loss function for deblurring defined as

$$\text{Loss}_{db} = \sum_{s=1}^{S} C(\boldsymbol{I}_s^{(i)} - \boldsymbol{X}_s^{(i)}), \tag{3.9.3}$$

where S is the adopted scale magnitude and $C(x)$ is the Charbonnier loss denoted as $C(z) = \sqrt{z^2 + \varepsilon}$, and ε is the hyperparameter set to

$1e^{-3}$. The commonly used MSE has a strong constraining effect on the PSNR, but it tends to lose high-frequency details, and the *L1 norm* has a better anti-noise and anti-jamming ability and is more robust and more favorable for high-frequency features and image reconstruction.

3.9.3 Experimental results and comparative analyses

These experiments were conducted based on the TensorFlow framework under the configuration of Intel i7-8700 and NVIDIA 1080Ti; a total of 2103 pairs of real clear images were captured by high-speed cameras and the corresponding simulated blurred images in the GOPRO dataset [45] were used as the training set, and the remaining 1111 pairs of the GOPRO dataset and the real motion blurred dataset [49] were used as the testing set. In terms of training parameters, a 256×256 image block was randomly cropped as the initial input, and the Adam optimization algorithm with an initial learning rate of minimizing the loss function was used to train 2000 epochs to achieve convergence with a batch size of 16, and the recovery results were compared and evaluated based on two objective indexes: the PSNR and the SSIM.

3.9.3.1 *Validation of the rationality of the network structure*

In order to verify the reasonableness of the proposed network structure, the U-net [50] structure was used as the benchmark framework model, i.e., in the encoding stage, the operation of maximum pooling was used after every two 3×3 convolutional layers, and the number of channels was doubled after each downsampling. In the decoding stage, the operation of up-sampling was followed by two 3×3 convolutional layers, the number of convolutional kernels corresponding to the structure of this proposed network, and the same number of channels of features were connected by skipping connection between layers, and the skipping connection was used between the layers, and finally, the recovered images were obtained by summing the initial input with the output of the 3×3 convolutional layers. The benchmark model was trained with the same parameters under the GOPRO dataset and different network modules were added

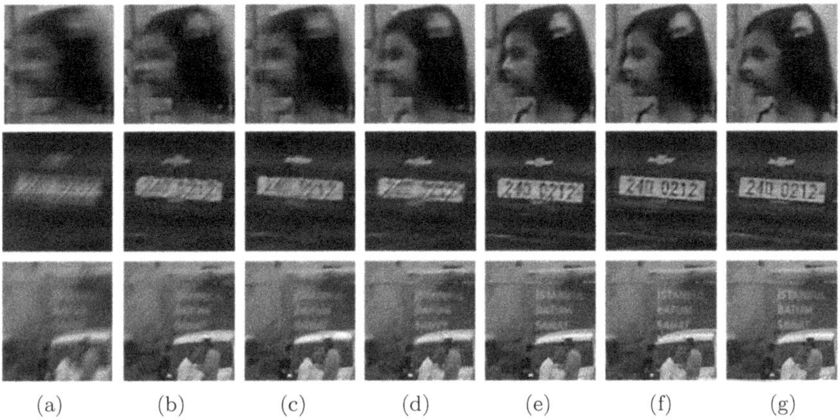

Figure 3.85. (a) Input blurred image. (b)–(d) Recovered images when scale S is 1,2,3, respectively. (e)–(g) Recovery results when DCB, ResBlock, and CLSTM are added sequentially at S = 3.

Table 3.11. Experimental results of different methods (corresponding to Figure 3.85(b)–(g), respectively).

Methodologies	PSNR	SSIM	Quantity of participant
$s = 1$ U	25.61	0.8361	0.59M
$s = 2$ U	26.01	0.8489	1.18M
$s = 3$ U	25.87	0.8474	1.77M
$s = 3$ U+D	27.56	0.8900	2.44M
$s = 3$ U+D+R	28.38	0.9072	2.84M
$s = 3$ U+D+R+A	28.53	0.9141	3.24M

sequentially to verify the feasibility of the structure. Figure 3.85 and Table 3.11 show the experimental examples and metrics evaluation under adding different modules, respectively.

In the recursive module, the state of the hidden layer in the CLSTM symbolizes the motion of the pictures, and the size of the kernel was related to the speed of the object's motion at the time of taking the picture, which was used to extract the motion information of the picture at the best resolution. At different scales of the same picture, the feature information was preserved from the high-resolution image and transferred to the cell state at the next scale, and Figure 3.86 shows that the feature layer with the addition of

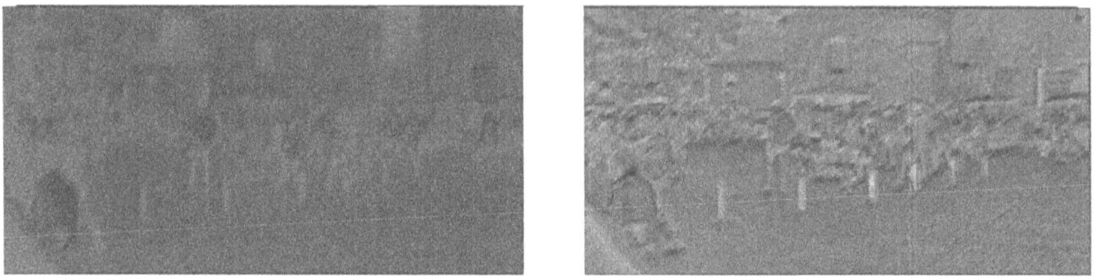

(a) (b)

Figure 3.86. Example of feature maps before and after passing CLSTM. (a) Feature map before CLSTM. (b) Feature map after CLSTM.

the recursive module not only removes some of the blurring but also enhances the region contour.

3.9.3.2 *Comparison with ResNet and DenseNet*

In order to analyze the effectiveness of the multi-scale feature fusion strategy, the experiment was conducted to analyze the network structure with DenseBlock, ResBlock, and DCB as the main body, respectively [51]. With the same hyperparameters, the experiments were first made based on ResNet (6 ResBlock) and DenseNet (12 DenseBlock), which were aiming at feature fusion, and referred to the framework of U-net and the framework, and experimental evaluations were analyzed under the GOPRO dataset. The depth of ResNet (18 ResBlock) and DenseNet (24 DenseBlock) was then increased. Figure 3.87 and Table 3.12 show the experimental examples and metrics evaluation of various network structures at different depths, respectively. Figure 3.87 shows that the deepening of the network depth can obviously improve on the performance of the proposed method, but as the number of parameters and the computational burden of the model grow dramatically, it will only produce a certain improvement in recovering the image edges, and this proposed method has a better recovery effect compared with the other two methods when the number of parameters is approximate.

3.9.3.3 *Multi-scale feature fusion methods*

For the fusion methods of different scale features, the experiment was done to compare the encoding tree (DCB-e), decoding tree (DCB-d),

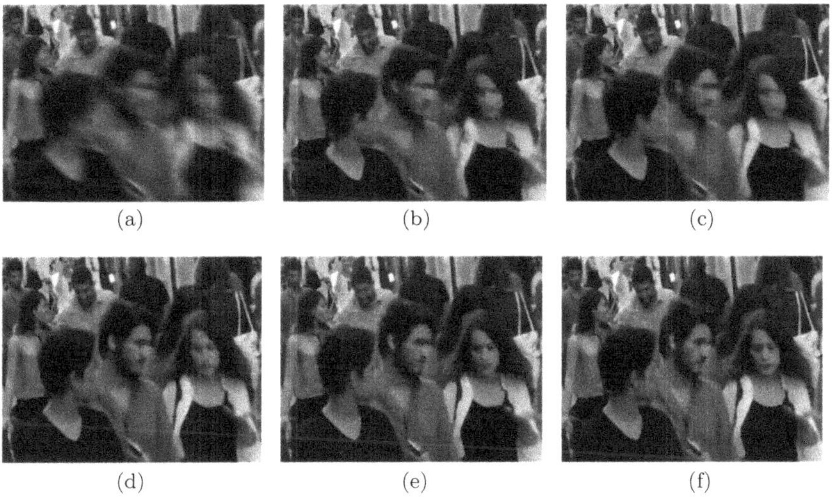

Figure 3.87. Comparison of the images of this section's method with DenseNet and ResNet on the GOPRP dataset. (a) Fuzzy input. (b) DenseNet (shallow). (c) DenseNet (deep). (d) ResNet (shallow). (e) ResNet (deep). (f) Methods of this section.

Table 3.12. Metrics comparison with ResNet and DenseNet on GOPRO dataset.

Norm	ResNet Shallow	ResNet Deep	DenseNet Shallow	DenseNet Deep	Methodology of this section
PSNR	24.31	28.69	26.89	27.48	28.38
SSIM	0.8004	0.9152	0.8754	0.8845	0.9072
Quantity of participant	3.42M	6.19M	3.20M	5.52M	3.24M

and DCB structure, respectively. Figures 3.88 and 3.89 show the schematic structure and metrics of the coding tree and decoding tree, respectively. Figures 3.88 and 3.89 show that feature extraction at multiple scales for the feature layer first performs better than layer-by-layer encoding. Simultaneously, the increase in the communication channels between information will also improve the performance of the network to a certain extent, so the DCB is chosen as the main module.

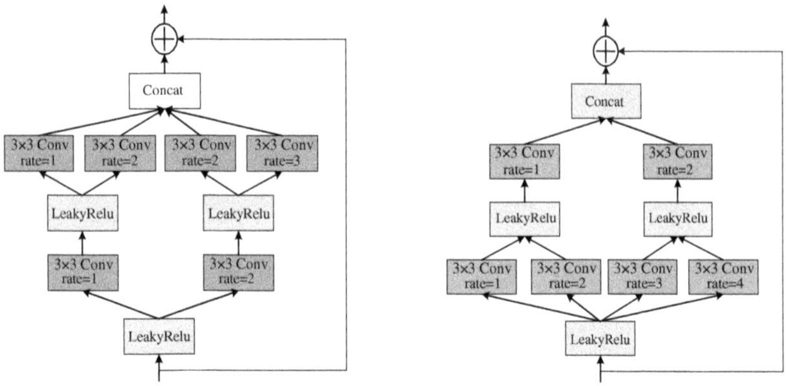

(a) (b)

Figure 3.88. Schematic structure of coding tree DCB-e and decoding tree DCB-d. (a) Encoding tree structure DCB-e. (b) Decoding tree structure DCB-d.

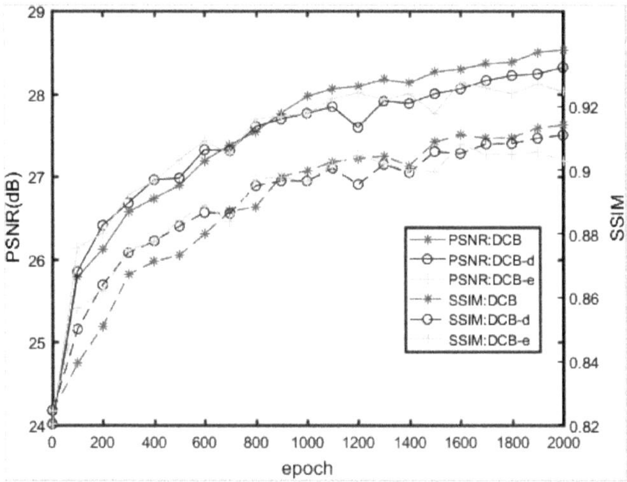

Figure 3.89. Schematic diagram of encoding tree DCB-e and decoding tree DCB-d structure.

3.9.3.4 *Expansion rate and number of blocks*

For the effect of expansion rate r and the number of blocks on the experimental results, the effect of r at 1,2,3,4 and 1,4,8,16 cases as well as the effect of the number of DCBs N1 and the number of Res-Blocks N2 on the experiments has been conducted. The experimental metrics for different hyperparameters are shown in Table 3.13.

Table 3.13. Experimental metrics with different expansion rate r, number of DCBs $N1$, and number of ResBlock $N2$.

	PSNR/SSIM($r = 1, 2, 3, 4$)	PSNR/SSIM($r = 1, 4, 8, 16$)
N1=1, N2=3	27.56/0.8900	26.89/0.8760
N1=2, N2=3	27.09/0.8803	26.64/0.8713
N1=1, N2=5	27.66/0.8939	27.06/0.8794
N1=1, N2=10	28.38/0.9072	27.30/0.8911
N1=1, N2=15	28.41/0.9088	27.39/0.8954

Table 3.13 shows that an increase in both r and $N1$ can lead to a decrease in the convergence degeneration performance of the network, and an increase in $N2$ only slightly improves the edge effect in the high depth case. $r = 1, 2, 3, 4$ and $N1 = 1$, $N2 = 10$ were finally chosen as the hyper-parameter choices in order to ensure the high efficiency of the network.

3.9.4 Simulation experiments and result analyses

This experiment compared the proposed method in this section with the existing methods for a large number of experimental results, and qualitative and quantitative analyses were done on the GOPRO dataset and the real motion blur dataset, respectively. The experimental examples and evaluation metrics of each method on the GOPRO dataset are shown in Figures 3.90, 3.91, and Table 3.14, respectively, and the experimental examples of each method on the real dataset are shown in Figure 3.89. Figures 3.90, 3.91, and Table 3.14 show that for the GOPRO testing set, the proposed methods in this section have good performance in terms of restoration effect and running time; the traditional algorithm [51] has fair performance in localized blur processing, but it cannot deal with the overall blurring caused by camera shake and the running time is too long. The CNN-based method [45] performs motion deblurring by predicting the blurring kernel of blurred images, and it has good performance in dealing with large blur areas and complex blur kernel scenes tend to produce artifacts. End-to-end CNN methods [52] are poor at detail restoration due to the limitations of the network's receptive field. Figure 3.92 show that the proposed method in this section has better deblurring results on real datasets.

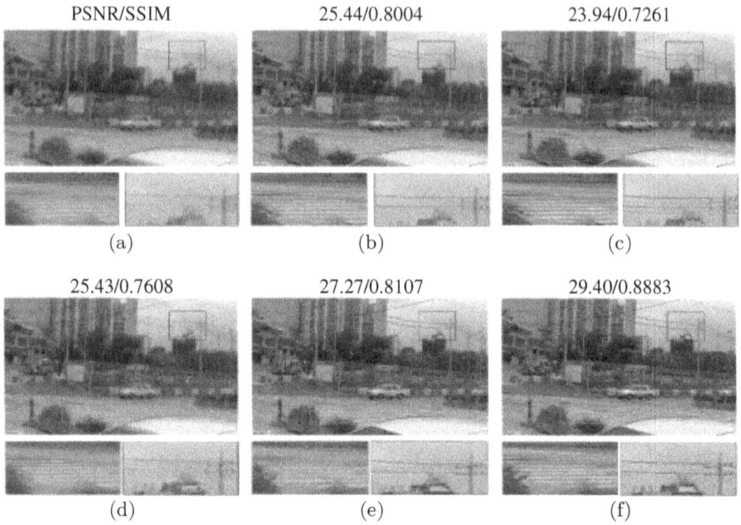

Figure 3.90. Comparison of test results between the methods in this section and other methods on the GOPRO dataset. (a) Fuzzy input. (b) Pan [51]. (c) Chakrabarti [52]. (d) Gong [53]. (e) Nah [45]. (f) Methodology of this section.

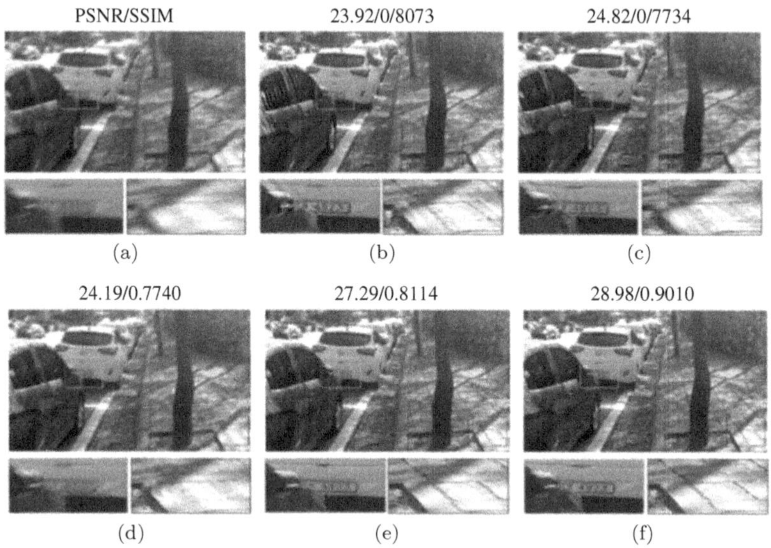

Figure 3.91. Comparison of test results between the methods in this section and other for license plate in the GOPRO dataset. (a) Fuzzy input. (b) Pan [51]. (c) Chakrabarti [52]. (d) Gong [53]. (e) Nah [45]. (f) Methodology of this section.

Table 3.14. Evaluation metrics vs. runtime for each method on the GOPRO dataset.

Method	PSNR	SSIM	Running time
Pan [106]	23.71	0.8216	1h
Chakrabarti [108]	24.18	0.8337	15min
Dong [107]	25.31	0.8549	20min
Nah [100]	28.34	0.9054	4.33s
Methodology of the chapter	28.53	0.9141	0.3s

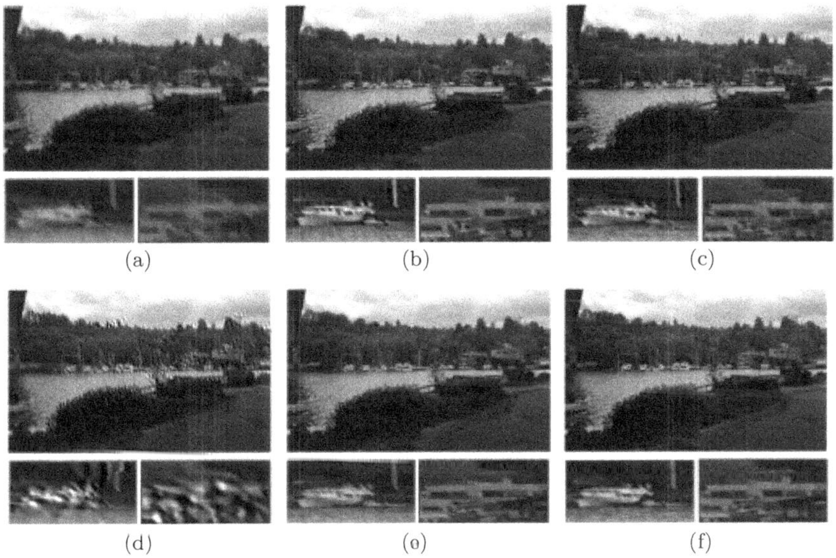

Figure 3.92. Comparison of test results between the proposed method in this section and other methods on real motion blur datasets. (a) Fuzzy input. (b) Pan [51]. (c) Chakrabarti [52]. (d) Gong [53]. (e) Nah [45]. (f) Methodology of this section.

In summary, the deep convolutional neural network is applied to the motion blur image restoration for the problem of motion blur images caused by camera shake, object motion, etc. The proposed method has better restoration effect, shorter running time, less number of parameters, and achieves better visual effect and evaluation index.

Chapter 4

Deep Generative Adversarial Networks

Summary

Based on the analyses of the principles and implementation architecture of Generative Adversarial Networks (GANs), the self-attention mechanism is introduced into the conditional GAN to form a self-attention mechanism conditional GAN, wavelet transform is also introduced into the deep convolutional GAN to obtain wavelet deep convolutional GAN and high-frequency wavelet GAN, and the multi-scale GAN based on turbulence-degraded images is analyzed and studied. Taking the motion-blurred image restoration based on deep residual GAN as an example, the method of motion-blurred image restoration using deep residual GAN is proposed, and the simulation experimental results are also discussed.

4.1 Introduction

Generative Adversarial Network (GAN) is a generative model proposed by Goodfellow *et al.* [54] in 2014 and also an unsupervised learning model. At present, the GAN has been used to solve many problems in the field of image and machine vision, such as face detection, segmentation image restoration, generation of high-resolution images from low-resolution images, speech processing, privacy protection, and virus detection. Deep Convolutional Generative Adversarial Network (DCGAN) proposed based on the GAN [55] is a completely new architecture, which is stable during the training process and can effectively achieve high-quality image generation and related generative model applications. Due to DCGAN's strong practicality, many subsequent GAN models are its improved versions. That is to say, the generative model and discriminative model in the GAN are replaced with two convolutional neural networks (CNNs) by the DCGAN. In the CNNs, an activation function is usually added after the convolution layer and then passed on to the next layer. When solving nonlinear problems, the activation function plays an important key role in enabling the neural network to better solve more complex problems.

4.2 Generative Adversarial Networks

4.2.1 Network structure

The Generative Adversarial Network (GAN) consists of a generator G and a discriminator D, as shown in Figure 4.1. The generator G generates an image $G(z)$ through random noise z, whereas the discriminator D is responsible for distinguishing the input samples and determining whether they are real data or the generated data $G(z)$. During the training process, the task of the generator G is to generate as many real data or images as possible to deceive the discriminator D, while the discriminator D needs to distinguish the data or images generated by the generator G from the real data or images as many as possible. The game between generator G and discriminator D continues until equilibrium is reached.

The objective function of the GAN is defined as

$$\min_G \max_D \mathrm{Loss}(D, G) = \mathbb{E}_{x\sim f_{\mathrm{dtat}(x)}}[\log D(x)] + \mathbb{E}_{z\sim f_z(\mathrm{z})}[\log(1 - D(G(z)))], \qquad (4.2.1)$$

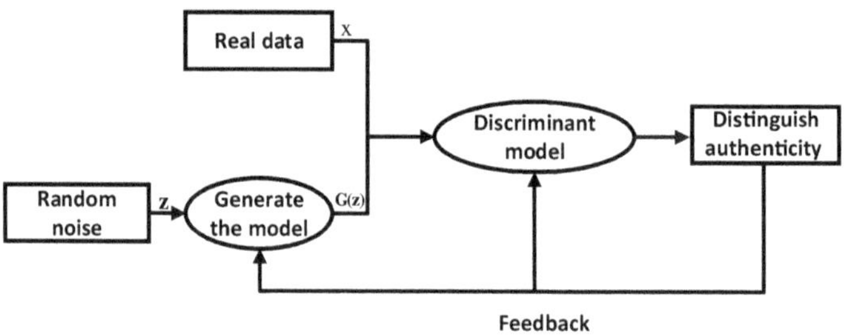

Figure 4.1. The working principle of the GAN.

where $f_z(z)$ represents the probability density of the random noise z and $f_{\text{data}}(x)$ is the probability density of the meter data x. $\log D(x)$, in the first term of Equation (4.2.1), represents the judgment result of the discriminator D on the real data, and the second term of Equation (4.2.1) denotes the synthesis and judgment of the data. Through the bilateral game of maximum and minimum values, the generator G and the discriminator D are alternately trained until the equilibrium is reached. For given any generator G, the training criterion of the discriminator$D$ is to maximize the objective function $\text{Loss}(D, G)$. Therefore, assuming that the generator G is fixed, the objective function that needs to be maximized can be written as

$$\begin{aligned}\text{Loss}(D,G) &= \int_x f_{\text{data}}(x)\log(D(x))\,\mathrm{d}x + \int_z f_z(z)\log(1-D(g(z)))\mathrm{d}z \\ &= \int_x f_{\text{data}}(x)\log(D(x)) + f_g(x)\log(1-D(x))\mathrm{d}x.\end{aligned} \tag{4.2.2}$$

The calculation formula of the optimal discriminator D is given by

$$D_{\text{G-opt}}(x) = \frac{f_{\text{data}}(x)}{f_{\text{data}}(x) + f_g(x)}. \tag{4.2.3}$$

After Equation (4.2.3) is substituted into Equation (4.2.2), the objective function of the generator G is obtained and written as

$$\begin{aligned}\text{Loss}(G) &= \max_D \text{Loss}(G,D) \\ &= \mathbb{E}_{x\sim f_{\text{data}}}[\log D_{G\text{-opt}}(x)] + \mathbb{E}_{z\sim f_z}[\log(1-D_{G\text{-opt}}(G(z)))] \\ &= \mathbb{E}_{x\sim f_{\text{data}}}[\log D_{G\text{-opt}}(x)] + \mathbb{E}_{x\sim f_g}(\log(1-D_{G\text{-opt}}(x))) \\ &\times \mathbb{E}_{x\sim f_{\text{data}}}\left[\log\frac{f_{\text{data}}(x)}{f_{\text{data}}(x)+f_g(x)}\right] \\ &+ \mathbb{E}_{x\sim p_g}\left[\log\frac{f_g(x)}{f_{\text{data}}(x)+f_g(x)}\right].\end{aligned} \tag{4.2.4}$$

In Equation (4.2.4), when $f_g(x) = f_{\text{data}}(x)$, and $D_{G\text{-opt}} = 0.5$, the discriminator D cannot make a distinction between real data and generated data, and the objective function of the generator G is given by $\text{Loss}(G) = -\log(4)$. If we represent the objective function $\text{Loss}(G)$ as the form of KL divergence, it can be written as

$$\text{Loss}(G) = -\log(4) + KL\left(f_{\text{data}} \middle\| \frac{f_{\text{data}} + f_B}{2}\right) + KL\left(f_g \middle\| \frac{f_{\text{data}} + f_g}{2}\right), \tag{4.2.5}$$

where the two KL divergences in Equation (4.2.5) can be denoted by the JS Divergences (JSD) as

$$\text{Loss}(G) = -\log(4) + 2JSD(f_{\text{data}} \| f_g). \tag{4.2.6}$$

Since the JSD between two distributions is always non-negative and is zero when and only when they completely overlap, the minimum value of the $\text{Loss}(G)$ is equal to $-\log(4)$, and the only solution is given by $f_g(x) = f_{\text{data}}(x)$. At this point, the generator G perfectly replicates the data generation process.

4.2.2 Implementation architecture

The specific steps to implement a GAN are as follows:

Step 1: Data generation and identification

The generator G receives a random noise z and generates data, and the discriminator D determines whether the generated data is "real".

Step 2: Calculate the loss function of the generator *G*

The loss function $\log(1 - D(G(z)))$ of the generator G is calculated, where $G(z)$ represents the output of the generator G and $D(G(z))$ denotes the probability that the discriminator D determines the probability that the data generated by the generator G is real.

Step 3: Calculate the loss function of the discriminator *D*

The loss function $\log(1-D(G(z)))+\log D(x)$ of the discriminator D is calculated, where x represents the input parameter, i.e., the real data, and $D(x)$ denotes the output of the discriminator D, namely the probability that the input x is the real data.

Step 4: Optimize loss function

The optimization of the loss function $Loss(D, G)$ can be written as

$$\min_G \max_D \text{Loss}(D, G) = \mathbb{E}_{X\sim f_{\text{data}}(x)}[\log D(x)] + \mathbb{E}_{Z\sim f_z(z)}[\log(1 - D(G(z)))]. \quad (4.2.7)$$

Step 5: Form a dynamic "game process"

During the training process, the generator G generates the real data as many as possible to deceive the discriminator D, while the discriminator D tries to distinguish the data generated by the discriminator G from the real data. Finally, a dynamic "game process" between the generator G and the discriminator D can be formed.

Step 6: Identify the data

Determine whether the discriminator D can distinguish that the data generated by the generator G is real data. If so, the trained data is obtained. Otherwise, the input parameter x needs to be adjusted, and the execution process needs to return to Step 2.

Step 7: Identify the data under ideal conditions

Under ideal conditions, the generator G generates $G(z)$ that is sufficient to "confuse the real with the fake". For the discriminator D, it is impossible to determine whether the data generated by the generator G is real, i.e., $D(G(z)) = 0.5$.

4.3 Conditional Generative Adversarial Networks

Conditional Generative Adversarial Networks (CGANs) are one of the early variants of the GAN or an improvement on the GAN. It introduces category labels into the GAN via adding additional information to the generator G and the discriminator D of the original GAN, where additional information is auxiliary information, such as a class label. The input of the generator G in the original GAN is random noise, while the generator G of the CGAN can combine the category labels with the random noise as the hidden layer input. The input of the discriminator D in the original GAN belongs to the image data, while the input of the discriminator D in the CGAN is the result of concatenating category labels and image

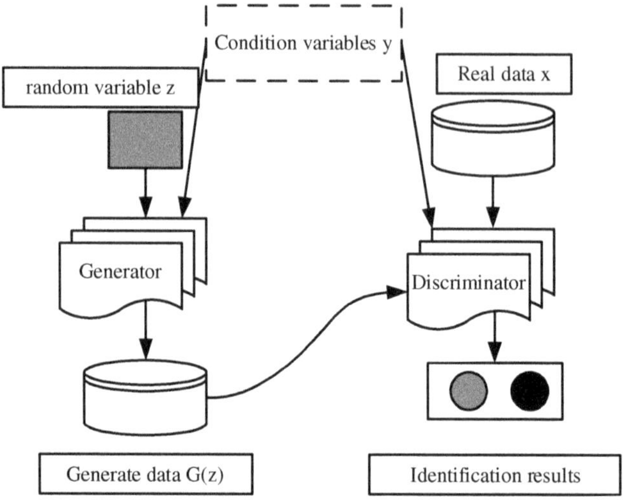

Figure 4.2. Working principle of the CGAN.

data, which serves as the basis for determining whether it is generated by the generator G or real data. Its working mechanism is shown in Figure 4.2 [56].

The biggest difference between the CGAN and the GAN lies in that the CGAN changes the probability in the objective function of the GAN to the conditional probability. Accordingly, the objective function of the CGAN can be defined as

$$\min\max \mathrm{Loss}(D, G) = \mathbb{E}_{x\sim f_{\mathrm{data(x)}}}[\log D(x|y)] + \mathbb{E}_{z\sim f_z(z)}[\log(1 - D(G(z|y)))], \tag{4.3.1}$$

where y is an additional information, which is used as part of the input of the generator G and the discriminator D. In the generator G, the prior input noise $f(z)$ and the additional information y together constitute the hidden layer input.

4.3.1 Conditional generative adversarial network based on self-attention mechanism

In order to further improve the quality of the images generated by the GAN, an effective is way to introduce the self-attention mechanism into the GAN, which can calculate the relationship between any two pixels in the images to obtain the global geometric features of the images. It is a mechanism that aligns internal details with external senses to improve the observation accuracy of certain areas, and it is an improvement on the attention mechanism, which can quickly extract important features of the sparse data, reduce dependence on the external information, and better capture the internal correlation of the data or the features.

In this section, the self-attention mechanism is introduced into the GAN to establish Self-Attention GAN (SA-GAN), which is expected to improve the quality of the generated images, then the Self-Attention mechanism CGAN (SA-CGAN) is proposed [57]. The main difference between the SA-CGAN and the CGAN is that after the self-attention mechanism is used to improve the correlation of long-distance features in the images, the SA-CGAN can better solve the problem that the GAN cannot generate a large number of images of different categories, and the quality of the generated images can be significantly improved.

4.3.1.1 *Self-attention mechanism*

Due to the limitation of the convolution kernel size, the generator G of the GAN can only capture the relationship of the local areas. Conventional convolutional GAN can generate high-resolution details and can only be used as a function of the local points in low-resolution feature maps. Since the self-attention mechanism can provide long-range dependency modeling based on self-attention for image generation tasks, the information from all feature positions can be used to generate detailed information. In addition, the discriminator D in the GAN can check whether the high-resolution detail features at a long distance in the image are consistent. At the same time, adjustments to the generator G in the GAN will affect the overall performance of the GAN. These fully demonstrate the necessity of introducing attention mechanisms. The working principle of self-attention mechanism is shown in Figure 4.3.

In Figure 4.3, the working principle of the self-attention mechanism is given as follows. First, the two 1×1 convolutions are used to perform linear transformation and channel compression on the convolution feature map. Second, the two tensors are converted into

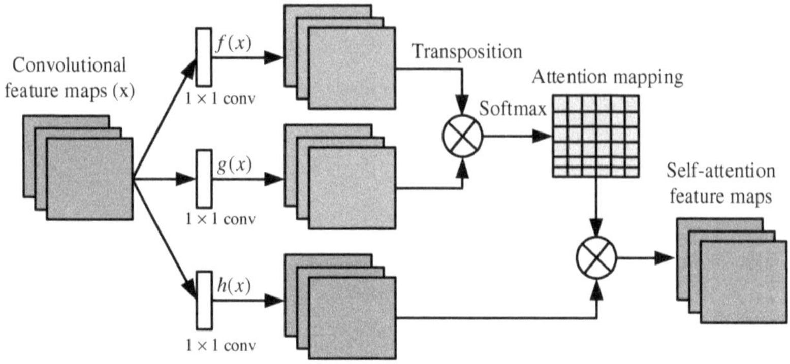

Figure 4.3. The working principle of self-attention mechanism.

matrix form, which is transposed and multiplied according to the matrix multiplication condition to obtain the attention map through the Softmax function. At the same time, after the original feature map is linearly transformed using 1×1 convolution, it is multiplied by the previously obtained attention map matrix and then added to obtain the self-attention feature map. Finally, the self-attention feature map and the original convolution feature map are weighted and summed as the final output. Among them, the self-attention feature map can be regarded as the product of the feature map and its own transposition. This operation can enhance the association among features that are far away in the images so that the dependency between any two pixels can be learned, and then the global feature can be obtained.

In Figure 4.3, $\boldsymbol{x} \in \boldsymbol{R}^{C \times N}$ represents the image features in the previous hidden layer and $f(\boldsymbol{x}) = \boldsymbol{W}_f\boldsymbol{x}$ and $g(\boldsymbol{x}) = \boldsymbol{W}_g\boldsymbol{x}$ denote the two feature spaces obtained by multiplying the image features with different weight matrices. After they are processed by the Softmax function, $f(\boldsymbol{x}) \otimes g(\boldsymbol{x})$ represents the participation (or correlation) of the image content area i in the model synthesis area j, where $\otimes$ denotes a convolution operation. Its mathematical expression is defined as follows:

$$\beta_{j,i} = \frac{\exp(s_{ij})}{\sum\limits_{i=1}^{N} \exp(s_{ij})}. \tag{4.3.2}$$

By integrating the global spatial information and the local information, an expression of the attention mechanism mapping can be obtained by

$$o_j = \sum_{i=1}^{N} \beta_{j,i} h(x_i), \tag{4.3.3}$$

where $o_1, o_2, \ldots, o_j, \ldots, o_N \in \boldsymbol{R}^{C \times N}$.

The final output of the attention layer is obtained by

$$y_i = \gamma o_i + x_i. \tag{4.3.4}$$

In order to take into account the correlation between neighborhood information and long-distance features, a parameter γ, which is initialized from zero, is introduced to let the network focus on the neighborhood information first, and then gradually distribute weights to other long-distance features.

At the same time, the loss functions Loss_G and Loss_D corresponding to the generator G and the discriminator D are respectively defined as

$$\text{Loss}_G = -\mathbb{E}_{z \sim f_z, y \sim f_{\text{data}}}[D(G(z), y)], \tag{4.3.5}$$

$$\begin{aligned} \text{Loss}_D = &-\mathbb{E}_{(x,y) \sim f_{\text{data}}}[\min(0, -1 + D(x, y))] \\ &- \mathbb{E}_{z \sim f_z \cdot y \sim f_{\text{data}}}[\min(0, -1 - D(G(z), y))]. \end{aligned} \tag{4.3.6}$$

Briefly speaking, since the receptive field of the convolution operations is limited, in order to make the generator G and the discriminator D be able to extract the high-resolution features, it is very necessary to introduce the global information into the convolution operation to make the connection between each part closer.

When generating images, conventional CGAN mostly uses convolution operations to simulate the dependencies between different regions of the images. Since each convolution operation has an acceptance region, the dependency relationship between long-range features in the images can only be determined after performing a convolution operation on many occasions. This will lead to inefficient learning of the generator G and the possibility of losing details. Whereas after introducing the self-attention mechanism to the CGAN, the SA-CGAN is proposed and it can improve the quality of the images generated by the CGAN via enhancing the relationships between different parts of the images.

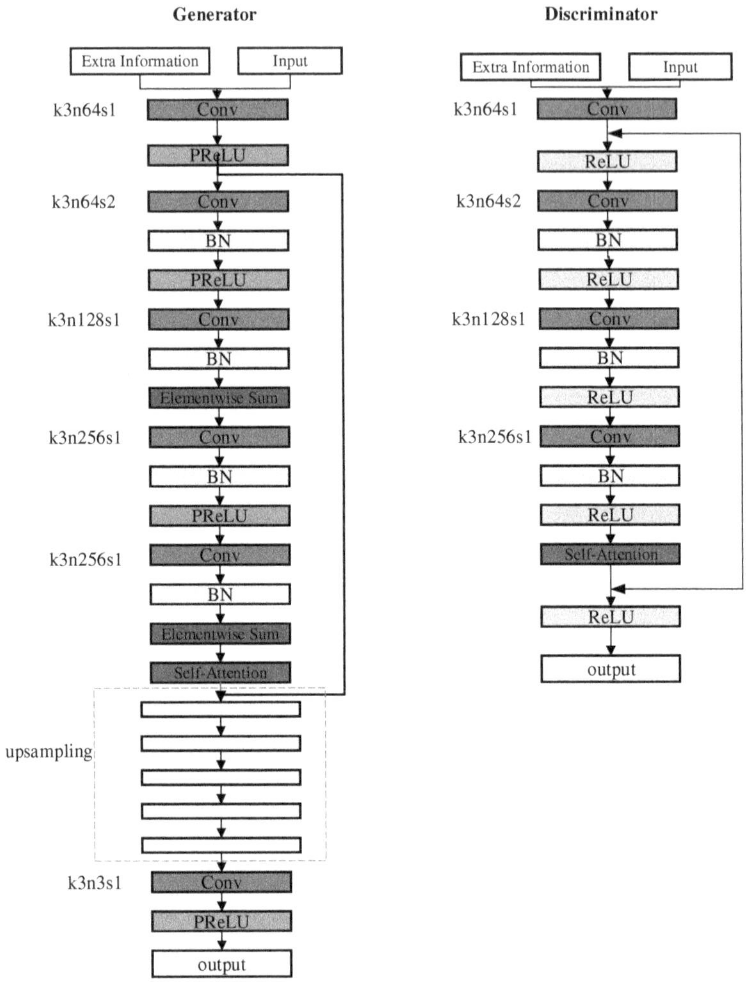

Figure 4.4. The network structure of the SA-CGAN.

Defining w as additional information, the objective function of the SA-CGAN can be described as

$$\min_G \max_D \text{Loss}(D, G) = \mathbb{E}_{x \sim f_{\text{data}}(x)}[\log D(x, y|w)] \\ + \mathbb{E}_{z \sim f_z(z)}[\log(1 - D(G((z|w), y)))]. \quad (4.3.7)$$

The network structure of the SA-CGAN is shown in Figure 4.4.

4.3.1.2 *CGAN architecture based on self-attention mechanism*

CGAN architecture based on the self-attention mechanism is given as follows:

For the iteration number of training do
For step t training do
A mini-batch m positive samples from the dataset $\{(c_1, x_1), (c_2, x_2), \ldots, (c_m, x_m)\}$.
A mini-batch m noise samples from the data distribution $\{z_1, z_2, \ldots, z_m\}$.
Let $\breve{x}_i = G(c_i, z_i)$ *generate dataset* $\{\breve{x}_1, \breve{x}_2, \ldots, \breve{x}_m\}$.
Mini-batch samples from the dataset $\{\widehat{x}_1, \widehat{x}_2, \ldots, \widehat{x}_m\}$ *are used to update the parameters*
θ_{d} *of the discriminator D.*
By maximizing $\breve{V}$, i.e.:

$$\breve{V} = \frac{1}{m}\sum_{i}^{m} \log D(c_i, x_i) + \frac{1}{m}\sum_{i}^{m} \log[1 - D(c_i, x_i)] + \frac{1}{m}\sum_{i}^{m} \log[1 - D(c_i, \widehat{x}_i)]$$

The parameter θ_{d} *is updated as follows:*

$$\theta_d \leftarrow \theta_d + \eta \nabla \breve{V}(\theta_d)$$

End for
Mini-batch m noise samples $\{z_1, z_2, \ldots, z_m\}$
Mini-batch m conditions $\{c_1, c_2, \ldots, c_m\}$
By maximizing $\breve{V}$, *i.e.:*

$$\breve{V} = \frac{1}{m}\sum_{i=1}^{m} \log D[G(c_i, z_i)]$$

The parameters θ_g *is updated as follows:*

$$\theta_g \leftarrow \theta_g + \eta \nabla V \breve{V}(\theta_g)$$

End for

4.3.2 Model-based reinforcement learning conditional generative adversarial networks

Deep Reinforcement Learning (DRL) [58] is a machine learning paradigm with the optimal strategy based on interacting with the environment and maximizing cumulative rewards via a trial-and-error mechanism. To obtain the optimal strategy, the DRL requires the intelligence agent to have knowledge and understanding of the surrounding environments and make decision actions that are in line with the environmental context according to task requirements. Based on whether the environmental model is available during the learning process, reinforcement learning can be divided into Model-based Reinforcement Learning (Mb-RL) and Model-free Reinforcement Learning (Mf-RL).

4.3.2.1 *Model-based reinforcement learning and model free reinforcement learning*

The environment model, also known as the system dynamics model, is a description of the state transition function. In the Model-free Reinforcement Learning (Mf-RL), the environment model is unknown, and the intelligence agent must interact extensively with the real environment to obtain sufficient training samples in order to ensure the decision-making performance of the intelligence agent. The collection process of the samples also places high demands on the system hardware configuration and there is even a risk of damaging the intelligent system. In addition, insufficient training samples can result in the intelligence agent being unable to extract useful information from a small number of training samples for accurate policy updates. Therefore, the sample utilization rate of the Mf-RL is low. In contrast, the Model-based Reinforcement Learning (Mb-RL) enables the intelligence agent to learn strategies without interacting with the real environment after accurately modeling the environment. It can directly interact with the environment model to generate the required training samples, thereby alleviating the problems of low learning efficiency and low sample utilization rate in practical applications of reinforcement learning to a certain extent.

The basic idea of the Mb-RL is to first establish the environment dynamic model and learn the parameters of the environment model. After the training algorithm of the model parameter converges to obtain a stable environment model, the intelligence agent can directly interact with the predicted environment model for policy learning [59]. Throughout the entire process, it is necessary for only a certain amount of the training samples to be used to learn the model parameters, and the sample demand is relatively small. However, due to numerous factors such as environmental noise and the dynamism of the system, it is usually difficult to make the predicted environmental model be able to accurately describe the real environment, i.e., there is a model error between the learned environment model and the real environment. When the environment models with model errors are used to generate data for policy learning, model errors will become greater, and ultimately lead to task failure. In order to reduce model errors and improve the accuracy of the environmental models, researchers have proposed a series of Mb-RL-based environmental models for different application scenarios. However, a key challenge in recent research is how to obtain accurate environmental models for large-scale complex dynamic environments.

A GAN-based environmental model reinforcement learning method was proposed [60] on the basis of making the most of the advantages of the GAN in data generation, which combined the CGAN with the Mb-RL model. It has been applied to learning state transition models, and combined the CGAN with a policy search method that is good at processing continuous action spaces, as well as further developed a new model-based policy search method based on the CGAN.

4.3.2.2 *Problem model*

Reinforcement learning refers to a learning paradigm in which an intelligent agent continuously interacts with the environment to learn optimal strategies in an unknown environment. An intelligent agent is a subject with decision-making ability, which interacts with the environment through state perception, action selection, and feedback reception. Usually, the interaction process between the intelligent agent and the environment can be modeled as a Markov Decision Process (MDP) [60]. A complete MDP is represented by a five-tuple (S, A, P, P0, R) consisting of state, action, state transition function, initial state, and reward. Among them, S represents the state space, which is the set of all states; s_k is the state at time k; A represents the action space, which is the set of all actions; a_k denotes the action selected at time k. P stands for the state transition probability, i.e., the environment model, based on whether the state transition probability is known, reinforcement learning methods are divided into Mb-RL and Mf-RL; $P0$ is regarded as the initial state probability, which is the probability of randomly selecting a certain initial state; R represents the cumulative reward of the intelligent agent; R_k is the instantaneous reward at time k.

At each time step k, the intelligent agent first observes the current state s_k of the environment and selects and takes actions a_k according to the current policy function. The taken actions are to interact with the environment on the one hand to realize state transition according to the state transition probability $p(s_{k+1}|s_k, a_k)$ and to obtain instantaneous rewards r_k on the other hand. After this process iterates T times until the final state is reached, a path $h^n = [s_1^n, a_1^n, \ldots, s_T^n, a_T^n]$ is presented.

The goal of reinforcement learning is to find the optimal strategy to maximize the expected cumulative rewards. Once a path is obtained, the cumulative rewards of this path can be calculated by

$$R(h) = \sum_{k=1}^{T} \gamma^{k-1} r(s_k, a_k, s_{k+1}), \tag{4.3.8}$$

where the parameter γ determines the time scale of the rewards, and $0 \leq \gamma < 1$.

The expected cumulative rewards are used to measure the quality of the strategy. The expected cumulative rewards are defined as

$$J_\pi = \int f(h) R(h) \mathrm{d}h, \tag{4.3.9}$$

where $f(h)$ denotes the probability density function of the occurrence path given by $f(h) = f(s_1) \prod_{k=1}^{T} f(s_{k+1}|s_k, a_k) \pi(a_k|s_k)$. The goal of reinforcement learning is to find the optimal strategy π_{opt} and maximize the expected rewards J_π. π_{opt} is represented as

$$\pi_{\text{opt}} = \arg\max_{\pi} J_\pi. \tag{4.3.10}$$

4.3.2.3 *Strategy search method*

The strategy search method is a strategy optimization method that can directly learn the strategy and is suitable for solving complex decision-making tasks with a continuous action space.

The learning goal of the strategy search method is to find the strategy parameter θ that can maximize the expected value $J(\theta)$ of the cumulative rewards, i.e., the optimal strategy parameter θ_{opt} is defined as

$$\theta_{\text{opt}} = \arg\max_{\pi} J(\theta), \tag{4.3.11}$$

where the expected cumulative reward $J(\theta)$ is a function of the strategy parameter θ and defined as

$$J(\theta) = \int f(h|\theta)R(h)\mathrm{d}h, \tag{4.3.12}$$

where $f(h|\theta)$ denotes the conditional probability density function and is represented by $f(h|\theta) = f(s_1)\prod_{k=1}^{T} f(s_{k+1}|s_k, a_k)\pi(a_k|s_k, \theta)$, and $R(h)$ determines the time scale of the return and is given by $R(h) = \sum_{k=1}^{T} \gamma^{k-1} r(s_k, a_k, s_{k+1})$, and $0 \leq \gamma < 1$.

At present, the most representative policy search algorithms include the PEGASUS [61], policy gradient method [62,63], and natural policy gradient method, etc. Among them, the policy gradient method is the simplest and most common method, which is used to find the optimal policy parameters. In view of the superior performance of the Proximal Policy Optimization (PPO) algorithm in the policy gradient methods, the PPO algorithm is introduced for policy learning.

4.3.2.3.1 *Implementation architecture of algorithm*

The Mb-RL needs to first learn an accurate state transition model, which is then used to generate the required samples and reduce the number of interactions between the intelligence agent and the environment in the policy learning phase. In the CGAN-based modeling policy search method, the CGAN is first employed for learning the sequence data $\{s_0, a_0, s_1, a_1, \ldots\}$ in order to obtain the prediction of the state transition function $\hat{f}(s_{k+1}|s_k, a_k)$. After that, when an input state–action pair is denoted as $[s_k, a_k]$, the learned state transition function $\hat{f}(s_{k+1}|s_k, a_k)$ can be directly used for predicting the next state s_{k+1} without waiting for real environment feedback.

The architecture of the CGAN-Mb-RL algorithm is given as follows:

Step 1: Collect samples

Real state transition samples $\{(s_k, a_k, s_{k+1})\}_{k=1}^{T}$ are collected.

Step 2: Establish and train model

The CGAN is used to model the state transition function $f_T(s_{k+1}|s_k, a_k)$ and these samples$\{(s_k, a_k, s_{k+1})\}_{k=1}^{T}$ are used for training the model.

Step 3: Implement strategy learning

The intelligence agent interacts with the state transition model $f_T(s_{k+1}|s_k, a_k)$ to obtain sufficient sample sequences $\{h_k\}_{k=1}^{N}$ for strategy learning.

Step 4: Update the parameters

The parameters in the policy model are updated until convergence, and the optimal strategy π_{opt} is ultimately obtained.

4.3.2.3.2 *CGAN-based environment model learning strategy*

In the Mb-RL, when the state transition model can fully simulate the real environment, the intelligent agent only needs to interact with the learned state transition model $\hat{f}(s_{k+1}|s_k, a_k)$ to obtain the next state $\hat{s}_{k+1}$, thereby resulting in reducing the interaction between the intelligent agent and the real environment. Therefore, the key issue for the Mb-RL method is how to obtain the state transition function of the real environment. Here, the CGAN is used to capture the distribution of the state transition function of the real environment, as shown in Figure 4.5 [55].

The next state s_{k+1} in the state transition function is limited by the current state s_k and the action a_k taken under the current state s_k. This state transition function is a conditional probability density model and is expressed as $f_T(s_{k+1}|s_k, a_k)$. Therefore, the current state s_k and the action a_k are regarded as the conditional variables y of the CGAN to add restrictions to both the generator G and the discriminator D at the same time, and guide the generation of the

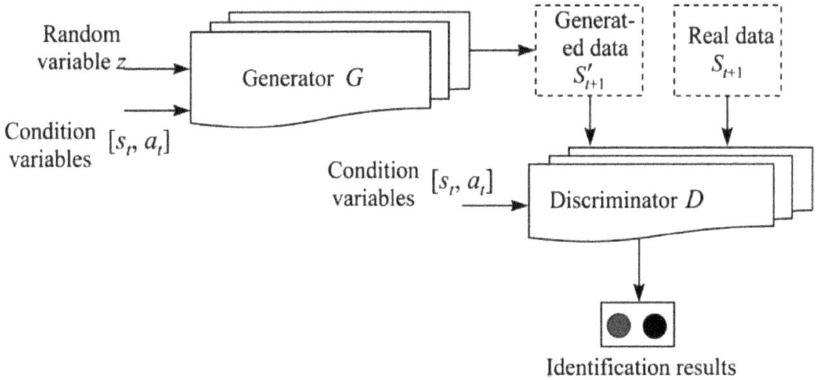

Figure 4.5. Environment model learning strategy based on the CGAN.

next state s_{k+1}. The conditional variable y and the random variable z are both used as inputs to the generator G. At this time, the output of the generator G is the next state s_{k+1} reached by the executed action a_k under the current state s_k. By inputting the output along with the real sample data and the conditional variable y into the discriminator D, the probability of the samples coming from the training data can be estimated.

The loss function of the above process can be expressed as

$$\begin{aligned}\min_{G}\max_{D}\mathrm{Loss}(D,G) = {} & \mathbb{E}_{s_{k+1}\sim f_{\mathrm{data}}(s_{k+1})}[\log D(s_{k+1}|(s_k,a_k))] \\ & + \mathbb{E}_{z\sim f_z(z)}[\log(1-D(G(z|(s_k,a_k))))].\end{aligned} \tag{4.3.13}$$

After the CGAN model is trained and stabilized, the trained generator G can be directly used as an environment prediction model to interact with the intelligent agent in order to generate a large amount of sample data for policy learning.

4.3.2.3.3 *Decision-searching method*

After obtaining a stable and efficient state transition model $\hat{f}(s_{k+1}|s_k,a_k)$, the classic Proximal Policy Optimization(PPO) algorithm, which is an improvement on the policy gradient method, is selected for policy learning. The traditional policy gradient method has the problem of low updating speed of the parameters and the need for resampling each update, whereas the PPO method with a simple structure can update policy parameters time and again with one sampling, and it has a high sample utilization rate and can automatically adjust the step size of the parameter space to achieve uniform changes in the policy space. Its expected cumulative reward is expressed as

$$\text{Loss}_{clip}(\rho) = \hat{\mathbb{E}}_k[\min(r_k(\rho)\hat{A}_k, clip(r_k(\rho), 1-\varepsilon, 1+\varepsilon)\hat{A}_k)], \quad (4.3.14)$$

where ρ is the policy parameter, $\hat{\mathbb{E}}_k$ denotes the expectation, ε is a constant, usually 0.1 or 0.2, r_k denotes the ratio of the new policy to the old policy, $\hat{A}_k$ represents the superiority at time k, the *clip* term ensures r_k that doesn't deviate from the interval defined by $[1-\varepsilon, 1+\varepsilon]$.

Based on the above discussions, in combining the CGAN with the PPO to find the optimal strategy, the state-action space is modeled by the CGAN as a state transition model $\hat{f}(s_{k+1}|s_k, a_k)$, and then the learned state transition model is applied to generating the samples for the strategy learning of the PPO algorithm, thereby the optimal strategy π_{opt} is obtained.

4.4 Wavelet Generative Adversarial Network

4.4.1 Deep convolutional generative adversarial network

In order to enable the Generative Adversarial Network (GAN) to adapt well to the CNN architecture, four architectural design rules were proposed by the DCGAN, namely:

(1) Use convolutional layers instead of pooling layers.
(2) Remove fully connected layers.
(3) Use batch normalization.
(4) Use appropriate activation functions. The ReLU function is used in the generator G and the LReLU function is used in the discriminator D.

The input of the generator G is a 100-dimensional random data z, which follows a uniform distribution in $[-1, 1]$. The first layer network of the generator G is a fully connected layer, whose task

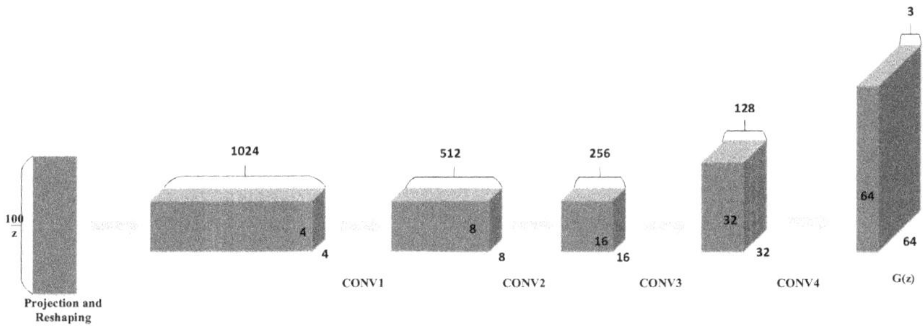

Figure 4.6. The structure of generator G in the DCGAN.

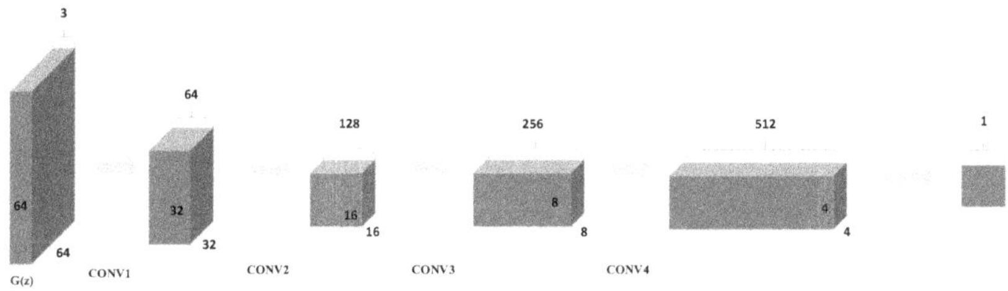

Figure 4.7. The structure of the discriminator D in the DCGAN.

is to convert the 100-dimensional noise vector into a $4 \times 4 \times 1024$-dimensional vector, apply stride convolution to perform upsampling operations starting from the second layer, and gradually reduce the number of channels. Finally, the final output is a $64 \times 64 \times 3$ image. The network structure of the generator G in the DCGAN is shown in Figure 4.6.

For discriminator D, its operation process is basically the reverse operation of the generator G, as shown in Figure 4.7. The input layer is a $64 \times 64 \times 3$ image data, which is reduced in dimension through a series of convolutions, and the final output of the discriminator D is a binary classification data.

In the DCGAN, after an image is processed by the discriminator D, the output of the discriminator D is the probability that it is a real image. After the discriminator D convolves the input image through several layers, it can extract the convolution features and input the obtained convolution features into the logistic function. Its output is only a probability.

4.4.2 Wavelet deep convolutional generative adversarial network

4.4.2.1 *Construction of wavelet functions*

Wavelet transform has the characteristics of multi-resolution analysis and the ability to characterize local features of the signals in both time and frequency domains. In this section, as for the generator G of the DCGAN, in order to obtain more local features of the original images, the activation function of the first convolution layer of generator G in the DCGAN is replaced by a nonlinear wavelet basis. In the selection of wavelet functions, some wavelet bases with similar properties to traditional activation functions may be considered.

There are many types of mother wavelets in wavelets, among which the Morlet wavelet has a smooth and continuous amplitude. If the Morlet wavelet is selected as the activation function to construct a wavelet deep convolutional GAN, its function expression is written as

$$\psi(t) = \mathrm{C}e^{-\frac{t^2}{2}}\cos(5t), \tag{4.4.1}$$

where C is the normalization constant during reconstruction. The Morlet wavelet has no scale function and is a non-orthogonal decomposition function. The time domain waveform is shown in Figure 4.8.

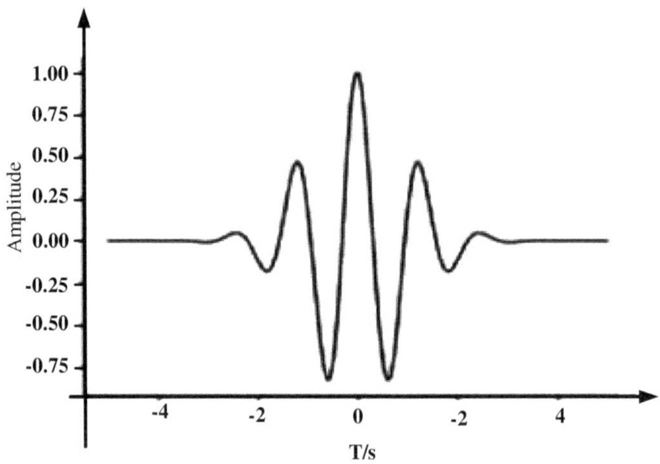

Figure 4.8. Time domain diagram of wavelet basis function.

In order to better adapt to the characteristics of the activation function, the Morlet wavelet function is modified as

$$\psi(t) = -\frac{7}{4}\sin\left(\frac{7}{4}t\right)e^{-\frac{t^2}{2}} - t\cos\left(\frac{7}{4}t\right)e^{-\frac{t^2}{2}}. \tag{4.4.2}$$

The corresponding time domain waveform is shown in Figure 4.9.

4.4.2.2 *Wavelet function generation model*

The modified wavelet function is added to the first convolutional layer in the generator G and is regarded as the activation function, as shown in Figure 4.10.

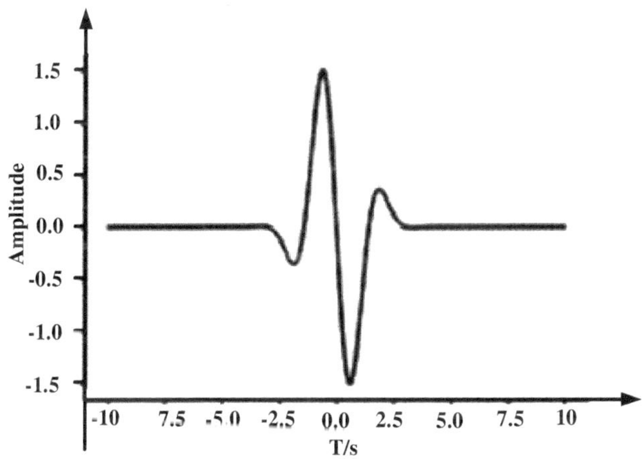

Figure 4.9. Time domain waveform of modified wavelet function.

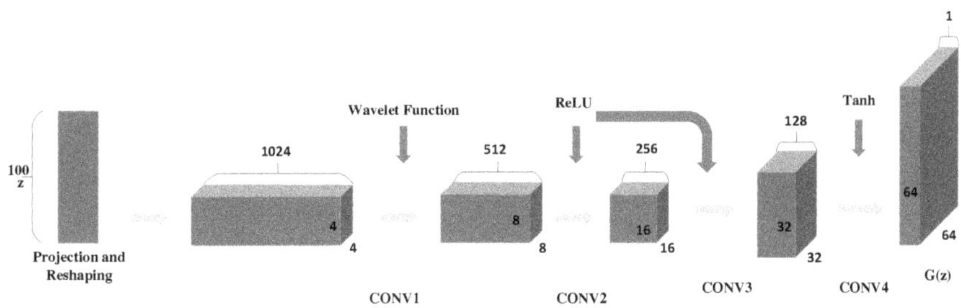

Figure 4.10. Generator G with wavelet function.

4.4.3 High-frequency wavelet generative adversarial network

Image completion has always been a hot topic field of computer vision research and application. For the occluded parts of the face, researchers have proposed various image completion algorithms, such as diffusion-based methods, statistical methods using local histograms, and sparsity-based methods, which are prone to poor results for images with large occluded areas or large areas of texture missing. In this section, the GAN and its various variants as the main image completion algorithms based on deep learning models are introduced.

In the Context Encoder (CE) algorithm, an encoder-to-decoder network was designed to generate the missing parts in the images, and the CE algorithm has also combined the training method of the GAN to improve the quality of the generated images. The Globally and Locally Consistent Image Completion (GLCIC) algorithm can increase the receptive field of the network via dilated convolution and improve the local consistency of the generated images via two global and local discriminators.

In response to the problem of image completion with occlusion, an image completion framework was designed based on the GAN to repair the image in the occluded areas. The dilated convolution was introduced in the network layer to expand the receptive field, and a high-frequency loss function based on wavelet decomposition was defined to improve the clarity of the generated images. The completion process of the occluded images is as follows. First, the original image is converted from the image space to the wavelet space and decomposed into high-frequency components and low-frequency ones. The blurriness of the repaired image is related to the lack of high-frequency information, so the L1 loss of the wavelet decomposition in the loss function is employed for training the network, which can effectively improve the quality of the generated image. The occluded image is completed into a full image by the generator G, then the ArcFace model [64] is provided as the image recognition framework to verify the image recognition of the repaired image.

4.4.3.1 *Image completion GAN*

4.4.3.1.1 *The structure of image completion generator*

In image recognition tasks, the occlusions covered the real image area, thereby resulting in missing recognition information and making it difficult for the network to make accurate judgments. Image completion can make up for this information loss to a certain extent and make the network be able to recognize the complete image information. If the occluded area is relatively fixed, the position of the occlusion can be predefined when training the network without considering the recognition and positioning of the occlusion. The image completion GAN consists of a generative network and a discriminative network. The goal of the generative network is to complete the occluded images into an integrated image. The structure of the image completion generator is shown in Figure 4.11.

In Figure 4.11, K is the size of the convolution kernel, N represents the number of channels, S denotes the stride of the convolution kernel, and D stands for the dilation rate. The input to image completion generator is a half-face image blocked by a mask, which passes through the encoding layer, the middle layer, and the decoding layer, and the ultimate output is a completed full-face image. The encoding layer is designed to reduce the spatial size of the input image to ensure that subsequent network operations can be performed in a smaller spatial size and effectively improve the operational efficiency of the entire network. Since the reduction of the spatial size is accompanied by a certain quantization error, the smaller the error,

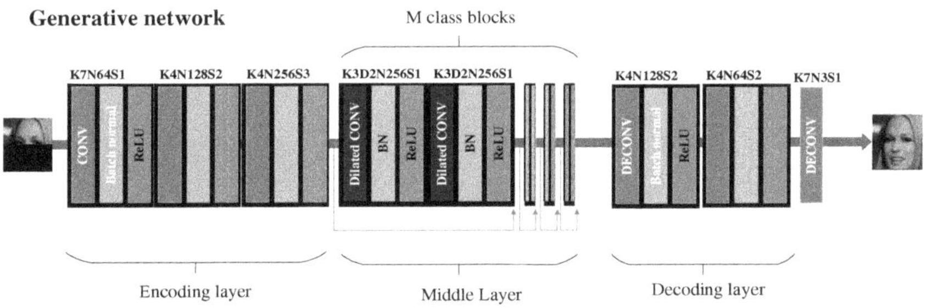

Figure 4.11. The structure of image completion generator.

the better the generated image quality. Therefore, the encoding layer first adopts the K7N64S1 structure, that is, the convolution kernel size is $7 \times 7 \times 64$, and the stride is set to 1. The convolution layer extracts the global features of the image without reducing the network spatial size. Then, two convolution layers with a convolution kernel size of 4×4 and a stride of 2 are used to reduce the network size by 4 times. The middle layer consists of M classes of ResNet blocks [65]. The deeper network can obtain better generalization ability and feature expression. At the same time, a dilated convolution with a dilation rate of 2 is replaced with a conventional convolution in the module, which can increase the network's receptive field and improve the network's ability to perceive global features. A class ResNet module is shown in Figure 4.12.

The decoding layer decodes the previously extracted features into a complete face image, mainly via enlarging the image size by four times through two deconvolution layers to restore the original image size. Finally, the channel is restored to three channels through a $7 \times 7 \times 3$ convolutional layer.

4.4.3.1.2 *Discriminator design*

The discriminator D is a binary classification network, as shown in Figure 4.13. Its input is an original complete image or the image generated by the generator G, and its output is a binary classification result that can determine whether it is the original image or the generated image.

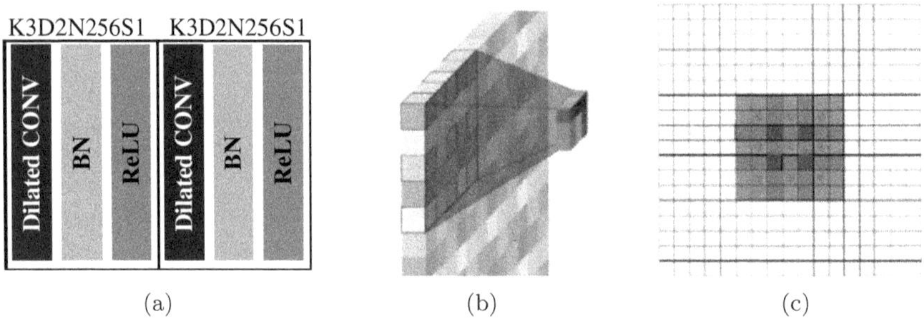

Figure 4.12. Class ResNet module. (a) A class ResNet blocks. (b) Dilated convolution with a dilation ratio of 2. (c) Receptive field of dilated convolution with a dilation ratio of 2.

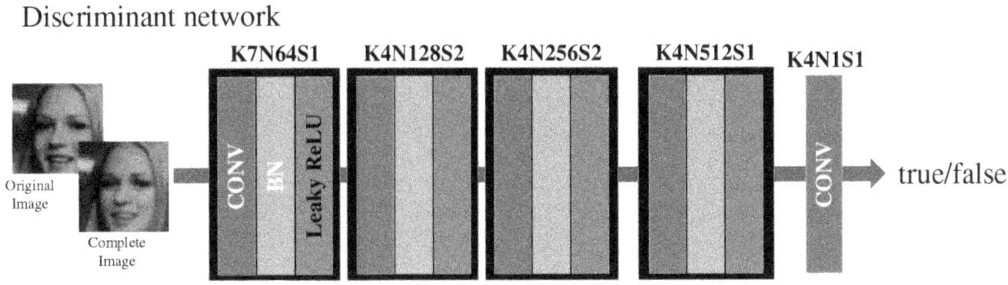

Figure 4.13. Structure of discriminator D.

4.4.3.2 *Loss function*

4.4.3.2.1 *Total loss function*

Let I_{gt} be the real sample image, and M be the mask, which is used to cover the real sample images in order to obtain the images with some missing areas, i.e.,

$$\widetilde{I} = I_{\text{gt}} \odot (1 - M), \tag{4.4.3}$$

where $\odot$ represents the point-by-point multiplication of the matrix elements. $\widetilde{I}$ is used as the input of the generator G, and a color image I_P with the filled missing area is formed as

$$I_P = G(\widetilde{I}) \tag{4.4.4}$$

where the resolution of the I_P is the same as that of the input image $\widetilde{I}$.

The total loss function of the network training is defined as

$$\text{Loss} = \alpha\,\text{Loss}_{l_1} + \beta\,\text{Loss}_{\text{adv}} + \gamma\text{Loss}_{\text{pere}} + \mu\text{Loss}_{\text{style}} + \delta\text{Loss}_{\text{wavelet}}, \tag{4.4.5}$$

where α, β, γ, μ, and δ are the coefficients of the defined loss function. Loss_{l_1} represents the reconstruction $L1$ loss.

In addition to the reconstruction $L1$ loss and in the process of network training, the definition of adversarial loss ensures the generation of adversarial processes and makes the completed image closer to the original image, i.e.,

$$\text{Loss}_{\text{adv}} = \mathbb{E}_{I_{\text{gt}} \sim f(I_{\text{gt}})}[\log D(I_{\text{gt}})] + \mathbb{E}_{I_{\text{comp}} \sim f(I_{\text{comp}})}[\log(1 - D(\widetilde{I}_p))], \tag{4.4.6}$$

where $f(I_{\text{gt}})$ is the distribution of the real samples, $D(\cdot)$ represents the output of the discriminator D, and $f(I_{\text{comp}})$ denotes the distribution of the generated samples after generating the completed image.

$\text{Loss}_{\text{pere}}$ is the perceptual loss, which uses the distance between the feature maps in the pretrained network as the penalty for the difference between the generated map $\widetilde{I}_p$ and the real map I_{gt} in perception, and is defined as

$$\text{Loss}_{\text{pere}} = \frac{1}{H_i W_i C_i} \|\phi^{(i)}(I_{\text{gt}}) - \phi^{(i)}(\widetilde{I}_p)\|_1, \tag{4.4.7}$$

where $\phi^{(i)}(\cdot)$ is the feature map output by the ith layer in the pretrained network, for instance, the feature map output from the ReLU layer of the VGG19 pretrained on ImageNet. At the same time, these feature maps are also demanded to calculate the style loss. For a given $H_i \times W_i \times C_i$ feature map, the style loss is calculated by the Gram matrix as

$$\text{Loss}_{\text{style}} = \|(\phi^{(i)}(I_{\text{gt}})^T \phi^{(i)}(I_{\text{gt}}) - \phi^{(i)}(\widetilde{I}_p)^T \phi^{(i)}(\widetilde{I}_p))\|_1. \tag{4.4.8}$$

In order to supplement the lack of high-frequency information in the generated images, a high-frequency loss function $\text{Loss}_{\text{wavelet}}$ is designed based on wavelet decomposition.

4.4.3.2.2 *Wavelet loss*

Pixel-based losses include $L1$ loss and Mean Square Error (MSE) loss [66]. However, while achieving a high Peak Signal-to-Noise Ratio (PSNR), there is often a lack of high-frequency information in the images, which results in excessively smooth textures in the generated images and visually unsatisfactory. Regarding this issue, a high-frequency wavelet loss is defined based on wavelet decomposition, as shown in Figure 4.14.

Suppose that a two-dimensional image is $f(u,v)$, where u and v are the width and the height of the two-dimensional image in pixels, respectively. Now, the wavelet function is utilized to perform one-layer decomposition on two-dimensional images $f(u,v)$ [67], we have

$$\begin{aligned} f(u,v) = &\frac{1}{\sqrt{MN}} \sum_{\tilde{u}} \sum_{\tilde{v}} W_{\varphi}(\tilde{u},\tilde{v}) \varphi_{\tilde{u},\tilde{v}}(u,v) \\ &+ \frac{1}{\sqrt{MN}} \sum_{\tilde{u}} \sum_{\tilde{v}} W_{\psi}^{\text{H}}(\tilde{u},\tilde{v}) \psi_{\tilde{u},\tilde{v}}^{\text{H}}(u,v) \end{aligned}$$

$$+\frac{1}{\sqrt{MN}}\sum_{\tilde{u}}\sum_{\tilde{v}}W_{\psi}^{\mathrm{V}}(\tilde{u},\tilde{v})\psi_{\tilde{u},\tilde{v}}^{\mathrm{V}}(u,v)$$

$$+\frac{1}{\sqrt{MN}}\sum_{\tilde{u}}\sum_{\tilde{v}}W_{\psi}^{\mathrm{D}}(\tilde{u},\tilde{v})\psi_{\tilde{u},\tilde{v}}^{\mathrm{D}}(u,v) \tag{4.4.9}$$

where MN represents the total number of the pixels of the input images and $\varphi_{\tilde{u},\tilde{v}}(u,v)$ denotes the scale function. $\psi_{\tilde{u},\tilde{v}}^{\mathrm{H}}(u,v)$, $\psi_{\tilde{u},\tilde{v}}^{\mathrm{V}}(u,v)$, and $\psi_{\tilde{u},\tilde{v}}^{\mathrm{D}}(u,v)$ are sensitive wavelets along the horizontal, vertical, and diagonal directions, respectively. The three-direction sensitive wavelet coefficients W_{ψ}^{H}, W_{ψ}^{V}, and W_{ψ}^{D} depict the details of the original image along the horizontal, vertical, and diagonal directions, respectively, i.e., the original two-dimensional image can be decomposed to obtain high-frequency information in the horizontal, vertical, and diagonal directions. Similarly, the scale coefficient $W_{\varphi}(\tilde{u},\tilde{v})$ presents the approximate decomposition of the original images, i.e.,

$$W_{\varphi}(\tilde{u},\tilde{v})=\frac{1}{\sqrt{MN}}\sum_{u}\sum_{v}f(u,v)\varphi_{\tilde{u},\tilde{v}}(u,v). \tag{4.4.10}$$

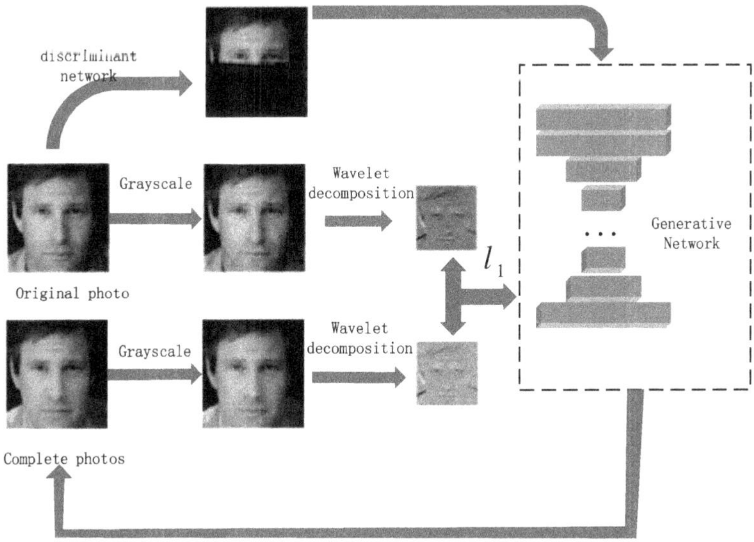

Figure 4.14. Wavelet loss.

In order to extract the high-frequency information in the horizontal and vertical directions, for a two-dimensional image I, $\xi(I)^j$ is redefined as the high-frequency information part of the wavelet decomposition of the original image I. According to the decomposition of the two-dimensional images in the wavelet, the wavelet loss is defined as the distance between the wavelet decomposition of the grayscale completed image $\tilde{I}_p$ and the wavelet $L1$ decomposition of the grayscale real image I_{gt}, that is,

$$\mathrm{Loss}_{\mathrm{wavelet}} = \|\zeta(I_{\mathrm{gt}})^j - \zeta(\tilde{I}_p)^j\|_1 \tag{4.4.11}$$

4.4.3.3 *Network training architecture*

The training architecture of the two networks is as follows:

Input: Initialize the parameters θ_d and θ_g of the discriminator D and generator G, respectively, and full image (unoccluded) data.

Output: Discriminator D and generator G.

Step 1: Sample and update Discriminator *D*

Sample $I_{gt1}, I_{gt2}, \ldots, I_{g\mathrm{tm}}$ from the full image data as a batch and the lower half of the image is covered to get $\tilde{I}_1, \tilde{I}_2, \ldots, \tilde{I}_m$. After they pass through the generator G, the completed full image data $\tilde{I}_{p1}, \tilde{I}_{p2}, \ldots, \tilde{I}_{\mathrm{pm}}$ will be obtained, where $\tilde{I}_{pi} = G(\tilde{I}_i)$ and $i = 1, 2, \ldots, m$. The full image data and the completed full image data are required to pass through the discriminator D to update the network parameters θ_d of the discriminator D.

Step 2: Sample and update generator *G*

Sample $I_{gt1}, I_{gt2}, \ldots, I_{\mathrm{gtm}}$ from the full image data as a batch, and the lower half of the image is blocked to obtain $\tilde{I}_1, \tilde{I}_2, \ldots, \tilde{I}_m$. After passing through the generator G to obtain the completed full image data $\tilde{I}_{p1}, \tilde{I}_{p2}, \ldots, \tilde{I}_{\mathrm{pm}}$, they are required to pass through the discriminator D to update the parameters θ_g of the generator G.

4.4.3.4 *Evaluation indicators*

General indicators, such as Structural SIMilarity (SSIM) index, Peak Signal-to-Noise Ratio (PSNR), and Mean Absolute Error(MAE), etc., of the image completion are employed for measuring the quality of images generated by the model. Assuming the independence between pixels, the SSIM index is calculated by

$$\mathrm{SSIM} = \frac{(2m_{I_{\mathrm{gt}}} m_{I_p} + c_1)(2C_{I_{\mathrm{gt}} l_p} + c_2)}{(m_{I_{\mathrm{gt}}}^2 + m_{I_p}^2 + c_1)(\sigma_{I_{\mathrm{gt}}}^2 + \sigma_{I_p}^2 + c_2)}, \tag{4.4.12}$$

where $m_{I_{\mathrm{gt}}}$ and $m_{\tilde{I}_p}$ represent the means of the real image I_{gt} and the completed image $\tilde{I}_p$, respectively; $\sigma_{I_{\mathrm{gt}}}$ and $\sigma_{\tilde{I}_p}$ stand for the standard deviations of I_{gt} and $\tilde{I}_p$, respectively; $\sigma_{I_{\mathrm{gt}}}$ and $\sigma_{\tilde{I}_p}$ denote the variances of I_{gt} and $\tilde{I}_p$, respectively; and $C_{I_{\mathrm{gt}}\tilde{I}_p}$ represents the covariance of I_{gt} and $\tilde{I}_p$, as well as c_1 and c_2 are constants and given by

$$\begin{aligned} c_1 &= (k_1 \times L)^2, \\ c_2 &= (k_2 \times L)^2. \end{aligned} \tag{4.4.13}$$

Assuming that C, W, and H are the number of channels, width, and height of the images, respectively, the PSNR is defined by the MSE index as

$$\mathrm{PSNR} = 20 \cdot \log_{10}\left(\frac{255}{\mathrm{MSE}}\right), \tag{4.4.14}$$

$$\mathrm{MSE} = \frac{1}{\mathrm{CWH}} \|I_{\mathrm{gt}} - \widetilde{I}_p\|^2. \tag{4.4.15}$$

At the same time, the MAE between the real image and the completed image is calculated by

$$\mathrm{MAE} = \frac{1}{\mathrm{CWH}} \|I_{\mathrm{gt}} - \widetilde{I}_p\|_1. \tag{4.4.16}$$

Precision and recall are calculated respectively by

$$\mathrm{Accuracy} = \frac{\sum TP + \sum TN}{\mathrm{AllSamples}}, \tag{4.4.17}$$

$$\mathrm{Recall} = \frac{TP}{TP + FN}, \tag{4.4.18}$$

where TP means the correct judgment is true, FP means the wrong judgment is true, TN means the correct judgment is false, and FN means the wrong judgment is false, as well as *Allsamples* means the number of samples.

4.5 Multi-scale Generative Adversarial Networks

In the optical wave transmission path, since the atmospheric refractive index is randomly affected by atmospheric turbulence, the geometric deformation and pixel blurring of the images captured by the optical imaging system, which seriously affect the image quality, are caused. Therefore, it is necessary to restore turbulence-degraded images. However, it is a difficult task to simultaneously remove deformation and blur caused by atmospheric turbulence in turbulence-degraded image restoration. In recent years, deep generative adversarial networks have been used to decompose the distortion caused by turbulence into blur and deformation components, which can be restored by a deblurring generator and a deformation correction generator, respectively. Although the restored images can be output via a fusion function, the complex training data needs to be prepared during the model training process, which introduces additional losses in the process of fuzzy decomposition and feature fusion.

In this section, a multi-scale GAN, which introduces a multi-scale attention feature extraction unit and a multi-level feature dynamic fusion unit to the U-Net in the GAN, is studied. The multi-scale attention feature extraction unit is embedded into the fully convolutional part of the U-Net to extract and encode the degraded image features. Then, the feature map is reconstructed in the upsampling part, and the coarse-to-fine feature fusion unit is used to restore the turbulence-degraded images [68].

4.5.1 Turbulence-degraded images

The degradation effect of atmospheric turbulence mainly includes the turbulence distortion operator and sensor optical blur. The mathematical model describing the turbulence degradation process is represented as

$$f(u) = D_u(H(I(u))) + n_u, \tag{4.5.1}$$

where $I(u)$ is the clear image to be restored, $f(u)$ is the turbulence-degraded image obtained by the imaging device, $u = (x, y)^{\mathrm{T}}$ represents the spatial position of the pixel in the images, H is the optical blur operator of the sensor, D_u is the turbulence distortion operator, which includes local deformation and spatial blur, and n_u is an additive noise. Since the turbulence distortion operator contains both blur kernels and deformation kernels, when a Convolution Neural Network (CNN) is required to extract pixel features, it is necessary to design a sufficiently large receptive field to cover the pixel areas. It helps focus on important information via dynamically adjusting the weight of the extracted features. In Equation (4.5.1), namely the expression of the min-max game of the GAN, $\mathbb{E}_{x\sim f_{\text{data}}}$ and $\mathbb{E}_z \sim f(z)$ denote the mathematical expectation when inputting the clear image x and the generated image z, respectively; $x \sim f_{\text{data}}(x)$ and $z \sim f(z)$ are the distributions of the clear image x and the generated image z, respectively. In conventional GAN [69], when the Sigmoid cross-entropy loss function is used as the adversarial loss, it is very easy to see the saturation of the gradient returning to zero, and this will result in model collapse, gradient disappearance, explosion, and other problems during training. In the discriminator D based on the least square GAN (LSGAN), the L2 loss function is used to measure the distance from the input x to the decision boundary and provide a gradient proportional to the distance, which helps further reduce the loss and generate higher quality images. At the same time, the LSGAN is not easy to reach saturation and has better training stability.

During the training process, the turbulence-degraded image is used as the input of the generator G, and the generated image and the clear image in the training data are used as the input of the discriminator D. The discriminator D and the generator G are trained separately and alternately, and the network parameters are updated by error feedback until the given number of iterations.

4.5.2 Multi-scale generative adversarial networks

4.5.2.1 *Multi-scale generative adversarial network structure*

In order to solve the turbulence image restoration task, it is necessary to remove geometric distortion and blur of the images at the same time. So, a multi-scale GAN is proposed, as shown in Figure 4.15. In Figure 4.15, the generator G is a symmetrical U-Net structure [50],

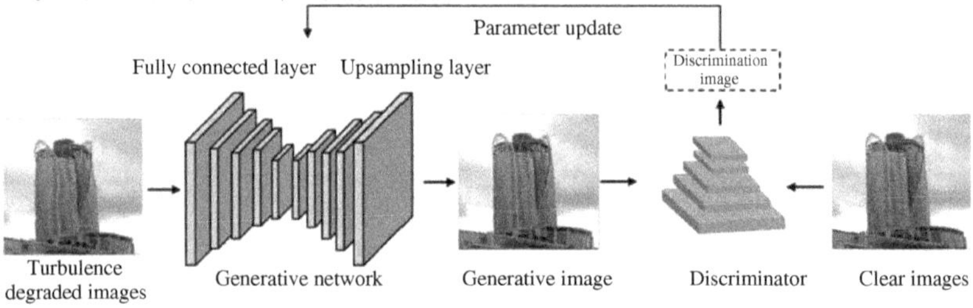

Figure 4.15. Multi-scale GAN [133].

and the discriminator D selects the PatchGAN structure [70, 71], which consists of four convolution layers with a convolution kernel size of 4×4. The multi-scale GAN model is manifested as follows: (1) In image feature extraction, the multi-scale attention feature extraction unit makes use of the convolution kernels of different sizes to extract multi-scale feature information within a larger receptive field. (2) In terms of model structure, multi-level feature fusion units dynamically fuse the feature maps of the different proportions by adjusting their weights and excavate semantic information at different levels.

The structure of the generator G is shown in Figure 4.16. From Figure 4.16, we can see that the full convolution part of the U-net consists of a pretrained convolution module and a multi-scale attention feature extraction unit. The pretrained convolution module uses the convolution layer and the maximum pooling layer in the Inception-ResNet-v2 network. The multi-scale attention feature extraction unit provides multi-scale feature information and uses feature attention to excavate channel correlation. The upsampling part includes convolution layers and an upsampling layer. Multiple proportional feature dynamic fusion units are inserted to upsample feature maps of different scales to the same input size and dynamically adjust the weights and connect them into a tensor to enhance information sharing of the feature maps of different scales. After passing through the full convolution layer, the input image is transformed into a feature map with a smaller spatial size and more compressed semantic information. The obtained feature map is fused and mapped in the upsampling part, and a higher spatial resolution from the semantically rich feature layer is reconstructed, and the size of the target images is gradually restored to the original one. Since the detailed features of the image

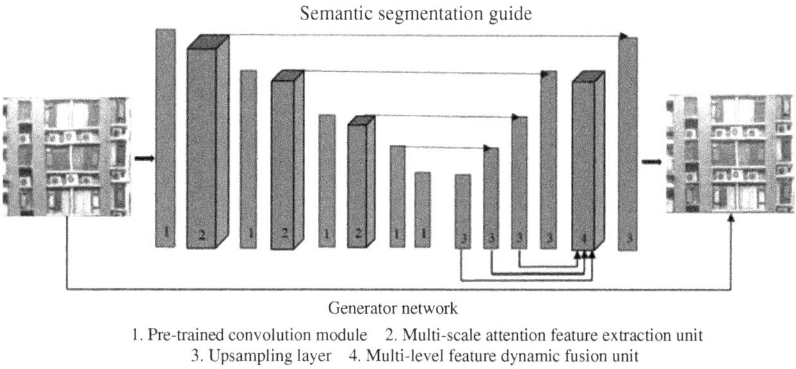

Figure 4.16. Generator structure of multi-scale GAN [130].

will be lost during the encoding and decoding process of the U-Net, the skip connections as hierarchical semantic guidance are added to the U-Net, and the shallow networks with more local information are combined with the corresponding deep network, and both the semantic information of the high-level features and the fine-grained features of the low-level features are made fuller use of to enhance the visual detail features of the reconstructed images. The generator introducing a skip connection directly from input to output can make the model focus on learning the residual.

4.5.2.2 *Multi-scale attention feature extraction unit*

In the turbulent image restoration algorithms, the geometric distortion and blur of the turbulent images have structural information at different scales. Conventional convolution for feature extraction isn't enough to fully restore the images. Therefore, a multi-scale attention feature extraction module is introduced to process feature information on the receptive field of different scales. The channel attention mechanism focuses on the relationship between channel features, and mining and learning the key contents of the images. The multi-scale attention feature extraction unit, as shown in Figure 4.17, is composed of multi-branch convolution layers and an attention layer. The multi-branch convolution layers correspond to the receptive fields of different sizes and can extract multiple features [72]. The attention layer [73] fully learns the important information in

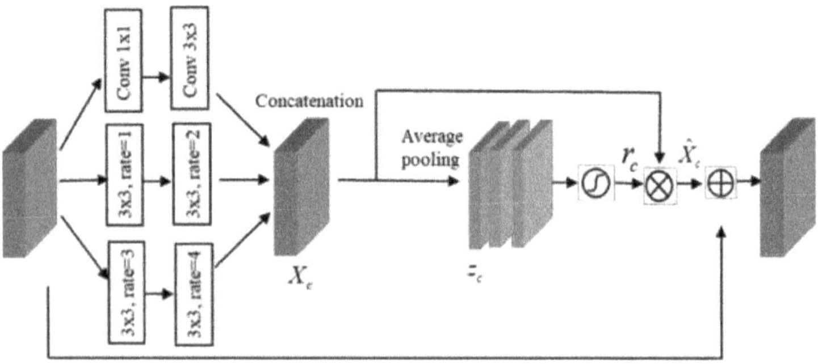

Figure 4.17. Multi-scale attention feature extraction unit [68].

the degraded images to ensure the accuracy and clarity of the reconstructed images.

The multi-branch convolutional layers are composed of parallel dilated convolutions of different sizes. The receptive fields of the three branches are 3×3, 7×7, and 15×15, respectively. Feature extraction is performed on the input feature map at the same time. After obtaining information feature maps of the different scales, the cascaded feature maps are readjusted to the input size through convolution operations.

In feature extraction, in order to distinguish between the low-frequency part (smooth or flat area) and the high-frequency part (such as lines, edges, textures) of the images, it is necessary to focus on and learn the key content of the images and to introduce the attention mechanism to generate different attention to each channel feature. First, based on the global context information of each channel, the global average pooling is selected to compress the spatial information of each channel. The compressed expression is written as

$$z_c = G_p(X_c) = \frac{1}{H \times W} \sum_{i=1}^{H} \sum_{j=1}^{W} X_c(i, j), \tag{4.5.2}$$

where X_c represents the aggregated convolutional feature map, whose size is $H \times W \times C$. z_c denotes the compressed global pooling layer, whose size is reduced to $1 \times 1 \times C$. The gating principle is implemented by using the ReLU activation function and the Sigmoid function to learn the nonlinear synergy and mutual exclusion relationships between channels. The attention mechanism can be expressed as

$$r_c = \text{sigmoid}(\text{Conv}(\text{ReLU}(\text{Conv}(z_c)))), \tag{4.5.3}$$

$$\widehat{X}_c = r_c \times X_c, \tag{4.5.4}$$

where r_c is the excitation weight and X_c represents the feature map after the attention mechanism is adjusted. According to Equation (4.5.3), first, the input z_c of the global pooling layer passes through the downsampling convolution layer and the ReLU activation function in turn. Second, the upsampling convolution layer restores the number of channels. Finally, the channel excitation weight r_c is obtained by using the Sigmoid function. According to Equation (4.5.4), the feature map X_c of the channels in the aggregated convolution layer is multiplied by different weights to obtain the output $\widehat{X}_c$ of adaptively adjusting the channel attention.

4.5.2.3 *Multi-level feature dynamic fusion unit*

In the upsampling part of the generator G, the feature maps at different levels contain different instance information [74]. In order to strengthen the information transmission between the feature maps at different levels, a multi-level feature fusion unit is adopted. There are large differences in the activation levels of the corresponding areas of the feature maps at different levels. When multi-level feature maps are fused, the conflict between features at different levels will interfere with information transfer and reduce the effectiveness of the feature fusion. To address the issue of conflicting features at different levels, a dynamic fusion network structure is presented, which assigns different weights to the spatial positions of feature maps and sifts effective features and filters conflicting information through learning. First, the feature maps of the different scales are upsampled to the same size, and spatial weights are set for feature maps at different levels during fusion to find the optimal fusion strategy, which can be specifically expressed as

$$F^* = w_1 \cdot F^{1\uparrow} + w_2 \cdot F^{2\uparrow} + w_3 \cdot F^{3\uparrow} + w_4 \cdot F^{4\uparrow}, \tag{4.5.5}$$

where $F^{i\uparrow}$ represents the standard feature map after the ith feature map is upsampled and adjusted to a uniform size, and the features of

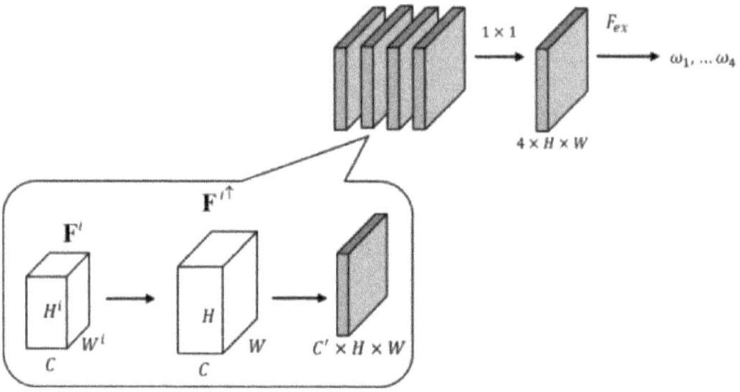

Figure 4.18. Multi-level feature dynamic fusion unit structure.

all levels are dynamically fused through adaptive weight allocation to output the final feature map F^*.

The learning method of weights w_i is shown in Figure 4.18. The spatial information of the standard feature map is compressed through the downsampling convolution layer, and the corresponding compressed convolution layers of the four different levels of feature maps are cascaded, and the feature information at the same position is mapped using 1×1 convolution. Finally, the parameters of the network are standardized by the Softmax function to obtain the spatial weight information, i.e.,

$$w_i = \frac{\exp(w_i')}{\sum \exp(w_i')}. \tag{4.5.6}$$

The learned feature map has adaptive spatial weights $w_i \in [0, 1]$ and the sum is equal to 1.

4.5.2.4 *Loss function*

During the training process, the loss function is employed for measuring the difference between the generated images and the real images. The loss function of the GAN includes adversarial loss and content loss. Among them, the adversarial loss adopts the adversarial loss in the LSGAN, i.e.,

$$\text{Loss}_{\text{GAN}} = \mathbb{E}[\|\log(D(I_{\text{gt}})) - 1\|_2^2] + \mathbb{E}[\|\log(D(I_{\text{gen}}))\|_2^2], \tag{4.5.7}$$

where I_{gt} represents the real images and I_{gen} represents the generated images. In terms of the content loss of image reconstruction, loss function Loss_{MSE} based on Mean Square Error (MSE) of the generated images and the real images is selected to obtain a higher PSNR. At the same time, in order to eliminate artifacts and promote the recovery of high-frequency details in the images, the reconstructed images must have a higher visual realism based on introducing the visual loss $\text{Loss}_{\text{pere}}$. Whereas sensory loss is achieved through a pretrained VGG19 model, which inputs the generated images and the real images into the VGG model, respectively, and then the Euclidean distance of the corresponding feature map after passing through the VGG model is calculated by

$$\text{Loss}_{\text{MSE}} = \frac{1}{M \times N} \sum_{i=1}^{m} \sum_{j=1}^{n} ||I_{gt,ij} - I_{\text{gen},ij}||^2, \tag{4.5.8}$$

$$\text{Loss}_{\text{pere}} = \sum_{i=1}^{L} \frac{1}{H_i W_i C_i} ||\phi^{(i)}(I_{\text{gt}}) - \phi^{(i)}(I_{\text{gen}})||_2^2, \tag{4.5.9}$$

where ϕ represents the pretrained VGG19 model. The overall loss function is defined as

$$\text{Loss} = \alpha \text{Loss}_{\text{MSE}} + \beta \text{Loss}_{\text{pere}} + \gamma \, \text{Loss}_{\text{adv}}, \tag{4.5.10}$$

where α, β, and γ are constant, and $0 < \alpha, \beta, \gamma < 1$.

4.6 Example 4: Motion Blurred Image Restoration Based on Deep Residual GAN

Early blurred image restoration problems often assumed that the degradation model was known and were solved by building mathematical models and algorithms. The degradation model of the general images is shown in Figure 4.19.

$$B = I \otimes K + N \tag{4.6.1}$$

In Figure 4.19, B and I represent the blurred images and the clear images, respectively. K is regarded as the blur kernel, $\otimes$ stands for the convolution operation, and N denotes the noises. Usually, according to whether the blur kernel is known, the image deblurring problem can be divided into two categories: non-blind deblurring of images

$$B = I \otimes K + N$$

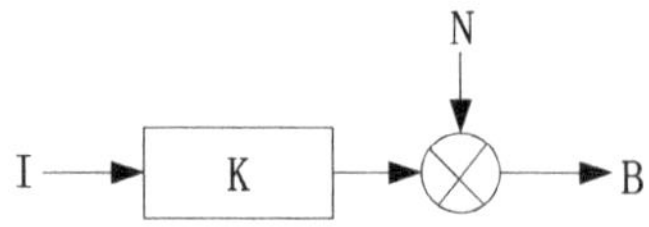

Figure 4.19. The degradation model of general images.

and blind deblurring. Non-blind deblurring of images means that it is necessary to know some prior information, such as the blur kernel. At this time, some classic algorithms, such as the Lucy–Richardson algorithm, Wiener filter, or Tikhonov filter, can be used to perform an estimate of the clear images via deconvolution operation so as to achieve image restoration. In the image process, when the motion trajectory of the camera or the object is unknown, the restoration process of the blurred images is called blind deblurring of images. Since blind deblurring of images lacks prior information about the images, it has a higher pathological property than non-blind deblurring. In Figure 4.19, all pixels in the clear images are convolved with the same blur kernel, which indicates that the blur degree of the images is globally consistent. However, the assumption of a globally consistent blur kernel, proposed by Levin *et al.* [75], is inconsistent with the motion blur caused by the camera or the object jitter. Most blurs in real life are non-uniform blurs. Therefore, in this section, the blind restoration problem of non-uniform blurred images is mainly studied.

It is a seriously ill-posed problem to find the blur kernel corresponding to each pixel in a non-uniformly blurred image, so most of the existing algorithms rely on the prior information of the images and assumptions on the blur sources, which solve the blur caused by camera shake by considering the uniformity of the blur image. For example, a representative traditional method for non-uniform blur was proposed by Whyte *et al.* [76], and the blur, assumed by Gupta *et al.* [77], is only caused by 3D camera motion. Kim and Lee [78] approximated the blur kernel as a local linear and proposed a method to jointly estimate the latent images and local linear motion. In recent years, the dark channel-based image deblurring method, proposed by Pan *et al.* [79], has achieved good results, but the estimated blur kernel is still inaccurate, especially in the case of sudden motion and occlusion. Inaccurate blur kernels will directly affect the quality of the restored image and easily cause ringing artifacts in the images. The previous research has shown that Convolutional Neural Network (CNN) has been applied to image deblurring and the result of image deblurring is good [80,81]. Due to the lack of real blurred images and real clear images in supervised learning, blurred images generated by convolving synthetic blur kernels with clear images are usually used. A classification CNN was proposed in Ref. [82] to estimate the local linear blur kernel. Therefore, the CNN-based model performs well in certain specific types of blurry images but is not suitable for common spatially varying blur [83].

Through the above image deblurring method, it has been found that in order to restore blurred images, it is essential to accurately estimate the blur kernel. However, the method of estimating the blur kernel is very sensitive to image noises, so the process of estimating the blur kernel needs to be carefully designed. In addition, the wrong blur kernel will directly affect the quality of the restored images and cause ringing; it is only effective for the CNN to remove image blur under specific models and a limited scope of use. Therefore, based on the CNN model of the GAN, an end-to-end approach is adopted. The input of the blurred image can directly obtain the restored images without the need for blur kernel estimation. At the same time, the use of the CNN for deblurring is no longer limited to a specific blur model.

4.6.1 Deep residual GANs

In the GAN model, the generator G captures the distribution of the sample data x, and the noise z that obeys a certain distribution (uniform distribution, Gaussian distribution, etc.) is used to generate a sample similar to the real training data. The effect is to make it as similar to the real sample as possible, and the discriminator D is a binary classifier that estimates the probability that the samples come from the training data (not the generated data). If the samples come from the real training data, the output of the discriminator D is a high probability. Otherwise, its output is a low probability. The deep residual GAN model designed by the classic GAN model is shown in Figure 4.20.

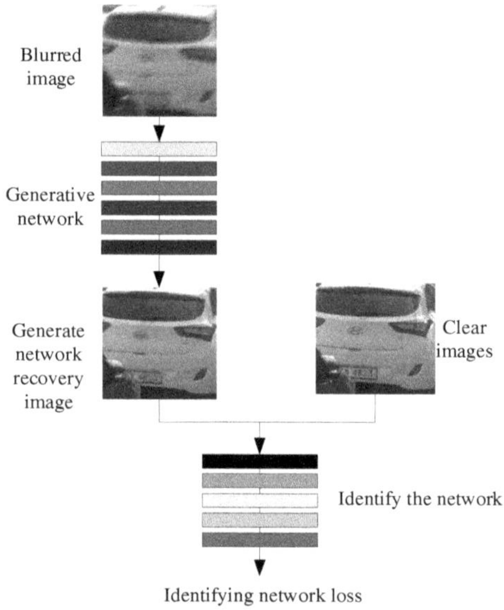

Figure 4.20. Deep residual generative adversarial network.

4.6.1.1 *Generative network*

In the generative network, the original blurred image is replaced with random noise and then input into the network, and the image dimension remains unchanged throughout the network. Although keeping the image dimension unchanged may consume a lot of computer memory, it can avoid the checkerboard effect caused by using deconvolution operations on the images.

The generative network designed in this section is shown in Figure 4.21. It adopts the network structure of U-net encoding-self-encoding, which mainly includes Resblock, Eblock, and Dblock.

The structure of the ResBlock in Figure 4.21 is similar to the traditional residual block structure, as shown in Figure 4.22. In the ResBlock, the Batch Normalization (BN) in the traditional residual block is removed, and due to the similarity between blurry and clear images, the mapping relationship between blurred images and clear images can be learned through the residual network. In addition, the rectified linear unit, after removing the shortcut connection of the original residual block, can improve the convergence speed during training.

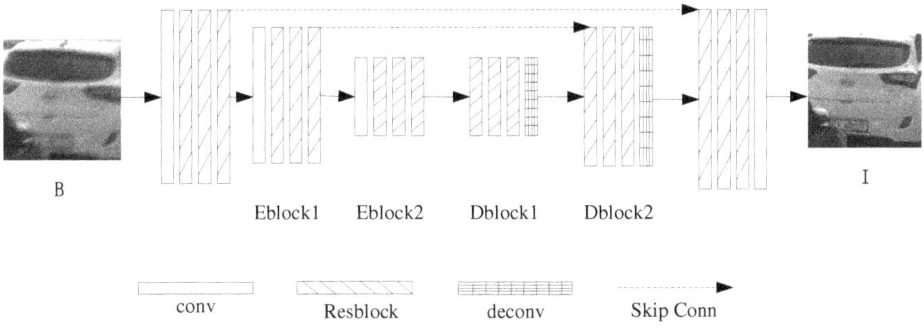

Figure 4.21. Generative network structure.

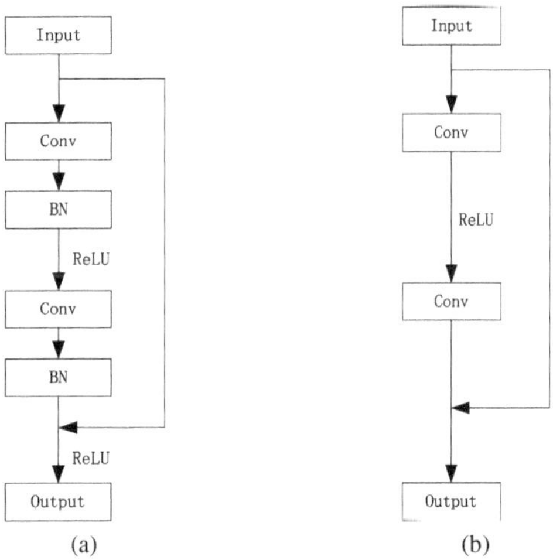

Figure 4.22. The conventional residual block and the Resblock in this section. (a) Traditional residual block. (b) Resblock in this section.

Both the Eblock and the Dblock form a U-net structure, where the Eblock is mainly composed of a convolution layer and three Resblocks. Whereas the Dblock corresponds to the Eblock and consists of three Resblocks and a deconvolution layer. By adopting the U-net structure, the information between feature maps can be combined using skip connections between Eblock and Dblock, which is beneficial for gradient propagation and accelerating model convergence.

The generative networks all use convolution kernels, and the number of convolution kernels in each layer is set to 32 for the first three layers, 64 for the Eblock1, 128 for the Eblock2, 128 for the Dblock1,

64 for the Dblock2, and 32 for the last four layers. In the generative network, there are two locations where skip connections are used to prevent gradient diffusion and degradation caused by increasing the number of network layers.

4.6.1.2 *Discriminative network*

The discriminative network is a binary classifier used to estimate the probability that the samples come from the training data (rather than generated data). The discriminative network structure is shown in Table 4.1.

The identification network in this section has a total of five convolutional layers, each of which contains Batch Normalization (BN), and the LReLU function is selected as the activation function. The first three layers use a convolution kernel size of 4×4 and a step size of 2. The convolution kernel size of the fourth layer remains unchanged, and the step size is set to 1. The fifth layer uses a convolution kernel size of 1×1 and a step size of 1, and then the Sigmoid activation function is used for calculating a class label of 0 or 1.

4.6.1.3 *Loss function*

The generator G in the GAN can generate clear and realistic images, and then the discriminator D can determine whether an image is "real". The output of the real images is used as the input of the discriminator D and is classified according to whether it is a clear image or a deblurred image. Throughout the training process, the

Table 4.1. Discriminative network structure.

Convolutional Layer	Number of convolution kernels	Convolution kernel size	Stride	Activation function
Conv1	64	4	2	Leaky-ReLU(LReLU)
Conv2	128	4	2	Leaky-ReLU(LReLU)
Conv3	256	4	2	Leaky-ReLU(LReLU)
Conv4	512	4	1	Leaky-ReLU(LReLU)
Conv5	512	1	1	Leaky-ReLU(LReLU)
Sigmoid	—	—	—	—

goal of the generator G is to generate as many real images as possible to deceive the discriminator D. The goal of the discriminator D is to distinguish between the fake images generated by the generator G and the real images as many as possible. The generator G and the discriminator D constitute a dynamic "game process" and eventually reach a balance point. Therefore, the loss function in this section is mainly composed of two parts: the generator loss and the discriminator loss.

The loss function of the discriminator D is defined as

$$\text{Loss}_D = -\lambda \times (\mathbb{E}[\log D(I)] + \mathbb{E}[\log(1 - D(G(B)))]), \tag{4.6.2}$$

where the λ value is equal to 0.001, I and $D(I)$ represent a clear real image and its corresponding to the probability that the discriminative network judges whether the image is "real", and B and $D(G(B))$ represent a blurred image and its corresponding to the probability that the discriminative network judges whether the image generated by the generation network is "real".

The loss function of the generator G is defined as

$$\text{Loss}_G = -\eta\mathbb{E}[\log(1 - D(B))] + \text{Loss}_{\text{MSE}}, \tag{4.6.3}$$

$$\text{Loss}_{\text{MSE}} = \frac{1}{\text{HWC}}||G(B) - I||_2^2, \tag{4.6.4}$$

where the η value is set to 0.001, $G(B)$ represents the image generated by the generator G, C, W, and H represent the number of channels, width, and height of the images, respectively. The loss function of the generator G consists of two parts, namely the loss function of the conventional discriminator D and the MSE function, which aims to reduce the ringing effect of the restored image.

4.6.2 Simulation experiments and their result analyses

4.6.2.1 *Experimental setup*

In this section, the proposed model was verified using the TensorFlow framework. All experiments were implemented on a workstation equipped with an Intel Core i7-7700 processor and an NVIDIA GTX 1070Ti graphics card.

The GoPro dataset, published by Nah *et al.*, includes 3214 pairs of blurry clear image pairs. Among them, 2103 pairs of the blurry clear images were used as the training dataset, and the remaining 1111 pairs of the blurry clear images were selected as the testing dataset. The input images in the network were the color image of size randomly cropped from the dataset.

The Adam algorithm was utilized to optimize the loss function during training, and the proposed model was converged by the stochastic gradient descent (SGD) algorithm. In the experiments, the global learning rate was set to 1e-4 and the momentum was set to 0.3. After 73,000 iterations, the CNN model converged.

4.6.2.2 *Experimental results*

In this section and other representative non-uniform image motion blur restoration methods, the GoPro dataset was all used. When training the network model, 2103 pairs of images were used as the training dataset and 1111 pairs of images were used as the testing set. The performance of the proposed model was carried out and compared with better image motion blur restoration effects in recent years, as shown in Figures 4.23 and 4.24. Among them, Figures 4.23(a) and 4.24(a) represent the blurred image, Figures 4.23(b) and 4.24(b) represent the original clear image, Figures 4.23(c) and 4.24(c) are the experimental result of the proposed model in this section, Figures 4.23(d) and 4.24(d) are the experimental result in Ref. [76], Figures 4.23(e) and 4.24(e) are the experimental result in Ref. [79], and Figures 4.23(f) and 4.24(f) are the experimental result in Ref. [52].

From Figures 4.23 and 4.24, it has been shown that the effect of the proposed model in removing image motion blur in this section is better than other methods. The ringing effect of the proposed model in the restored images [76] and [52] is more obvious, yet the restoration of the image details isn't very good. The ringing effect of the proposed model in the restored image [79] is not very obvious, and the restoration of the details is also poor. The PSNR and the SSIM are used as evaluation indicators, as shown in Table 4.2. Table 4.2 shows that the image deblurring effect of the proposed model in this section is better.

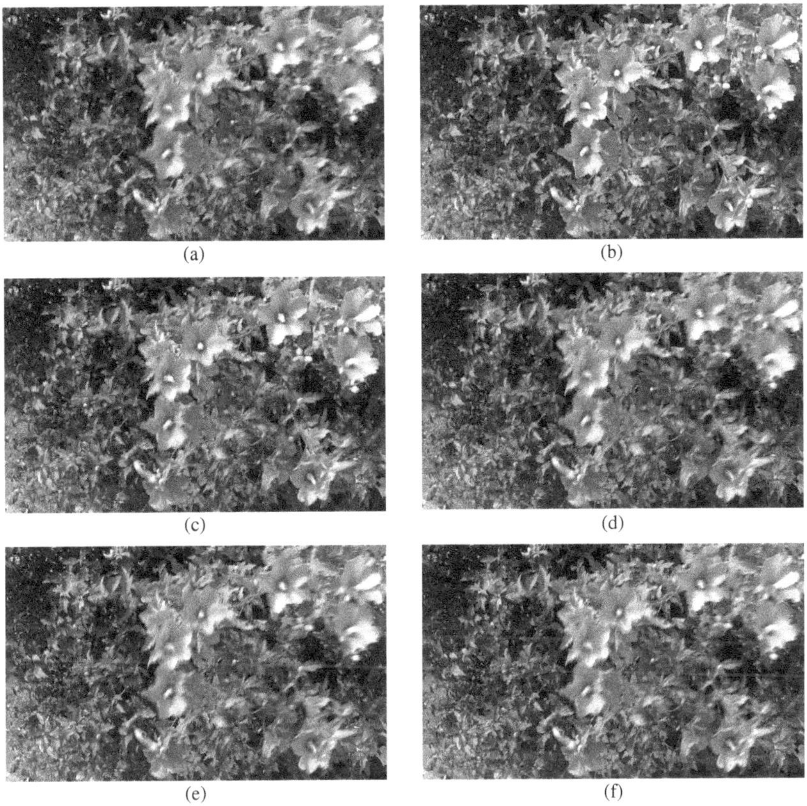

Figure 4.23. Image (flower). (a) Blurred image. (b) Original image. (c) The restored image in this section. (d) The experimental results in Ref. [76]. (e) Experimental results from Ref. [79]. (f) Experimental results in Ref. [52].

In this section, the GAN is composed of two convolutional neural networks, the generator G, and the discriminator D. In order to show that the GAN is better than the direct use of the CNN, the impact of the discriminator D on the experimental results is studied. The discriminator D is removed to make the generator G alone as an experiment and compared with the complete discriminator D, as shown in Figure 4.25. Among them, Figure 4.25(a) is the original image, Figure 4.25(b) is the blurred image, Figure 4.25(c) shows the

Figure 4.24. Image (road). (a) Blurred image, (b) Original image. (c) This section restores the image. (d) Experimental results in Ref. [76]. (e) The experimental results of Ref. [41]. (f) Experimental results in Ref. [52].

Table 4.2. Image restoration quality evaluation results.

	Evaluation indicators	Literature [76]	Literature [79]	Literature [52]	Methods in this section
Image Flower	PSNR(dB)	24.0860	23.4903	23.4616	27.3588
	SSIM	0.7669	0.7836	0.7587	0.8905
Image road	PSNR(dB)	23.6357	22.5833	24.5398	26.9535
	SSIM	0.7631	0.7414	0.7387	0.9194

Figure 4.25. Comparison of experimental results with and without discriminator D. (a) Original image. (b) Fuzzy image. (c) Experimental results without discriminator D. (d) There are experimental results with discriminator D.

Table 4.3. Image restoration quality evaluation results.

Evaluation indicators	Model without discriminator D	The proposed model (i.e., with discriminator D)
PSNR(dB)	29.8398	32.4774
SSIM	0.9232	0.9506

experimental results of the original image without the discriminator D, and Figure 4.25(d) shows the experimental results with the discriminator D, i.e., the proposed model. Table 4.3 shows the two indicators for evaluating images, namely the PSNR and the SSIM. From the visual effect and evaluation indicators, it can be seen that the proposed model in this section can restore the image details more clearly, while the structure without the discriminator D and only the generator G has a poor restoration effect.

In summary, compared with existing models, the proposed model in this section is an end-to-end structure, which avoids the problem of poor image restoration effect caused by blur kernel estimation bias in traditional models. Based on the GAN model, the designed deep residual GAN can effectively remove image motion blur and obtain clear images.

Chapter 5

Deep Restricted Boltzmann Machine

Summary

Starting from the structure, principles, and searching mechanism of the Boltzmann machine, this chapter analyzes the working mechanisms of the Restricted Boltzmann Machine (RBM), Sparse RBM (SRBM), Competitive SRBM (CSRBM), and classification RBM. By introducing Principal Component Analysis (PCA) method into the RBM, the performance and characteristics of 2D-RBM and image classification RBM networks are proposed. Take the RBM as a tool, gait feature extraction and recognition methods are studied, and simulation experiments and result analyses are provided.

5.1 Introduction

The Deep Boltzmann Machine (DBM) is a deep learning model based on the Restricted Boltzmann Machine (RBM), which is essentially a specially constructed neural network. The DBM is formed by stacking multiple layers of the RBMs, with bidirectional connections between adjacent layers.

5.2 Boltzmann Machine

The Boltzmann Machine (BM) is a stochastic neural network that incorporates simulated annealing. Its main differences from other neural networks are as follows:

(1) During the learning (training) phase, the stochastic network does not adjust weights based on deterministic algorithms but needs to modify them according to a probability distribution.
(2) During the operation (prediction) phase, the stochastic network does not follow deterministic state transition equations but determines the state transitions based on a probability distribution. The net input of a neuron cannot determine whether its state takes 1 or 0, but it can determine the probability of its state taking 1 or 0. This is the basic concept of an algorithm for random neural networks.

5.2.1 Boltzmann machine and operation principles

5.2.1.1 *Structure of Boltzmann machine*

The structure of the BM network is between the fully interconnected Discrete Hopfield Neural Network (DHNN) and the hierarchical structure of the Backpropagation (BP) network. Formally, it is similar to the single-layer feedback DHNN, with symmetrical weights $w_{ij} = w_{ji}$ and no feedback of its own, i.e., $w_{ii} = 0$. Functionally, similar to a three-layer BP network, it consists of input neurons (or nodes), output nodes, and hidden layer neurons. Generally, input and output nodes are referred to as visible neurons, and hidden nodes are referred to as invisible neurons. During training, input and output nodes collect training samples, while hidden nodes mainly play an auxiliary role in achieving the connection between input and output so that the training set can be reproduced in visible units, as shown in Figure 5.1. There is no clear hierarchy between the three types of neurons in the BM network.

5.2.1.2 *Principles of Boltzmann machine*

Each neuron in the BM network exhibits stochastic excitation or inhibition, with probabilities determined by the input neurons. A formal description of a single neuron in the BM network is shown in Figure 5.2.

The net input of a single neuron in the BM network is given by

$$y_j = \sum_i w_{ji} x_i + b_j, \tag{5.2.1}$$

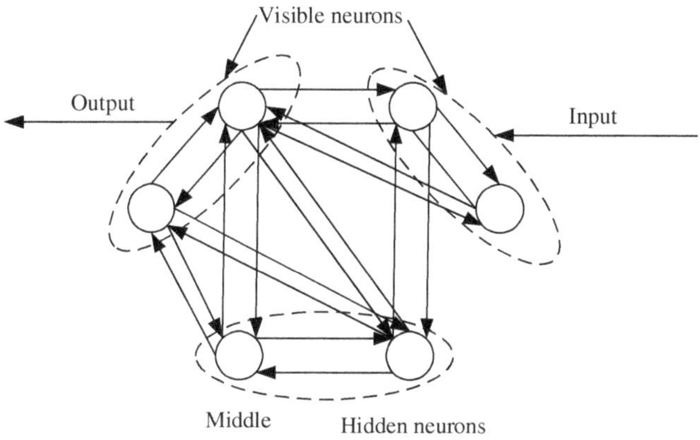

Figure 5.1. Structure of the BM network.

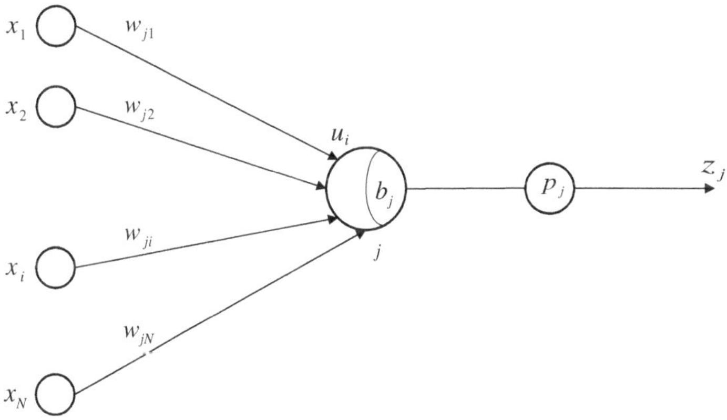

Figure 5.2. Single neuron.

where x_i is the ith input of the input layer, w_{ji} is the connection weight between the jth neuron and the ith neuron, and b_j is the bias of the jth neuron.

Unlike the DHNN, the net input cannot directly obtain a deterministic output state through the sign transfer function. The actual output state will occur according to a certain probability, and the jth neuron will output z_j according to the probability of taking 1 or 0. The probability of taking one is defined as

$$p_j(z_j = 1) = \frac{1}{1 + \mathrm{e}^{-y_j/T}}. \tag{5.2.2}$$

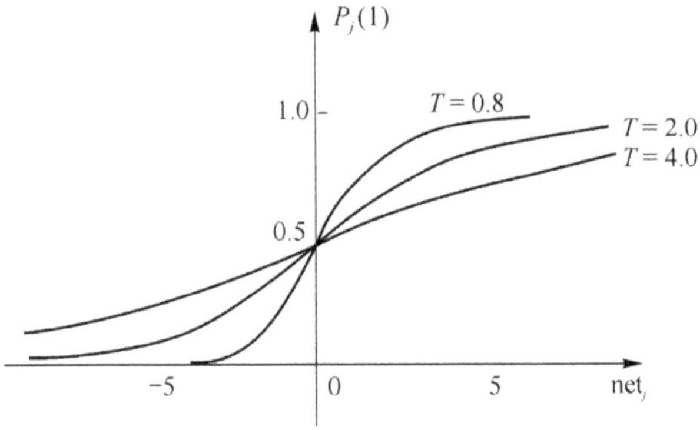

Figure 5.3. Effect of temperature (T) on activation probability.

Equation (5.2.2) gives the probability that the jth neuron outputs the state of taking one, while the probability outputting the state of zero can be subtracted from 1 to the probability of being in state 1. The larger the net input, the higher the probability of the neuron state of taking 1. The smaller the net input, the higher the probability of the neuron state of taking 0.

The temperature parameter (T) controls the steepness of the probability curve, as illustrated in Figure 5.3.

Figure 5.3 shows that at higher temperatures T, the probability curve is smoother, resulting in smaller differences between the probabilities of states of taking 0 and 1 for the same net input. At lower temperature T, the curve becomes steeper, the difference between probabilities becomes greater. When temperatures $T = 0$, the probability function degenerates into a sign function, and the output state of the neuron will have no randomness.

5.2.2 Network energy function and searching mechanism

The BM network employs the same energy function as the DHNN for describing the network state, i.e.,

$$E(k) = -\frac{1}{2}\boldsymbol{x}^{\mathrm{T}}(k)\boldsymbol{w}\boldsymbol{x}(k) + \boldsymbol{x}^{\mathrm{T}}(k)\boldsymbol{b} = -\frac{1}{2}\sum_{i=1}^{N_v}\sum_{j=1}^{J} w_{ji}x_i x_j + \sum_{j=1}^{J} b_j x_j. \tag{5.2.3}$$

Assuming that the BM network operates asynchronously, the jth neuron changes state each time, according to the following energy change:

$$\Delta E(k) = -\Delta x_j(k) y_j(k). \tag{5.2.4}$$

We discuss the following:

(1) When the net input is greater than 0, the probability of the neuron state taking 1 is greater than 0.5. If the original neuron state $x_j = 1$, then its state change quantity $\Delta x_j = 0$, as a result, leading to its energy change quantity $\Delta E = 0$. If the original neuron's state $x_j = 0$, then its state change quantity $\Delta x_j = 1$, thus resulting in its energy change quantity $\Delta E < 0$, and the energy decreases.
(2) When the net input is less than 0, the probability of the neuron state taking state 1 is smaller than 0.5. If the original neuron state $x_j = 0$, then its state change quantity $\Delta x_j = 0$, thereby leading to its energy change quantity $\Delta E = 0$. If the original neuron state $x_j = 1$, then its state change quantity $\Delta x_j = -1$, resulting in its energy change quantity $\Delta E < 0$, and the energy decreases.

The above various possible scenarios indicate that for the BM network, as the network state evolves, the energy of the network always changes in a decreasing direction in a probabilistic sense. This means that although the overall trend of network energy is evolving toward a decreasing direction, it cannot be ruled out that some neuron states may take on a small probability value, temporarily increasing network energy. It is precisely because of this possibility that the BM network has the ability to climb mountains and jump out of locally small valleys, which is the fundamental difference between the BM network and DHNN in terms of energy changes. Due to the use of a probabilistic random selection of neuron states, the BM network has the ability to continuously jump out of high position valleys and search for new low position valleys. This operating mode is called the searching mechanism, which means that the network continuously searches for lower energy minima during operation until it reaches the global minimum of energy. From the principle of simulated annealing, it can be seen that the continuous decrease in temperature T

can weaken the "climbing" ability of the network, which is an effective measure to ensure that the BM network of the function can successfully search for the global minimum energy.

5.2.3 Boltzmann distribution

In this section, we discuss Boltzmann distribution. When $x_j = 1$, the corresponding network energy is E_1, and when $x_j = 0$, the network energy is E_0. When x_j changes from one to zero, $\Delta x_j = -1$, so $E_0 - E_1 = \Delta E = y_j$. The probabilities of the corresponding state being 1 or 0 are as follows:

$$p_j(z_j = 1) = \frac{1}{1 + \mathrm{e}^{-y_j/T}} = \frac{1}{1 + \mathrm{e}^{-\Delta E/T}}, \tag{5.2.5}$$

$$p_j(z_j = 0) = 1 - p_j(z_j = 1) = \frac{\mathrm{e}^{-\Delta E/T}}{1 + \mathrm{e}^{-\Delta E/T}}, \tag{5.2.6}$$

$$\frac{p_j(z_j = 0)}{p_j(z_j = 1)} = \mathrm{e}^{-\Delta E/T} = \mathrm{e}^{-(E_0 - E_1)/T} = \frac{\mathrm{e}^{-E_0/T}}{\mathrm{e}^{-E_1/T}}. \tag{5.2.7}$$

Now, the above equations are extended by us to the relationship between the probability of any two states appearing in the network and their corresponding energy. We can get the relationship between their probabilities and energies as follows:

$$\frac{p(\alpha)}{p(\beta)} = \frac{\mathrm{e}^{-E_\alpha/T}}{\mathrm{e}^{-E_\beta/T}}. \tag{5.2.8}$$

This is the famous Boltzmann distribution. Equation (5.2.7) indicates that the probability of the BM network being in a certain state mainly depends on the energy in this state, and the lower the energy, the greater the probability. The probability of the BM network being in a certain state also depends on the temperature parameter T. The higher the temperature, the closer the probabilities of different states appearing, and making it easier for the network energy to jump out of the local minimum and search for the global minimum. The lower the temperature, the greater the difference in probabilities of different states appearing, and the less likely the network energy is to change, thus making the network search converge. This is precisely why the simulated annealing method is used to search for the global minimum.

When using the BM network for optimization calculation, the objective function can be constructed as the energy function of the network. To prevent the objective function from falling into local optima, the above simulated annealing algorithm is used to search for the optimal solution. At the beginning, the temperature is set very high, and the probability of the neuron state being 1 or 0 is almost equal. Therefore, the network energy can reach any possible state, including local minimum or global minimum. When the temperature drops, the probability of different states changes; the probability of low-energy states occurring is higher, while the probability of high-energy states occurring is lower. When the temperature gradually drops to 0, each neuron can either only take 1 or 0, and the state of the network freezes around the global minimum of the objective function. The corresponding network state is the optimal solution to the optimization problem.

When using the BM network for association, the probability of training samples appearing can be simulated by learning the probability of stable network states. According to the type of learning, the BM network can be divided into self-association and hetero-association. These two types of structures, as shown in Figure 5.4, can have zero hidden nodes and some lines are unidirectional.

Figure 5.4(a) shows a self-associative BM, while Figure 5.4(b) shows a hetero-associative BM network. The visible node V in the self-associative BM network is similar to the nodes in the DHNN, serving as both input and output nodes. The number of hidden nodes H is determined by the learning needs and can be at least 0. The

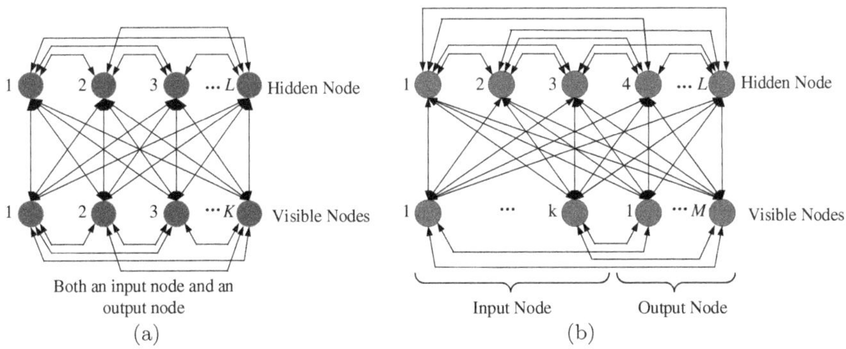

Figure 5.4. BM network types. (a) Autoassociation. (b) Hetero-association.

visible nodes V in the heterogeneous BM network need to be divided into the input node group I and output node group O according to their functions.

5.2.4 Boltzmann learning algorithm

5.2.4.1 *Learning process*

Through supervised learning, the BM networks can simulate the probability distributions of various patterns in the training set, thereby achieving associative memory. The learning objective is to reproduce the patterns in the training set with the same probability in the network state by adjusting the weights. The learning process can be divided into two stages: (1) positive learning stage or input period and (2) reverse learning stage or free activity period.

5.2.4.1.1 *Positive learning stage*

A pair of input–output patterns is input into the network, and the state of the network's input–output nodes is clamped to the desired state, as well as the hidden nodes are allowed to move freely to capture the corresponding patterns between the pattern pairs.

5.2.4.1.2 *Negative learning stage*

For heteroassociative learning, the input nodes are constrained while the hidden nodes and output nodes are allowed to move freely. For self-associative learning, both visible and hidden nodes can be freely active, and reflect in the corresponding rules of input and output of the network. This correspondence law is manifested as the average probability that the connected node states are all equal to one at the same time when the network reaches thermal equilibrium. The difference between the expected corresponding law and the simulated corresponding law is manifested as the difference between the average probabilities corresponding to the two learning stages, which serves as the basis for weight adjustment.

If let the BM network have M hidden nodes and N visible nodes, visible states X that nodes can express have 2^N (for heteroassociative tasks, X includes input and output patterns). Assuming that the training set provides L pairs of patterns, generally $L < N$, the training set uses a set of probability distributions to represent the probability of each pattern appearing denoted as $p(x_1), p(x_2), \ldots, p(x_L)$, which is exactly the expected network state probability distribution in the forward learning stages. The free-running network's distribution is $p'(x_1), p'(x_2), \ldots, p'(x_L)$. Training aims to align these distributions.

5.2.4.2 *Network thermal equilibrium state*

To estimate these probabilities, the BM network must repeatedly execute the simulated annealing algorithm until thermal equilibrium is reached. The detailed procedure is as follows:

Step 1: Clamp nodes

In the positive learning phase, a pair of training patterns x_L is used to clamp onto the visible nodes. In the negative learning phase, only the input portion of the training pattern is used to clamp onto the input nodes of the visible layer.

Step 2: Update the state of nodes

A free node j is randomly selected and its state is updated according to

$$x_j(k+1) = \begin{cases} 1, & x_j(k) = 0 \\ 0, & x_j(k) = 1 \end{cases}. \tag{5.2.9}$$

Step 3: Calculate the energy change

The energy change ΔE_j of the node j in the network is calculated by $\Delta E_j = -\Delta x_j(k) y_j(k)$.

Step 4: Determine whether the state continues to be updated

If the energy change $\Delta E_j < 0$, the state continues to update. If $\Delta E_j > 0$ and $p[x_j(k+1)] > \rho$, the new state is accepted at that time, otherwise the original state will be maintained. Here, $\rho \in (0, 1)$ is a predefined parameter that gradually decreases during the simulated annealing process. As shown in Equation (5.2.8), for a given constant ρ, if the condition $p[x_j(k+1)] > \rho$ must be met, it is necessary to make ΔE_j continuously decrease in training, so that the ability of the network to climb mountains is constantly decreasing

Step 5: Repeat execution

Steps from 2 to 4 are repeated until all free nodes have been sampled once.

Step 6: Set the initial temperature

According to the pre-selected cooling equation, the cooling law of the annealing algorithm is not uniformly specified, and the initial temperature is generally required.

5.2.5 BM operation steps

After training is completed, the process of obtaining output based on input data is called the running state, and the weights remain unchanged during the running process. Its operating steps are very similar to the simulated annealing algorithm, but the difference is that the simulated annealing algorithm needs to define different cost functions for different problems, while the cost function of the Boltzmann machine is an energy function.

Step 1: Initialization

The number of neurons in the BM network is set to N. Connection weight w_{ji} between neuron j and neuron i and all neuron states are initialized. We set the initial temperature T_0 and the termination temperature T_{final}.

Step 2: Update state at temperature T_m

The state of neuron j at temperature T_m is calculated by

$$y_j = \sum_{i=1}^{N} w_{ji} x_i + b_j. \tag{5.2.10}$$

If $y_i > 0$, i.e., $\Delta E_j < 0$, it indicates that energy has a reduction tendency; at this point, the next state of neuron j is set to 1.

If $y_j < 0$, next state can be determined based on probability, i.e.,

$$p_j(z_j = 1) = \frac{1}{1 + \mathrm{e}^{-y_j/T_m}}. \tag{5.2.11}$$

If $p_j(z_j = 1)$ is greater than or equal to a given threshold, $z_j = 1$ is taken as the next state value of neuron j, otherwise it remains the next state value of neuron j. During this process, the states of other nodes remain unchanged.

Step 3: Check the termination conditions

The same temperature T_m in the small loop is used. If the current state has reached thermal equilibrium, the running of the program goes to step 4 for cooling, otherwise it goes to step 2 and continues to randomly select a neuron for iteration. Otherwise, it proceeds to the next step.

Step 4: Decrease temperature

Temperature is gradually decreased according to the specified pattern and the termination conditions of the large cycle are checked. If the temperature has reached the termination temperature, the algorithm ends. Otherwise, it goes to Step 2 to continue the calculation.

In this section, the selection method of the initial temperature T_0 is as follows: N neurons in the network can be randomly selected to take the variance of their energy, or some neurons can be randomly selected and the maximum difference between their energies can be taken.

5.3 Sparse Restricted Boltzmann Machine and Competitive Learning

5.3.1 Restricted Boltzmann machine and sparse restricted Boltzmann machine

5.3.1.1 *Restricted Boltzmann machine*

Restricted Boltzmann Machine (RBM) is a two-layer neural network composed of unit connections within the layers of the BM network. As an undirected graph model, in the RBM, the visible unit layer $\boldsymbol{v}$ is the observed data, and the hidden unit layer $\boldsymbol{h}$ is the feature detector, as shown in Figure 5.5.

The energy of an RBM configuration is represented as

$$E(\boldsymbol{v}, \boldsymbol{h}|\boldsymbol{\theta}) = -\sum_{i=1}^{N_v} b_i v_i - \sum_{j=1}^{J} c_j h_j - \sum_{i=1}^{N_v}\sum_{j=1}^{J} v_i w_{ji} h_j, \tag{5.3.1}$$

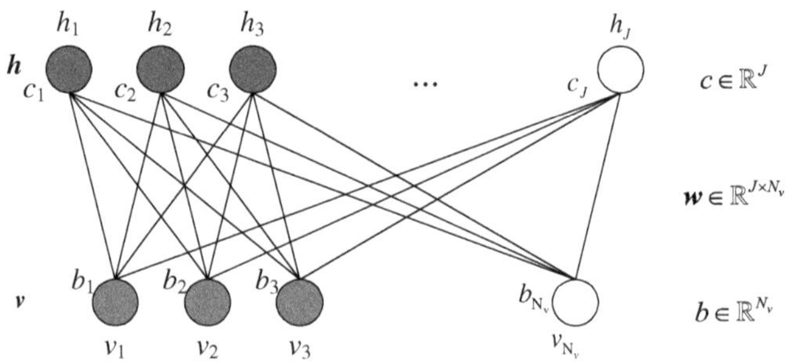

Figure 5.5. Structure of RBM.

where $\boldsymbol{\theta} = \{w_{ji}, a_i, b_j\}$ represents the parameters of the RBM, w_{ji} denotes the weights of the edges between the hidden unit h_j and the visible unit v_i, b_i is the bias for the ith visible unit, and c_j is the bias for the jth hidden unit h_j. Given the joint configuration energy of $\boldsymbol{v}$ and $\boldsymbol{h}$, the joint probability of $\boldsymbol{v}$ and $\boldsymbol{h}$ can be calculated by

$$p_{\boldsymbol{\theta}}(\boldsymbol{v}, \boldsymbol{h}) = \frac{1}{Z(\boldsymbol{\theta})} \exp(-E(\boldsymbol{v}, \boldsymbol{h}|\boldsymbol{\theta})), \tag{5.3.2}$$

where $Z(\boldsymbol{\theta}) = \sum \exp(-E(\boldsymbol{v}, \boldsymbol{h}|\boldsymbol{\theta}))$ is the partition function and the marginal probability.

Substituting Equation (5.3.1) into Equation (5.3.2), we obtain

$$p_{\boldsymbol{\theta}}(\boldsymbol{v}, \boldsymbol{h}) = \frac{1}{Z(\boldsymbol{\theta})} \exp\left(\sum_{j=1}^{J}\sum_{i=1}^{N_v} w_{ji} v_i h_j + \sum_{i=1}^{N_v} v_i b_i + \sum_{j=1}^{J} h_j c_j\right). \tag{5.3.3}$$

In practical applications, the focus is on the probability distribution $p_{\boldsymbol{\theta}}(\boldsymbol{v})$ of the observed data $\boldsymbol{v}$, and its marginal distribution over $p_{\boldsymbol{\theta}}(\boldsymbol{v}, \boldsymbol{h})$ is given by

$$\begin{aligned} p_{\boldsymbol{\theta}}(\boldsymbol{v}) &= \sum_{h} p_{\boldsymbol{\theta}}(\boldsymbol{v}, \boldsymbol{h}) \\ &= \frac{1}{Z(\boldsymbol{\theta})} \sum_{h} \exp(-E(\boldsymbol{v}, \boldsymbol{h}|\boldsymbol{\theta})) \\ &= \frac{1}{Z(\boldsymbol{\theta})} \sum_{h} \exp(\boldsymbol{v}^{\boldsymbol{T}} \boldsymbol{w} \boldsymbol{h} + \boldsymbol{c}^{\boldsymbol{T}} \boldsymbol{h} + \boldsymbol{b}^{\boldsymbol{T}} \boldsymbol{v}). \end{aligned} \tag{5.3.4}$$

By maximizing $p_{\boldsymbol{\theta}}(\boldsymbol{v})$, we can obtain the parameters of the RBM. Maximizing the $p_{\boldsymbol{\theta}}(\boldsymbol{v})$ is equivalent to maximizing $\log(p_{\boldsymbol{\theta}}(\boldsymbol{v}))$. The loss function is defined as

$$Loss(\boldsymbol{\theta}) = \frac{1}{N_v}\sum_{i=1}^{N_v} p_{\boldsymbol{\theta}}(v_i). \tag{5.3.5}$$

The parameters can be optimized by the gradient of the loss function $Loss(\boldsymbol{\theta})$ with respect to the weight coefficient w_{ji}, i.e.,

$$\begin{aligned}\frac{\partial Loss(\boldsymbol{\theta})}{\partial w_{ji}} &= \frac{1}{N_v}\sum_{i=1}^{N_v}\left(\frac{\partial}{\partial w_{ji}}\log\left(\sum_h \exp(\boldsymbol{v_i^T w h} + \boldsymbol{c^T h} + \boldsymbol{b^T v_i})\right)\right.\\ &\quad - \frac{\partial}{\partial w_{ji}}\log Z(\boldsymbol{\theta})\\ &= -\frac{1}{d}\sum_{k=1}^{d} v_{ik}p(\boldsymbol{h_j}|\boldsymbol{v}(k)) + \sum_{v,h} v_i h_j p_{\boldsymbol{\theta}}(\boldsymbol{v},\boldsymbol{h}).\end{aligned} \tag{5.3.6}$$

Since there are no connections between units within the same layer, the posterior probability distributions of the hidden units and visible units can be conveniently derived as

$$p(h_j = 1|v) = \frac{1}{1+\exp\left(-\sum_i w_{ji}v_i - c_j\right)}, \tag{5.3.7}$$

$$p(v_i = 1|h) = \frac{1}{1+\exp\left(-\sum_j w_{ji}v_j - b_i\right)}. \tag{5.3.8}$$

In Equation (5.3.6), the first term on the right-hand side is referred to as the positive phase (PP), which can be directly calculated by substituting the training data into Equations (5.3.7) and (5.3.8). The second term is called the negative phase (NP), where the expectation $p_\theta(\boldsymbol{v},\boldsymbol{h})$ cannot be derived analytically. Therefore, Hinton proposed the Contrastive Divergence (CD) algorithm [84], which takes each training data as the initial state, performs several state transitions through block Gibbs sampling and then uses the transitioned state as a sample to estimate the mean of the NP. Hinton also demonstrated through experiments that in practical applications, even a single-state transition is sufficient to ensure good estimation performance.

In a multi-layer RBM network, all adjacent two-layer structures can be considered as one RBM, while the lower-level hidden layer is used as the input layer for the adjacent higher-level hidden layer. A greedy layer-by-layer training algorithm is adopted to train one RBM from bottom to top each time based on image feature vectors as input and can preliminarily determine the spatial parameters $\{w_{ji}, b_j, c_i\}$ of the entire RBM network. Afterwards, it is necessary to perform overall fine-tuning and optimization of the parameters between all layers based on the reverse BP neural network. After repeatedly training, the spatial parameters between layers are continuously adjusted to achieve a balanced state of the network.

5.3.1.2 *Sparse restricted Boltzmann machine*

The Sparse Restricted Boltzmann Machine (SRBM) optimizes the training objective of the RBM by adding a sparse penalty factor to the maximum likelihood objective function of the RBM so that the average activation probability of all hidden units approaches a small constant p (i.e., sparse objective). When given a training sample V, the optimization problem of sparse RBM becomes

$$\underset{\{w_{ji},b_i,c_j\}}{\text{minimize}} \left\{ -\sum_{l=1}^{N_h} \log \sum_{h} p(\boldsymbol{v}^{(l)}, \boldsymbol{h}^{(l)}) + \lambda R \right\}, \tag{5.3.9}$$

$$R = \sum_{j=1}^{J} \left| p - \frac{1}{N_h} \sum_{l=1}^{N_h} \mathbb{E}[h_j^{(l)} | \boldsymbol{v}^{(l)}] \right|^2, \tag{5.3.10}$$

where $\mathbb{E}[\cdot]$ represents the conditional expectation when the observed data are known, λ is a regularization constant, p is a constant controlling the sparsity of hidden units, and N_h and J denote the number of hidden units and training samples, respectively.

5.3.1.3 *Improved sparse restricted Boltzmann machine*

In sparse encoding, the *L0 norm* should be used to measure sparsity (i.e., the number of non-zero elements in the parameter vector) for the sparse regularization term R. However, the problem of solving the *L0 norm* is non-convex and an NP hard problem; it is very difficult to solve it directly. In recent years, many approximation algorithms have

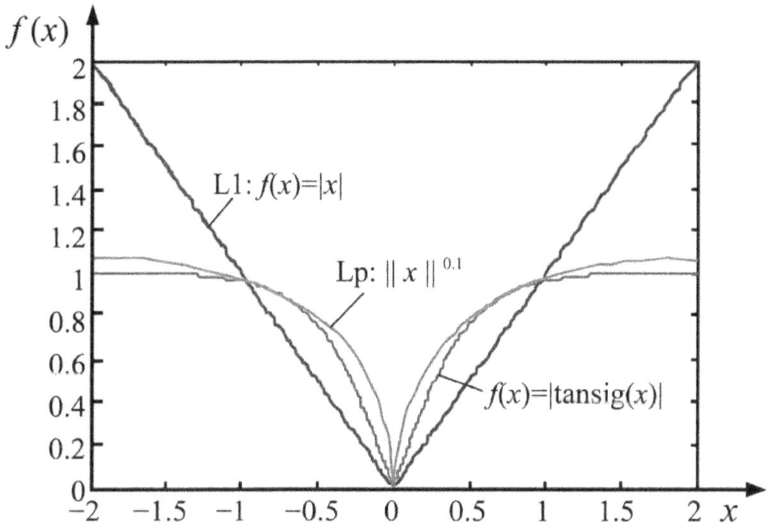

Figure 5.6. Comparison of different norms for $||x||$.

emerged, such as the *Lp norm* sparse constraint algorithm, which replaces the $||x||_p^p(0 < p, 1)$ norm with the $||x||_0^0$ norm. According to compressive sensing theory, the *L1 norm* provides more effective sparsity performance, and under certain conditions, the solution with the minimum *L1 norm* is the sparseness solution of the function. Compared with the *L1 norm*, the slope of the tan-sigmoid function is closer to zero, which can provide more effective sparse induction performance, as shown in Figure 5.6. The penalty term for using the tan-sigmoid function as the likelihood function is formulated as

$$R = \sum_{l=1}^{N_h} \sum_{j=1}^{J} (1 - 2/(1 + \mathrm{e}^{2E|[h_j^{(l)}|v^{(l)}]|/T})), \qquad (5.3.11)$$

where T is a scaling factor that controls the similarity between the tan-sigmoid function and the *L0 norm*. When T approaches 0, the tan-sigmoid function approaches the *L0 norm*. The regularization term R in Equation (5.3.11) does not limit the sparsity of each hidden unit but can be automatically obtained according to different tasks. That is, the sparsity level of each hidden unit can be calculated based on the input data rather than forcing each hidden unit to have the same sparsity (sparsity target p) by adding regularization terms.

In Ref. [85], sparse representation can be implemented based on the sparsity of each hidden unit, and the method of implementing sparse representation is to make the hidden units as inactive as possible (i.e., only a small part of the hidden units can be activated). Here, sparsity is generated by directly limiting the activation probability of all hidden units through regularization terms. The former generates sparse representations by limiting the activation time of each hidden unit on N training samples. And here, sparse representation is achieved by limiting the number of activities of hidden units. This is the most natural way to induce sparse representation.

When given a training sample $\boldsymbol{v}$, the sparse RBM optimization problem becomes

$$\begin{aligned}\underset{\{w_{ji},b_i,c_j\}}{\text{minimize}} \Bigg\{ & \sum_{l=1}^{N_h} \log \sum_h p(\boldsymbol{v}^{(l)}, \boldsymbol{h}^{(l)}) \\ & +\lambda \sum_{l=1}^{N_h} \sum_{j=1}^{J} (1 - 2/(1 + \mathrm{e}^{2|\mathrm{E}[\boldsymbol{h}_j^{(l)}|\boldsymbol{v}^{(l)}]|/T})) \Bigg\}. \end{aligned} \tag{5.3.12}$$

In principle, the gradient descent method should be used to solve this optimization problem, but it is time-consuming to calculate the gradient of the logarithmic likelihood function. Therefore, the CD fast learning algorithm is first used to calculate the gradient approximation of the logarithmic likelihood function [85], and then the regularization term R is subjected to gradient descent until the parameters converge. And the gradient of the parameters on the regularization term R is given by

$$\begin{aligned}\frac{\partial R}{\partial w_{ji}} &= \sum_{l=1}^{N_h} \frac{\partial}{\partial w_{ji}} \left(1 - \frac{2}{1 + \mathrm{e}^{\frac{2|E[\boldsymbol{h}_j^{(l)}|\boldsymbol{v}^{(l)}]|}{T}}} \right) \\ &= \sum_{l=1}^{N_h} \frac{\partial}{\partial w_{ji}} \left(1 - \frac{2}{1 + \mathrm{e}^{\frac{2p(h_j^{(l)}=1|\boldsymbol{v}^{(l)})}{T}}} \right) \\ &= \frac{1}{T} \sum_{l=1}^{N_h} (1 - f^2(l)) p_j^{(l)} (1 - p_j^{(l)}) \boldsymbol{v}_i^{(l)}, \end{aligned} \tag{5.3.13}$$

$$\begin{aligned}\frac{\partial R}{\partial c_j} &= \sum_{l=1}^{N_h} \frac{\partial}{\partial c_j}\left(1 - \frac{2}{1+\mathrm{e}^{\frac{2|E[h_j^{(l)}|\boldsymbol{v}^{(l)}]|}{T}}}\right) \\ &= \sum_{l=1}^{N_h} \frac{\partial}{\partial c_j}\left(1 - \frac{2}{1+\mathrm{e}^{\frac{2p(h_j^{(l)}=1|\boldsymbol{v}^{(l)})}{T}}}\right) \\ &= \frac{1}{T}\sum_{l=1}^{N_h}(1-f^2(l))p_i^{(l)}(1-p_i^{(l)}), \end{aligned} \tag{5.3.14}$$

where $f(l) = 1 - \frac{2}{1+\mathrm{e}^{\frac{2p(h_j^{(l)}=1|v(l))}{T}}}$, $p_i^{(l)} = \sigma(\sum_j v_j^{(l)} w_{ji} + b_i)$.

When training the model, the Contrastive Divergence (CD) fast-learning algorithm is first applied to calculate the gradient approximation of the logarithmic likelihood function, and then gradient descent on the regularization term R is performed according to Equations (5.3.13) and (5.3.14). In order to improve the computational efficiency of minimizing the regularization term R, the implementation details mentioned in Ref. [85] are adopted, where only the bias term of the hidden layer is updated. The specific weight update steps are outlined in Algorithm 5.1.

Algorithm 5.1: Sparse RBM training algorithm

Step 1: Update weights

These weights are updated by the Contrastive Divergence (CD) fast-learning algorithm as follows:

$$\Delta w_{ji} \leftarrow \Delta w_{ji} + \eta(\langle v_i h_j\rangle_{\text{data}} - \langle v_i h_j\rangle_{\text{recon}}),$$

where η denotes the learning rate and $\langle\cdot\rangle_{\text{recon}}$ represents the data reconstructed via Gibbs sampling.

Step 2: Update the bias terms

The bias terms Δb_i and Δc_j for the regularization term R are updated using Equation (5.3.14) by

$$\Delta b_i \leftarrow \Delta b_i + \eta(\langle v_i\rangle_{\text{data}} - \langle v_i\rangle_{\text{recon}}),$$
$$\Delta c_j \leftarrow \Delta c_j + \eta(\langle h_j\rangle_{\text{data}} - \langle h_j\rangle_{\text{recon}}).$$

Step 3: Perform repeatedly

Steps 1–2 are carried out repeatedly until all parameters converge.

5.3.2 Competitive learning

Competitive neural networks have many specific forms and different learning algorithms, but their main feature is reflected in the competition between neurons in the competition layer, where only one neuron ultimately wins to adapt to the training sample. Among them, Self-Organizing Map Network (SOMN) is one of the most widely applied competitive neural networks. The SOMN can automatically find similarities between training data and configure similar data nearby in the network. The training steps are as follows:

Step 1: Network initialization

The connection weights $\boldsymbol{w}$ between the input layer and the mapping layer are initialized with random numbers.

Step 2: Calculate the distance

The distance between the weight vectors of the mapping layer and the input vector, and the Euclidean distance between the weight vectors of each neuron in the network and the input vector are calculated to obtain the neuron j with the minimum distance as the optimal neuron.

Step 3: Weight learning

The weights of the output neuron and its neighboring neurons are modified based on the optimal neuron, i.e.,

$$\Delta w_{ji} = w_{ji}(k+1) - w_{ji}(k) = \eta(x_j(k) - w_{ji}(k)), \tag{5.3.15}$$

where w_{ji} denotes the connection weight between input unit j and map unit i at the kth training iteration and $\boldsymbol{x}_j(k)$ represents the training data corresponding to unit i in the kth iteration.

5.3.3 Competitive deep sparse restricted Boltzmann machine

In response to the limitation of previous regularization factors not being able to adaptively adjust based on the sparsity of hidden units during training, the Competition Deep SRBM (CD-SRBM) [86] was proposed to improve the sparsity of hidden units and model training efficiency.

5.3.3.1 *Sparse penalty mechanism based on competition*

The CDSRBM adopts a neuron competition mechanism similar to the SOMN for sparse hidden units. During the model training process, the CDSRBM first selects the optimal matching hidden unit based on the training samples, then sparsely suppresses other hidden units based on the activation state of the optimal matching hidden unit, and finally performs parameter updates. The specific mechanism is as follows:

5.3.3.1.1 *Distance measurement*

The RBM network maps the raw data from the original dimensional space to the multi-dimensional 0-1 space through model-connected weights, and the 0-1 sequence generated by the samples is the corresponding multi-feature combination. Given that the connection weights of the RBM network are the product of the visible unit dimension and the hidden unit dimension, i.e., the number of columns of the connection weight is equal to the number of hidden units, and the connection weight is not consistent with the sample on the unit scale. Therefore, the CDSRBM does not use the Euclidean distance commonly used in the SOMN as a measurement standard but instead uses the cosine value of the angle between the neuron weight vector and the input vector to evaluate these two similarities, i.e., the cosine similarity $S_{\cos(i,j)}$ between sample i and hidden unit j is defined as

$$S_{\cos(i,j)} = \frac{\boldsymbol{v}_i \cdot \boldsymbol{w}_{\cdot j}}{\|\boldsymbol{v}_i\|\|\boldsymbol{w}_{\cdot j}\|}, \tag{5.3.16}$$

where $\boldsymbol{v}_i$ denotes the ith training sample and $\boldsymbol{w}_{\cdot j}$ represents the jth column of the model's connection weight matrix.

5.3.3.1.2 *Selection of the best-matching hidden unit*

Based on the cosine similarity between sample i and all hidden units, the Best-Matching hidden Unit (BMU) for sample i can be determined as the hidden unit $h_{\cos-\max}$ with the highest similarity to the sample. We have

$$S_{\cos(i,h_{\cos-\max})} = F_{\max}(S_{\cos(i,j)}), \tag{5.3.17}$$

where $F_{\max}$ represents the function for finding the argument that maximizes the similarity, J denotes the total number of hidden units, and $j = 1, 2, \ldots, J$.

5.3.3.1.3 *Sparse suppression of the best-matching neuron*

The CDSRBM determines the sparsity level of other units based on the state of the BMU. The sparse suppression of the BMU depends on the cosine similarity between two columns of the connection weight matrix. The implementation process of this suppression is as follows:

Step 1: Compute the cosine similarity

The cosine similarity between the weight column $\boldsymbol{w}_{\cdot\cos-\max}$ (corresponding to the BMU) and all other columns $\boldsymbol{w}$ is calculated to yield a similarity vector $\boldsymbol{S}_{\cos}$.

Step 2: Normalize process

A similarity vector $\boldsymbol{S}_{\cos}$ is normalized to obtain $\overline{\boldsymbol{S}}_{\cos}$.

Step 3: Set sparsity penalty degree

Each element of $\overline{\boldsymbol{S}}_{\cos}$ is set to the sparsity penalty degree p for the corresponding hidden unit.

5.3.3.2 *CD-SRBM training procedure*

The training of an RBM network is unsupervised, with the objective of maximizing the likelihood probability of the training data. The employed training method is the Contrastive Divergence (CD) algorithm.

For the CD-SRBM, the updating rules for parameters $\boldsymbol{w}_{\cdot j}$ and bias c_j of the hidden unit under its competitive sparsity mechanism are given by

$$\Delta w_{\cdot j} = (\overline{S}_{\cos,j} - p(h_j|v_n)) \cdot (v_n)^{\mathrm{T}}, \tag{5.3.18}$$

$$\Delta_1 c_j = \overline{S}_{\cos,j} - P(h_j|\boldsymbol{v}_n), \tag{5.3.19}$$

where $\overline{S}_{\mathrm{cos,j}}$ represents the jth element of the vector, which is the normalized cosine similarity between $\boldsymbol{w}_{\cdot j}$ and $\boldsymbol{w}_{\cdot\cos-\max}$.

In summary, the training procedure of the CD-SRBM is as follows:

Input: Learning rate η, connection weight of network $\boldsymbol{w}$, bias of visible unit b, bias of hidden unit c.

Output: Updated W, b, c.

Training Steps:

Step 1: Update parameters

The parameters $\boldsymbol{w}, b, c$ are updated by the CD algorithm as follows:

$$\begin{aligned}
w_{ji}(k+1) &= w_{ji}(k+1) + \eta(\langle v_j(k)h_i(k)\rangle_{\text{data}} \\
&\quad - \langle v_j(k)h_i(k)\rangle_{\text{recon}}), \\
b_i(k+1) &= b_i(k) + \eta(\langle v_i(k)\rangle_{\text{data}} - \langle v_i(k)\rangle_{\text{recon}}), \\
c_j(k+1) &= c_j(k) + \eta(\langle h_j(k)\rangle_{\text{data}} - \langle h_j(k)\rangle_{\text{recon}}).
\end{aligned}$$

Step 2: Locate the best-matching hidden unit

The Best-Matching hidden Unit (BMU) h_p for the current sample p is found according to Equation (5.3.17).

Step 3: Compute and update w and c

The parameters $\boldsymbol{w}$ and $\boldsymbol{c}$ are calculated according to Equations (5.3.18) and (5.3.19) and the sparse suppression procedure for the BMU.

Step 4: Rerun program

Steps 1–3 are carried out repeatedly until the model converges or the maximum training iterations are reached.

Deep Boltzmann Machine (DBM) is a deep learning model based on the RBM, which is similar to the information processing mechanism of the human brain and a structural system composed of multiple stacked RBMs together. Considering that in the DBM training process, the first step is to perform greedy layer-by-layer initialization training with stacked RBMs, so the sparse penalty mechanism of the CDSRBM is introduced into the construction of the DBM to form the Competition Sparse Deep Boltzmann Machine (CDSDBM).

5.4 Classification Restricted Boltzmann Machine

Classification restricted Boltzmann machine (ClassRBM or CFRBM) [87] is an undirected graph model based on energy function, which is a self-labeled random neural network model used to solve classification problems.

The label layer of the CFRBM uses one neuron to represent a category. Therefore, the number of neurons in the label layer is consistent with the number of categories in the data. The label layer neurons are always sparse, and each neuron can only provide very little information for the model parameters, which may lead to overfitting [87]. Thereby, it is necessary to improve the classification RBM by representing a category with K neurons to provide more information for the model parameters and improve the classification performance of the model.

5.4.1 Classification restricted Boltzmann machine

The Classification Restricted Boltzmann Machine (ClassRBM) model can be characterized as a three-layer stochastic neural network model, as illustrated in Figure 5.7. In Figure 5.7, the first layer is the visible layer and consists of N_v neurons representing input data v. The second layer is called as hidden layer, which consists of J neurons to denote data $\boldsymbol{h}$. The third layer represents the labels y of the input data and is regarded as the label layer, where $y \in \{1, 2, \ldots, C\}$. Fully connected weights between the visible and hidden layers are denoted as $\boldsymbol{w}$. Fully connected weights between the label and hidden layers *are* recorded as U. There are no connections among neurons of each layer.

Now, we are considering a binary element model, but we can also consider Gaussian elements, polynomial elements, correctable linear

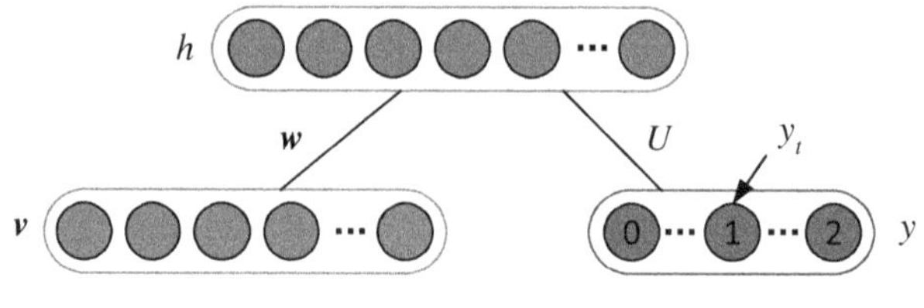

Figure 5.7. Basic RBM model.

elements, and so on. The joint probability distribution of binary CFRBM with labels is defined as

$$p(\boldsymbol{y}, \boldsymbol{v}, \boldsymbol{h}) = \frac{1}{\mathrm{Z}(\boldsymbol{\theta})} e^{-E(\boldsymbol{y}, \boldsymbol{v}, \boldsymbol{h}|\boldsymbol{\theta})}, \tag{5.4.1}$$

where $Z(\boldsymbol{\theta}) = \sum e^{-E(\boldsymbol{y}, \boldsymbol{v}, \boldsymbol{h}|\boldsymbol{\theta})}$, and it is called the partition function, which ensures that the joint probability distribution is effective.

The energy function $E(\boldsymbol{y}, \boldsymbol{v}, \boldsymbol{h}|\theta)$ is defined as

$$\begin{aligned} E(\boldsymbol{y}, \boldsymbol{v}, \boldsymbol{h}|\boldsymbol{\theta}) = &-\sum_{i=1}^{N_v}\sum_{j=1}^{J} \boldsymbol{w}_{ji} h_j v_i - \sum_{i=1}^{N_v} b_i v_i \\ &- \sum_{j=1}^{J} c_j h_j - \sum_{t=1}^{C}\sum_{j=1}^{J} U_{tj} y_t h_j - \sum_{t=1}^{C} d_t y_t, \end{aligned} \tag{5.4.2}$$

where θ denotes the vector of real-valued parameters b_i, c_j, $\boldsymbol{w}_{ji}$, U_{ti} and d_t, v_i, $h_j \in \{0, 1\}$. If and only if the label is t, $y_t = 1$. Otherwise, $y_t = 0$. $\boldsymbol{w}_{ji}$ is the connection weight between neuron v_i and h_j, U_{tj} is the connection weight between neuron y_t and h_j, b_i is the bias of the ith visible neuron, c_j is the bias of the jth hidden neuron, and d_t is the bias of the tth neuron of the label layer.

For classification tasks, the posterior probability $p(y_t|\boldsymbol{v})$, called conditional probability, is calculated by

$$p(y_t|\boldsymbol{v}) = \frac{e^{d_t + \sum_{j=1}^{J} \mathrm{softplus}\left(c_j + \sum_{i=1}^{N_v} \boldsymbol{w}_{ji} v_i + \sum_{t=1}^{C} U_{tj} y_t\right)}}{\sum_{t^*=1}^{C} e^{d_t^* + \sum_{j=1}^{J} \mathrm{softplus}\left(c_j + \sum_{j=1}^{N_v} \boldsymbol{w}_{ji} v_j + \sum_{t^*=1}^{C} U_{t^* j} y_t^*\right)}}, \tag{5.4.3}$$

where $\mathrm{softplus}(x) = \log(1 + e^x)$ and t^* represents the label of the input data, where $t^* \in \{1, 2, \ldots, C\}$.

The training of the CFRBM typically adopts an objective function similar to that of the RBM, where the generative model serves as the training target. Given the joint probability $p(\boldsymbol{y}, \boldsymbol{v})$, the parameters are optimized by maximizing the log-likelihood function $Loss(\boldsymbol{\theta}) = \sum_{n=1}^{N} \log p(\boldsymbol{y}_n, \boldsymbol{v}_n; \boldsymbol{\theta})$ over the training dataset, where N is the number of training samples for classification.

To update the parameters θ, the key step involves computing the partial derivatives of $Loss(\boldsymbol{\theta}) = \log p(\boldsymbol{y}_n, \boldsymbol{v}_n; \boldsymbol{\theta})$ with respect to the model parameters. For the nth sample, the gradient of the log-likelihood $\boldsymbol{\theta}$ [87] is calculated by

$$\frac{\partial Loss(\boldsymbol{\theta})}{\partial \boldsymbol{\theta}} = -\sum_{\boldsymbol{h}} p(\boldsymbol{h}|\boldsymbol{y}_n, \boldsymbol{v}_n) \frac{\partial E(\boldsymbol{y}_n, \boldsymbol{v}_n, \boldsymbol{h}|\boldsymbol{\theta})}{\partial \boldsymbol{\theta}} + \sum_{\boldsymbol{y},\boldsymbol{v},\boldsymbol{h}} \boldsymbol{p}(\boldsymbol{y}, \boldsymbol{v}, \boldsymbol{h}) \frac{\partial E(\boldsymbol{y}, \boldsymbol{v}, \boldsymbol{h}|\boldsymbol{\theta})}{\partial \boldsymbol{\theta}}, \tag{5.4.4}$$

where the first term is straightforward to compute. Due to the presence of the partition function $Z(\boldsymbol{\theta})$, the second term's computational complexity is high. To avoid computational complexity, there are currently various algorithms that approximate gradients, such as CD algorithm, Persistent Contrastive Divergence (PCD) algorithm [88], and Parallel Tempering (PT) algorithm [89]. Among them, the CD algorithm is a commonly used algorithm for completing CFRBM training.

In the CFCBM model, $\boldsymbol{w}$ learns data features with labeled information. When performing classification tasks, the CFRBM distinguishes categories through U to determine the labels of the data. Therefore, U is a very important parameter for controlling different categories of information. The label layer of the CFRBM model only uses one neuron to represent a specific category, and the neurons are always sparse. Moreover, the category information carried by a single neuron is limited, which can affect the classification performance. Reference [87] studied a modified RBM model.

5.4.2 Enhanced CFRBM model

5.4.2.1 *Description of the enhanced model*

Let L neurons represent a specific category, the information of which is carried by the neurons and increased, thereby the classification accuracy of which can be improved. To this end, a Label ClassiFication RBM (L-Classification RBM, L-CFRBM) is established, which is the same as the CFRBM structure except for the label part. In the

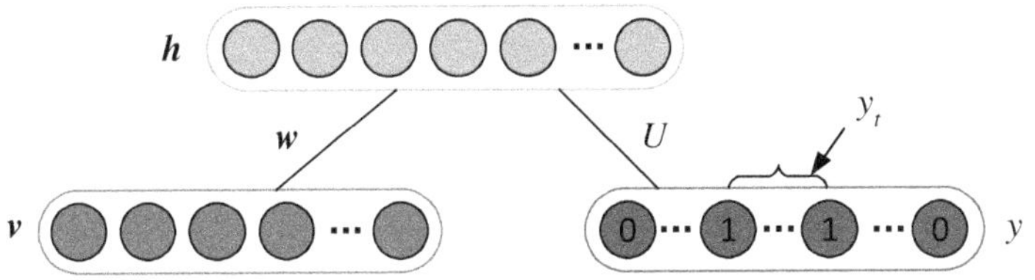

Figure 5.8. RBM model with $C_L$labeled neurons.

label section, C_L neurons are used, and each category uses consecutive L neurons, as shown in Figure 5.8. If the data category is a class t, the neuron $y_{1t}, y_{2t}, \ldots, y_{Lt}$ takes the value 1, and the remaining neurons take the value $0, t \in \{1, 2, \ldots, C\}$. Similarly, $\boldsymbol{w}$ is the connection weight between neurons of the visible layer and the hidden layer, and U is the connection weight between neurons of the label layer and the hidden layer.

The energy function of the labeled binary L-CFRBM model is calculated by

$$\begin{aligned} \mathrm{E}(\boldsymbol{y}_L, \boldsymbol{v}, \boldsymbol{h}|\boldsymbol{\theta}) = &-\sum_{i=1}^{N_v}\sum_{j=1}^{J} w_{ji}h_jv_i - \sum_{i=1}^{N_v} b_iv_i - \sum_{j=1}^{J} c_jh_j \\ &-\sum_{j=1}^{J}\sum_{t=1}^{C}\sum_{l=(t-1)L+1}^{tL} U_{lj}y_lh_j - \sum_{t=1}^{C}\sum_{l=(t-1)L+1}^{tL} d_ly_l. \end{aligned} \tag{5.4.5}$$

Similar to the CFRBM, the L-CFRBM uses θ to represent the real parameters b_i, c_j, w_{ji}, U_{lj} and d_l, $v_i, h_j \in \{0, 1\}$. However, there is a difference in the label part. L depends on the class label, so $t \in \{1, 2, \ldots, C\}, l \in \{(t-1)L+1, (t-1)L+2, \ldots, tL\}$. The weights of the l consecutive neuron connections between the hidden layer and the label layer and the biases of the l consecutive neurons are different. Since there are no connections between the neurons in each layer of the L-CFRBM, the conditional probability formulas $p(\boldsymbol{h}|\boldsymbol{y}_L, \boldsymbol{v}) = \prod_{j=1}^{N_h} p(h_j|\boldsymbol{y}_L, \boldsymbol{v})$ and $p(\boldsymbol{v}|\boldsymbol{h}) = \prod_{i=1}^{N_v} p(v_i|\boldsymbol{h})$ hold.

When the visible layer data and the corresponding label are given, the probability of the ith hidden layer unit being activated is calculated by

$$p(\boldsymbol{h}|\boldsymbol{y}_L,\boldsymbol{v}) = \frac{p(\boldsymbol{y}_L,\boldsymbol{v},\boldsymbol{h})}{\sum\limits_{h} \boldsymbol{p}(\boldsymbol{y}_L,\boldsymbol{v},\boldsymbol{h})} = \frac{e^{-E(\boldsymbol{y}_L,\boldsymbol{v},\boldsymbol{h})}}{\sum\limits_{h} e^{-E(\boldsymbol{y}_L,\boldsymbol{v},\boldsymbol{h})}}$$

$$= \prod_{j=1}^{J} \frac{e^{(c_j+\sum_{i=1}^{N_v} \boldsymbol{w}_{ji}v_i+\sum_{t=1}^{C}\sum_{l=(t-1)\mathrm{L}+1}^{tL} U_{lj}y_l)h_j}}{\sum\limits_{h_j} e^{(c_j+\sum_{i=1}^{N_v} \boldsymbol{w}_{ji}v_i+\sum_{t=1}^{C}\sum_{l=(t-1)\mathrm{L}+1}^{tL} U_{lj}y_l)h_j}}. \quad (5.4.6)$$

Since $h_j \in \{0,1\}$, the conditional probability of $\boldsymbol{h}_j = 1$ is defined as

$$p(h_j = 1|\boldsymbol{y}_L,\boldsymbol{v}) = \mathrm{sigmoid}\left(c_j + \sum_{i=1}^{N_v} w_{ji}v_i + \sum_{t=1}^{C}\sum_{l=(t-1)L+1}^{tL} U_{lj}y_l\right). \quad (5.4.7)$$

Given a neuron of the hidden layer, the conditional probability that the state of the jth neuron in the visible layer is 1 can be written as

$$p(v_i = 1|\boldsymbol{h}) = \mathrm{sigmoid}\left(b_i + \sum_{j=1}^{J} w_{ji}h_j\right). \quad (5.4.8)$$

Given the hidden layer data, the conditional probability corresponding to the neuron of the class t is calculated by

$$p(y_t|\boldsymbol{h}) = \frac{\sum\limits_{v} p(y_L,\boldsymbol{v},\boldsymbol{h})}{\sum\limits_{v,y_L} p(y_L,\boldsymbol{v},\boldsymbol{h})} = \frac{\sum\limits_{v} e^{-E(y_L,\boldsymbol{v},\boldsymbol{h})}}{\sum\limits_{v,y_L} e^{-E(y_L,\boldsymbol{v},\boldsymbol{h})}}$$

$$= \frac{e^{\sum_{l=(t-1)L+1}^{tL} d_l+\sum_{j=1}^{J}\sum_{l=(t-1)L+1}^{tL} U_{lj}h_j}}{\sum\limits_{t^*=1}^{C} e^{\sum_{l=(t^*-1)L+1}^{t^*} d_l+\sum_{l=(t^*-1)L+1}^{t^*L} U_{lj}h_j}}. \quad (5.4.9)$$

When performing a classification task, the classification of the data can be inferred using the posterior probability $p(y_t|\boldsymbol{v})$, which can be calculated by

$$p(y_t|\boldsymbol{v}) = \frac{\sum\limits_{h} p(y_L,\boldsymbol{v},\boldsymbol{h})}{\sum\limits_{h,y_L} p(y_L,\boldsymbol{v},\boldsymbol{h})} = \frac{\sum\limits_{h} e^{-E(y_L,\boldsymbol{v},\boldsymbol{h})}}{\sum\limits_{h,y_L} e^{-E(y_L,\boldsymbol{v},\boldsymbol{h})}}$$

$$= \frac{e^{\sum_{l=(t-1)L+1}^{tL} d_l+\sum_{j=1}^{J} \mathrm{softplus}\left(c_j+\sum_{i=1}^{N_v} w_{ji}v_i+\sum_{t=1}^{C}\sum_{l=(t-1)L+1}^{tL} U_{lj}y_L\right)}}{\sum\limits_{t^*=1}^{c} e^{\sum_{l=(t^*-1)L+1}^{t^*} d_l+\sum_{j=1}^{J} \mathrm{softplus}\left(c_j+\sum_{j=1}^{N_v} w_{ji}v_i+\sum_{t^*=1}^{C}\sum_{l=(t^*-1)L+1}^{t^*L} U_{lj}y_L\right)}}. \quad (5.4.10)$$

5.4.2.2 *Improvement of model training*

The improved model uses the generative model as the training target. Given the joint probability $p(\boldsymbol{y_L}|v)$, the model is trained using the stochastic gradient ascent method and the CD algorithm and maximizing the log-likelihood function $Loss(\boldsymbol{\theta})$ of the L-CFRBM on the training dataset.

The log-likelihood function $Loss(\boldsymbol{\theta})$ and the updating formula of the parameter θ are respectively given by

$$Loss(\boldsymbol{\theta}) = \sum_{n=1}^{N} \log p(\boldsymbol{y}_{Ln}, \boldsymbol{v}_n; \boldsymbol{\theta}), \tag{5.4.11a}$$

$$\boldsymbol{\theta}(k+1) = \boldsymbol{\theta}(k) + \eta \Delta \boldsymbol{\theta}(k), \tag{5.4.11b}$$

where η is the learning rate. The updated formulas of the gradient of the model parameter are as follows:

$$\begin{aligned}\Delta w_{ji} = {} & \text{sigmoid}\left(c_j + \sum_{i=1}^{N_v} w_{ji} v_{in} + \sum_{t=1}^{C} \sum_{l=(t-1)L+1}^{tL} U_{lj} y_{ln}\right) v_{in} \\ & - \text{sigmoid}\left(c_j + \sum_{i=1}^{N_v} w_{ji} v'_{in} + \sum_{t=1}^{C} \sum_{l=(t-1)L+1}^{tL} U_{lj} y'_{ln}\right) v'_{ln},\end{aligned} \tag{5.4.12}$$

$$\begin{aligned}\Delta U_{lj} = {} & \text{sigmoid}\left(c_j + \sum_{i=1}^{N_v} w_{ji} v_{in} + \sum_{t=1}^{C} \sum_{l=(t-1)L+1}^{tL} U_{lj} y_{ln}\right) y_{ln} \\ & - \text{sigmoid}\left(c_j + \sum_{i=1}^{N_v} w_{ji} v'_{in} + \sum_{t=1}^{C} \sum_{l=(t-1)L+1}^{tL} U_{lj} y'_{ln}\right) y'_{ln},\end{aligned} \tag{5.4.13}$$

$$\begin{aligned}\Delta c_j = {} & \text{sigmoid}\left(c_j + \sum_{i=1}^{N_v} w_{ji} v_{in} + \sum_{t=1}^{C} \sum_{l=(t-1)L+1}^{tL} U_{lj} y_{ln}\right) \\ & - \text{sigmoid}\left(c_j + \sum_{i=1}^{N_v} w_{ji} v'_{in} + \sum_{t=1}^{C} \sum_{l=(t-1)L+1}^{tL} U_{lj} y'_{ln}\right),\end{aligned} \tag{5.4.14}$$

$$\Delta b_i = v_{in} - v'_{in}, \tag{5.4.15}$$

$$\Delta d_l = y_{ln} - y'_{ln}, \tag{5.4.16}$$

where v'_i is the reconstruction of v_{in} and $y'_{ln}$is the reconstruction of y_{ln}.

The specific training steps of the L-CFRBM are as follows:

Step 1: Initialize model parameters

The connection weights are initialized into a set of random values that satisfy the normal distribution and are very small, and the biases of each layer are initialized to zero. Training data $(\boldsymbol{y}_{Ln}, \boldsymbol{v}_n)$ are used as input, the learning rate η needs to be set, and the training end condition is also given.

Forward phase: The probability distribution $p(h_j = 1|\boldsymbol{y}_{Ln}, \boldsymbol{v}_n)$ is calculated according to the training data to obtain sampling data from the distribution, and the first term of the gradient of the model parameter is calculated.

Backward phase: The probability distributions $p(y_t|\boldsymbol{h}_n)$ and $p(v_i = 1|\boldsymbol{h}_n)$ are calculated to obtain sample $\boldsymbol{y'}_{Ln}, \boldsymbol{v'}_n$ from these two distributions. The probability distribution $p(h_j = 1|\boldsymbol{y'}_{Ln}, \boldsymbol{v'}_n)$ is calculated to obtain sample $\boldsymbol{h'}_n$, and the second term of the gradient of the model parameter is also calculated.

Step 2: Update parameters

The parameters are updated as follows:

$$\begin{aligned}
\boldsymbol{w}(k+1) &= \boldsymbol{w}(k) + \eta\Delta\boldsymbol{w}(k),\\
\boldsymbol{U}(k+1) &= \boldsymbol{U}(k) + \eta\Delta\boldsymbol{U}(k),\\
\boldsymbol{b}(k+1) &= \boldsymbol{b}(k) + \eta\Delta\boldsymbol{b}(k),\\
\boldsymbol{c}(k+1) &= \boldsymbol{c}(k+1) + \eta\Delta\boldsymbol{c}(k),\\
\boldsymbol{d}(k+1) &= \boldsymbol{d}(k) + \eta\Delta\boldsymbol{d}(k).
\end{aligned}$$

The forward and backward phases and the update of parameters are continuously performed until the training end conditions are met.

5.4.2.3 *Analysis of the improved model*

The improved model increases the number of neurons in the label layer of the classification model so that the neurons carry more category information. The calculation formula of the improved model is slightly different from that of the classification model. The difference is that each category is identified by L label neurons. In order to better analyze the impact of the added neurons on the model parameters and the improvement of the final classification performance, the change of the improved model parameters is introduced by taking parameter U as an example. The gradient of the CFRBM is calculated by

$$\Delta U_{tj}(\text{CFRBM}) = \text{sigmoid}\left(c_j + \sum_{i=1}^{N_v} w_{ji}v_i + \sum_{t=1}^{C} U_{tj}y_t\right) y_t - \text{sigmoid}\left(c_j + \sum_{i=1}^{N_v} w_{ji}v_i' + \sum_{t=1}^{C} U_{ti}y_t'\right) y_t'. \tag{5.4.17}$$

The gradient of the L-CFRBM is calculated by

$$\Delta U_{lj}(\text{L-CFRBM}) = \text{sigmoid}\left(c_j + \sum_{i=1}^{N_v} w_{ji}v_i + \sum_{t=1}^{C}\sum_{l=(t-1)L+1}^{tL} U_{lj}y_l\right) y_l - \text{sigmoid}\left(c_j + \sum_{i=1}^{N_v} w_{ji}v_i' + \sum_{t=1}^{C}\sum_{l=(t-1)L+1}^{tL} U_{lj}y_l'\right) y_l'. \tag{5.4.18}$$

The above formula shows that the update formulas of the connection weight for the two models differ in $\sum_{l=(t-1)L+1}^{tL} U_{lj}y_l$ and $\sum_{l=(t-1)L+1}^{tL} U_{lj}y_l'$. Of course, the reconstruction labels used by them are also different, but they are all attributed to the model parameter U. In order to better describe the difference between the two models, only the different parts in the gradient formula are compared.

We define:

$$\text{sum}_{\text{CFRBM}} = \sum_{t=1}^{C} U_{tj}y_t,$$

$$\text{sum}_{\text{L-CFRBM}} = \sum_{t=1}^{C}\sum_{l=(t-1)L+1}^{tL} U_{lj}y_l.$$

Assuming that the label of the training data is the first category, then we have

$$\mathrm{sum}_{\text{CFRBM}} = U_{1j},$$
$$\mathrm{sum}_{\text{L-CFRBM}} = U_{1j} + U_{2j} + \cdots + U_{Li}.$$

Therefore, the L-CFRBM has more parameters than the CFRBM to control the data categories, and numerically, $\mathrm{sum}_{\text{CFRBM}} \neq \mathrm{sum}_{\text{L-CFRBM}}$. It is precisely this difference that leads to differences in the classification performance of the L-CFRBM.

5.5 (2D)^{2}PCA Restricted Boltzmann Machine

In image classification, principal component analysis (PCA), also known as K-L transform, is one of the most successful Linear Discriminant Analysis (LDA) methods. The traditional PCA method first converts the image matrix into an image vector and then uses the image vector as the original feature for the LDA [90–92]. The PCA method has many advantages, such as fast speed, easy implementation, and high image recognition rate, but it is easily affected by factors such as lighting, expression, and posture. The estimated accuracy expression of the PCA and 2DPCA (2-dimensional PCA) sample covariance matrix was given using mean square error as the metric [93], and thus, we can obtain the judgment condition that the 2DPCA image feature is better than the PCA one. The 2DPCA method was used to extract features from block difference images, which improved the facial recognition performance [94]. A 2DPCA dimensionality reduction method based on L1 paradigm was proposed and applied to unsupervised learning. An improved weighted 2DPCA algorithm that can effectively improve the efficiency of target recognition was proposed [94, 95]. Two-dimensional linear discriminant analysis (2DLDA) is a parallel extension of the LDA in matrix mode, equivalent to the PCA segmented by rows [96–99], but manually designing features requires a lot of experience and debugging workload.

We now introduce the RBM image classification algorithm based on (2D)^{2}PCA, which has been parallelized on the Hadoop platform. Compared with traditional RBMs, this introduced algorithm can effectively improve the processing speed of high-resolution images and has good parallelism [100].

5.5.1 $(2D)^2$PCA image classification RBM

Assuming the training image is recorded as $I_i \in \boldsymbol{R}^{m\times n}$ $(i = 1, 2, \ldots, M)$, the average image of all training images is defined as

$$\bar{I} = \frac{1}{M}\sum_{i=1}^{M} I_i. \tag{5.5.1}$$

Its covariance matrix C_{I} is defined as

$$C_I = \frac{1}{M}\sum_{i=1}^{M}(\boldsymbol{I}_i - \bar{\boldsymbol{I}})^{\mathrm{T}}(\boldsymbol{I}_i - \bar{\boldsymbol{I}}). \tag{5.5.2}$$

Let $\{x_1, x_2, \ldots, x_D\}$ be the standard orthogonal eigenvectors corresponding to the D largest eigenvalues of the matrix C_{I}; it can be proved that the optimal projection matrix is given by

$$\boldsymbol{x}_{\mathrm{opt}} = [x_1, x_2, \ldots, x_D]. \tag{5.5.3}$$

Let $\boldsymbol{I}_i = [(\boldsymbol{I}_{i1})^{\mathrm{T}}, (\boldsymbol{I}_{i2})^{\mathrm{T}}, \ldots, (\boldsymbol{I}_{im})^{\mathrm{T}}]^{\mathrm{T}}$ and $\bar{\boldsymbol{I}} = [(\bar{\boldsymbol{I}}_1)^{\mathrm{T}}, (\bar{\boldsymbol{I}}_2)^{\mathrm{T}}, \ldots, (\bar{\boldsymbol{I}}_m)^{\mathrm{T}}]^{\mathrm{T}}$, where $\boldsymbol{I}_{ij}$ and $\bar{\boldsymbol{I}}_{\boldsymbol{j}}$ are the jth row vectors of $\boldsymbol{I}_i$ and $\bar{\boldsymbol{I}}$, respectively, then Equation (5.5.2) can be rewritten as

$$C_I = \frac{1}{M}\sum_{i=1}^{M}\sum_{j=1}^{m}(\boldsymbol{I}_{ij} - \bar{\boldsymbol{I}}_j)^{\mathrm{T}}(\boldsymbol{I}_{ij} - \bar{\boldsymbol{I}}_j). \tag{5.5.4}$$

Similarly, let $\boldsymbol{I}_i = [(\boldsymbol{I}_{i1}), (\boldsymbol{I}_{i2}), \ldots, (\boldsymbol{I}_{in})]$ and $\bar{\boldsymbol{I}} = [(\bar{\boldsymbol{I}}_1), (\bar{\boldsymbol{I}}_2), \ldots, (\bar{\boldsymbol{I}}_m)]$, where $\boldsymbol{I}_{ij}$ and $\bar{\boldsymbol{I}}_j$ represent the jth column vector of $\boldsymbol{I}_i$ and $\bar{\boldsymbol{I}}$, respectively. Similarly, the covariance matrix of the image is defined as

$$C_{\mathrm{I}}' = \frac{1}{M}\sum_{i=1}^{M}\sum_{j=1}^{m}(\boldsymbol{I}_j - \bar{\boldsymbol{I}}_j)^{\mathrm{T}}(\boldsymbol{I}_{ij} - \bar{\boldsymbol{I}}_j)^{T}. \tag{5.5.5}$$

Let $\{z_1, z_2, \ldots, z_Q\}$ be the standard orthogonal eigenvectors corresponding to the Q largest eigenvalues of the matrix C_I', so the optimal projection matrix is written as

$$\boldsymbol{Z}_{\text{opt}} = [z_1, z_2, \ldots, z_Q]. \tag{5.5.6}$$

The matrix $\boldsymbol{A}$ is projected onto $\boldsymbol{X}$ and $\boldsymbol{Z}$ at the same time, and the characteristic matrix of $Q \times D$ is calculated by

$$\boldsymbol{C} = \boldsymbol{z}^{\boldsymbol{T}} \boldsymbol{A}\boldsymbol{x}. \tag{5.5.7}$$

First, the image of each training set $\boldsymbol{I}_i$ is projected onto the X and Z axes at the same time to obtain the feature matrix $\boldsymbol{C}_i$, then the feature matrix $\boldsymbol{C}$ of the given testing images is calculated, and the images are classified using the RBM-based deep learning method.

The implementation steps of the image classification algorithm based on $(2\text{D})^2$PCA extraction and the RBM network are as follows:

Input: Image data $I_i \in \boldsymbol{R}^{m \times n}$.

Step 1: Read and calculate images

The images in the dataset are read and the average graph of all training samples is calculated according to Equation (5.5.1).

Step 2: Calculate the covariance matrix

For each image, the covariance matrix C_I is calculated according to Equation (5.5.4).

Step 3: Take eigenvectors and obtain optimal vector

The standard orthogonal eigenvectors corresponding to the D largest eigenvalues of the matrix C_{I} are taken to generate $\boldsymbol{x}_{\text{opt}} = [x_1, x_2, \ldots, x_d]$.

Step 4: Calculate the covariance matrix

The covariance matrix C'_I is calculated according to Equation (5.5.5).

Step 5: Take eigenvectors and obtain optimal vector

The standard orthogonal eigenvectors corresponding to the Q largest eigenvalues of the matrix C'_I are taken to generate $\boldsymbol{z}_{\text{opt}} = [z_1, z_2, \ldots, z_D]$.

Step 6: Project the image and calculate feature matrix

The image I is projected in the X and Z directions at the same time, and the feature matrix C of $Q \times D$ is calculated according to Equation (5.5.7).

Step 7: Train the RBM network

The characteristic matrix C is used as the input layer data v_i to train the RBM network, where the number of neurons in the input layer of the network is determined by the dimension of the projection matrix C.

Step 8: Calculate PP

PP in Equation (5.3.6) is calculated according to Equation (5.3.7) and input layer data v_i.

Step 9: Calculate NP

NP in Equation (5.3.6) is calculated according to the CD algorithm.

Step 10: Calculate the parameter and output classification result

The gradient descent method is used to calculate the parameter $\boldsymbol{\theta} = \{w_{ji}, b_i, c_j\}$ in Equation (5.3.6), and it is input to the Softmax classifier to obtain the image classification result.

Output: Image classification result.

5.5.2 (2D)^{2}PCA image classification RBM parallel implementation

In this section, the MapReduce distributed programming mode on the Hadoop [101–103] platform was used to implement an image classification RBM based on (2D)^{2}PCA extraction. First, the entire dataset was divided into several small datasets (assuming that there are N small dataset, $N > 0$). In the Map stage, each "mapper" entity processed each small "split" training set, i.e., the principal components of the (2D)^{2}PCA were extracted and the RBM for the training set, which is responsible for, is trained. In the reduction process, the calculation results of the mapper stage were received and output to the file system. The parallel programming model of the algorithm is shown in Figure 5.9 [100].

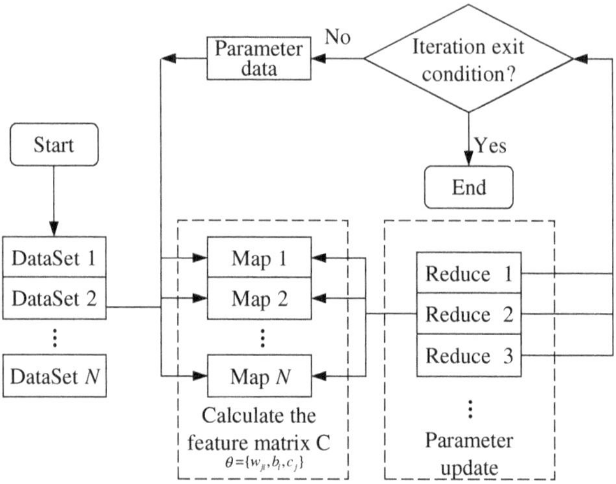

Figure 5.9. $(2D)^2$PCA classification RBM parallel implementation model.

5.6 Example 9: Gait Feature Extraction and Recognition Using Restricted Boltzmann Machine

Gait recognition is a biometric technology that uses gait features for long-distance perception. It is non-invasive, can be obtained from a long distance, and is difficult to disguise. At present, the methods for extracting gait features can be mainly divided into two categories: one is a model-based method and the other is a holistic method (also called a non-model-based method). The model-based method mainly divides the human body area into several parts based on the physiological characteristics of human gait and extracts gait features from them. While the holistic method does not need to build a model, but it must consider the gait features from a holistic perspective and utilize mathematical methods to describe the gait features. The more commonly used method is the features of the gait profile [103–106]. Researches have shown that the accuracy of gait recognition is often affected by walking speed, clothing changes, and perspective changes. Among them, the perspective changes will greatly affect the generalization ability of the recognition method. At present, most gait recognition research focuses on extracting features from gait video sequences and using traditional PCA and Linear Discriminant Analysis (LDA) methods. However, a difficult problem in gait recognition is how to automatically extract effective gait features. In this section, we analyze the method of gait feature extraction and recognition using restricted Boltzmann machines [107].

5.6.1 Gait feature extraction based on restricted Boltzmann machine

5.6.1.1 *Training restricted Boltzmann machine*

For the RBM model, the purpose of training is to obtain the value of the parameter by learning the training samples so that the distribution represented by the corresponding RBM can fit the given gait sample training data as much as possible. The use of the CD algorithm can quickly train the RBM network. When initializing the RBM, the CD algorithm only needs to use one-step Gibbs sampling on the experimental samples to obtain sufficiently good results. During the training process, $\langle\cdot\rangle_{\text{data}}$ represents a training sample input into the network at each step of training. The activation probability of neurons of the hidden layer is calculated using Equation (5.3.7). For each neuron of hidden layer, the activation probability is compared with a random number sampled from a uniform distribution. If it is greater than the random number, it is set to an activated state, otherwise it is set to an inhibited state. After determining the activation state of all neurons of the hidden layer, the activation probability of the neurons of the visible layer is calculated using Equation (5.3.8), i.e., the reconstruction of the visible layer for the training samples is obtained. The update formula of the parameters in the RBM is thus

$$\begin{aligned}
\Delta w_{ji} &= \eta(\langle v_i h_j\rangle_{\text{data}} - \langle v_i h_j\rangle_{\text{recon}}),\\
\Delta b_i &= \eta(\langle v_i\rangle_{\text{data}} - \langle v_i\rangle_{\text{recon}}),\\
\Delta c_i &= \eta(\langle h_j\rangle_{\text{data}} - \langle h_j\rangle_{\text{recon}}),
\end{aligned} \tag{5.6.1}$$

where η is the learning rate and, usually, a small number.

The gait energy map is input into the trained RBM, and the activation probability of the neurons of the hidden layer is obtained by Equation (5.3.7). This value is the new obtained feature after feature extraction.

5.6.1.2 *Feature extraction*

Gait recognition consists of two parts: feature extraction and classification recognition. Feature extraction mainly extracts gait features and represents them through samples using selected features, including background segmentation, normalization, calculation of gait cycles, and generation of gait energy maps. While classification recognition mainly uses classifier classification methods for recognition.

5.6.1.2.1 *Background segmentation*

The segmentation of moving targets is to extract the changing area from the background image in the image. However, due to the dynamic changes that may exist in the background image, such as weather changes, color temperature changes, shadows, and object occlusion, motion detection becomes a more difficult task. At present, motion segmentation algorithms mainly include background subtraction method, time domain difference method, and spatiotemporal gradient method. Background subtraction method is employed for extracting the side profile of pedestrians.

The specific steps are as follows:

Step 1: Background estimation

The mixed Gaussian model is used to model the background to obtain the background image in the image sequence.

Step 2: Target detection and segmentation

The background subtraction method is used to detect the moving target in the sequence image to perform binary segmentation on the image under the condition of setting a threshold.

Step 3: Filling processing

Since the contour image after binary segmentation will produce some noise and holes, it is necessary to use the open operation processing on the binary image, and then use the connected domain analysis to fill the residual noise area. Figure 5.10 shows the generation process of a pedestrian gait contour map [107].

5.6.1.2.2 *Normalization of contour*

Figure 5.10(c) shows that the target outline only occupies a small part of the original image. Since the camera is fixed at an angle

(a) (b) (c)

Figure 5.10. Generation of pedestrian silhouette image. (a) Original image. (b) Background image. (c) Contour image.

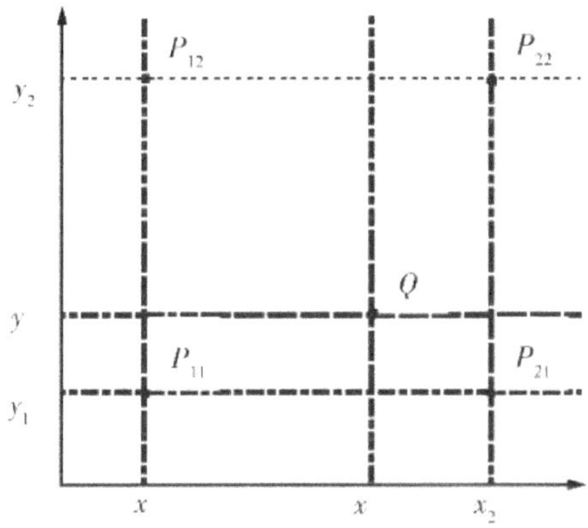

Figure 5.11. Bilinear interpolation method.

when shooting, the distance between the pedestrian and the camera changes during walking, which directly affects the size of the silhouette. The silhouette image is cut to obtain the target silhouette image, which is then normalized and scaled to a fixed-size template. Bilinear interpolation method is used here, and the image scale is transformed using Equation (5.6.2), as shown in Figure 5.11. $L_{11} = (x_1, y_1), L_{12} = (x_1, y_2), L_{21} = (x_2, y_1)$, and $L_{22} = (x_2, y_2)$ are the positions of the four pixel points, $Q = (x, y)$, and $f(P)$ is the pixel value of point P:

$$f(x, y) = \frac{1}{(x_2 - x_1)(y_2 - y_1)} \times [x_2 - x \;\; x - x_1] \times \begin{bmatrix} f(L_{11}) & f(L_{12}) \\ f(L_{21}) & f(L_{22}) \end{bmatrix} \begin{bmatrix} y_2 - y \\ y - y_1 \end{bmatrix}. \tag{5.6.2}$$

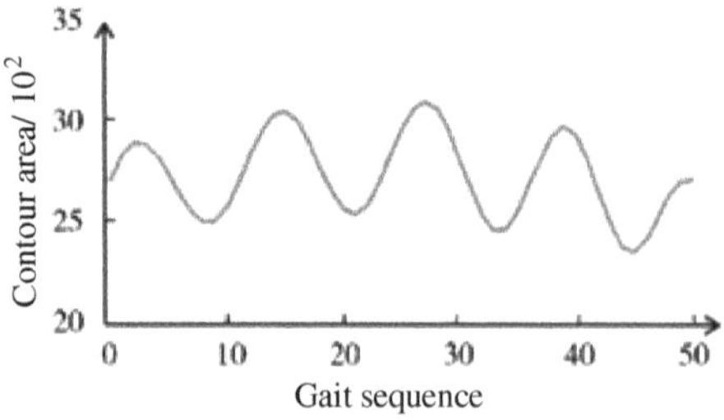

Figure 5.12. Curve of contour area changing with time.

5.6.1.2.3 *Calculation of the gait period*

Since the area of the moving human body contour changes periodically over time, the area curve on the time axis has obvious peaks and troughs, as shown in Figure 5.12. To this end, the sequences contained between two consecutive peaks are classified into the same period, and then the entire sequence is divided into smaller period sequences to provide a basis for obtaining the gait energy diagram.

5.6.1.2.4 *Gait feature extraction*

Gait energy graph is a commonly used method for obtaining features in gait recognition. It is constructed by superimposing binary contour images on the center-normalized gait cycle sequence images. Assuming that the contour image set is $S = \{s_1, s_2, \ldots, s_{N_T}\}$, s_k represents the contour image at time k, and N_T is the length of this periodic sequence, the gait energy graph is defined as

$$\mathrm{GEI} = G(x, y) = \frac{1}{N_T} \sum_{t=1}^{N_T} s_k(x, y), \tag{5.6.3}$$

where $s_k(x, y)$ represents the gray value at (x, y) coordinate at time k, $G(x, y)$ is the pixel value at (x, y) after the gait contour image is superimposed, and G is the corresponding gait energy map. The gait energy map used in this section is shown in Figure 5.13.

The obtained gait energy graph is vectorized and then trained using the RBM to obtain the characteristic representation of the gait.

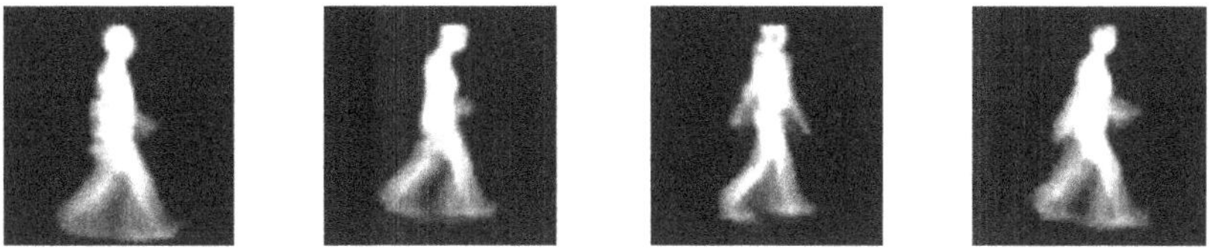

Figure 5.13. Example of gait energy graph.

5.6.2 Simulation experiments and result analyses

5.6.2.1 *Experimental data and methods*

The data from the CASIA Dataset A were selected, and a gait recognition database of the Chinese Academy of Sciences was used as the testing set. The database contains 20 pedestrian samples, each of which consists of 12 different image sequences, of which the first 4 are normal walking sequences taken at an angle of 90° viewing angle, and each image sequence has about 70 original images. First, the original gait images taken at an angle of 90° viewing angle in CASIA Dataset A are subjected to human body region detection, normalization, and converted into gait energy maps with a size of 64×64. Then, 2/3 of the gait energy map samples of each pedestrian were randomly selected as the training set, and the remaining 1/3 were used as the testing set. The RBM was mainly used as the feature extractor and is compared with the feature extraction methods of the PCA, LDA, and CNN. The classifiers used Support Vector Machine (SVM), Twin SVM (TSVM), K-Nearest Neighbor (K-NN), and Artificial Neural Network (ANN) for gait recognition. For different combinations of feature extractors and classifiers, the corresponding parameters must be adjusted and the recognition accuracy was recorded. The above experimental steps were repeated 5 times and the average recognition rate of each set of parameters was taken as the final output.

5.6.2.2 *Experimental results and analyses*

5.6.2.2.1 *Gait feature extraction and recognition based on the RBM*

In the gait feature extraction based on the RBM, the value of the number of hidden layers was an integer from 50 to 500, the values of the parameter c in the SVM and TSVM were set to range from

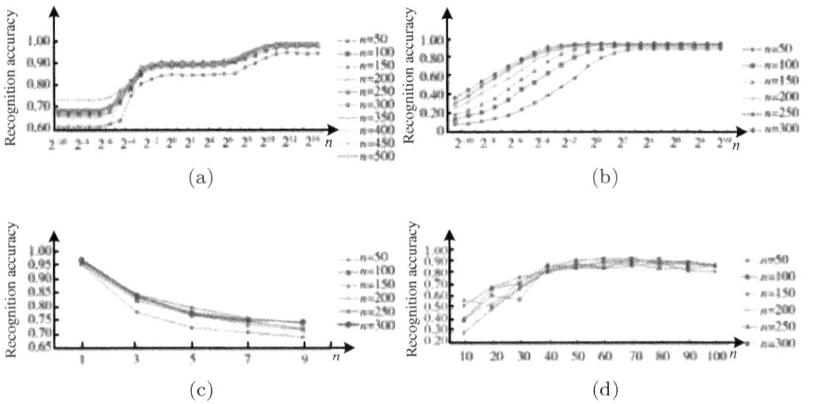

Figure 5.14. RBM gait feature extraction and recognition rate of different classification methods. (a) SVM penalty. (b) TSVM penalty. (c) Number of K-nearest neighbors. (d) Number of hidden neurons in ANN.

2^{-10} to 2^{14}, respectively, and the multi-class classification adopted a one-to-many method. The values of K in the K-NN were 1, 3, 5, 7, and 9, respectively. The number of neurons in each hidden layer of the ANN was from 10 to 100, respectively. The experimental results are shown in Figure 5.14 [107]. Among them, from Figure 5.14(a) to Figure 5.14(d), it can be seen the recognition results of using the SVM, the TSVM, the K-NN, and the ANN, respectively. It can also be seen that when using the SVM and the TSVM classifiers, as the value of parameter c increases, the recognition rate gradually increases, but the recognition rate of the TSVM is lower than that of the SVM. For the K-NN classifier, the gait recognition rate using the nearest neighbor method is higher than that of $K(K \geq 2)$ nearest neighbors, but its recognition rate is lower than that of the SVM classifier. Similarly, for the ANN, as the number of features and the number of neurons in the hidden layer increase, its gait recognition rate continues to improve. When the number of neurons exceeds 70, its recognition rate begins to decrease.

5.6.2.2.2 *Gait feature extraction and recognition based on the PCA and LDA method*

(1) Feature extraction and recognition based on the PCA method. The number of principal components ranges from 50 to 300, and

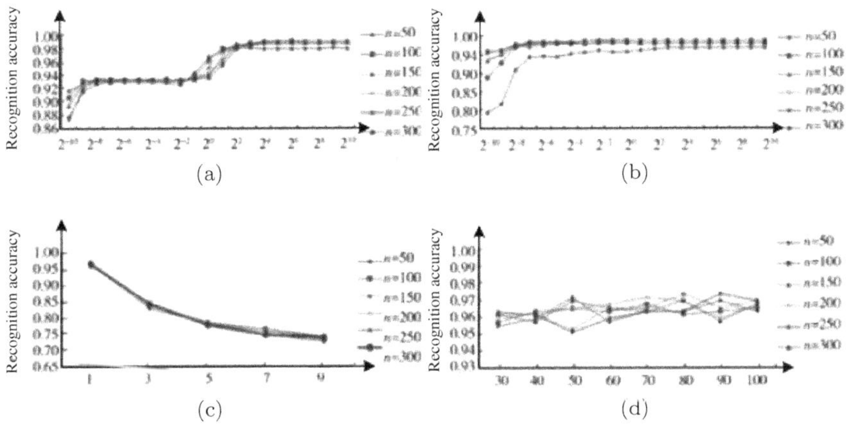

Figure 5.15. PCA gait feature extraction and recognition rate of different classification methods. (a) SVM penalty. (b) TSVM penalty. (c) Number of K-nearest neighbors. (d) Number of hidden neurons in ANN.

the experimental results are shown in Figure 5.15 [107]. Figure 5.15 shows that for the PCA method, when using the SVM and the TSVM classifiers, a higher recognition rate can be achieved.

(2) Feature extraction and recognition based on the LDA method. Since the LDA method uses class labels as prior knowledge, fewer feature space dimensions can be used to represent samples. The number of features in this experiment ranges from 1 to 19, and the experimental results are shown in Figure 5.16 [107]. Figure 5.16 shows that the recognition rate using K-NN as a classifier is higher than that of the SVM, the TSVM and the ANN classifiers, while the recognition rates of the SVM, the TSVM, and the ANN classifiers are roughly the same.

(3) Gait feature extraction and recognition based on CNN. In the experiment, the CNN consists of two convolutional layers, two pooling layers, a fully connected layer, and the SVM, the TSVM, the K-NN, and the ANN classifiers. The sizes of the convolution kernels are 3 and 5, respectively. The experimental results are shown in Figure 5.17.

(4) Comparison of experimental results of gait feature extraction using different methods. In order to compare the performance of different methods, four feature extraction and classification methods were experimentally studied. The experimental results are shown in

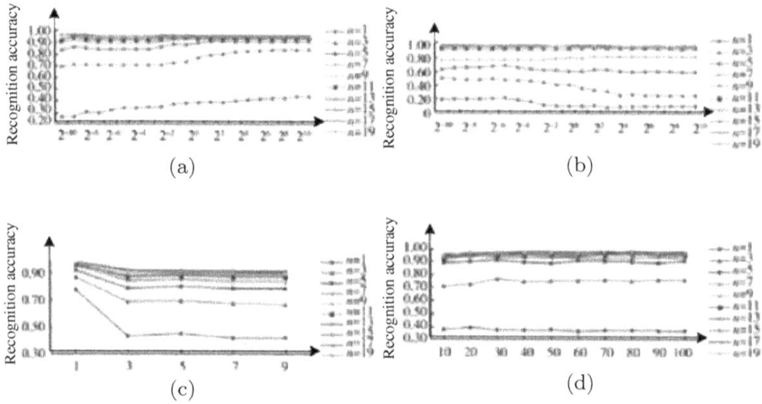

Figure 5.16. LDA Gait feature extraction and recognition rate of different classification method. (a) SVM penalty. (b) TSVM penalty. (c) Number of K-nearest neighbors. (d) Number of hidden neurons in ANN.

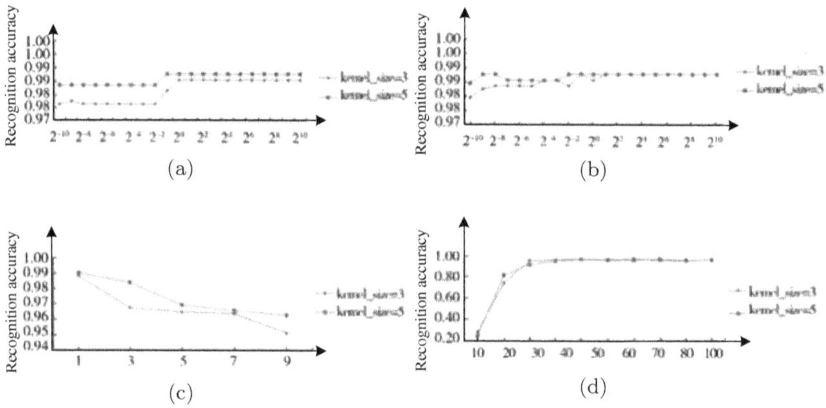

Figure 5.17. CNN gait feature extraction and recognition rate of different classification methods.

Table 5.1 [107], where the content in brackets is the parameter value when a better recognition accuracy is obtained. In general, for different feature extraction methods, when the SVM classifier is used, its gait recognition rate is better than that of the TSVM, K-NN,

Table 5.1. Experimental results of feature extraction and recognition using different methods.

Feature extraction methods	SVM	TSVM	K-NN	ANN
RBM	(250, 2^{-11})	(300, 2^{0})	(300, 1)	(200, 70)
	0.989 7	0.956 9	0.971 3	0.937 4
PCA	(250, 2^{4})	(150, 2^{-1})	(250, 1)	(250, 90)
	0.991 8	0.991 8	0.967 2	0.973 3
LDA	(19, 2^{-9})	(17, 2^{-4})	(17, 1)	(19, 70)
	0.976 4	0.979 5	0.985 6	0.979 5
CNN	(5, 2^{5})	(5,2^{-1})	(5, 1)	(3, 70)
	0.987 7	0.987 7	0.989 7	0.981 5

and ANN methods. In addition, the recognition rate of the feature extraction using the RBM network is higher than that of the LDA method and CNNs, and slightly lower than that of the PCA method, but the RBM network achieves automatic extraction of gait features.

In summary, for the problem of gait recognition, the RBM is used to extract and identify gait features, and the gait energy map is used as the input of the RBM. The gait features are automatically extracted by the RBM. For the CASIA gait database, support vector machine, twin support vector machine, neural network, and K-nearest neighbor recognition methods are selected to extract the features of the RBM for gait recognition. At the same time, the feature extraction of principal component analysis (PCA), linear discriminant analysis (LDA) and convolutional neural network (CNN) is compared. The results are as follows:

(1) For the automatic feature extraction method, when the SVM is used as a classifier, the recognition rate of the RBM network is better than that of the CNN, while when the TSVM, K-NN, and ANN are used as classifiers, the recognition rate of the CNN is better than that of the RBM network.
(2) For different feature extraction methods, when the SVM is used as a classifier, the recognition rate of the RBM method is better than that of the LDA method and CNN, and slightly lower than that of the PCA method.

5.7 Example 5: Speech DOA Estimation Based on RBM Data Compression

5.7.1 Introduction

Direction of Arrival (DOA) estimation is an important research direction in the field of array signal processing and has a large number of practical application scenarios. For example, radar, sonar, microphone arrays, and communication systems [108, 109]. Sensor arrays receive single or multiple source signals to obtain the array signal, and DOA estimation method is the technique to obtain the direction of incidence of the signals through the array signals. Most of the early DOA estimation methods are based on beamforming algorithms, such as Capon [110] algorithm, which are simple to implement and widely used, but perform poorly when there are multiple sources or sources with similar angles. Modern DOA estimation methods are mainly represented by subspace class methods and sparse reconstruction methods. The subspace class methods include MUltiple SIgnal Classification (MUSIC) [111] and Estimation of Signal Parameters via Rotational Invariance Techniques (ESPRIT) [112], which have good performance and are widely used, but at the same time there are problems such as the need for information about the number of sources. Sparse reconstruction methods include L1-SVD [113] and Sparse Bayesian Learning (SBL) [114], which do not need information about the number of sources and have excellent discriminative ability for neighboring sources but at the same time require considerable computational cost.

In recent years, due to the development of computing power and the rise of artificial intelligence, the DOA estimation methods [115, 116] based on Artificial Neural Networks (ANN) have appeared, which first obtain a large number of array signals through software simulation, do simple feature extraction on the array signals to construct a dataset, and use the dataset to train the ANN to complete the nonlinear mapping between the array signal features and the DOA estimation results. The effect of the ANN-based DOA estimation method is significantly improved with the increase in the dataset size and the number of layers of the ANN, but at the same time, it will increase a large number of network's parameters, which makes the training process of the network too complicated.

For the training problems of the above ANN, in this section, a data compression method based on the RBM [108] is proposed, which reduces the size of the ANN's parameters by compressing signal features, thus achieving the effect of accelerating network training. The data compression of the array signals is accomplished by training an RBM-based autoencoder, which retains only the Auto-Encoder (AE), to encode the array signal features. In the process of training the auto-codec, noise-pure data pairs are used so that the encoding results of the AE on the noise-containing signals are closer to the encoding results of the pure signals, which can improve the performance of the whole DOA estimation system in a noisy environment. Compared with the DOA estimation method without RBM data compression, the method with data compression significantly reduces the network's parameters and improves the estimation performance in low SNR environments.

5.7.2 Array model

Let the uniform linear microphone array have M array element with a spacing distance of d. Each array element is an identical omnidirectional microphone, and the far-field signals are incident at θ. Assuming that the noise is independent and identically distributed Gaussian white noise and the incident signals have the feature with mean zero and variance σ^2, the output of the array at the time k can be expressed as

$$\boldsymbol{x}(k) = \boldsymbol{a}(\theta)\boldsymbol{s}(k) + \boldsymbol{n}(k), \tag{5.7.1}$$

where $\boldsymbol{a}(\theta)$ denotes the $M \times 1$-dimensional array flow matrix, $\boldsymbol{s}(k)$ denotes the $1 \times K$-dimensional target source complex amplitude vector, and $\boldsymbol{n}(k)$ denotes the $M \times K$-dimensional additive noise complex vector:

$$\boldsymbol{a}(\theta) = \frac{1}{\sqrt{M}}[1 \ e^{j\varphi(\theta)} \ \cdots \ e^{j(M-1)\varphi(\theta)}], \tag{5.7.2}$$

where $\varphi(\theta) = \frac{2\pi d}{\lambda}\sin\theta$, $\theta \in \left[-\frac{\pi}{2}, \frac{\pi}{2}\right]$.

5.7.3 RBM-based data compression

In machine learning-based DOA estimation algorithms, and the feature extraction method in estimating wideband signals, there are the covariance matrix and the Generalized Cross-Correlation (GCC) matrix. The GCC matrix increases the dimension of the extracted features dramatically when the speech sampling rate is large or the number of array elements is large. The covariance matrix encounters the same problem when the signal's bandwidth is wide or the number of array elements is large. Too many input features will lead to a corresponding increase in the number of parameters that need to be trained in the subsequent machine learning algorithm, and the size of the required training set will also increase accordingly, which significantly increases the computational load of the training process.

To address this problem, the RBM-based AE is introduced to compress the input features. The results of the GCC-based feature extraction are first fed into the AE, and then the encoding results are used as features to construct the dataset of the ANN. Since the dimensionality of the processed data is reduced, the number of parameters in the ANN is also reduced, which reduces the training difficulty of the ANN. Since the dataset used in the training process of the AE is pure-noise-containing speech pairs, the DOA estimation performance is enhanced in the environment with lower Signal-to-Noise Ratios (SNRs).

5.7.3.1 *RBM-based AE*

The RBM is introduced as a basis to build the AE; a complete AE with a single hidden layer is shown in Figure 5.18.

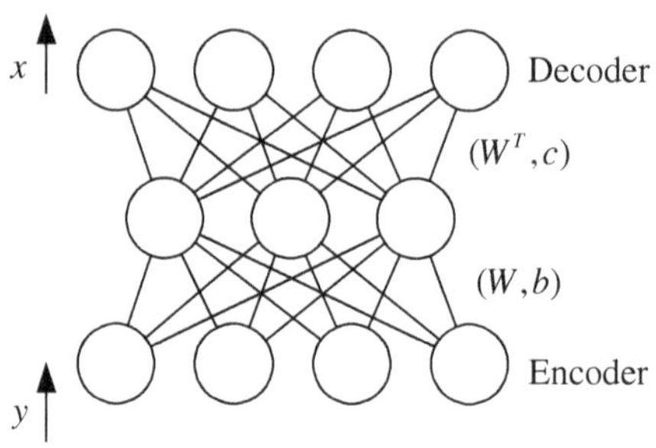

Figure 5.18. Structure of auto-encoder.

Here, y is the input noisy speech and x is the corresponding pure speech. A complete autocoder consists of an autoencoder (AE) part and a decoder part. The formulas in the autocode are as follows:

$$h(\boldsymbol{y}_i) = \sigma(\boldsymbol{W}\boldsymbol{y}_i + \boldsymbol{b}), \tag{5.7.3}$$

$$\boldsymbol{x}_i = \boldsymbol{W}^{\mathrm{T}} h(\boldsymbol{y}_i) + \boldsymbol{c}. \tag{5.7.4}$$

Equation (5.7.3) is the encoder part, where $\boldsymbol{W}$ is the connection weight matrix between neurons, $\boldsymbol{b}$ is the bias between neurons, $\boldsymbol{y}_i$ is the ith noisy speech sample, and $\sigma(x) = (1 + e^{-x})^{-1}$, called as activation function. Equation (5.7.4) is the decoder part, $\boldsymbol{c}$ is the bias between neurons, and $\boldsymbol{x}_i$ is the output of the ith sample after the decoder. The connection weights between neurons are set to $\boldsymbol{W}^{\mathrm{T}}$ to reduce the number of parameters and prevent overfitting, where T denotes as transposition operation.

The cost function of the network is set to l_2 norm, i.e.,

$$Loss(\boldsymbol{W}, \boldsymbol{b}, \boldsymbol{c}) = \sum_i |\boldsymbol{X}_i - \boldsymbol{x}_i|_2^2, \tag{5.7.5}$$

where $\boldsymbol{X}_i$ is the pure speech corresponding to the ith sample. A regularization method is introduced to prevent overfitting and the cost function is obtained by

$$J(\boldsymbol{W}, \boldsymbol{b}, \boldsymbol{c}) = L(\boldsymbol{W}, \boldsymbol{b}, \boldsymbol{c}) + \alpha|\boldsymbol{W}|_2^2. \tag{5.7.6}$$

The parameters can be obtained by optimizing the cost function, i.e.,

$$(\boldsymbol{W}, \boldsymbol{b}, \boldsymbol{c}) \arg\min_{(\boldsymbol{W}, \boldsymbol{b}, \boldsymbol{c})} J(\boldsymbol{W}, \boldsymbol{b}, \boldsymbol{c}). \tag{5.7.7}$$

There are many optimization methods that can be used to solve Equation (5.7.7). Here, the proposed Newton [117] method is used.

The connection between the AE and the ANN is shown in Figure 5.19. Only the coding part of the trained autocodec is retained, and the output of the hidden layer is fed into the ANN as input features to complete the DOA estimation of speech.

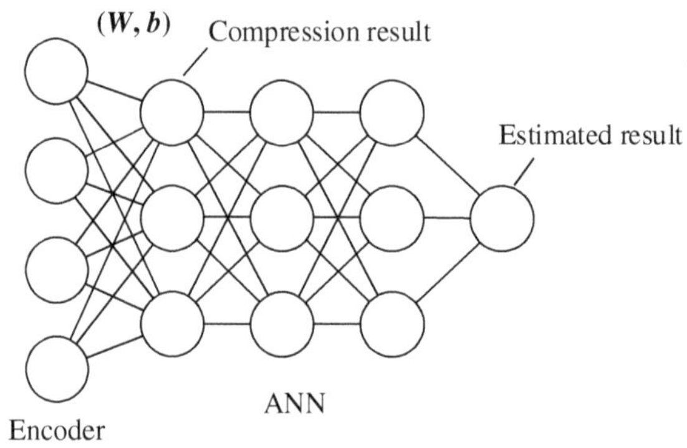

Figure 5.19. Connection diagram of the AE and the ANN.

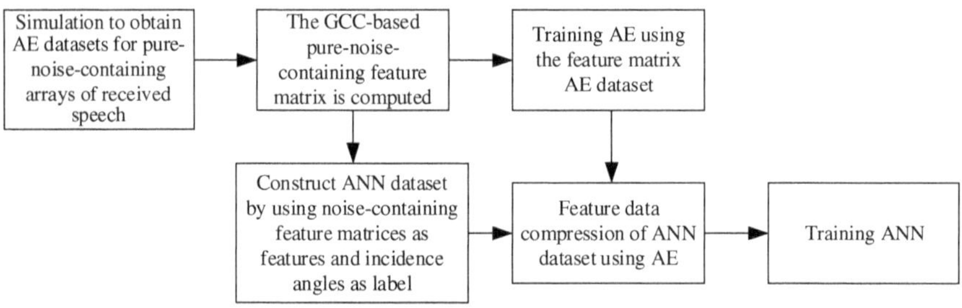

Figure 5.20. Flowchart of training and estimation of DOA.

5.7.3.2 *DOA estimation system*

The training and estimation flow of the DOA estimation system is shown in Figure 5.20, where the final obtained AE dataset is characterized by a noise-containing GCC matrix labeled as pure GCC matrix, and the ANN dataset is characterized by a compressed noise-containing GCC matrix labeled as incidence angle.

The system that completes the training only needs to complete the encoder part of the AE and the forward propagation of the ANN in turn after computing the GCC matrix from the received signals. The DOA estimation results can be obtained.

5.7.4 GCC-based feature extraction

The GCC-PHAT is selected as the basis of input features. The speech signals received by the microphone array are divided into subframes of 0.1s in length, and PHAT weighting is done after calculating the GCC matrix for each subframe. Based on the array model parameters and environment parameters, the possible appearance locations of the GCC peaks are calculated, and only the GCC data of that segment are retained to obtain the sub-feature matrix. Finally, the sub-feature matrices are all weighted and summed to get the feature matrix of the speech signals.

5.7.4.1 *GCC-PHAT*

Let the signals received by the mth and nth microphone array elements in the array model be $x_m(k)$ and $x_n(k)$ after discrete sampling, respectively:

$$x_m(k) = a_m s(k - \tau_{k_m}) + n_{em}(k) + n_{rm}(k), \tag{5.7.8}$$

$$x_n(k) = a_n s(k - \tau_{k_n}) + n_{en}(k) + n_{rn}(k), \tag{5.7.9}$$

where $n_{em}(k)$ and $n_{en}(k)$ are the additive noise in the environment, $n_{rm}(k)$ and $n_{rn}(k)$ are the multipath reflected noise received by the two microphones, a_m and a_n are the attenuation factors of signal amplitudes, τ_{k_m} and τ_{k_n} are the time delays for the propagation of the acoustic source signal to the two microphones, and $s(k)$ is the acoustic source signal.

If neglecting the effects of reverberation and noises, the correlation function between $x_m(k)$ and $x_n(k)$ can be represented as

$$R_{x_m x_n}(\tau_k) = E[x_m(k) x_n(k - \tau_k)] \tag{5.7.10}$$

Substituting Equations (5.7.8) and (5.7.9) into Equation (5.7.10), the following calculation formula is given by

$$\begin{aligned} R_{x_m x_n}(\tau_k) &= E\{[a_m s(k - \tau_{k_m}) + n_{em}(k)] \\ &\quad \times [a_n s(k - \tau_{k_n} - \tau_k) + n_{en}(k - \tau_k)]\} \\ &= a_m a_n R_{ss}[\tau_k - (\tau_{k_m} - \tau_{k_n})] + a_m R_{sn_{en}}(\tau_k - \tau_{k_m}) \\ &\quad + a_n R_{sn_{em}}(-\tau_k - \tau_{k_n}) + R_{n_{em} n_{en}}(\tau_k). \end{aligned} \tag{5.7.11}$$

Let $s(k)$, $n_{em}(k)$, and $n_{en}(k)$ be uncorrelated, then Equation (5.7.11) can be written as

$$R_{x_m x_n}(\tau_k) = a_m a_n R_{ss}(\tau_k - \tau_{k_{mn}}), \tag{5.7.12}$$

where $\tau_{k_{mn}} = \tau_{k_m} - \tau_{k_n}$ and $R_{ss}(\tau_k)$ are the autocorrelation functions of the sound source $s(k)$.

When $\tau_k - \tau_{k_{mn}} = 0$, $R_{x_m x_n}(\tau_k)$ takes the maximum value. Therefore, the maximum value of $R_{x_m x_n}(\tau_k)$ can be used to estimate the time delay sampling point $\tau_{k_{mn}}$ of the received signals of the two microphone arrays, which in turn gives the time delay τ_{mn}.

From the relationship between the cross-correlation function and the mutual power spectrum, the following equation is obtained by

$$R_{x_m x_n}(\tau_k) = \frac{1}{N}\sum_{n=0}^{N-1} G_{x_m x_n}(n) e^{j\frac{2\pi}{N}\tau_k n} = \frac{1}{N}\sum_{n=0}^{N-1} X_m(n) X_n^*(n) e^{j\frac{2\pi}{N}\tau_k n}. \tag{5.7.13}$$

The GCC matrix adds a weighting function to Equation (5.7.13) to get

$$R_{x_m x_n}(\tau_k) = \frac{1}{N}\sum_{n=0}^{N-1} W_{x_m x_n}(n) G_{x_m x_n}(n) e^{j\frac{2\pi}{N}\tau_k n}, \tag{5.7.14}$$

where $W_{x_m x_n}(n)$ is the weighting function and $G_{x_m x_n}(n)$ is the cross power spectrum of the two signals. The PHAT weighting function is given by $W_{x_m x_n}(n) = \frac{1}{|G_{x_m x_n}(n)|}$. The PHAT weighting eliminates the amplitude information of the signals and highlights its phase information, the effect of which is equivalent to the whitening filter.

5.7.4.2 *Feature matrix construction*

Let the uniform line array consist of M array element, the array spacing distance is d, there are C_M^2 pairs of microphone pairs between two combinations of microphones of the array element, the GCC matrix is calculated for every 0.1s signal and the phase transform (GCC-PHAT) is done, the maximum spacing distance between two microphone arrays is $(M-1)d$, and the maximum time delay between two microphone arrays is $\tau = \frac{(M-1)d}{c}$ (set the speed of sound to be c), and let the incident signal be sampled at a rate of f_s Hz, and the peaks of the GCC waveforms that represent the time delay must appear in the middle $f_s \times \tau$ points. Therefore, the input matrix dimension is calculated by $C_M^2 \times f_s \times \tau$.

Due to the characteristics of the speech signals, there may be some non-speech frames when the speech signals are processed in frames.

For more robustness weighting, each speech frame can be expressed by the following equation as

$$\boldsymbol{o} = \sum_{m=1}^{L} \omega_m \boldsymbol{o}_m, \tag{5.7.15a}$$

where

$$\omega_m = \frac{\sum_{d=1}^{D} |o_m(d)|^{\alpha}}{\sum_{m=1}^{L} \sum_{d=1}^{D} |o_m(d)|^{\alpha}}, \tag{5.7.15b}$$

where $\boldsymbol{o}_m$ is the mth speech frame, D is the number of elements of the GCC matrix, $|\cdot|$ is the absolute value, α is the control parameter, and if $\alpha = 0$, then the average value of the GCC vector is obtained. A large α is used to effectively reduce the effect of mute frames on the GCC matrix.

5.7.5 System performance indicators

Root Mean Square Error (RMSE) is used to compare the accuracy and stability of the Multi-Layer Perceptron (MLP) [115] and the Least Squares TDOA (LS-TDOA) algorithm before and after data compression. The RMSE is defined as follows:

$$\text{RMSE} = \sqrt{\frac{\sum_{r=1}^{R} (\theta_S - \hat{\theta}_s^r)^2}{R}}, \tag{5.7.16}$$

where R is the number of samples used for the test, θ_S is the true angle of incidence of the signal, and $\hat{\theta}_s^r$ is the estimated angle of incidence derived from the DOA system.

5.7.6 Experiments and analyses

5.7.6.1 *Simulation conditions*

The simulation conditions are shown in Figure 5.21. The simulation program simulated a room with the dimensions of room, with uniform line arrays placed at a height of 5.5 m × 3.3 m × 2.3 m, and the array spacing distance of the arrays d was 0.2 m. The reflection coefficients

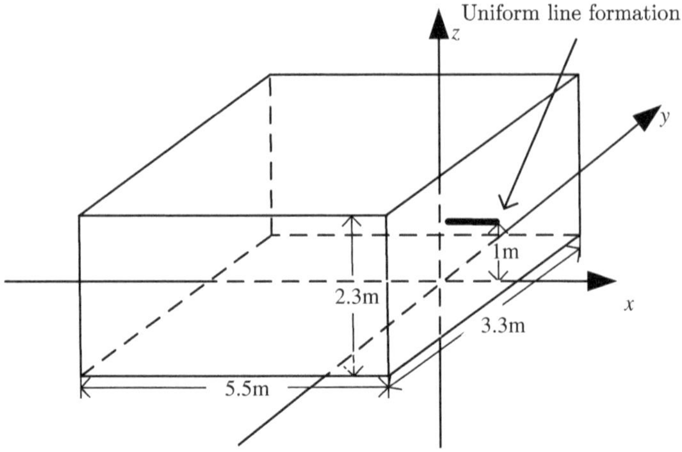

Figure 5.21. The simulation condition.

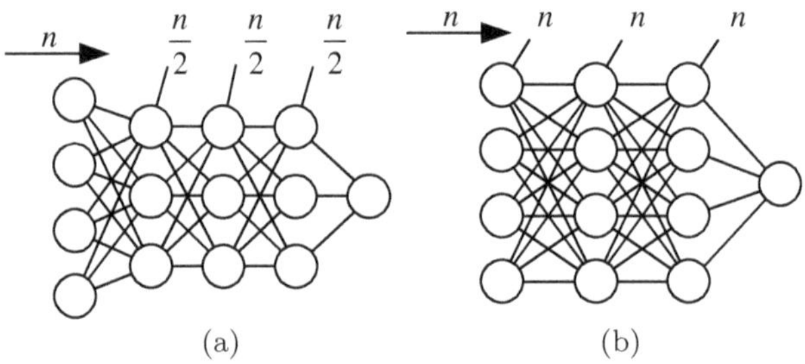

Figure 5.22. The network parameters of the DOA estimation system.

of the room's four walls and ceiling were set to 0.95 (ordinary lime walls), and the reflection coefficient of the floor was set to 0.90. The sound sources were traversed to position range from -90° to 90° with the sound source at 0.5° to form a dataset, and the dataset was divided into the training set and testing set in a 7:3 ratio. The dataset size was approximately 70000.

The parameters of the DOA estimation system are shown in Figure 5.22. Figure 5.22(a) shows the network with the addition of data compression, Figure 5.22(b) shows the original network, n in Figure 5.22 is the input dimension of the network, and the number of hidden layers in the MLP all are 2.

Figure 5.23. The anechoic chamber.

5.7.6.2 *Experimental conditions*

Figure 5.23 shows the anechoic chamber of 5.5 m × 3.3 m × 2.3 m. The sound was set at different angles of the reference array element as the sound source, and the experimental data were collected by an 8-element linear microphone array with a spacing distance of 0.2 m between the array elements.

5.7.6.3 *Analyses of results*

5.7.6.3.1 *Effect of the number of array elements on the performance*

When SNR = 10 dB, the effect of the number of array elements on the performance of the algorithms is shown in Figure 5.24.

From Figure 5.24, it can be seen that when the number of array elements is higher, the algorithm after adding RBM data compression has some improvement compared to the original MLP algorithm. When the number of array elements is less than 6, the performance of the RBM-MLP algorithm decreases significantly. The reason for this problem is that the accuracy of the RBM-based AE for modeling the input features decreases when the input dimension is too small. Considering that data compression is only required for the parameter size of the ANN when the number of array elements is large, this issue does not affect the normal use of the RBM-MLP algorithm.

5.7.6.3.2 *Impact of signal-to-noise ratio on performance*

When the number of array elements is 8, the effect of signal-to-noise ratio on the performance of the algorithm is shown in Figure 5.25.

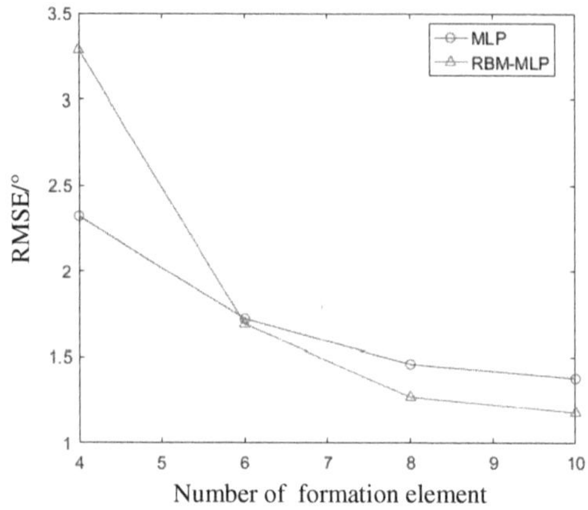

Figure 5.24. The effect of the number of array elements on the performance of the algorithms.

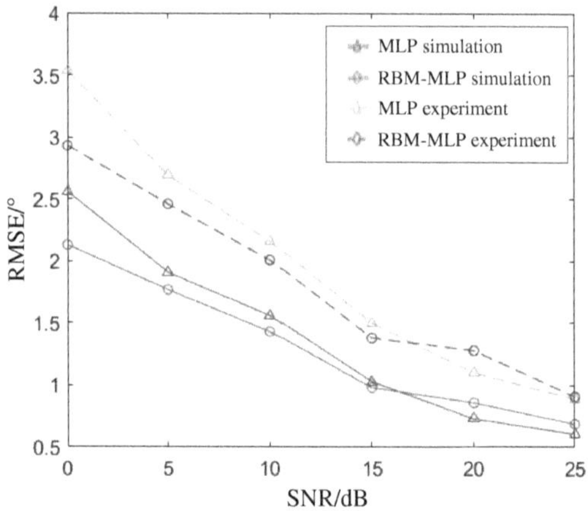

Figure 5.25. The effect of the number of array elements on the performance of the algorithm.

It can be seen that when the signal-to-noise ratio is low, the performance of the MLP algorithm with the addition of the RBM data compression is improved over the original MLP algorithm. However, when the signal-to-noise ratio is high, the performance decreases. This may be due to the fact that the noise immunity introduced by using the noise-containing-pure dataset when training the AE is not effective at higher signal-to-noise ratios, and there is an unavoidable loss of information in the process of data compression. Due to

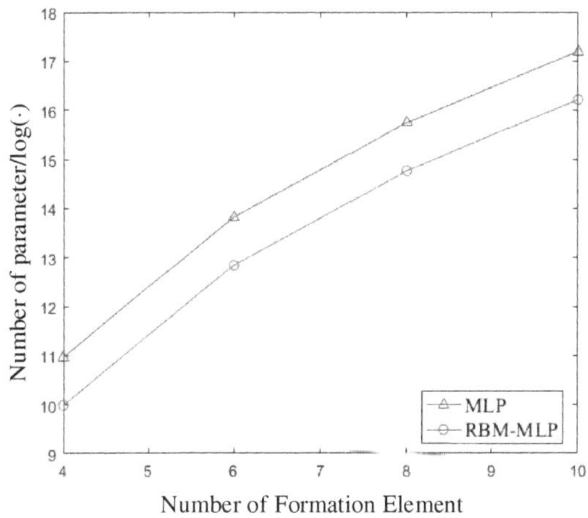

Figure 5.26. The network parameter scales of the two algorithms.

the electrical noise of experimental equipment and inevitable errors introduced during measurement, both algorithms showed an increase in the RMSE under same experimental conditions, but the basic conclusion remained unchanged.

5.7.6.3.3 *Changes in total parameters of network before and after adding data compression*

The network parameter scales of the two algorithms are shown in Figure 5.26 when the array spacing distance $d = 0.2\,\text{m}$, the speech sampling rate $f_s = 16000\,\text{Hz}$, and the sound speed $c = 340\,\text{m/s}$.

Let the input feature dimension be n and the number of hidden layers in the MLP be m, then the number of weights of the original MLP is

$$N_{MLP} = mn^2 + n. \tag{5.7.17}$$

The number of weights for the RBM-MLP is calculated by

$$\begin{aligned} N_{RBM\text{-}MLP} &= \frac{1}{2}n^2 + m\left(\frac{1}{2}n\right)^2 + \frac{1}{2}n \\ &= \left(\frac{1}{2} + \frac{m}{4}\right)n^2 + \frac{1}{2}n. \end{aligned} \tag{5.7.18}$$

When $m = 2$, the parameter size of the MLP algorithm with the addition of the RBM data compression is half of the original MLP algorithm, and the results are shown in Figure 5.26.

In summary, the RBM-based AE is used for the first time to compress the input feature data of the ANN. The use of noise-containing-pure datasets in the training phase of the AE introduces a certain degree of noise immunity, which ensures the performance of the algorithm at low signal-to-noise ratios while reducing the number of parameters in the network, and solves the problem of training difficulties caused by the large size of the ANN parameters when the network depth is large or the number of array elements is large.

Chapter 6

Deep Belief Networks

Summary

Starting from the conventional structure and principles of Deep Belief Networks (DBNs), this chapter discusses sparse DBN architectures and provides a detailed analysis of the Gamma-DBN structure, including layer-wise activation responses and the Gibbs sampling process in both upward and downward directions. It further explores adaptive DBNs by integrating the Adam algorithm with the Nesterov method. Additionally, Kernel Principal Component Analysis (KPCA) is used for feature extraction from sampled data to obtain a KPCA-based DBN model. By analyzing hyperparameters and trainable parameters of the DBN, the training of parameters is adaptively adjusted to form a fully parameter-dynamic learning DBN. Finally, the optimization of the DBNs is achieved through chaotic immune algorithm and Particle Swarm Optimization (PSO) algorithm.

6.1 Introduction

Deep Belief Network (DBN) is one of the foundational architectures in Deep Learning (DL), a multi-layer neural network that integrates feature learning and deep learning and serves as a probabilistic machine learning model combining unsupervised and supervised learning across multiple layers. The DBNs can automatically learn and extract features from original data and enable a more essential and abstract representation of the original input. Moreover, the layer-wise pretraining strategy of the DBNs effectively addresses the difficulties encountered in training deep neural networks. The introduction of the DBNs has opened a new research direction in the field of deep learning.

Before the emergence of Deep Belief Networks (DBNs), the training of deep neural networks remained an unresolved challenge. Some issues, such as gradient instability, overfitting caused by supervised learning, and the sensitivity of gradient descent methods to initialization, often led to convergence to local optima, and so on, making it difficult to train parameters in deep architectures effectively. As a result, deep networks often underperformed compared to shallow networks. The DBN model, proposed by Hinton [1] in 2006, along with a greedy layer-wise unsupervised learning algorithm, effectively addresses the training difficulties of deep learning networks and significantly improves the learning efficiency of deep networks.

This chapter focuses on the study of deep belief networks.

6.2 Deep Belief Networks

6.2.1 Conventional DBN

6.2.1.1 *DBN structure*

Taking a deep belief neural network (DBN-DNN) with three hidden layers as an example, the network consists of three stacked Restricted Boltzmann Machines (RBMs), as shown in Figure 6.1. Each RBM comprises two layers: a visible layer (input layer) and a hidden layer. When constructing a DBN, the hidden layer of the previous RBM serves as the visible layer (input layer) of the next RBM. By stacking the RBMs in this way, a basic DBN architecture is formed. Finally, an additional output layer is appended to complete the DBN-DNN structure.

In the RBM, connections exist only between the neurons of the hidden layer and the visible layer, and there are no connections among neurons within the visible layer or within the hidden layer. The neurons of the hidden layer are usually binary-valued and follow a Bernoulli distribution, while the neurons of the visible layer can be either binary or real-valued depending on the type of input data. Based on the value types of the visible layer ($\boldsymbol{v}$) and the hidden layer ($\boldsymbol{h}$), the RBMs can be categorized into two main types. If both $\boldsymbol{v}$ and $\boldsymbol{h}$ follow binary distributions, the model is referred to as a

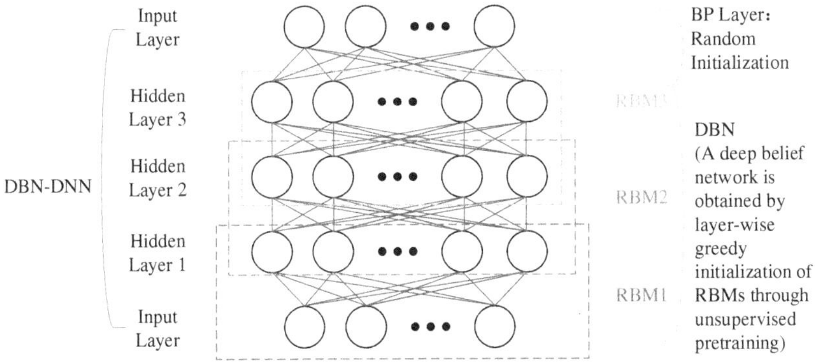

Figure 6.1. Structure of deep belief network (DBN-DNN).

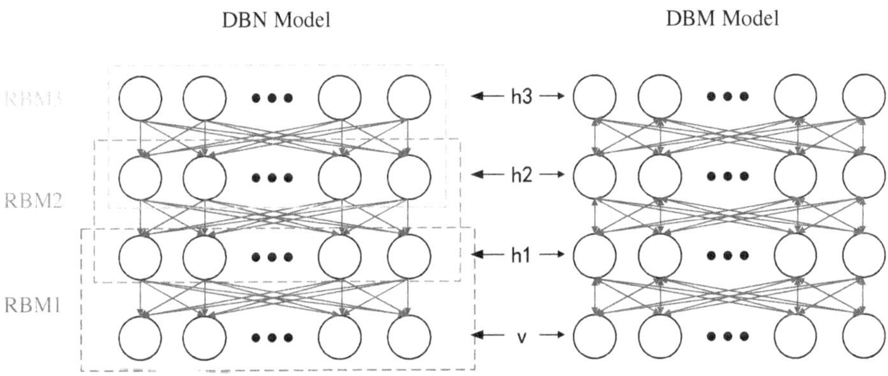

Figure 6.2. Two types of models constructed by RBM.

Bernoulli–Bernoulli RBM network. If $\boldsymbol{v}$ takes real values, such as in the case of speech features, and $\boldsymbol{h}$ is binary, the model is known as a Gaussian–Bernoulli RBM network. Therefore, in Figure 6.1, RBM1 is a Gaussian–Bernoulli RBM, while RBM2 and RBM3 are both Bernoulli—Bernoulli RBMs.

The DBN and Deep Boltzmann Machine (DBM) models constructed by the RBMs are shown in Figure 6.2. When the DBN model is pretrained layer by layer by stacking RBM, the distribution of a certain layer is only determined by the previous layer. For example, in a DBN, the distribution of layer $\boldsymbol{v}$ depends on $h1$, and $h1$ depends only on $h2$. This means that the distribution of $h1$ is not influenced by $\boldsymbol{v}$, once the distribution of $\boldsymbol{v}$ is fixed, $h1$ is determined solely by $h2$. In contrast, a DBM is an undirected graphical model, meaning that the $h1$ layer in the DBM is influenced by both $h2$ and

$\boldsymbol{v}$, and it is a bidirectional relationship. In terms of performance, the DBM structure generally provides better robustness than the DBN structure, but it requires all layers to be trained jointly and can lead to high computational complexity and limited practicality. From the perspective of borrowing the RBM network layer-by-layer pretraining method, the DBN structure is much more convenient and efficient, making it easier to apply widely.

6.2.1.2 *DBN training and backpropagation tuning*

6.2.1.2.1 *Unsupervised pretraining based on the RBM*

The weights are initialized using the Contrastive Divergence (CD-k) algorithm, and Hinton found that when k was set to 1, unsupervised pretraining based on the RBM can yield good learning results. The specific steps are as follows:

Step 1: Initialize parameters

The weights $\{\boldsymbol{w}, \boldsymbol{b}, \boldsymbol{c}\}$ may be randomly initialized, where $\boldsymbol{w}$ is the weight vector, $\boldsymbol{b}$ is the bias vector for the visible layer, and $\boldsymbol{c}$ is the bias vector for the hidden layer. These are initialized to small values (which can be 0).

$$\boldsymbol{x} = \boldsymbol{v} = \begin{bmatrix} v_1 \\ v_2 \\ \vdots \\ v_N \end{bmatrix}, \tag{6.2.1a}$$

$$\boldsymbol{h} = \begin{bmatrix} h_1 \\ h_2 \\ \vdots \\ h_J \end{bmatrix}, \tag{6.2.1b}$$

$$\boldsymbol{w} = \begin{bmatrix} w_{1,1} & w_{2,1} & \cdots & w_{M,1} \\ w_{1,2} & w_{2,2} & \cdots & w_{M,2} \\ \vdots & \vdots & \vdots & \vdots \\ w_{1,J} & w_{2,J} & \cdots & w_{N,J} \end{bmatrix}, \tag{6.2.1c}$$

$$\boldsymbol{b} = \begin{bmatrix} b_1 \\ b_2 \\ \vdots \\ b_N \end{bmatrix}, \tag{6.2.1d}$$

$$\boldsymbol{c} = \begin{bmatrix} c_1 \\ c_2 \\ \vdots \\ c_J \end{bmatrix}, \tag{6.2.1e}$$

where N is the number of visible units and J is the number of hidden units. $\boldsymbol{w}$, can be initialized as random numbers from a normal distribution $N(0,0.01)$ and initialized as $b_i = \log \frac{p_i}{1-p_i}$, where p_i represents the proportion of the ith samples in the training data in the activated state (i.e., where the value is 1) and $\boldsymbol{c}$ can be initialized directly as 0. The formulas for calculating the values of hidden and visible units are written as

$$\boldsymbol{v} = \left(\boldsymbol{w}^T \cdot \boldsymbol{h} + \boldsymbol{b}\right) = \begin{bmatrix} w_{1,1} \cdot h_1 & + & w_{1,2} \cdot h_2 & + & \cdots & + & w_{1,J} \cdot h_J \\ w_{2,1} \cdot h_1 & + & w_{2,2} \cdot h_2 & + & \cdots & + & w_{2,J} \cdot h_J \\ \vdots & & \vdots & & & \vdots & \\ w_{N,1} \cdot h_1 & + & w_{N,2} \cdot h_2 & + & \cdots & + & w_{N,J} \cdot h_J \end{bmatrix}$$

$$+ \begin{bmatrix} b_1 \\ b_2 \\ \vdots \\ b_N \end{bmatrix} - \begin{bmatrix} v_1 \\ v_2 \\ \vdots \\ v_N \end{bmatrix}, \tag{6.2.2}$$

$$\boldsymbol{h} = (\boldsymbol{w} \cdot \boldsymbol{x} + \boldsymbol{c}) = \begin{bmatrix} w_{1,1} \cdot v_1 & + & w_{2,1} \cdot v_2 & + & \cdots & + & w_{N,1} \cdot v_N \\ w_{1,2} \cdot v_1 & + & w_{2,2} \cdot v_2 & + & \cdots & + & w_{N,1} \cdot v_N \\ \vdots & & \vdots & & & & \vdots \\ w_{1,J} \cdot v_1 & + & w_{2,J} \cdot v_2 & + & \cdots & + & w_{N,J} \cdot v_N \end{bmatrix}$$

$$+ \begin{bmatrix} c_1 \\ c_2 \\ \vdots \\ c_J \end{bmatrix} = \begin{bmatrix} h_1 \\ h_2 \\ \vdots \\ h_J \end{bmatrix}. \tag{6.2.3}$$

Step 2: Calculate the probability

Input vector x is assigned to the visible layer $\boldsymbol{v}^{(0)}$ and the probability that the neurons of the activated hidden layer are calculated. For Gaussian or Bernoulli visible units, the probability is given by

$$p(h_j^{(0)} = 1|\boldsymbol{v}^{(0)}) = \sigma(\boldsymbol{w}_J \cdot \boldsymbol{v}^{(0)} + c_j), \tag{6.2.4}$$

where the superscript distinguishes between different vectors and the subscript distinguishes between different dimensions within the same vector.

Step 3: Perform one step of Gibbs sampling according to the computed probability distribution

For each unit in the hidden layer, sample from {0,1} to obtain the corresponding value $\boldsymbol{h}^{(0)} \sim p(\boldsymbol{h}^{(0)}|\boldsymbol{v}^{(0)})$. When a random number $r_j \in [0, 1]$ is generated, it determines the value of h_j as

$$h_j = \begin{cases} 1, & \text{if } p(h_j^{(0)} = 1|\boldsymbol{v}^{(0)}) > r_j \\ 0, & \text{otherwise} \end{cases}. \tag{6.2.5}$$

Step 4: Reconstruct the visible layer using $\boldsymbol{h}^{(0)}$, calculate the probability density, and then perform Gibbs sampling again

For Bernoulli visible units, the probability is calculated by

$$p(v_i^{(1)} = 1|\boldsymbol{h}^{(0)}) = \sigma(\boldsymbol{w}_I^T \boldsymbol{h}^{(0)} + b_j). \tag{6.2.6}$$

For Gaussian visible units, the probability is calculated by

$$p(v_j^{(1)} = 1|\boldsymbol{h}^{(0)}) = N(\boldsymbol{v}^{(0)}; \boldsymbol{w}_I^T \cdot \boldsymbol{h}^{(0)} + b_i, I), \tag{6.2.7}$$

where N represents the normal distribution function.

Step 5: Reconstruct neurons

Based on the calculation probability distribution, another step of Gibbs sampling is performed to extract corresponding values from {0,1} for the neurons in the hidden layer for sampling reconstruction. When a random number in [0,1] is generated, the value of $\boldsymbol{v}_i$ can be determined by

$$v_i = \begin{cases} 1, & \text{if } p(v_i^{(1)} = 1 \mid \boldsymbol{h}^{(0)}) > r_j \\ 0, & \text{otherwise} \end{cases}. \tag{6.2.8}$$

Step 6: Calculate probability again

The visible units (reconstructed ones) are used to calculate the probability that the neurons of the hidden layer are activated.

For Gaussian or Bernoulli visible layer units, the probability is written as

$$p(h_j^{(1)} = 1|\boldsymbol{v}^{(1)}) = \sigma(\boldsymbol{w}_J\boldsymbol{v}^{(1)} + c_j). \tag{6.2.9}$$

Step 7: Update the weights and biases

The weights and biases are updated by using the following formulas:

$$\boldsymbol{w}(k+1) = \boldsymbol{w}(k) + \eta[p(h^{(0)} = 1 \mid \boldsymbol{v}^{(0)})\boldsymbol{v}^{(0)T} - p(h^{(1)} = 1|\boldsymbol{v}^{(1)})\boldsymbol{v}^{(1)T}], \tag{6.2.10a}$$

$$\boldsymbol{b}(k+1) = \boldsymbol{b}(k) + \alpha[\boldsymbol{v}^{(0)} - \boldsymbol{v}^{(1)}], \tag{6.2.10b}$$

$$\boldsymbol{c}(k+1) = \boldsymbol{c}(k) + \beta[p(h^{(0)} = 1|\boldsymbol{v}^{(0)}) - p(h^{(1)} = 1|\boldsymbol{v}^{(1)})], \tag{6.2.10c}$$

where η is the learning rate.

Note: The training of the RBM essentially seeks to find the probability distribution that can generate the best training samples. In other words, the goal is to find the distribution for which the probability of the training samples is maximized. Since the determining factor of the distribution lies in the weight vector $\boldsymbol{w}$, the training objective is to find the optimal weight vector. Therefore, the Contrastive Divergence (CD) algorithm is used to find the optimal weight vector.

In Figure 6.1, when pretraining is performed using the CD algorithm, it only requires iterating to calculate the values of $\boldsymbol{w}$, $\boldsymbol{b}$, and $\boldsymbol{c}$ for the three RBM units: RBM1, RBM2, and RBM3. For the final BP unit, the values of $\boldsymbol{w}$ and $\boldsymbol{c}$ are directly initialized randomly. Typically, the structure formed by RBM1, RBM2, and RBM3 is called the DBN structure (Deep Belief Network structure). After adding an output layer (BP layer), it forms the standard DNN structure, which is called DBN-DNN.

6.2.1.2.2 *Supervised fine-tuning based on the RBM*

In supervised fine-tuning training, the forward propagation algorithm is first used to obtain certain output values from the input, and then the backpropagation (BP) algorithm is applied to update the network's weights and bias values.

(1) Forward propagation

① The pretrained $\boldsymbol{w}$ and $\boldsymbol{c}$ using the CD algorithm are utilized to determine the activation (or deactivation) of each hidden unit.

The activation value of each hidden unit is calculated by

$$\boldsymbol{h}^{(l)} = \boldsymbol{w}^{(l)} \cdot \boldsymbol{v} + \boldsymbol{c}^{(l)}, \tag{6.2.11}$$

where l is the index for the number of hidden layers and $\boldsymbol{w}$ and $\boldsymbol{b}$ are the weights and biases, respectively, and

$$\boldsymbol{w} = \begin{bmatrix} w_{1,1} & w_{2,1} & \cdots & w_{N,1} \\ w_{1,2} & w_{2,2} & \cdots & w_{N,2} \\ \cdots & \cdots & \cdots & \cdots \\ w_{1,J} & w_{2,J} & \cdots & w_{N,J} \end{bmatrix}, \quad \boldsymbol{c} = \begin{bmatrix} c_1 \\ c_2 \\ \vdots \\ c_J \end{bmatrix}, \tag{6.2.12}$$

where $w_{i,j}$ represents the weight from the ith visible unit to the jth hidden unit.

② The signals spread upwards layer by layer, the excitation value of each element in the hidden layer by layer is calculated and standardized by using the following Sigmoid function:

$$\sigma^{(l)}(h_j) = \frac{1}{1 + e^{-h_j}} \tag{6.2.13}$$

③ Finally, the activation and output of the output layer are calculated by

$$\boldsymbol{h}^{(l)} = \boldsymbol{w}^{(l)} \cdot \boldsymbol{h}^{(l-1)} + \boldsymbol{c}^{(l)}, \tag{6.2.14}$$

$$\hat{\boldsymbol{z}} = f(\boldsymbol{h}^{(l)}), \tag{6.2.15}$$

where $f(\cdot)$ is the activation function of the output layer and $\hat{\boldsymbol{z}}$ represents the output layer's value.

(2) Backpropagation

① The backpropagation algorithm based on the minimum Mean Square Error (MSE) criterion is used to update the entire network's parameters, and the cost function is defined as

$$Loss = \frac{1}{N}\sum_{i=1}^{N}(\hat{z}_i(\boldsymbol{w}^{(l)}, \boldsymbol{c}^l) - z_i)^2, \tag{6.2.16}$$

where *Loss* is the average square error for the DNN and $\hat{z}_i$ and z_i represent the output and ideal output from the output layer, respectively. i denotes the sample index, and $(\boldsymbol{w}^{(l)}, \boldsymbol{c}^{(l)})$ refers to the learnable weights and bias parameters at layer l.

② The gradient descent method is used to update the weights and biases of the network:

$$(\boldsymbol{w}^{(l)}, \boldsymbol{c}^{(l)})(k+1) = (\boldsymbol{w}^{(l)}, \boldsymbol{c}^{(l)})(k) - (\eta, \beta)\frac{\partial Loss}{\partial(\boldsymbol{w}^{(l)}, \boldsymbol{c}^{(l)})}, \tag{6.2.17}$$

where (η, β) is the learning rate.

The above describes the two key steps in constructing the entire DBN-DNN architecture: Unsupervised pretraining and supervised fine-tuning. By selecting the appropriate number of hidden layers, number of neural units in each layer, and learning rate, and iterating a certain number of times for training, we can obtain the DNN mapping model to be ultimately wanted.

From the perspective of unsupervised learning, its goal is to preserve the characteristics of the original features as much as possible while reducing feature dimensionality. From the perspective of supervised learning, its goal is to minimize classification error. Regardless of whether it is supervised or unsupervised learning, the essence of the DBN lies in feature learning, i.e., how to obtain better feature representations.

6.2.2 Sparse deep belief network

6.2.2.1 *Conventional sparse deep belief network*

The sparse deep belief network (sparse DBNs) is composed of multiple layers of sparse RBM models, each layer of which captures highly correlated associations from the hidden units of the preceding layer, as shown in Figure 6.3.

The learning process of the sparse DBN model mainly consists of two steps:

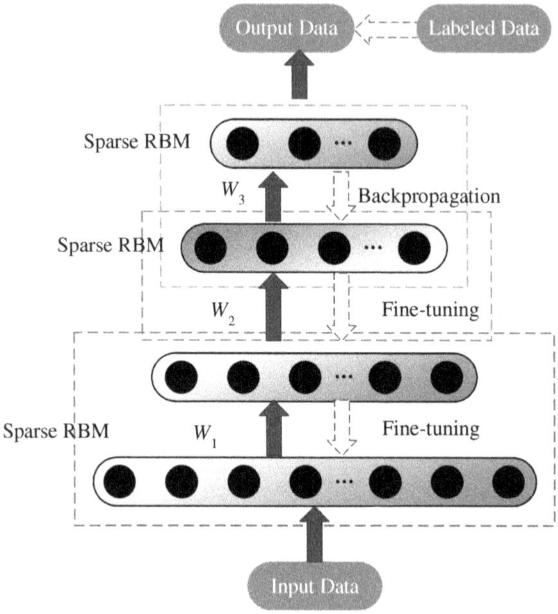

Figure 6.3. Structure of sparse DBNs.

Step 1: Pretraining

Each sparse RBM model is trained layer-by-layer using the Contrastive Divergence (CD) algorithm to obtain the connection weights between the visible and hidden layers.

Step 2: Fine-tuning

After pretraining, in order to enhance the model's feature representation capability, the labeled data is used and the model's discriminative performance is optimized using the conjugate gradient method.

Experiments in Ref. [118] demonstrate that sparse DBNs can extract features such as contours and corners from natural images. This is highly similar to the receptive field functions of the V2 area in the visual cortex and shows a qualitative improvement over the previous linear sparse representation method. The addition of the regularization term R can ensure that each stochastic unit in the hidden layer maintains the same activation rate, which implies that, under certain conditions, all hidden units represent the input data features in a consistent manner.

6.2.2.2 *Improvement of sparse DBNs*

A new sparse Deep Belief Network (sparse DBN) is constructed by stacking multiple layers of the improved sparse RBM models. During training network, the improved sparse RBM algorithm is used to train the bottom-layer sparse RBM model to obtain the parameters w_{ji}, b_i, and c_j, which are then used as the input for training the next sparse RBM layer. The training architecture for the sparse DBM with L layers is as follows:

Step 1: Train the first sparse RBM layer to obtain the weight matrix $\boldsymbol{w}^{(1)}$.

Step 2: Use the hidden layer data from the previous layer to train the next sparse RBM in the same manner and to fix the connection weights $\boldsymbol{w}^{(l)}$, and repeat until layer L-1.

Step 3: Initialize the weights $\boldsymbol{w}^{(l)}$, and use the label values of the data as the output layer.

Step 4: Use the conjugate gradient method to optimize the weights $\{\boldsymbol{w}^{(1)}, \boldsymbol{w}^{(2)}, \ldots, \boldsymbol{w}^{(\mathrm{L})}\}$.

6.3 Gamma Deep Belief Network

Based on the DBN architecture, the Gamma distribution is introduced to model the hidden units to make the network connect different layers via binary units and infer multi-layer representations of feature vectors without bias. Unlike traditional deep networks such as the DBN, which require binary units for inference and necessitate tuning both the width (number of hidden units) and depth (number of layers) of each layer, the Gamma DBM performs joint training of all hidden layers. This is achieved by performing Gibbs sampling: Dirichlet-distributed connection weights are sampled upward, and Gamma-distributed hidden units are sampled downward. This greedy layer-wise learning approach enables the model to determine the width of each hidden layer under a given fixed width of the first layer. This is the Gamma deep belief network [119].

6.3.1 Structure of the Gamma deep belief network

The Gamma DBN [119] is composed of several layers from bottom to top: the observation layer, the first hidden layer, ..., the $(L-1)$th hidden layer, and the Lth top layer. The bottom-most layer is the observation layer, which consists of the observed units $\boldsymbol{y}_n^{(1)}$. By employing Poisson factor analysis [118], it can be represented as the product of the connection weights and the next-layer hidden units:

$$\boldsymbol{y}_n^{(1)} \sim \text{Pois}(\boldsymbol{\Omega}^{(1)}\boldsymbol{h}_n^{(1)}), \tag{6.3.1}$$

where $\boldsymbol{\Omega}^{(1)}$ denotes the connection weights, $\boldsymbol{y}_n^{(1)}$ denotes the first hidden layer, and $\boldsymbol{x} \sim \text{Pois}(\lambda)$ indicates that x follows a Poisson distribution with parameter λ.

From the observation layer to the top layer, the 1st, ..., lth, ..., Lth hidden layers are arranged in order. From the 1st hidden layer to $(L-1)$th hidden layer, $\boldsymbol{h}_n^{(1)}$, $\boldsymbol{h}_n^{(2)}, \ldots, \boldsymbol{h}_n^{(L-1)}$ can be expressed respectively as

$$\begin{cases} \boldsymbol{h}_n^{(1)} : \text{Gam}\left(\boldsymbol{\Omega}^{(2)}\boldsymbol{h}_n^{(2)}, \frac{\varphi_n^{(2)}}{1-\varphi_n^{(2)}}\right) \\ \boldsymbol{h}_n^{(l+1)} : \text{Gam}\left(\boldsymbol{\Omega}^{(l+2)}\boldsymbol{h}_n^{(l+2)}, \frac{1}{\lambda_n^{(l+2)}}\right), \\ \boldsymbol{h}_n^{(L-1)} : \text{Gam}\left(\boldsymbol{\Omega}^{(L)}\boldsymbol{h}_n^{(L)}, \frac{1}{\lambda_n^{(L)}}\right) \end{cases} \tag{6.3.2}$$

where $\boldsymbol{\Omega}^{(l)}$ is the connection weight matrix, $\boldsymbol{h}_n^{(l)}$ is the parameters of the lth hidden layer at level l, and $\varphi_n^{(2)}$ is the probability parameter satisfying

$$\frac{\varphi_n^{(2)}}{1-\varphi_n^{(2)}} = \frac{1}{\lambda_n^{(2)}}. \tag{6.3.3}$$

$x \sim \text{Gam}\left(\alpha, \frac{1}{\lambda}\right)$ indicates that $\boldsymbol{x}$ follows a Gamma distribution with shape parameter α and scale parameter $\frac{1}{\lambda}$. Its probability density function is calculated by

$$f(x;\alpha,\lambda) = \begin{cases} \frac{\lambda^\alpha x^{\alpha-1}e^{-\lambda x}}{\Gamma(\alpha)}, & x > 0 \\ 0, & x \leq 0 \end{cases}. \tag{6.3.4}$$

Similarly, the Lth layer can be expressed as

$$\begin{cases} \boldsymbol{h}_n^{(L)} : \text{Gam}\left(\boldsymbol{\alpha}, \frac{1}{\lambda_n^{(L+1)}}\right) \\ \boldsymbol{\alpha} = (\alpha_1, \alpha_2, \ldots, \alpha_{M_L}) \end{cases}. \tag{6.3.5}$$

For the top layer $\boldsymbol{h}_n^{(L)}$, $\boldsymbol{\alpha}$ is the shared shape parameter of Gamma distribution, and $\frac{1}{\lambda_n^{(L+1)}}$ is the scale parameter.

To limit network complexity and facilitate parameter inference, each column of $\boldsymbol{\Omega}^{(l)} \in \boldsymbol{R}^{M_{l-1} \times M_l}$ is subject to L1 regularization. For $l \in 1, 2, \ldots, L-1$, ensure that

$$\begin{cases} \omega_i^{(l)} : \text{Diri}\left(\eta^{(l)}, \ldots, \eta^{(l)} | \alpha_i\right) \\ \alpha_i : \ \text{Gam}\left(\frac{\gamma_0}{M_L}, \frac{1}{\lambda_0}\right) \end{cases}, \tag{6.3.6}$$

where Diri($\boldsymbol{p}|\boldsymbol{\alpha}$) indicates that $\boldsymbol{p}$ follows a Dirichlet distribution with parameter $\alpha = (\alpha_1, \alpha_2, \ldots, \alpha_L)$. Its probability density function is given by

$$\text{Diri}(\boldsymbol{p}|\boldsymbol{\alpha}) = \frac{1}{\boldsymbol{B}(\boldsymbol{\alpha})} \prod_{l=1}^{L} p_l^{\alpha_l - 1}. \tag{6.3.7}$$

$B(\alpha)$ represents the normalization constant of the Dirichlet distribution:

$$B(\boldsymbol{\alpha}) = \int \prod_{l=1}^{L} p_l^{\alpha_l - 1} \text{d}p. \tag{6.3.8}$$

In Equation (6.3.6), $\boldsymbol{\omega}_i^{(l)} \in \boldsymbol{R}^{M_{l-1}}$ is the ith column of $\boldsymbol{\Omega}^{(l)}$, and both λ_0 and γ_0 follow Gamma distributions:

$$\begin{cases} \lambda_0 : \text{Gam}\left(a_0, \frac{1}{b_0}\right) \\ \gamma_0 : \text{Gam}\left(c_0, \frac{1}{d_0}\right). \end{cases} \tag{6.3.9}$$

For $l = 3, \cdots, T+1$, we have

$$\begin{cases} \lambda_n^{(l)} : \text{Gam}\left(a_0, \frac{1}{b_0}\right) \\ \varphi_n^{(2)} : \text{Beta}(c_0, d_0) \end{cases}, \tag{6.3.10}$$

where $x \sim \text{Beta}(\alpha, \beta)$ indicates that x follows a Beta distribution with parameters α, β [174], and its probability density function is calculated by

$$f(x; \alpha, \beta) = \frac{1}{B(\alpha, \beta)} x^{\alpha-1}(1-x)^{\beta-1}. \tag{6.3.11}$$

Therefore, the relationship between the hidden units $\boldsymbol{h}_1^{(l)}, \boldsymbol{h}_2^{(l)}, \ldots, \boldsymbol{h}_n^{(l)}$ in row M_l can be represented by the column vector $\boldsymbol{\omega}_i^{(l+1)}$ $(i = 1, 2, \ldots, M_L)$ of $\boldsymbol{\Omega}^{(l+1)}$.

Since the conjugate prior of the Gamma distribution's shape parameter within the network is unknown, it is difficult to derive the network structure by computing the conditional posterior. Therefore, the data augmentation algorithm proposed in Ref. [120] is employed for simplifying the computation and derivation. It is found that in a single-layer network ($T = 1$), the hidden units at each layer are independent of the prior. However, in a deep network ($T \geq 2$), the model can effectively capture the correlation among hidden units. For $l = 1, 2, \ldots, L$, we have

$$\begin{cases} \varphi_n^{(1)} = 1 + \frac{1}{e} \\ \varphi_n^{(l+1)} = \frac{-\ln\left(1-\varphi_n^{(l)}\right)}{\lambda_n^{(l+1)} - \ln\left(1-\varphi_n^{(l)}\right)}. \end{cases} \tag{6.3.12}$$

Then, $\boldsymbol{u}_n^{(l)}$ ($l = 1$ for the observation layer, $l = 2, 3, \ldots, L$ for the hidden layers) can be represented by the product of $\boldsymbol{\Omega}^{(l)}$ and $\boldsymbol{h}_n^{(l)}$ under the Poisson probability for the l th layer as follows:

$$\boldsymbol{u}_n^{(l)} : \text{Pois}(-\boldsymbol{\Omega}^{(l)} \boldsymbol{h}_n^{(l)} \ln(1 - \varphi_n^{(l)})). \tag{6.3.13}$$

Equation (6.3.13) holds for $l = 1$, while for $l = 2, 3, \ldots, L$, we have

$$\begin{cases} u_{mn}^{(l)} = \sum_{i=1}^{M_L} u_{mni}^{(l)} \\ u_{mni}^{(l)} : \text{Pois}(-\omega_{mi}^{(l)} \boldsymbol{h}_{in}^{(l)} \ln(1 - \varphi_n^{(l)})) \end{cases}. \tag{6.3.14}$$

Let $q_{in}^{(l)(l+1)} = u_{ni}^{(l)} = \sum_{m=1}^{M_l} u_{mmi}^{(l)}$ represent the number of occurrences of the factor $i \in \{1, 2, \ldots, M_l\}$ in the nth observation unit in

kth layer, where $q_n^{(l)(l+1)} = (u_{n1}^{(l)}, u_{n2}^{(l)}, \ldots, u_{nM_l}^{(l)})'$. Then, marginalizing $\boldsymbol{\Omega}^{(l)}$, we obtain

$$q_n^{(l)(l+1)} : \text{Pois}(-\boldsymbol{h}_n^{(l)} \ln(1 - \varphi_n^{(l)})). \tag{6.3.15}$$

From the above Poisson probability, after marginalizing the Gamma distribution $\boldsymbol{o}_n^{(l)}$, we get

$$q_n^{(l)(l+1)} \sim \text{NB}(\boldsymbol{\Omega}^{(l+1)} \boldsymbol{h}_n^{(l+1)}, \boldsymbol{\varphi}_n^{(l+1)}), \tag{6.3.16}$$

where $x \sim \text{NB}(r, p)$ indicates that x follows the negative binomial distribution with parameters r and p.

From Equations (6.3.10) and (6.3.12), the $l + 1$th layer can be derived from the lth layer's hidden variable $u_{mn}^{(l)}$:

$$\begin{aligned} &\{(u_{mn1}^{(l)}, \ldots, u_{mnM_k}^{(l)}) | u_{mn}^{(l)}, \phi_m^{(l)}, \boldsymbol{h}_n^{(l)}\} \\ &\qquad \sim \text{Multi}\left(u_{mn}^{(l)}, \frac{\omega_{m1}^{(l)} \boldsymbol{h}_{1n}^{(l)}}{\sum_{i=1}^{M_k} \omega_{mi}^{(l)} \boldsymbol{h}_{in}^{(l)}}, \ldots, \frac{\omega_{mi}^{(l)} \boldsymbol{h}_{in}^{(l)}}{\sum_{i=1}^{M_k} \omega_{mi}^{(l)} \boldsymbol{h}_{in}^{(l)}} \right), \end{aligned} \tag{6.3.17}$$

$$\{u_{in}^{(l+1)} | q_{in}^{(l)(l+1)}, \omega_{i:}^{(l+1)}, \boldsymbol{h}_n^{(l+1)}\} \sim \text{CRT}(q_{in}^{(l)(l+1)}, \omega_{i:}^{(l+1)}, \boldsymbol{h}_n^{(l+1)}), \tag{6.3.18}$$

where *Multi* represents the multinomial distribution and *CRT* stands for the Chinese Restaurant Process (CRP) [121].

6.3.2 Gibbs upward and downward sampling

Gibbs sampling is an algorithm for approximately drawing a sample sequence from a multi-component probability distribution when direct sampling from the samples is difficult. In the DBN, the RBM uses two-dimensional Gibbs sampling to approximate parameter estimation and maps the visible vector values to hidden units and then reconstructs the visible units based on the hidden units. This process is repeated layer by layer during training. Similarly, in a Gamma DBN, N-dimensional Gibbs sampling can be used to estimate hidden variables and jointly train all layers of the network. In each iteration, one layer of the network is sampled, with upward sampling of the connection weights that follow a Dirichlet distribution and downward sampling of the hidden units that follow a Gamma distribution. This is referred to as Gibbs upward and downward sampling.

For each layer of the Gamma DBN, the iterative sampling process is as follows:

(1) Sampling $u_{mni}^{(l)}$

From Equation (6.3.17), sampling for $u_{mni}^{(l)}$ is performed across all layers. For the first hidden layer, the observation unit $u_{mn}^{(1)}$ can be considered as the sequence of the mth feature in the nth state. The $\{m_{nj}\}_{j=1,\ldots,u_n}$ values are sequentially assigned to the hidden factors, and $\boldsymbol{\Omega}^{(1)}$ and $\boldsymbol{h}_n^{(1)}$ are marginalized. For $i \in \{1, 2, \ldots, M_{1\,\max}\}$, we have

$$p(l_{nj} = i \mid -) \propto \frac{\delta^{(1)} + u_{m_{nj}\ gi}^{(1)-nj}}{M\delta^{(1)} + u_{ggi}^{(1)-nj}} (u_{gni}^{(1)^{-nj}} + \boldsymbol{\omega}_i^{(2)} \boldsymbol{h}_n^{(2)}), \qquad (6.3.19)$$

where l_{nj} is the feature label of m_{nj}, and the symbol represents the summation for the corresponding label, such as $u_{\cdot ni}^{(1)} = \sum_m x_{mni}^{(1)}$, where x^{-nj} represents the count sequence for feature j in the nth state. To simplify the model, a truncation step is introduced: If $L = 1$, the number of hidden units is constrained to $M_{1\,\max}$, and $\alpha_i \sim \text{Gam}\left(\frac{\gamma_o}{M_{1\,\max}}, \frac{1}{\lambda_o}\right)$.

(2) Sampling $\omega_i^{(l)}$

$$(\omega_i^{(l)} \mid -) = \text{Diri}(\eta^{(l)} + u_1^{(l)}, \ldots, \eta^{(l)} + u_{M_{l-1}gl}^{(l)}) \qquad (6.3.20)$$

(3) Sampling $u_{mn}^{(l+1)}$

From Equation (6.3.19), sampling for $\boldsymbol{u}_n^{(l+1)}$ is performed, and $\boldsymbol{\Omega}^{(L+1)} \boldsymbol{h}_n^{(L+1)}$ is replaced by $\boldsymbol{\alpha} = (\alpha_1, \alpha_2, \ldots, \alpha_{M_L})^{\mathrm{T}}$.

(4) Sampling $\boldsymbol{h}_n^{(l)}$

Based on the conjugacy between Gamma and Poisson distributions from Equation (6.3.13), sampling for $\boldsymbol{h}_n^{(l)}$ is performed according to

$$(\boldsymbol{h}_n^{(l)} \mid -) = \text{Gam}(\boldsymbol{\Omega}^{(l+1)} \boldsymbol{h}_n^{(l+1)} + q_n^{(l)(l+1)}, (\lambda_n^{(l+1)} - \ln(1 - \varphi_n^{(l)}))^{-1}). \qquad (6.3.21)$$

(5) Sampling $\boldsymbol{\alpha}$

$$(\alpha_m|-) = \text{Gam}\left(\frac{\gamma_0}{M_L} + u_{m\cdot}^{(L+1)}, \lambda_0 - \sum_n \ln(1-\varphi_n^{(L+1)})^{-1}\right). \tag{6.3.22}$$

(6) Sampling $\lambda_n^{(l)}$

First, sampling for $\varphi_n^{(2)}$ is performed according to

$$(\varphi_n^{(2)}|-) = \text{Beta}(c_0 + q_{\cdot n}^{(1)(2)}, d_0 + \ \boldsymbol{h}_{\cdot n}^{(2)}). \tag{6.3.23}$$

Then, from Equation (6.3.17), we obtain

$$(\lambda_n^{(l)}|-) = \text{Gam}(a_0 + \boldsymbol{h}_{\cdot n}^{(l)}, (b_0 + \boldsymbol{h}_{\cdot n}^{(l-1)})^{-1}), \tag{6.3.24}$$

where $\boldsymbol{h}_{\cdot n}^{(l)} = \sum_{i=1}^{M_k} \boldsymbol{h}_{in}^{(l)} (l = 1, 2, \ldots, L), \boldsymbol{h}_{\cdot n}^{(L+1)} = \alpha$.

6.4 Adaptive Deep Belief Network

In the field of deep learning, the choice of optimization algorithm is of paramount importance. Even with the same dataset and model architecture, different optimization algorithms can lead to vastly different training effectiveness. In traditional DBN, the CD algorithm is typically used to solve the negative gradient of the log-likelihood function of Equation (6.3.2) layer by layer to obtain the optimal solution $\boldsymbol{\theta}$ of each layer's RBM [12, 122]. In this section, the Nadam optimization algorithm, which improves the prediction error of the DBN and provides better adaptability [123], is introduced.

6.4.1 Momentum update rule

Classical momentum accumulates the sum of decay of previous gradients (along with decay constant γ) into a momentum vector $\boldsymbol{m}$, and it is replaced with the true gradient. After adding this feature to the gradient, it can make the gradient faster in the dimension where the direction remains unchanged and slower in the dimension where the direction has changed; thus, this approach can accelerate convergence and reduce oscillation.

The updating formulas of the gradients and momentum term are written as

$$\boldsymbol{g}(k) = \nabla_{\boldsymbol{\theta}(k)} Loss(\boldsymbol{\theta}(k)), \tag{6.4.1}$$

$$\boldsymbol{m}(k) = \gamma \boldsymbol{m}(k) + \eta \boldsymbol{g}(k), \tag{6.4.2}$$

$$\boldsymbol{\theta}(k+1) = \boldsymbol{\theta}(k) - \boldsymbol{m}(k). \tag{6.4.3}$$

where $\boldsymbol{g}(k)$ is the gradient vector, $Loss(\boldsymbol{\theta}(k))$ is the loss function, $\boldsymbol{m}$ is the momentum vector (with initial value 0), γ is the decay constant, η is the learning rate, and k represents the number of iterations. After Equations (6.4.1)–(6.4.3) are expanded, we get

$$\boldsymbol{\theta}(k+1) = \boldsymbol{\theta}(k) - (\gamma \boldsymbol{m}(k-1) + \eta \boldsymbol{g}(k)). \tag{6.4.4}$$

As seen, the momentum term takes a step along the direction of the previous momentum vector and the current gradient.

6.4.2 Nadam algorithm for DBN optimization

Nadam algorithm is similar to the Adam algorithm with a Nesterov momentum term, where the Nesterov momentum term is added to the Adam algorithm, the previous momentum vector can effectively replace the earlier momentum vector. Therefore, the updating formulas for the Adam algorithm becomes

$$\boldsymbol{m}(k) = \beta_1 \boldsymbol{m}(k-1) + (1-\beta_1)\, \boldsymbol{g}(k), \tag{6.4.5}$$

$$\boldsymbol{v}(k) = \beta_2 \boldsymbol{v}(k-1) + (1-\beta_2)\, \boldsymbol{g}(k) e \boldsymbol{g}(k), \tag{6.4.6}$$

$$\hat{\boldsymbol{m}}(k) = \frac{\boldsymbol{m}(k)}{1-\beta_1^k}, \tag{6.4.7}$$

$$\hat{\boldsymbol{v}}(k) = \frac{\boldsymbol{v}(k)}{1-\beta_2^k}, \tag{6.4.8}$$

$$\boldsymbol{\theta}(k+1) = \boldsymbol{\theta}(k) - \frac{\eta}{\sqrt{\hat{\boldsymbol{v}}(k)} + \varepsilon} \hat{\boldsymbol{m}}(k), \tag{6.4.9}$$

where $\boldsymbol{m}(k)$ and $\boldsymbol{v}(k)$ are the first-order and second-order moment estimates of the gradient, which can be considered estimates of the expectations of $|\boldsymbol{g}(k)|$ and $|\boldsymbol{g}(k)|^2$, respectively. β_1, β_2, and ε are correction parameters. Extensive experiments show that the good default parameters for the tested machine learning problems [12] are set to $\beta_1 = 0.9$, $\beta_2 = 0.999$, and $\varepsilon = 10^{-8}$, where ε prevents the denominator from becoming zero. Equation (6.4.9) is expanded as

$$\boldsymbol{\theta}(k+1) = \boldsymbol{\theta}(k) - \frac{\eta}{\sqrt{\hat{\boldsymbol{v}}(k)} + \varepsilon}\left(\frac{\beta_1 \boldsymbol{m}(k-1)}{1-\beta_1^k} + \frac{1-\beta_1}{1-\beta_1^k}\boldsymbol{g}(k)\right). \tag{6.4.10}$$

The first term in the parentheses, i.e., $\frac{\beta_1 \boldsymbol{m}(k-1)}{1-\beta_1^k}$, is merely the bias-corrected estimate of the momentum vector from the previous time step, denoted by $\hat{\boldsymbol{m}}(k)$, i.e.,

$$\boldsymbol{\theta}(k+1) = \boldsymbol{\theta}(k) - \frac{\eta}{\sqrt{\hat{\boldsymbol{v}}(k)} + \varepsilon}\left(\boldsymbol{m}(k-1) + \frac{1-\beta_1}{1-\beta_1^k}\boldsymbol{g}(k)\right). \tag{6.4.11}$$

Equation (6.4.11) is quite similar to the expanded momentum term in Equation (6.4.4). When incorporating Nesterov momentum term, the look-ahead momentum vector is directly applied to update the parameters. This is achieved by replacing the bias estimate of momentum vector $\hat{\boldsymbol{m}}(k-1)$ at the previous time step with bias estimate $\hat{\boldsymbol{m}}(k)$ at the current momentum vector. Thus, the Nadam updating formulas becomes

$$\boldsymbol{\theta}(k+1) = \boldsymbol{\theta}(k) - \frac{\eta}{\sqrt{\hat{\boldsymbol{v}}(k)} + \varepsilon}\left(\boldsymbol{m}(k) + \frac{1-\beta_1}{1-\beta_1^k}\boldsymbol{g}(k)\right). \tag{6.4.12}$$

Traditional stochastic gradient descent (SGD) method holds a single learning rate to update all weights during training. In contrast, the Nadam algorithm is utilized to compute first-order and second-order moment estimates of the gradient to design independent and adaptive learning rates for different parameters. As a result, the Nadam algorithm imposes a stronger constraint on the learning rate and has a more direct influence on updating gradient. Generally speaking, the Nadam algorithm can be employed in most cases where the RMSprop or Adam algorithm with momentum is used and often achieves better performance. By incorporating second-order information of the loss function, the Nadam algorithm can capture the relative change of the gradient between two iterations—essentially an approximation of the second derivative—thereby, it can enhance the adaptivity of the optimization process.

6.5 KPCA-Based Deep Belief Network Model

6.5.1 Kernel principal component analysis

Kernel Principal Component Analysis (KPCA) method is an effective nonlinear dimensionality reduction technique that eliminates nonlinear dependencies among variables and reduces data dimensionality. After the data are transformed by the KPCA method, the extracted data along with their corresponding labels are input into the DBN model to make the network fully learn and extract key features from the samples [123–125].

6.5.1.1 *Statistical smoothing method*

Due to various influencing factors, observational data are often subject to measurement errors. These erroneous data can significantly compromise the accuracy of model training and diagnostics. The statistical smoothing method, based on mathematical statistics, can reduce such measurement errors. By applying this method, the reliability and precision of the data can be effectively enhanced.

The statistical smoothing method is defined as

$$\overline{x}(k) = \frac{y(k) + y(k-1) + \cdots + y(k-N+1)}{N}, \tag{6.5.1}$$

where $\overline{x}(k)$ is the data processed by the smoothing method at time k and $y(k)$ is the observed value at time k.

In a set of original data, points that significantly differ from the others are labeled as outliers. Since outliers are distinctly different from other points, they may contain important information, so they are not subjected to smoothing.

The N observation data points are arranged in chronological order to form a random dataset that follows a normal distribution.

The standard deviation of the samples is given by

$$\sigma_k = \sqrt{\frac{1}{N-1}\sum_{i=1}^{N}(y_i - \overline{x}(k))}.$$

Based on the confidence criterion, if the kth data point in the random sequence falls within the 99.7% confidence interval $[\bar{x}_{(k)} - 3\sigma_k, \bar{x}_{(k)} +$

$3\sigma_k]$, smoothing is applied to the data processing. If the data do not fall within $[\bar{x}(k) - 3\sigma_k, \bar{x}(k) + 3\sigma_k]$, it is considered an outlier, and the original data is retained without change.

6.5.1.2 *Normalization processing*

In practice, different observed parameters often have different units and scales. To eliminate the influence of inconsistent dimensionalities among parameters, the smoothed data must be normalized. The normalization is performed according to the following equation:

$$x_i' = \frac{x_i - x_{\min}}{x_{\max} - x_{\min}}, \tag{6.5.2}$$

where x_i denotes the original measured value, $x_{\max}$ and $x_{\min}$ represent the maximum and minimum values in the sample data, respectively, and x_i' is the normalized value. After normalization, all values are scaled to the range $[0, 1]$, which facilitates subsequent data analysis and modeling.

6.5.1.3 *Analysis of Kernel principal component*

In complex and dynamic experimental environments, nonlinear correlations often exist among observed parameters. To eliminate such nonlinear dependencies and reduce the dimensionality of the dataset, the Kernel Principal Component Analysis (KPCA) method is introduced. The core idea of the KPCA method is to use a nonlinear mapping function to project the correlated data into a high-dimensional feature space. Then, traditional Principal Component Analysis (PCA) method is applied in this transformed space, where the inner product matrix is replaced by a kernel matrix, enabling effective nonlinear feature extraction and dimensionality reduction.

6.5.1.3.1 *Kernel function*

Let the function $\varphi(\cdot)$ be a nonlinear mapping that projects the correlated low-dimensional monitoring data into a high-dimensional feature space. A vector $\boldsymbol{x}_i$ in the low-dimensional feature space, mapped through function $\varphi(\cdot)$, becomes the vector $\boldsymbol{\varphi}(\boldsymbol{x}_i)$. If there exists a function $k(\cdot)$ in the low-dimensional space such that $k(\boldsymbol{x}_i, \boldsymbol{x}_j) = \boldsymbol{\varphi}(x_i)^{\mathrm{T}} \cdot \boldsymbol{\varphi}(\boldsymbol{x}_j)$, then this function $k(\cdot)$ is called a kernel function. If the matrix $\boldsymbol{K} = \boldsymbol{x}^{\mathrm{T}}\boldsymbol{x}$ in the high-dimensional space satisfies Equation (6.5.3), i.e., all elements $\varphi(x_i)^{\mathrm{T}} \cdot \varphi(\boldsymbol{x}_j)$ in matrix $\boldsymbol{K}$ are represented using the kernel function, then matrix $\boldsymbol{K}$ is called a kernel matrix:

$$\boldsymbol{K} = \boldsymbol{\Phi}^{\mathrm{T}}\boldsymbol{\Phi} = \begin{bmatrix} \varphi(\boldsymbol{x}_1)^{\mathrm{T}} \\ \vdots \\ \varphi(\boldsymbol{x}_N)^{\mathrm{T}} \end{bmatrix} [\varphi(\boldsymbol{x}_1) \cdots \varphi(\boldsymbol{x}_N)]$$
$$= \begin{bmatrix} k(\boldsymbol{x}_1, \boldsymbol{x}_1) & \cdots & k(\boldsymbol{x}_1, \boldsymbol{x}_N) \\ \vdots & & \vdots \\ k(\boldsymbol{x}_N, \boldsymbol{x}_1) & \cdots & k(\boldsymbol{x}_N, \boldsymbol{x}_N) \end{bmatrix}, \tag{6.5.3}$$

where $k(\cdot)$ represents the kernel function, $\boldsymbol{\Phi}$ denotes the sample matrix mapped into the high-dimensional feature space, and N indicates the number of samples.

Since the exact form of the function $\varphi(\cdot)$ is unknown, when extracting principal components in the high-dimensional feature space, the kernel matrix is used to replace the inner product matrix of that space. Thus, it is only necessary to analyze $\boldsymbol{K}$. The most commonly used kernel functions are as follows:

① Linear Kernel function

$$k(\boldsymbol{x}_i, \boldsymbol{x}_j) = \boldsymbol{x}_i^{\mathrm{T}}\boldsymbol{x}_j, \tag{6.5.4}$$

② Gaussian Kernel function

$$k(\boldsymbol{x}_i, \boldsymbol{x}_j) = \exp\left(-\frac{\|\boldsymbol{x}_i - \boldsymbol{x}_j\|^2}{2\sigma^2}\right), \tag{6.5.5}$$

③ Polynomial Kernel function

$$k(\boldsymbol{x}_i, \boldsymbol{x}_j) = (\boldsymbol{x}_i^{\mathrm{T}}\boldsymbol{x}_j)^d, \tag{6.5.6}$$

④ Laplacian Kernel function

$$k(\boldsymbol{x}_i, \boldsymbol{x}_j) = \exp\left(-\frac{\|\boldsymbol{x}_i - \boldsymbol{x}_j\|}{\sigma}\right), \tag{6.5.7}$$

⑤ Sigmoid Kernel function

$$k(\boldsymbol{x}_i, \boldsymbol{x}_j) = \tanh\left(\beta\boldsymbol{x}_i^{\mathrm{T}}\boldsymbol{x}_j + \alpha\right), \tag{6.5.8}$$

where σ, β, and α are parameters in the function expressions and d represents the exponent.

6.5.1.3.2 *Computation of Kernel principal component*

Kernel principal component computation involves transforming the extraction of kernel principal components into the problem of calculating the eigenvalues and corresponding eigenvectors of the kernel matrix. The calculation process of the KPCA method is as follows:

Step 1: N observations of observation parameter vector $\boldsymbol{x} = [x_1, x_2, \ldots, x_N]^T$ are performed to obtain the sample matrix $\boldsymbol{X} \in \boldsymbol{R}^{n\times m}$. An appropriate kernel function is selected based on analyses, and the corresponding kernel matrix $\boldsymbol{K}$ is calculated by using the selected kernel function and the sample matrix:

$$\boldsymbol{K} = \begin{bmatrix} k(\boldsymbol{x}_1, \boldsymbol{x}_1) & \cdots & k(\boldsymbol{x}_1, \boldsymbol{x}_N) \\ \vdots & & \vdots \\ k(\boldsymbol{x}_N, \boldsymbol{x}_1) & \cdots & k(\boldsymbol{x}_N, \boldsymbol{x}_N) \end{bmatrix}. \tag{6.5.9}$$

Step 2: KPCA method is performed under the assumption that the vector $\varphi(x_i)$ has zero mean. Since the explicit form of the function $\varphi(\cdot)$ cannot be given, the kernel matrix cannot be directly centralized. The matrix $\boldsymbol{K}$ is centralized using the following formula:

$$\overline{\boldsymbol{K}} = \boldsymbol{K} - \frac{1}{N}\boldsymbol{KI} - \frac{1}{N}\boldsymbol{IK} + \frac{1}{N^2}\boldsymbol{IKI}, \tag{6.5.10}$$

where $\boldsymbol{I}$ is an $N \times N$ matrix that all elements equal to 1 and $\overline{\boldsymbol{K}}$ is the centralized kernel matrix obtained by applying Equation (6.5.10).

Step 3: The eigenvalues λ_i and the corresponding eigenvectors $\boldsymbol{v}_i (i = 1, 2, \ldots, I)$ of $\overline{\boldsymbol{K}}$ are calculated.

Step 4: The kernel principal component vectors of $\boldsymbol{\Phi}$ are calculated according to the following equation:

$$\boldsymbol{t}_i = (\lambda_i)^{-\frac{1}{2}} \boldsymbol{\Phi}\boldsymbol{\Phi}' v_i = (\lambda_i)^{-\frac{1}{2}} \overline{\boldsymbol{K}}\boldsymbol{v}_i. \tag{6.5.11}$$

Step 5: The variance contribution rate and cumulative contribution rate are calculated by

$$\begin{cases} \xi_i = \dfrac{\lambda_i}{\sum\limits_{j=1}^{J} \lambda_j} \\ \eta_r = \sum\limits_{i=1}^{r} \xi_i = \dfrac{\sum\limits_{i=1}^{r} \lambda_i}{\sum\limits_{j=1}^{J} \lambda_j} \end{cases}, \tag{6.5.12}$$

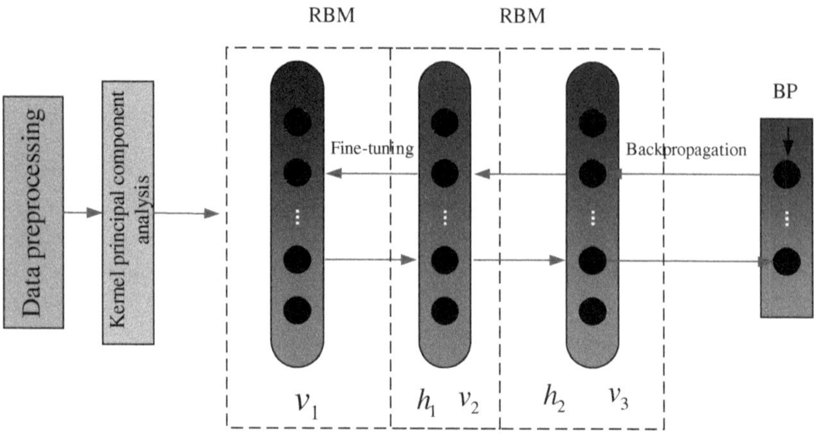

Figure 6.4. Network model based on KPCA and DBN.

where λ_i represents the variance of the ith principal component, ξ_i denotes the variance contribution rate of the ith principal component, and η_r is the cumulative variance contribution rate of the first r principal components.

6.5.2 DBN based on KPCA analysis

After normalizing the observed original data, the KPCA method is applied to extract the main features of the data and reduce its dimensionality. The reduced-dimensional data is then sequentially fed into the visible layer of the first RBM. By using a greedy layer-wise learning strategy to train the RBM, all RBMs in the network are progressively trained. Finally, a neural network is added at the top layer of the DBN to perform classification tasks. The backpropagation (BP) algorithm is used to fine-tune the parameters of the entire DBN using labeled samples. The DBN model based on the KPCA method [126] is shown in Figure 6.4.

6.6 Fully Parametric Dynamic Learning Deep Belief Network

The training process of the DBN mainly consists of two steps: First, each layer of the RBM is trained using an unsupervised learning method, where the input of each RBM is the output of the previous one. Each RBM is trained individually to ensure that the feature vector retains as much information as possible when being mapped into different feature spaces. Second, the final layer—typically a back-propagation (BP) neural network—receives the output of the last RBM and is trained in a supervised manner to fine-tune the overall network.

In a typical DBN composed of three stacked RBMs (as shown in Figure 6.1), during the unsupervised forward-stacking RBM learning phase, the vector $\boldsymbol{v}$ is generated in the visible layer and transmits the input data from the visible layer to the hidden layer. During this process, the input in the visible layer is randomly sampled to attempt reconstruction of the original input signals. Then, the neuron activation units of the newly obtained visible layer continue to propagate forward to obtain $\boldsymbol{h}$ via reconstructing the neuron activation units of the hidden layer. These repeated forward and backward steps constitute Gibbs sampling [126]. Throughout this process, the update of the weights is primarily based on the difference in correlation between the activation units of the hidden layer and the visible input signals.

For all units of the hidden layer, we calculate

$$p\left(h_j^1 = 1|v^1\right) = \text{sigmoid}\left(c_j + \sum_i v_i^1 w_{ji}\right), \tag{6.6.1}$$

where $h_j^1 \in \{0,1\}$ and $j = 1, 2, \ldots, J$.

For all units of the visible layer, we calculate

$$p(v_i^2 = 1|h^1) = \text{sigmoid}\left(b_i + \sum_j h_j^1 w_{ji}\right), \tag{6.6.2}$$

where $v_i^2 \in \{0,1\}$ and $i = 1, 2, \cdots, N$.

For all units of the hidden layer, we compute

$$p(h_j^2 = 1|v^2) = \text{sigmoid}\left(c_j + \sum_i v_i^2 w_{ji}\right), \tag{6.6.3}$$

where $h_j^2 \in \{0,1\}$ and $j = 1, 2, \ldots, J$.

The updating formulas of parameters are written as

$$\begin{cases} \boldsymbol{w}(k+1) = \boldsymbol{w}(k) + \eta(p(h^1 = 1|\boldsymbol{v}^1)(\boldsymbol{v}^1)^{\mathrm{T}} - p(h^2 = 1|\boldsymbol{v}^2)(\boldsymbol{v}^2)^{\mathrm{T}}) \\ \boldsymbol{b}(k+1) = \boldsymbol{b}(k) + \alpha(\boldsymbol{v}^1 - \boldsymbol{v}^2) \\ \boldsymbol{c}(k+1) = \boldsymbol{c}(k) + \beta(p(h^1 = 1|\boldsymbol{v}^1) - p(h^2 = 1|\boldsymbol{v}^2)) \end{cases}. \tag{6.6.4}$$

By following the above steps, iterative updates are completed, and the next RBM network is trained sequentially. Eventually, the final updated parameters of the DBN are obtained.

After the unsupervised forward stacking RBM training is completed, the parameters of each RBM layer can be initialized, which essentially provides prior knowledge of the input data for subsequent supervised learning. Then, a supervised backpropagation fine-tuning algorithm [127] is used to adjust the weights of the DBN. The output error is used to estimate the error between the output layer and the preceding layer. Similarly, layer-by-layer backpropagation training is performed to obtain the error between other layers. Finally, the batch gradient descent method is applied to compute and update the weights of each node until the output error meets the required threshold. The fine-tuning process starts from the last layer of the DBN. The fine-tuning formula is as follows:

$$(\boldsymbol{w}^{(l)}, \boldsymbol{c}^{(l)})(k+1) = (\boldsymbol{w}^{(l)}, \boldsymbol{c}^{(l)})(k) - (\eta, \beta)\frac{\partial Loss}{\partial(\boldsymbol{w}^{(l)}, \boldsymbol{c}^{(l)})}. \tag{6.6.5}$$

The BP algorithm in the DBN only requires a local search in the weight parameter space. This weight fine-tuning method overcomes the drawbacks of traditional BP networks, which are prone to getting stuck in local minima and taking too long to train due to randomly initializing weight parameters. By fine-tuning within a known parameter space, the convergence time for parameter optimization is significantly reduced. Furthermore, the use of the Contrastive Divergence (CD) algorithm enables rapid training by decomposing the DBN into multiple RBM structures, thereby avoiding the high complexity of training the entire DBN directly. This training method, followed by traditional backpropagation algorithm for fine-tuning of the network, greatly enhances the modeling capability of the network and accelerates convergence to the optimal solution.

In the DBN model, two processes involve the use of a learning rate: the forward stacking of the RBMs and the backward fine-tuning process. The learning rate significantly influences the learning progress of the network. An appropriate learning rate is essential to ensure that the parameters $\boldsymbol{\theta} = (\boldsymbol{w}, \boldsymbol{b}, \boldsymbol{c})$ can be optimized to their best state.

In the DBN model, parameter optimization—namely, the updating of weights and biases—follows the general update formula:

$$\boldsymbol{\theta}(k+1) = \boldsymbol{\theta}(k) - \eta(k)g(\boldsymbol{\theta}(k)), \tag{6.6.6}$$

where $\boldsymbol{\theta}(k+1)$ represents the parameter value after $k+1$ iterations, $\theta(k)$ is the parameter value after k iterations, $\eta(k)$ denotes the learning rate (step size), and $g(\boldsymbol{\theta}(k))$ is recorded as the gradient of the loss function defined on the dataset.

According to the distinct characteristics and functions of connection weights and biases, a fully parameterized dynamic learning strategy is proposed [128], mathematically expressed as follows:

(1) Parameter learning strategy during the forward stacking of the RBM networks is given by

$$\eta(k+1) = \frac{\eta(k)}{\sqrt{\varepsilon + D(k)}}, \tag{6.6.7}$$

$$D(k) = \xi \mathrm{d}^2(k-1) + \mathrm{d}^2(k), \tag{6.6.8}$$

$$\alpha(k+1) = \alpha(k)\left(1 + \frac{k}{T_{\max}}\right)^{-q}, \tag{6.6.9}$$

$$\beta(k+1) = \beta(k)\left(1 + \frac{k}{T_{\max}}\right)^{-q}, \tag{6.6.10}$$

where $\eta(k+1)$ is the learning rate for the next round of the connection weights, $\eta(k)$ is the learning rate for the current round of connection weights, ε is set to 1, $D(k)$ is a certain proportion of the square sum of the previous gradient and the current gradient, and ξ is the decay factor, with a value of 0.9. $\alpha(k+1)$ and $\beta(k+1)$ are the learning rates for the visible and bias of the hidden unit in the kth iteration. $\alpha(k)$ and $\beta(k)$ are the learning rates for the visible and bias of the hidden unit in the $k+1$th iteration. A power-exponential function with a decreasing trend is used. $T_{\max}$ is the maximum number of iterations and q is set to 0.75.

(2) Parameter learning strategy during the backpropagation fine-tuning process is given by

$$\mu(k+1) = \frac{\mu(k)}{\sqrt{\varepsilon + D(k)}}, \tag{6.6.11}$$

where $\mu(k+1)$ is the learning rate for the next round of the connection weights during the backpropagation fine-tuning process and $\mu(k)$ is the learning rate for the current round of connection weights.

The idea of this learning strategy is to adaptively adjust the learning rate for the next round by using the current learning rate and the sum of the squares of the last two gradients of weights. The sum of squares of the last two gradients only is used to reduce the tedious calculation of historical gradients. At the same time, the learning rate dynamically changes with the number of iterations, which accelerates the convergence speed of the model. For bias, from the perspective of reducing computational complexity, a power exponential function that is only related to the current learning rate is set to accelerate the convergence speed of the model.

6.7 Optimization Algorithms of Deep Belief Network

6.7.1 Deep belief network based on chaotic immune optimization

The traditional parameter training methods for DBNs exhibit certain limitations, which affect their feature extraction capability and convergence speed to some extent. First, the random initialization of parameters of the network often leads the shallow network layers into local optima during the training process, thereby impairing the DBN's ability to extract features. Second, when extracting low-level features from high-dimensional data, the DBNs require the high-dimensional data to be directly input into the network, which significantly increases the number of parameters in the network and slows down the training convergence. To overcome these shortcomings, the Particle Swarm Optimization (PSO) algorithm can be employed for determining the optimal structure, connection weights, and biases of the DBN. Alternatively, the dropout technique can be used during training DBN, where a portion of the nodes of the hidden layer is randomly dropped in each iteration to prevent potential overfitting. However, this random node removal may introduce some degree of error. The advantage of the DBNs lies in their powerful feature extraction capabilities, which are directly dependent on the network's parameters. The parameters of the DBN include hyperparameters and trainable parameters. Hyperparameters refer to the number of hidden layers and nodes, learning rate, momentum, etc., while trainable parameters are the connection weights and biases through learning and training of the network.

The Artificial Immune Algorithm (AIA), known for its strong global search capability and fast convergence, has been widely used in optimizing neural networks to enhance their convergence speed and generalization performance. This section introduces a method for optimizing the parameters of the DBN using an improved chaotic immune algorithm.

6.7.1.1 *Clone selection algorithm*

The Clone Selection Algorithm (CSA) draws inspiration from the immune mechanism of antibody clone selection in artificial immune systems and has advantages such as strong global searching ability and fast optimization speed. Compared with other intelligent algorithms, the CSA can generate optimal solutions that are more conducive to complex optimization problems. When applying the CSA to complex optimization problems, the solution of the problem to be optimized is mapped to antibodies, the objective function of the problem to be optimized is mapped to affinity, and the degree of matching between the optimized solution and the objective function is mapped to antigen and antibody affinity. Antibodies with higher affinity are cloned and selected, and the diversity of antibodies is maintained through high-frequency mutation and concentration suppression. The implementation process is shown in Figure 6.5.

6.7.1.2 *Improved chaotic immune algorithm*

The basic CSA algorithm tends to fall into local optima during the optimization process due to the randomness and lack of directionality in antibody mutation, which affects the convergence speed of the algorithm. In this section, adaptive mutation is employed for

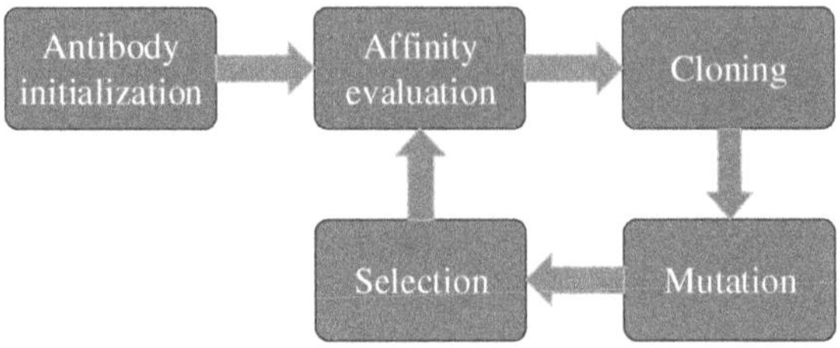

Figure 6.5. Implementation flow of the CSA.

enhancing the algorithm's global and local search capabilities. Firefly optimization mutation is used to conduct directional search on the antibody population, chaotic mutation performs global searching, and boundary mutation controls the population's searching range. In addition, the time complexity of the CSA is denoted as $O(T \times Ab)$ (where T is the number of generations and Ab is the antibody population size). Therefore, for low-dimensional optimization problems, the small antibody population leads to rapid convergence, whereas for optimization problems of high-dimensional parameters, the required antibody population becomes extremely large, and the convergence of the algorithm slows down. A variable selection operator is proposed here, where the antibody selection scale gradually decreases with the number of generations, thus the optimization speed is accelerated.

(1) Chaotic Initialization: A chaotic algorithm is introduced to initialize the antibody population of the CSA. The Logistic map is used as a chaotic formula to enable rapid search for antibodies. The chaotic formula is given by

$$z_{j+1} = uz_j(1 - z_j) \quad j = 1, 2, \ldots, m. \tag{6.7.1}$$

(2) Affinity Calculation: The affinity between antigen and antibody is measured by the output error of the DBN. A smaller output error of the DBN corresponds to a higher matching degree and thus higher affinity. The calculation formula is defined as

$$f(ab_i, ab_j) = \frac{1}{N}\sum_{n=1}^{N}(\hat{y}_n - y_n)^2, \tag{6.7.2}$$

where $\hat{y}_n$ is the actual output of the nth training sample, y_n is the expected output of the nth training sample, and N is the total number of training samples.

(3) Adaptive Mutation: The adaptive combination of firefly optimization mutation, chaotic mutation, and boundary mutation helps prevent the algorithm from falling into local optima while ensuring that antibodies evolve in a beneficial direction. In the firefly algorithm, each individual is regarded as a firefly with a certain perception ability. Within the searching space, the position of each firefly is updated based on the brightness and mutual attractiveness. Fireflies with weaker brightness are attracted to brighter ones, and searching for the optimal solution is equivalent to searching for the brightest firefly. Here, the firefly algorithm is introduced into the mutation process. Each antibody is treated as a firefly, and the affinity between antigen and antibody is considered the firefly's brightness. Through position optimization of the fireflies, directional mutation of antibodies can be achieved. The updating formula of the antibody is as follows:

$$ab_i(k+1) = ab_i(k) + \gamma_0 e^{-\beta r_{ij}^2} + \zeta\varepsilon_i, \tag{6.7.3}$$

where $ab_i(k+1)$ and $ab_i(k)$ represent the position of the ith antibody in the $(k+1)$th and kth generations, respectively. ζ is a random number within [0,1], ε_i is a random factor that follows a normal distribution within $[0,1]$, γ_0 is the maximum attraction, β is the absorption coefficient, and r_{ij} is the distance between antibody$i$and antibody j. In the firefly optimization mutation process, a local searching counter l is introduced, and a new updating mechanism of antibody is proposed as follows: Calculate the difference in affinity between the antibody obtained in each iteration and the original antibody. If the difference is greater than a set threshold χ, the current antibody is updated; Otherwise, the local searching counter $l = l+1$. If no change or a minimal change occurs in the antibodies for several generations, and the local searching counter reaches a certain value, it indicates the algorithm is trapped in the local searching process. At this point, Chaotic mutation is adopted to quickly escape the local optimum. Meanwhile, to prevent the antibodies from deviating from the searching space during optimization, boundary mutation is also considered. When an antibody crosses the boundary, boundary mutation is performed. The updating formula of antibody is as follows:

$$ab_i(k) = \begin{cases} x_{\max} \times (1 - c \times \text{rand}()), ab_i(k) > x_{\max} \\ x_{\min} \times (1 - c \times \text{rand}()), ab_i(k) > x_{\min}, \end{cases} \tag{6.7.4}$$

where $x_{\max}$ represents the maximum searching range, $x_{\min}$ represents the minimum searching range, and $c = 0.01$.

The improved adaptive mutation algorithm is as follows:

Algorithm 6.1. Mutation $(N_0, m, \gamma_0, \beta, x, \varsigma)$// Adaptive Mutation Algorithm [129]

Input: Initial antibody population N_0, antibody population size M, maximum attraction γ_0, absorption coefficient β, antibody searching range x, local searching threshold χ, and critical value ζ.

Output: Mutated antibody population T

For $t = 1$ to $T_{\max}$

  For $i = 1$ to M //The ith antibody begins adaptive mutation

   $ab_i(k) = ab_i(k-1) + \gamma_0 e^{-\beta r_{ij}^2} + \zeta\varepsilon_i$ //Perform firefly optimization mutation for the antibody

  If $ab_i(k) - ab_i(k-1) < \chi$

   $l = l + 1$

   If $l > \varsigma$ // Perform a chaotic mutation jumping out of local searching

    $ab_i(k) = \chi ab_i(k-1) \times (1 - ab_i(k-1))$

   End If

  End If

  If $ab_i(k) > x_{\max}$ // Boundary variations when crossing the border

   $ab_i(k) = x_{\max} \times (1 - c \times \text{rand}())$

  Else if $ab_i(k) < x_{\min}$

   $ab_i(k) = x_{\min} \times (1 - c \times \text{rand}())$

  End if

End for

End for

---

(4) Variable selection operator: When adaptive mutation is used, the range of antibody optimization will keep shrinking to the optimal antibody. At this time, if we continue to keep a fixed value to select the antibody, it is not only unhelpful to optimization but also slows down the convergence speed of the algorithm. Now, the selection

operator S_K is improved by replacing the fixed value with a variable value. At the early stage of evolution, the selection operator S_K is larger, and the antibody is selected on a larger scale, and as the evolution gradually reaches convergence, the selection operator S_K becomes smaller, and the antibody is selected on a smaller scale. The advantage of this approach is that it reduces the time complexity and speeds up the optimization of the algorithm. The adjustment formula of selection operator S_K is as follows:

$$S_K = \begin{cases} k_0, k, \frac{T_{\max}}{2} \\ \left[k_0 \cos\left(\frac{\pi}{2} \times \frac{2(T_{\max}-k)}{T_{\max}}\right)\right], & k > \frac{T_{\max}}{2}, \end{cases} \tag{6.7.5}$$

where k and $T_{\max}$ are the current and maximum evolutionary generations, respectively, and k_0 is the initial selection scale.

6.7.1.3 *Improved chaos immunity algorithm to optimize parameters in the DBN*

On the basis of the traditional DBN training method, an improved chaos immune algorithm is added to optimize the parameters of the DBN. Based on the improved chaotic immune algorithm, the DBN's parameters obtained from pretraining are first globally optimized, and then the traditional BP algorithm is locally fine-tuned to obtain the optimal parameters. The main framework of the improved chaotic immune algorithm is the clone selection algorithm, in which, first, the DBN connection weights and biases obtained from pretraining are used as the antibodies and the antibody population is initialized using the chaotic formula. Then, each antibody is used as a network parameter to determine a unique DBN, obtain its output response value, and then calculate the affinity of each antibody. According to the affinity of each antibody, different degrees of clonal mutation are performed on the antibody to continuously obtain a new antibody population. Finally, the parameters of the DBN are updated according to the optimal antibodies.

The optimization of parameters in the DBN using the improved chaotic immune algorithm [130] is as follows:

Algorithm 6.2. ICIM-DBN $(\boldsymbol{\theta}, x, y, \kappa)$ // Improved chaos immunity algorithm optimizing DBN's parameters

Input: Parameter vector $\boldsymbol{\theta} = \{\boldsymbol{w}, \boldsymbol{b}, \boldsymbol{c}\}$ with connection weight and bias of the DBN, training data x and desired output y, and parameters κ of the improved chaotic immune algorithm.
Output: Global optimal parameters of the DBN $\boldsymbol{\theta}' = \{\boldsymbol{w}', \boldsymbol{b}', \boldsymbol{c}'\}$.
Begin

$Ab_0 = \theta$ //Initialization antibody of DBN connection weights and bias
For $i = 1$ to $opt.M$ //Initialization antibody population of Chaos algorithm
$ab_i = opt.u \times ab_0 \times (1 - ab_0)$
End for
For $t = 1$ to $opt.num$
For $i = 1$ to M // Calculate the affinity of each antibody
$fit_i^t = dbn(ab_i^t)$
End for
If $t \leq \frac{opt.tnum}{2}$ // Variable selection operator assignment
$S = opt.\ k$
Else

$$S = \left[opt.k \times \cos\left(\frac{\pi}{2} \times \frac{2(opt.tnum - 1)}{opt.tnum}\right)\right]$$

End if
$Ab = \text{sort}\left(ab_0^t, ab_1^t, \cdots, ab_S^t\right)$ //Select the first S antibodies with lower affinity and sort them in descending order of affinity.
T = [ ] // Initialize temporary antibody library
For $h = 1$ to S
$C_h = opt.cfactor \times opt.m$
$T = [T, ab_{h1}^t, ab_{h2}^t, \ldots, ab_{hC_h}^t]$ // Add the cloned antibody to the temporary antibody library.
End for

$Ab =$ mutation $(T\ , opt.\gamma_0, opt.\beta, opt.\ x)$
// Adaptively mutate the antibodies in the temporary antibody library.
End for
$\theta' = ab_1^{\text{opt.thum}}$ // Update connetion weights and bias of the DBN according to the optimal antibody
End

6.7.2 Particle swarm algorithm for optimizing deep belief networks

The network structure, number of hidden layers, and learning rate of Deep Belief Networks (DBNs) can have a significant impact on the classification results of DBNs. At present, DBNs mostly rely on experience or spend a lot of time adjusting parameters multiple times to determine the network structure. Based on this, this section presents a DBN based on Particle Swarm Optimization (PSO) algorithm. This proposed model uses PSO algorithm to learn and train several important parameters of the DBN and then assigns the optimal parameters obtained from the training to the DBN network. The DBN with the optimal structure is used to extract features from the data. This proposed model uses the adaptive time estimation method (Adam) to effectively fine-tune all connection weights, which further improves the classification accuracy of the DBN [129].

The optimized DBN-based algorithm consists of two main parts: 1. Initialization of the DBN and 2. PSO algorithm-based optimization of the DBN.

6.7.2.1 *Initialization of the DBN model*

The DBN is stacked by multiple RBMs, and its training process of the DBN is accomplished by sequential training of each RBM in turn, which can be divided into two phases:

The first phase is a learning process of the forward stacked RBM; the second phase is the backward fine-tuned learning process of the DBN. While the learning process in the first stage considers only a single RBM layer at a time for unsupervised training, the supervised

Adaptive moment (Adam) estimation [129] fine-tuning of the parameters in the second stage considers all the layers simultaneously.

In order to make the probability distribution under the RBM structure as consistent as possible with the training samples, the parameter fine-tuning method in Ref. [131] was proposed to maximize the log-likelihood function generated during the RBM training process, so as to obtain the appropriate parameters $\boldsymbol{\theta}$.

Without loss of generality, the probability of the visible layer v_i is calculated by

$$p(v_i|\boldsymbol{\theta}) = \frac{f(v_i|\boldsymbol{\theta})}{Z(\boldsymbol{\theta})}, \tag{6.7.6}$$

where $\boldsymbol{\theta}$ can be obtained as a suitable value based on finding the maximum value of the maximum likelihood function, and the loss function and its gradient are given by

$$Loss(\boldsymbol{\theta}) = \sum_{i=1}^{N} \log p\left(v_i|\boldsymbol{\theta}\right) = \sum_{i=1}^{N} \left[\log f\left(v_i|\boldsymbol{\theta}\right) - \log Z(\boldsymbol{\theta})\right], \tag{6.7.7}$$

$$\frac{\partial Loss(\boldsymbol{\theta})}{\partial \boldsymbol{\theta}} = \sum_{i=1}^{N} \left[\frac{\partial \log f\left(v_i|\boldsymbol{\theta}\right)}{\partial \boldsymbol{\theta}} - \left\langle \frac{\partial \log f(v|\boldsymbol{\theta})\mathrm{d}v}{\partial \boldsymbol{\theta}} \right\rangle_{p(v|\boldsymbol{\theta})}\right], \tag{6.7.8}$$

where $\langle\cdot\rangle_{p(v|\boldsymbol{\theta})}$ represents the expectation of the partial derivatives under the distribution $p(\boldsymbol{v}|\boldsymbol{\theta})$, is not easy to find [130], and can only be approximated by some sampling method. v_i is the expectation of both the partial derivatives under the given distribution and the visible state of the sample data of the positive direction, while it cannot be calculated in the negative direction due to the linearity of the partition function. In this case, Gibbs sampling is often used to estimate the negative phase.

In summary, the training process of the RBM can be summarized as follows: The training data are first provided to the neurons v_i of the visible layer, which are then used by the following equation to find the probability that a unit h_j in the hidden layer is activated:

$$p(h_j = 1|v) = \mathrm{ReLU}\left(h_j + \sum_{i=1}^{N} v_i w_{ji}\right). \tag{6.7.9}$$

The process is repeated to update the neurons in the visible layer, and then the neurons in the hidden layer further "reconstruct" the states of v_i and h_j. As the gradient of the joint likelihood function of the data changes, the updating rule for the weight w_{ji} between the visible and hidden layers is written as

$$\Delta w_{ji} = \eta(\langle v_i h_j \rangle_{\text{data}} - \langle v_i h_j \rangle_{\text{recon}}), \tag{6.7.10}$$

where $\langle \cdot \rangle_{\text{data}}$ denotes the expectation of the training data, $\langle \cdot \rangle_{\text{recon}}$ denotes the expectation of the data under the distribution of the reconstructed model, η denotes the learning rate, and $\eta \in (0, 1)$. When the learning rate is larger, the algorithm converges faster, but it may be instability. When the learning rate is smaller, the instability can be avoided, but the convergence becomes slower so as to affect the computation time.

To solve these problems, the Minibatch Stochastic Gradient Descent (MSGD) method is generally used for updating parameters, which associates the parameters for the current round of training process with those for the previous round of training process and can bring a good training speed. This algorithm can solve the problem of unstable convergence to a certain extent, but it is easy to converge to the local minima, and there is a possibility of being trapped in the saddle point.

Therefore, Ref. [129] adopted the Adam algorithm for updating parameters. The advantage of Adam algorithm mainly lies in the fact that after bias correction, there is a definite range of learning rate in each iteration, which makes the parameters relatively smooth. The algorithm is formulated from Equations (6.4.5) to (6.4.9).

6.7.2.2 *PSO algorithm-based optimization of the DBN*

Numerous studies have shown that DBN models containing multiple hidden layers are much better than those with only one layer. The classification error rate of the deep neural network can decrease as the number of hidden layers increases, but when the number of hidden layers is increased to four or more, its classification error rate can rise and the generalization performance can decrease.

Particle Swarm Optimization (PSO) algorithm is a population-based stochastic optimization algorithm. In the PSO algorithm, the solution of each optimization problem is a particle in the searching space. All particles have a fitness value determined by an optimized function, and each particle also has a velocity V that determines its direction and distance of flight. The PSO algorithm initializes a swarm of particles and then searches for the optimal solution in the solution space based on the current optimal particle in the swarm. In each iteration, particles update themselves by tracking two "extremum values": One is the optimal solution found by the particle itself, called as individual extremum (pbest), the other extremum is the optimal solution found by the entire population, known as the global extremum (gbest). The PSO algorithm requires fewer parameters to be adjusted and is simple and easy to implement, suitable for dynamic, multi-objective optimization problems and has a faster computational speed and better global searching capability than the traditional algorithms.

As for a hidden layer with three layers in the DBN model, there are J_1, J_2, and J_3 neurons in each layer, respectively, and a learning rate $\eta \in [0, 1)$. To encode the particle swarm, each particle in the PSO algorithm is set to be a four-dimensional vector $\boldsymbol{x}(J_1, J_2, J_3, \eta)$. The number of particle populations is N, and N is generally between 10 and 20. The maximum number of iterations of the PSO algorithm is recorded as $T_{\max}$, PSO optimization of the DBN process is shown as Figure 6.6.

Step 1: Pre-processing data

In order to ensure that the original data is relatively undistorted, it is necessary to preprocess the acquired signals. Preprocessing formula of the data is given by

$$\hat{x}_i = (x_i - x_{\min})/(x_{\max} - x_{\min}). \tag{6.7.11}$$

If the original signal is a high-dimensional signal, the training time and convergence speed are greatly affected, so it is necessary to carry out dimensionality reduction. In preprocessing, it is also necessary to employ the principal component analysis (PCA) method for dimensionality reduction.

Step 2: Divide the dataset

The dataset D, whose dimensionality is reduced by the PCA method, is divided into two mutually exclusive sets, one of which is used as the training set and the other as the testing set. After training the model

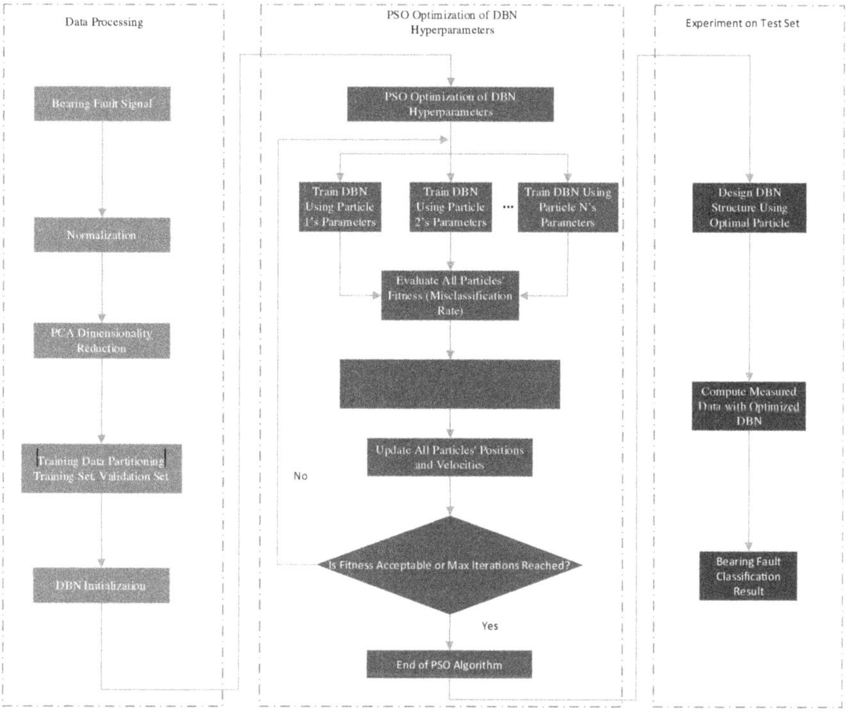

Figure 6.6. PSO optimization of the DBN process.

on the training set, the testing set is used to evaluate its testing error as an estimate of the generalization error.

Step 3: Initialize the particle swarm

The particle swarm is generated based on the parameters of the DBNs (connection weights and hidden node values) and the positions and velocities of these particles are initialized, i.e., the positions $X_I(k)$ and velocities $V_I(k)$ of the particles are initialized into their initial positions $X_I(0)$ and initial velocities $V_I(0)$, respectively.

Step 4: Calculate adaptation value

After obtaining the output response value of the DBN based on the initialized positions and velocities of particles, the adaptation function of the particle swarm is defined as

$$fitness = \frac{1}{2N}\sum_{i=1}^{N}\sum_{j=1}^{M}(\hat{y}_{ij} - y_{ij})^2, \tag{6.7.12}$$

where N is the training sample, M is the number of output neurons, and $\hat{y}_{ij}$ and y_{ij} denote the output value and the desired output value of the jth component of the ith sample, respectively.

Step 5: Update the velocities and positions of the particles

Based on the fitness value of the particle swarm calculated in Step 4, the optimal particle $\boldsymbol{X}_{gbest}(k)$ of the particle swarm in the current round and the optimal particle $\boldsymbol{X}_{pbest}(k)$ in the search history must be found. The velocities and positions of particles are calculated by

$$\boldsymbol{X}_I(k+1) = \boldsymbol{X}_I(k) + \boldsymbol{V}_I(k+1), \tag{6.7.13}$$

$$\begin{aligned}\boldsymbol{V}_I(k+1) &= \omega \boldsymbol{V}_I(k) + c_1 r_1 \left(\boldsymbol{X}_{ipbest}(k) - \boldsymbol{X}_I(\mathrm{k})\right) \\ &\quad + c_2 r_2 \left(\boldsymbol{X}_{igbest}(k) - \boldsymbol{X}_I(k)\right),\end{aligned} \tag{6.7.14}$$

where ω denotes the inertia weight, its value is taken in [0,1], and generally $\omega = 0.9$. c_1 and c_2 denote the acceleration parameter, and they are generally equal and take the value of the range of [0,4]. After a number of experiments carried out by Shi and Eberhart, they recommended that $c_1 = c_2 = 2$ were in order to balance the role of stochastic factors [132]. r_1 and r_2 change within the range of [0,1] in the value of a random value.

Step 6: Determine whether the operation is terminated or not

If the misclassification rate of the training samples satisfies the set condition or the number of iterations is equal to M, the optimization process of PSO algorithm ends, otherwise it goes to Step 4 with $k = k+1$ and repeat Step 5 and Step 6 until the termination condition is satisfied.

Step 7: Further test and verify, and output the results

The trained DBN is used to train the testing data and output the classification results of signals.

6.8 Example 6: Diagnosis of Attention Deficit Hyperactivity Disorder by Deep Belief Network Based on Greedy Approach

Attention Deficit Hyperactivity Disorder (ADHD) is a behavioral and developmental disorder in which children usually do not have the ability to focus on a problem and their learning is slow. They have unusual activities and this disorder is associated with lack of attention, hyperactivity, impulsive behavior, or a combination of these behaviors. Many of these children suffer from one or several other behavioral disorders, as well as psychological problems, such as depression or bipolar disorder. Doctors believe that people with this disease lack sufficient amounts of specific chemicals, namely neurotransmitters, in their brains. These chemicals help control the behavior of children. Early diagnosis of the disease is important because the nature of the disease is not yet fully understood, and it is difficult to diagnose and categorize the disease based on the problems that humans have with timely diagnosis. Currently, the diagnosis of the AHAD has attracted the attention of many researchers [133–135]. In 2015, Vaslui Suchnio [136] proposed an effective classification scheme based on Meta-Cognitive Neuro-Fuzzy Interface System (McIFS). In this scheme, the feature selection mechanism is based on Binary Coded Genetic-Extreme Learning Machine (BCG-ELM). In 2014, Kuang *et al.* [137] utilized the frequency features of the brain vectors as feature vectors and used deep learning for feature extraction and classification. In 2012, Igual *et al.* [138] made use of the information from fMRI at rest and introduced a G-algorithm-based algorithm to construct the entire brain network. In 2014, Alexander Tenve *et al.* [139] proposed a hybrid machine learning approach based on Electroencephalogram (EEG) for the classify adults with the AHAD, four SVM classifiers were used in the training phase, and the output containing the obtained logical expressions, with significantly improved results.

In 2012, Abibullaev *et al.* [140] proposed a decision support system for the diagnosis of the ADHD by means of brainwave signals, with the main goal of providing a supportive diagnostic tool and machine learning-based diagnostic algorithms, where the diagnostic tool contains a signal processing process for determining the features, and where a semi-supervised approach is used for efficient updating of the training set with a higher level of accuracy. In 2012, Eloyan *et al.* [141] established an automated method for diagnosing the ADHD through the use of imaging and biolabeling, which developed the model based on the ADHD symptoms, and focus on structural imaging of the brain's "Resting State Functional Connectivity (RS-FC)".

Currently, deep network models are used to learn representations and nonlinear relationships for the corresponding problems, which are based on feature engineering and nonlinear modeling, with fully automatic, unsupervised, and semi-supervised methods for feature extraction. Among them, Deep Belief Network (DBN) is used to diagnose ADHD in Ref. [142]. This method first transforms a large number of problem features, including fMRI features, diagnostic status, Attention Deficit Hyperactivity Disorder (ADHD) measurements, secondary symptoms, age, gender, IQ, and lifetime status of medication, into composite features. Then, in the greedy algorithm, the structure of the deep neural network, the number of hidden layers, and the number of neurons are determined. Finally, to calculate the probability distribution of the output, Softmax layer is used as the last layer. This method uses two standard datasets introduced by the ADHD-200 global competition, Neuro Image (Mennes, Biswal, Castellanos, & Milham, 2013) and NYU (Mennes *et al.*, 2013) datasets, and compares them with state-of-the-art well-known algorithms. The results show that this method outperforms other known systems in terms of accuracy, checking completeness, f-value and precision, etc. The prediction accuracy on the Neuro Image and NYU datasets has an improvement of about +12.04 and +27.81, respectively, comparison with the best method presented in the ADHD-200 global competition.

6.8.1 Deep belief networks based on greedy methods

A greedy algorithm is used to determine the setup parameters of the DBN model.

6.8.1.1 *Deep confidence networks*

The DBN is actually a network with several RBMs connected together in a bottom-up manner. The RBM is a type of generative stochastic artificial neural network that can learn the probability distribution over its input set, and a Boltzmann machine with the constraint that the neurons must form a bipartite graph, the 'visible' and 'hidden' layers must be made up of two groups of neurons, respectively. And there are no connections among the nodes within one group. In order to improve the training efficiency, a gradient-based contrast dispersion algorithm is used. The RBM is shown in Figure 6.7. The connections among the neurons of the visible and

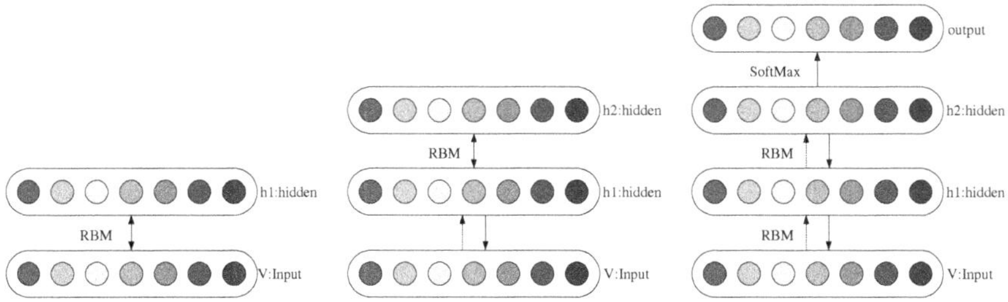

Figure 6.7. Deep belief network.

hidden layers have a higher reliability compared to conventional networks.

$\boldsymbol{v} = (v_0, v_1, \ldots, v_{N-1})$ and $\boldsymbol{h} = (h_0, h_1, \ldots, h_{J-1})$ denote the input vector and output vector, respectively. The DBN consists of a bunch of RBMs, as shown in Figure 6.7. Figure 6.7 explains the training phase of the network.

The probability distribution of the RBM used to generate the output is defined as

$$p(\boldsymbol{v},\boldsymbol{h}) = \frac{1}{Z(\boldsymbol{v},\boldsymbol{h})} e^{E(\boldsymbol{v},\boldsymbol{h})}, \tag{6.8.1}$$

where $Z(\boldsymbol{v},\boldsymbol{h})$ and $p(\boldsymbol{v},\boldsymbol{h})$ are called the normalization factor and energy function, respectively, and

$$Z(\boldsymbol{v},\boldsymbol{h}) = \sum_{v} \sum_{h} e^{E(\boldsymbol{v},\boldsymbol{h})} \tag{6.8.2}$$

and

$$E(\boldsymbol{v},\boldsymbol{h}) = -\boldsymbol{bv} - \boldsymbol{ch} - \boldsymbol{vwh}, \tag{6.8.3}$$

where $\boldsymbol{b}$ and $\boldsymbol{c}$ are the visible and invisible threshold variables, $\boldsymbol{v}$ and $\boldsymbol{h}$ denote the visible and hidden variables, and $\boldsymbol{w}$ is the weight of the connections between nodes. The update criterion for calculating the weights of the RBM network based on the logarithmic gradient is given by

$$w_{ji} = E_{data}(v_i, h_j) - E_{model}(v_i, h_j), \tag{6.8.4}$$

where $E_{data}(v_i, h_j)$ is the expected value of the training sample and $E_{model}(v_i, h_j)$ is the expected value of the model.

6.8.1.2 *A greedy training algorithm*

In order to build a deep model with RBMs, a multi-layer deep network is now trained using the greedy algorithm. First, the RBM network is learned as the first layer of the DBN model with a weight matrix $\boldsymbol{w}$. Then, the weights $(\boldsymbol{w}^2 = (\boldsymbol{w}^1)^{\mathrm{T}})$ of the second layer are initialized to ensure that the two hidden layers of the DBN are at least equal to the RBM. By generating the output of the first layer $\boldsymbol{h}^1$, the DBN can be improved by modifying the weight matrix $\boldsymbol{w}^2$. The vector $\boldsymbol{h}^2$ obtained from the second RBM can also be used to learn the third layer of the RBM. By initializing $\boldsymbol{w}^3 = (\boldsymbol{w}^2)^{\mathrm{T}}$, it ensures that the lower bound of the log-likelihood function is improved. During the iterative process, the implementation steps of the depth hierarchical model are shown as follows:

Step 1: Determine the parameters

The parameters $\boldsymbol{w}^1$ of the first hidden layer are determined to further determine the CD training algorithm.

Step 2: Train the features for the next layer

The parameter $\boldsymbol{w}^1$ is fixed and $Q(\boldsymbol{h}^1|\boldsymbol{v}) = p(\boldsymbol{h}^1|\boldsymbol{v}; \boldsymbol{w}^1)$ is used as the input vector to train the features for the next layer.

Step 3: Train the features in the third layer

The parameter $\boldsymbol{w}^2$, which defines the second layer of the attribute, is fixed and the first sample from $Q(\boldsymbol{h}^2|\boldsymbol{h}^1) = p(\boldsymbol{h}^2|\boldsymbol{h}^1; \boldsymbol{w}^2)$ is used as data for training the features in the third layer.

In the last layer, in order to compute the probability distribution of the inputs, a Softmax layer is used. The Softmax function is calculated by

$$p_j = \frac{e^{x_j}}{\sum\limits_{i=1}^{k} e^{x_i}}, \tag{6.8.5}$$

where p_j is the possible value of neuron j and x_i is the value entering each neuron.

6.8.2 Simulation experiments and results analyses

6.8.2.1 *Datasets*

The training and testing datasets were provided by the ADHD-200 global competition. The NYU training dataset consisted of 222

Table 6.1. Characteristics of deep confidence networks.

Description	Features
Sex of the patient	Gender
Patient's preference for one hand over the other	Handedness
Individual's IQ in different states	IQ measure
State of the patient after medication	medication status
Control of the individual's behaviors and states	quality control
Characteristics from the FMRI images	FMRI Features
Degree of inattention of the individual	Inattentive
Degree of hyperactivity of the individual	Hyper/Impulsive
Verbal or spoken IQ	Verbal IQ
IQ of performance function	Performance IQ
Depressive state after medication	med status
First quality control of people's behavior and emotions at rest (least active)	QC_Rest_1
Second quality control of people's behavior and emotions at rest (least active)	QC_Rest_2
Third quality control of people's behavior and emotions at rest (least active)	QC_Rest_3
Fourth quality control of people's behavior and emotions at rest (least active)	QC_Rest_4
First quality control of anatomical state	QC_Anatomical_1
Second quality control of anatomical state	QC_Anatomical_2

training samples and 41 testing samples. The NeuroImage dataset consisted of 48 training samples and 25 testing samples. The participants in the study ranged in age from 7 to 21 years old. The features used for the deep belief network are shown in Table 6.1. Functional magnetic resonance imaging (fMRI) was used to obtain the required data.

6.8.2.2 *Comparison of the methods*

The greedy training algorithm was utilized to learn from the training data and evaluated on the testing data. The method in Ref. [142] was evaluated on standard datasets from NYU and Neuron Image, respectively, and a binary categorization task, where positive categorization includes the ADHD patients and negative categorization includes non-ADHD patients. The results of the evaluation are shown in Table 6.2. Note that one of the most important problems in diagnosing ADHD is imbalance. To address this problem, the Synthetic

Table 6.2. Evaluation results of precision, recall, and F-value for the NYU and Neuroimage datasets.

	Precision		Recall		F Score ($\alpha = 0.5$)	
Datasets	Positive	Negative	Positive	Negative	Positive	Negative
NYU	72%	42%	75%	38%	74%	40%
NeuroImage	64%	73%	75%	62%	69%	67%

Table 6.3. Comparison results between NYU and NeuroImage datasets in terms of accuracy.

	Presion	
	NYU Datasets	NeuroImage Datasets
Proposed methodology	63.68	69.83
ADHD-200	35.19	59.96
RBF SVM	64.2	64.2
RBF	51.9	48.2
SVM	55.1	61.2
Decision Tree	51.9	54.1

Minority Oversampling TEchnique (SMOTE) method was used to artificially synthesize the oversampling of the minority category. The minority category includes the ADHD patients. There are a number of methods that can be used to oversample a dataset in a typical classification problem. The most common technique is known as the SMOTE. It is important to note that the dataset becomes balanced after applying the SMOTE to the dataset to mitigate the imbalance.

Table 6.3 shows the results of the methods from Ref. [142] and the best methods introduced by the ADHD-200 Global Competition as well as the SVM, RBF, RBF-SVM, and Decision Tree algorithms (Brown *et al.*, 2012).

Table 6.3 shows that the method in Ref. [142] improves by +12.04 and 27.81% in comparison with NYU and NeuroImage datasets, respectively. The performance of the proposed method with a deep learning network and Softmax classifier outperforms that of other methods.

In conclusion, Attention Deficit Hyperactivity Disorder (ADHD) has become one of the most common disorders and its early diagnosis is of great importance. Reference [142] proposed a deep belief network-based approach that utilized a greedy algorithm to construct and train the network structure. Experimental results conducted on two standard datasets show that the proposed method in this section has significant advantages over existing methods.

Chapter 7

Deep Sparse Autoencoder

Summary

This chapter begins with the principles and training methods of autoencoders, then it delves into the theoretical foundations and architectural implementations of several extended models, including variational autoencoder, stacked variational autoencoder, deep convolutional variational autoencoder, autoencoder-based echo state network, deep classical correlation sparse autoencoder, and dual adversarial autoencoder network. By integrating autoencoders into practical applications, this chapter explores some models, such as soft sensing model based on mutual information sparse autoencoder, fuzzy inference model based on deep autoencoding networks, fast sparse autoencoding model based on feature clustering, and extreme learning machine model based on stacked denoising sparse autoencoders. Finally, the feature extraction and modulation recognition algorithms are investigated by integrating an improved latent Dirichlet allocation algorithm with autoencoders, accompanied by simulation experiments and result analyses.

7.1 Introduction

In 1986, Rumelhart proposed the concept of autoencoder (AE) and used it for high-dimensional complex data processing, which promoted the development of neural networks. In 2006, Hinton improved the structure of the prototype autoencoder to obtain Deep Auto-Encoder (DAE). In 2007, Benjio proposed the concept of a sparse autoencoder to further deepen the depth of the DAE. In 2008, Vincent proposed a denoising reduction autoencoder to prevent the overfitting phenomenon. In 2009, Benjio used a stacked autoencoder to construct deep learning neural network. In 2010, Salah proposed the shrinkage autoencoder, which restricted the process of dimensionality upgrading and dimensionality downgrading. In 2011, Jonathan proposed to construct convolutional neural networks with convolutional autoencoders. Taylor constructed different types of deep structures using autoencoders, in 2012. Hinton, Bengio, and Vincent compared the performance of structures such as prototype autoencoders, sparse autoencoders, de-noising autoencoders, shrinkage autoencoders, convolutional autoencoders, and RBMs, which will be useful for future practice and scientific research. In 2013, Telmo studied the performance of the trained DAE with different cost functions, which showed the direction for the development of cost function optimization strategies.

This chapter describes the research results of the structure, principles, implementation, and application of autoencoding network.

7.2 Autoencoder

7.2.1 Autoencoding principle

7.2.1.1 *Encoding–decoding*

The principle of encoding and decoding is illustrated in Figure 7.1.

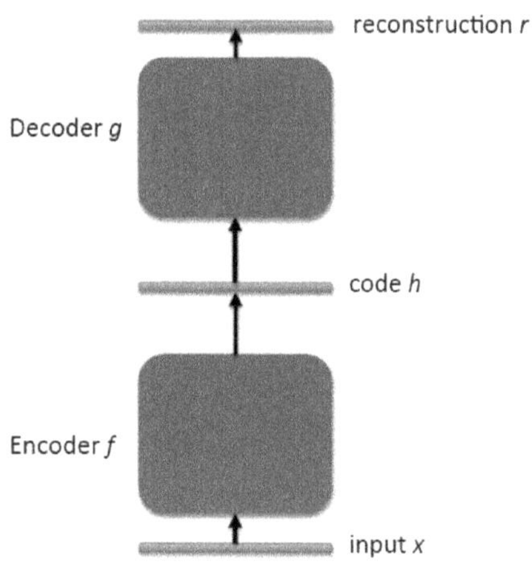

Figure 7.1. Encoding and decoding principle.

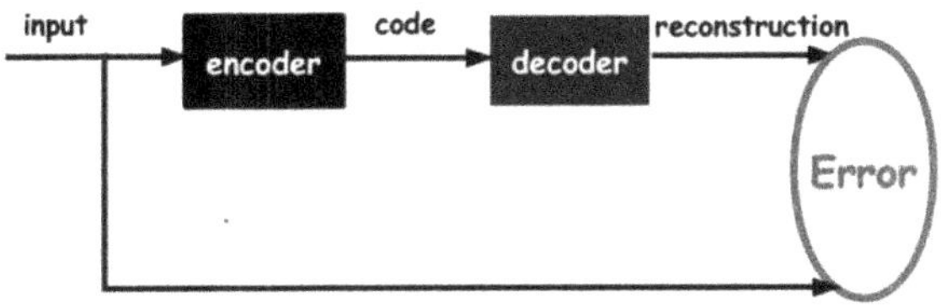

Figure 7.2. Errors in process of reconstruction.

In Figure 7.1, input $\boldsymbol{x}$ is passed through the encoder f, and the encoded result is written as

$$\boldsymbol{h} = \boldsymbol{f}(\boldsymbol{x}).$$

The encoding $\boldsymbol{h}$ is decoded or said to be reconstructed by the decoder g as

$$\hat{\boldsymbol{x}} = g(\boldsymbol{h}).$$

The closeness of the reconstructed $\hat{\boldsymbol{x}}$ to the input $\boldsymbol{x}$ can be measured by a loss function. This loss function, recorded as $Loss(\boldsymbol{x}, \hat{\boldsymbol{x}})$, is used to measure the quality of reconstruction, with the goal of minimizing the expected value of $Loss(\boldsymbol{x}, \hat{\boldsymbol{x}})$. $Loss(\boldsymbol{x}, \hat{\boldsymbol{x}})$ can be defined in a variety of ways, which of them is the Mean Square Error (MSE) being one of the most common forms, and as

$$Loss(\boldsymbol{x}, \hat{\boldsymbol{x}}) = \frac{1}{2}||\hat{\boldsymbol{x}} - \boldsymbol{x}||^2.$$

Figure 7.2 illustrates the reconstruction error that exists in the decoding process.

7.2.1.2 *Autoencoder*

An autoencoder is an unsupervised learning model in deep learning. It first maps high-dimensional features to a low-dimensional hidden representation via an encoder and then reconstructs the input features through a decoder, as shown in Figure 7.3.

The encoder maps the high-dimensional original input feature vector $\boldsymbol{x} = [x_1, x_2, \ldots, x_N]$ into a low-dimensional hidden vector $\boldsymbol{z}$ (of dimension M), and the decoder then maps $\boldsymbol{z}$ to an N-dimensional output layer, and, therefore, the autoencoder achieves the reproduction of the original input feature quantity.

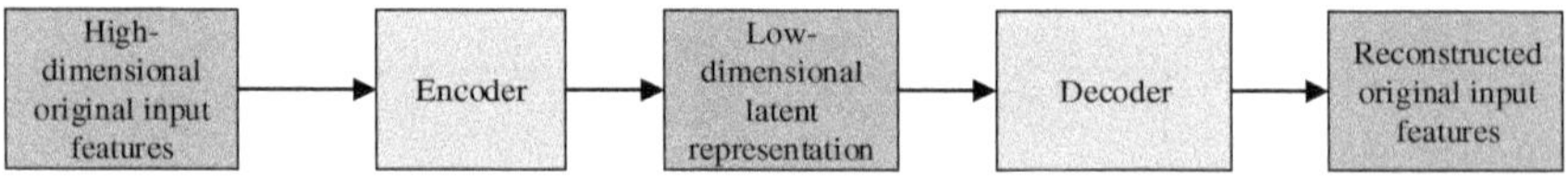

Figure 7.3. Structure of an autoencoder.

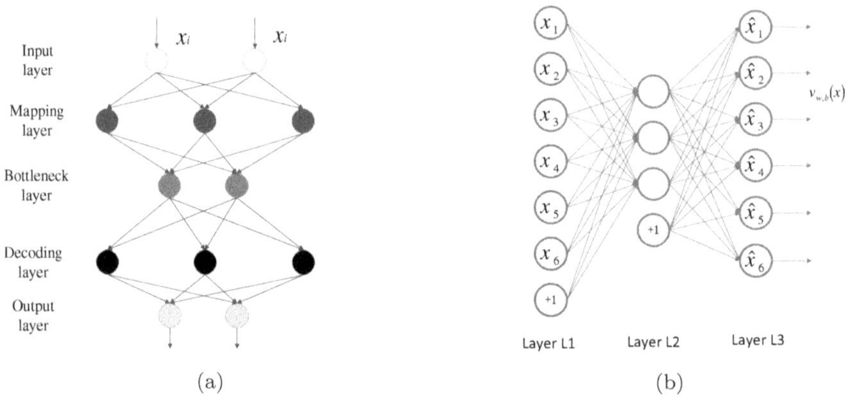

Figure 7.4. Autoencoder. (a) Five-layer structure. (b) Three-layer structure.

Figure 7.4(a) shows a five-layer structure with an input layer, a mapping layer (encoding layer), a bottleneck layer, an inverse mapping layer (or denoted as decoding layer), and an output layer, while Figure 7.4(a) is simplified into a three-layer structure, as shown in Figure 7.4(b). Suppose that the input vector of the input layer is marked as $\boldsymbol{x} = [x_1, x_2, \ldots, x_N]^T$, the encoding function of the encoding layer is denoted as $\boldsymbol{f}(x) = [f_1(x), f_2(x), \ldots, f_{N_e}(x)]$, the output layer is represented as $\hat{\boldsymbol{x}} = [\hat{x}_1, \hat{x}_2, \ldots, \hat{x}_N]^T$, and the decoding function is recorded as $\boldsymbol{h} = [h_1(x), h_2(x), \ldots, h_{N_h}(x)]^T$. N is the dimensionality of the input and output samples, and N_h is the dimensionality of the hidden layer. The mapping relationship between the hidden layer and the input layer is defined as

$$\boldsymbol{h} = \boldsymbol{f}(\boldsymbol{x}) = s_f(\boldsymbol{w}^{(1)}\boldsymbol{x} + \boldsymbol{b}^{(1)}), \tag{7.2.1}$$

where s_f is a linear or nonlinear activation function, $\boldsymbol{w}^{(1)} \in \boldsymbol{R}^{N_h \times N}$ is the weight matrix, and $\boldsymbol{b}^{(1)} \in \boldsymbol{R}^{N_h}$ is the bias vector of the hidden layer.

Similarly, the mapping from the hidden layer to the output layer can also be represented by a function $\boldsymbol{g}$ as follows:

$$\hat{\boldsymbol{x}} = \boldsymbol{g}(\boldsymbol{h}) = s_g(\boldsymbol{w}^{(2)}\boldsymbol{h} + \boldsymbol{b}^{(2)}), \tag{7.2.2}$$

where s_g is the activation function, $\boldsymbol{w}^{(2)} \in R^{N\times N_h}$ is the weight matrix, and $\boldsymbol{b}^{(2)} \in R^N$ is the bias vector of the output layer. The basic idea of the AE is to learn a function from the input layer to the output layer such that $p_\theta(x) = g(f(x)) \approx x$. The activation function is typically chosen as the sigmoid function and defined as

$$f(x) = \frac{1}{1+\exp(-x)}. \tag{7.2.3}$$

Since the range of $f(x)$ is between 0 and 1, the data needs to be normalized by

$$x = \frac{x - x_{\min}}{x_{\max} - x_{\min}}. \tag{7.2.4}$$

The parameters of the autoencoder include the weights and bias vectors of the network, i.e., $\boldsymbol{\theta} = \{\boldsymbol{w}^{(1)}, \boldsymbol{w}^{(2)}, \boldsymbol{b}^{(1)}, \boldsymbol{b}^{(2)}\}$, which can be solved by minimizing the loss function $L(x, \hat{x})$. Assuming that the training samples are $\boldsymbol{x} = \{x_1, x_2, \ldots, x_N\}$, N is the number of samples, and $\boldsymbol{x} \in \boldsymbol{R}^d$, then the loss function may be defined as

$$J_{\text{AE}} = L(\boldsymbol{x}, \hat{\boldsymbol{x}}) = \frac{1}{2N}\sum_{i=1}^{N} ||x_i - \hat{x}_i||^2 = \frac{1}{2N}\sum_{i-1}^{N}\sum_{j=1}^{D}[x_{id} - \hat{x}_{id}]^2. \tag{7.2.5}$$

Based on the above assumptions, the steps of the backpropagation algorithm are as follows:

Step 1: Calculate the activation values

The activation values of the lth neurons in each layer of forward propagation are calculated by

$$h_{\boldsymbol{w},\boldsymbol{b}}^{(l)} = f(\boldsymbol{w}^{(l-1)}\boldsymbol{x}^{(l-1)} + \boldsymbol{b}^{(l-1)}),$$

where $l = L-1, L-2, \ldots, 2$.

Step 2: Calculate the gradient of neuron

The gradient of the output of the ith neuron in the lth layer (output layer) is calculated by

$$g_i^{(l)} = \frac{\partial J}{\partial z_i^{(l)}} = \frac{\partial}{\partial z_i^{(l)}} \frac{1}{2} ||\boldsymbol{x} - \hat{\boldsymbol{x}}||^2 = -(x_i - h_{(w,b)i}^{(l)}) f'(z_i^{(l)}). \quad (7.2.6)$$

The gradient of the output of the ith neuron in layer $l - 1$ (hidden layer) is calculated by

$$\begin{aligned} g_i^{(l-1)} &= \frac{\partial J}{\partial z_i^{(l-1)}} = \frac{\partial J}{\partial z_i^{(l)}} \frac{\partial z_i^{(l)}}{\partial z_i^{(l-1)}} \\ &= g_i^{(l)} \frac{\partial z_i^{(l)}}{\partial z_i^{(l-1)}} = g_i^{(l)} \frac{\partial}{\partial z_i^{(l-1)}} \sum_{j=1}^{N_{l-1}} w_{ij}^{(l-1)} f\left(z_i^{(l-1)}\right) \\ &= \sum_{j=1}^{N_{l-1}} w_{ij}^{(l-1)} g_i^{(l)} f'(z_i^{(l-1)}), \end{aligned} \quad (7.2.7)$$

where

$$g_i^{(l)} = \sum_{j=1}^{N_l} w_{ij}^{(l)} g_i^{(l+1)} f'(z_i^{(l)}). \quad (7.2.8)$$

Step 3: Calculate the partial derivatives

The partial derivatives in the final network are calculated by

$$\begin{aligned} \nabla_{w^{(l)}} J &= h_{(w,b)i}^{(l)} g_i^{(l+1)}, \\ \nabla_{b^{(l)}} J &= g_i^{(l+1)}. \end{aligned} \quad (7.2.9)$$

7.3 Sparse Adaptive Encoder

An autoencoder requires the output to be as close as possible to the input, and their hidden layers must satisfy certain sparsity constraints, which means the hidden layers should not carry too much information. Therefore, the hidden layer compresses the

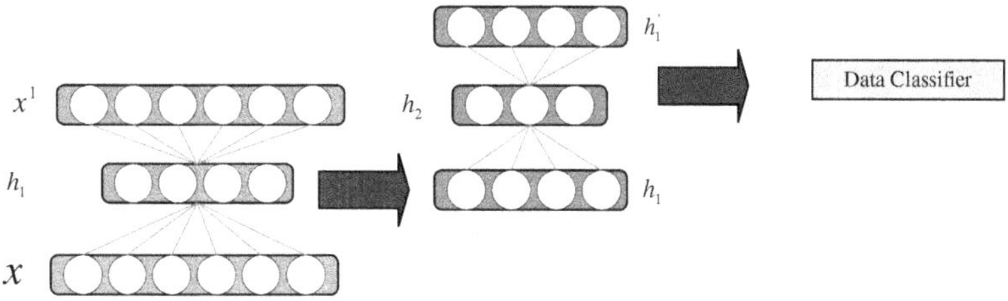

Figure 7.5. Structure of sparse autoencoding network.

input, and the output layer decompresses it. The entire process will inevitably lead to some information loss, but the training aims to minimize the loss of information. A sparse autoencoding network builds upon the basic autoencoding network by adding sparsity constraints to the hidden layer and can stack multiple autoencoding networks [143, 144]. Figure 7.5 shows a sparse autoencoding network that stacks two autoencoding networks. After training the first autoencoding network, the outputs of its hidden layer are used as the inputs and the expected outputs for the next autoencoding network. This process is repeated until the desired depth of network is reached. Finally, the fine-tuning process of the network begins, where the input layer x, the first hidden layer h_1, the second hidden layer h_2, and all subsequent hidden layers are combined into a new neural network. A data classifier is then connected, and the parameters of the network are readjusted using all labeled data in a supervised manner.

Since the commonly used activation function in networks is the sigmoid function, whose outputs are between 0 and 1. Therefore, the average activation of the jth neuron in the lth hidden layer for all neurons in the $l-1$th hidden layer $\hat{\rho}_j^{(l)}$ is calculated by

$$\hat{\rho}_j^{(l)} = \frac{1}{N_{l-1}} \sum_{i=1}^{N_{l-1}} \left(h_{(w,b)j}^{(l)} \left(h_{(w,b)i}^{(l-1)} \right) \right). \tag{7.3.1}$$

This is close to a small real number ρ, i.e., $\hat{\rho}_j^{(l)} = \rho$, where ρ represents the sparsity target and ensures the sparsity of the network's hidden layer. To make these two values as close as possible, the Kullback–Leibler divergence (KL divergence) is introduced and defined as

$$J_{\mathrm{KL}}(\rho\|\hat{\rho}) = \sum_{l=1}^{N_h} J_{\mathrm{KL}}\left(\rho\|\hat{\rho}_j^{(l)}\right)$$
$$= \sum_{l=1}^{N_h}\sum_{j}^{N_l}\left(\rho\log\frac{\rho}{\hat{\rho}_j^{(l)}} + (1-\rho)\log\frac{1-\rho}{1-\hat{\rho}_j^{(l)}}\right), \quad (7.3.2)$$

where N_h represents the number of nodes in the hidden layer. The total cost function of the Sparse AutoEncoder (SAE) is represented as

$$J_{\mathrm{SAE}} = \frac{1}{N}\sum_{i=1}^{N}\frac{1}{2}\|\hat{\boldsymbol{x}}_i - \boldsymbol{x}_i\|^2 + \frac{\lambda}{2}\sum_{l=1}^{N_l}\|\boldsymbol{w}^{(l)}\|^2 + \beta J_{KL}(\rho\|\hat{\rho}), \quad (7.3.3)$$

where β is the sparsity penalty term. By minimizing the cost function, the optimal parameters $\boldsymbol{w}^{(l)}$ and $\boldsymbol{b}^{(l)}$ can be obtained. Since the cost function has an additional term, the expression for the gradient also changes.

For simplicity, we only calculate the sparse penalty term for the first-layer parameters, and let

$$S(\boldsymbol{w}^{(l)}, \boldsymbol{b}^{(l)}) = \beta J_{KL}(\rho) = \beta\sum_{l=1}^{N_h} J_{\mathrm{KL}}(\rho\|\hat{\rho}_j^{(l)}). \quad (7.3.4)$$

Thus,

$$\frac{\partial S\left(\boldsymbol{w}^{(1)}, \boldsymbol{b}^{(1)}\right)}{\partial w_{ij}^{(1)}}$$
$$= \sum_{n=1}^{N}\frac{\partial S\left(\boldsymbol{w}^{(1)}, \boldsymbol{b}^{(2)}\right)}{\partial z_n^{(2)}(x_n)} \times \frac{\partial z_n^{(2)}(x_n)}{\partial w_{ij}^{(1)}}$$
$$= \sum_{n=1}^{N} h_j^{(l)}(x_n) \times \beta\frac{\partial}{\partial z_i^{(2)}(x_n)}\left[\rho\log\frac{\rho}{\hat{\rho}_i} + (1-\rho)\log\frac{1-\rho}{1-\hat{\rho}_i}\right]$$

$$
\begin{aligned}
&= \sum_{n=1}^{N} h_j^{(1)}(x_n) \times \beta \left(-\frac{\rho}{\hat{\rho}_i} + \frac{1-\rho}{1-\hat{\rho}_i} \right) \frac{\partial \hat{\rho}_n}{\partial z_i^{(2)}(x_n)} \\
&= \sum_{n=1}^{N} h_j^{(1)}(x_n) \times \beta \left(-\frac{\rho}{\hat{\rho}_i} + \frac{1-\rho}{1-\hat{\rho}_i} \right) \left(\frac{1}{N} \frac{\partial h_i^{(2)}(x_n)}{\partial z_i^{(2)}(x_n)} \right) \\
&= \frac{1}{N} \sum_{n=1}^{N} h_j^{(1)}(x_n) \times \beta \left(-\frac{\rho}{\hat{\rho}_i} + \frac{1-\rho}{1-\hat{\rho}_i} \right) f'(z_i^{(2)}(x_n)). \qquad (7.3.5)
\end{aligned}
$$

Therefore,

$$
\begin{aligned}
&\frac{\partial J_{\text{sparse}}(\boldsymbol{w}, \boldsymbol{b})}{\partial w_{ij}^{(1)}} \\
&= \frac{1}{N} \sum_{n=1}^{N} h_j^{(1)}(x_n)\, g_i^{(2)}(x_n) + \lambda w_{ij}^{(1)} + \frac{\partial S\left(\boldsymbol{w}^{(1)}, \boldsymbol{b}^{(1)}\right)}{\partial w_{ij}^{(1)}} \\
&= \frac{1}{N} \sum_{n=1}^{N} a_j^{(1)}(x_n) \left[g_i^{(2)}(x_n) + \beta \left(-\frac{\rho}{\hat{\rho}_i} + \frac{1-\rho}{1-\hat{\rho}_i} \right) f'(z_i^{(2)}(x_n) \right] \\
&\quad + \lambda w_{ij}^{(1)} \\
&= \frac{1}{N} \sum_{n=1}^{N} h_j^{(1)}(x_n) \left[\left(\sum_{r=1}^{s_3} g_r^{(3)}(x_n)\, w_{ri}^{(2)} \right) \right. \\
&\quad \left. + \beta \left(-\frac{\rho}{\hat{\rho}_i} + \frac{1-\rho}{1-\hat{\rho}_i} \right) f'(z_i^{(2)}(x_n) \right] + \lambda w_{ij}^{(1)}, \qquad (7.3.6)
\end{aligned}
$$

where

$$
g_i^{(2)} = \left(\sum_{j=1}^{N_3} g_j^{(3)} w_{ji}^{(2)} \right) f'(z_i^{(2)}) \qquad (7.3.7)
$$

and can become

$$
g_i^{(2)} = \left[\sum_{j=1}^{N_3} g_j^{(3)} w_{ji}^{(2)} + \beta \left(-\frac{\rho}{\hat{\rho}_i} + \frac{1-\rho}{1-\hat{\rho}_i} \right) \right] f'(z_i^{(2)}). \qquad (7.3.8)
$$

7.4 Variational Autoencoder

7.4.1 Variational autoencoder theory

Variational AutoEncoder (VAE) [145] is a type of autoencoder which can extract high-dimensional raw features into low-dimensional high-order features while retaining as much of the original information as possible. Unlike general autoencoders, the VAE is based on variational Bayesian inference. By finding the Gaussian distribution that satisfies high-order latent variables, the high-order features obtained through mapping have stronger robustness. This helps enhance the generalization ability of the classifiers and reduce interference caused by noise. The VAEs typically consist of three layers of neural networks: the input layer, the hidden layer, and the output layer. By encoding the input $\boldsymbol{x} \in \boldsymbol{R}^{D\times N}$ (where D is the sample's dimension and N is the number of samples), the hidden layer output $\boldsymbol{h} \in \boldsymbol{R}^{D_h\times N}$ (with D_h being the space dimension of the hidden layer) is obtained. Then, by decoding, the output of the hidden layer is reconstructed back to the original spatial dimension of the sample to obtain the reconstructed sample $\hat{x}$. The training of the autoencoder aims to make the output $\hat{x}$ approach the input x and thus obtain latent features that can represent the characteristics of the input samples.

As a generative model, the basic structure of the VAE is shown in Figure 7.6. The VAE uses the latent variable z to represent the distribution of the original data $\boldsymbol{x}$. By optimizing the generation parameters $\boldsymbol{\theta}$, the latent variable z is used to generate data $\hat{\boldsymbol{x}}$, which is highly similar to the original data $\boldsymbol{x}$, i.e., the following marginal distribution can be maximized:

$$f_{\boldsymbol{\theta}}(\boldsymbol{x}) = \int f_{\boldsymbol{\theta}}(\boldsymbol{x}|z) f_{\boldsymbol{\theta}}(z) dz, \tag{7.4.1}$$

where $\boldsymbol{f}_{\boldsymbol{\theta}}(\boldsymbol{x}|z)$ represents the reconstruction of the original data $\boldsymbol{x}$ using the latent variable z and $f_{\theta}(z)$ represents the prior distribution of the latent variable z, which follows a Gaussian distribution $N(\boldsymbol{0}, \boldsymbol{I})$. Since there is no label corresponding to z, it is impossible to make the samples generated via using z correspond to the original samples. Therefore, $f_{\boldsymbol{\theta}}(\boldsymbol{x}|z)$ is used to represent the latent variable z

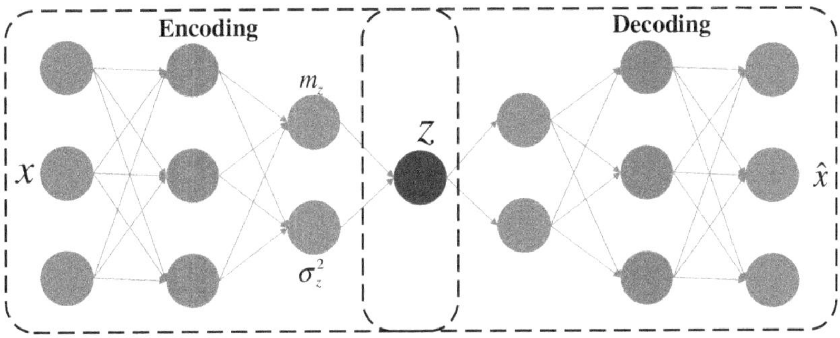

Figure 7.6. Structure of the VAE.

learned from the original data to establish the relationship between z and $\boldsymbol{x}$. Since the true posterior distribution $f_{\boldsymbol{\theta}}(\boldsymbol{x}|z)$ is difficult to be calculated, an approximate posterior distribution $f_{\varphi}(z|\boldsymbol{x})$, which follows a Gaussian distribution, is used to replace the true posterior. The Kullback–Leibler divergence between these two distributions is given by

$$\begin{aligned} J_{KL}[f_{\boldsymbol{\varphi}}(z|\boldsymbol{x})\|f_{\theta}(z|\boldsymbol{x})] &= E_{f_{\boldsymbol{\varphi}}(z|\boldsymbol{x})}[\log f_{\boldsymbol{\varphi}}(z|\boldsymbol{x}) - \log f_{\theta}(z|\boldsymbol{x})] \\ &= E_{f_{\boldsymbol{\varphi}}(z|\boldsymbol{x})}[\log f_{\boldsymbol{\varphi}}(z|\boldsymbol{x}) - \log f_{\theta}(\boldsymbol{x}|z) \\ &\quad - \log f_{\theta}(z)] + \log f_{\theta}(\boldsymbol{x}) \end{aligned} \tag{7.4.2}$$

By transforming Equation (7.4.2), we obtain

$$\begin{aligned} &\log f_{\boldsymbol{\theta}}(\boldsymbol{x}) - J_{KL}\left[f_{\boldsymbol{\varphi}}(z|\boldsymbol{x})\|f_{\boldsymbol{\theta}}(z|\boldsymbol{x})\right] \\ &\quad = E_{f_{\varphi}(z|\boldsymbol{x})}[\log f_{\boldsymbol{\varphi}}(z|\boldsymbol{x}) - \log f_{\boldsymbol{\theta}}(\boldsymbol{x}|z) - \log f_{\boldsymbol{\theta}}(z)]. \end{aligned} \tag{7.4.3}$$

Since the KL divergence is non-negative, let the right-hand side of Equation (7.4.3) equal to $Loss(\boldsymbol{\theta}, \boldsymbol{\varphi}; \boldsymbol{x})$; we can get $\log f_{\boldsymbol{\theta}}(\boldsymbol{x}) \geq Loss(\boldsymbol{\theta}, \boldsymbol{\varphi}; \boldsymbol{x}) \cdot \log f_{\boldsymbol{\theta}}(\boldsymbol{x})$ is the log-likelihood function that needs to be maximized, while the approximate posterior distribution $f_{\varphi}(z|\boldsymbol{x})$ is desired to be close to the true posterior distribution $f_{\boldsymbol{\theta}}(z|\boldsymbol{x})$ so that $J_{KL}\left[f_{\omega}(z|\boldsymbol{x})\|f_{\boldsymbol{\theta}}(z|\boldsymbol{x})\right]$ is close to zero. Here, $Loss(\boldsymbol{\theta}, \boldsymbol{\varphi}; \boldsymbol{x})$ is said to be the variational lower bound of $\log f_{\boldsymbol{\theta}}(\boldsymbol{x})$. To optimize $\log f_{\boldsymbol{\theta}}(\boldsymbol{x})$ and $J_{KL}\left[f_{\omega}(z|\boldsymbol{x})\|f_{\boldsymbol{\theta}}(z|\boldsymbol{x})\right]$, the loss function of the VAE can be defined by the variational lower bound of the likelihood function as

$$
\begin{aligned}
&Loss(\boldsymbol{\theta}, \boldsymbol{\varphi}; \boldsymbol{x}^{(i)}) \\
&\quad = E_{f_{\boldsymbol{\varphi}}(z|\boldsymbol{x})}[\log f_{\boldsymbol{\varphi}}(z|\boldsymbol{x}^{(i)}) - \log f_{\boldsymbol{\theta}}(\boldsymbol{x}^{(i)}|z) - \log f_{\boldsymbol{\theta}}(z)] \\
&\quad = -J_{KL}(f_{\varphi}(z|\boldsymbol{x}^{(i)})||p_{\theta}(z)) + E_{f_{\boldsymbol{\varphi}}(z|\boldsymbol{x}^{(i)})}[\log f_{\theta}(\boldsymbol{x}^{(i)}|z)], \quad (7.4.4)
\end{aligned}
$$

where $J_{KL}(f_{\boldsymbol{\varphi}}(z|\boldsymbol{x}^{(i)})|f_{\boldsymbol{\theta}}(z))$ denotes the regularization term and $E_{f_{\boldsymbol{\varphi}}(z|\boldsymbol{x}^{(i)})}[\log f_{\boldsymbol{\theta}}(\boldsymbol{x}^{(i)}|z)]$ denotes the reconstruction error. Similar to an autoencoder, $f_{\boldsymbol{\varphi}}(z|\boldsymbol{x})$ can be represented as an encoder with a variational parameter $\boldsymbol{\varphi}$, and $f_{\boldsymbol{\theta}}(\boldsymbol{x}|z)$ can be represented as a decoder with a generative parameter $\boldsymbol{\theta}$.

Assuming that $f_{\theta}(\boldsymbol{z})$ obeys a Gaussian distribution $N(\mathbf{0}, \boldsymbol{I})$ and $f_{\varphi}(z|\boldsymbol{x})$ obeys $N(m, \sigma^2)$, the first term on the right-hand side of Equation (7.4.4) can be calculated by

$$
-J_{KL}(f_{\boldsymbol{\varphi}}(z|\boldsymbol{x}^{(i)})||f_{\boldsymbol{\theta}}(z)) = \frac{1}{2}\sum_{j=1}^{N}[1 + \log(\sigma_j^{(i)})^2 - (m_j^i)^2 - (\sigma_j^{(i)})^2]. \quad (7.4.5)
$$

The second term on the right-hand side of Equation (7.4.4) can be calculated by

$$
E_{f_{\boldsymbol{\varphi}}(z|x^{(i)})}[\log f_{\theta}(\boldsymbol{x}^{(i)}|z)] = \frac{1}{S}\sum_{s=1}^{S}\log f_{\theta}(\boldsymbol{x}^{(i)}|z^{(s)}) = \log f_{\boldsymbol{\theta}}(\boldsymbol{x}^{(i)}|z), \quad (7.4.6)
$$

where S denotes the number of times that$f_{\boldsymbol{\varphi}}(z|\boldsymbol{x})$ is sampled, typically $S = 1$. Due to the non-differentiability of the sampling process, in order to avoid the inability to directly derive z and update the parameters of the network through the gradient descent method, the reparameterization technique is used to reparameterize the random variable z, i.e.,

$$
\boldsymbol{z} = \boldsymbol{m} + \boldsymbol{\sigma} e \varepsilon, \quad (7.4.7)
$$

where ε is a randomly sampled value from an N-dimensional independent standard Gaussian distribution, e is the elemental product, m is the mean, and σ^2 is the variance. In order to calculate Equation (7.4.6), $f_{\boldsymbol{\theta}}(x|z)$ is generally chosen to be either a Bernoulli distribution or a Gaussian distribution. If there is an input signal to the network, which is non-binary data, the distribution $f_{\boldsymbol{\theta}}(x|z)$ is Gaussian distribution, i.e.,

$$f_\theta(\boldsymbol{x} \mid z) = \frac{1}{\prod_{i=1}^{N} \sqrt{2\pi\sigma_i^2}} \exp\left(-\frac{1}{2}\left\|\frac{\boldsymbol{x}_i - m_i}{\sigma_i}\right\|^2\right). \tag{7.4.8}$$

Taking the logarithm on both sides of Equation (7.4.8), we obtain

$$\log f_\theta(\boldsymbol{x} \mid z) = -\sum_{i=1}^{N}\left(\left(\frac{1}{2}\left\|\frac{\boldsymbol{x}_i - \boldsymbol{m}_i}{\sigma_i}\right\|\right) + \log(\sqrt{2\pi}\sigma_i)\right). \tag{7.4.9}$$

Based on Equations (7.4.5) and (7.4.8), $Loss(\boldsymbol{\theta}, \boldsymbol{\varphi}; \boldsymbol{x})$ can be calculated to yield the loss function of the VAE.

From Equations (7.4.7) to (7.4.9), the N-dimensional standard deviation vector may be represented as $\boldsymbol{\sigma} = [\sigma_1, \sigma_2, \ldots, \sigma_N]$, and the N-dimensional mathematical expectation vector may be denoted as $\boldsymbol{m} = [m_1, m_2, \ldots, m_n]$. At this point, Figure 7.6 can be redrawn as Figure 7.7.

The training goal of the VAE model is to minimize the reconstruction error and to make the $f(z|\boldsymbol{x})$ as close as possible to the

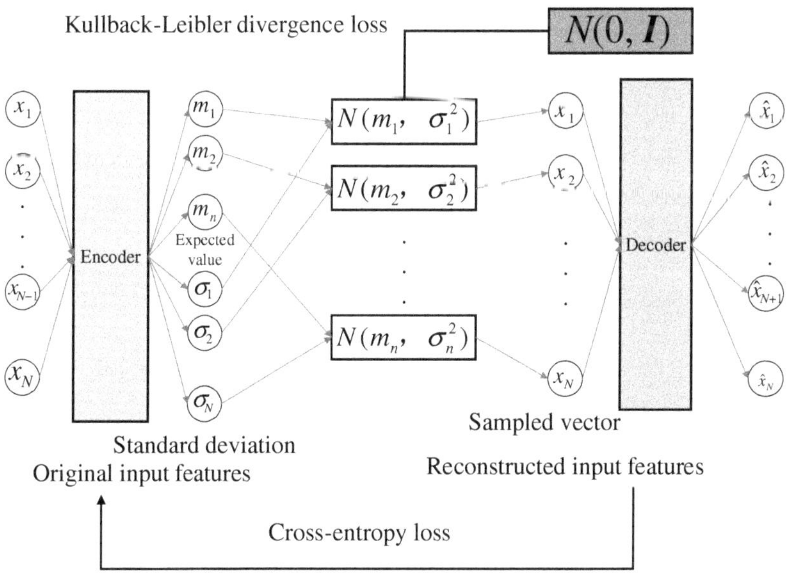

Figure 7.7. VAE structure.

standard multivariate Gaussian distribution. The loss functions of the VAE are given by

$$Loss_{\text{VAE}} = \frac{1}{2}(Loss_{\text{xent}} + Loss_{KL}), \quad (7.4.10)$$

$$Loss_{\text{xent}} = -\frac{1}{N}\sum_{i=1}^{N}[x_i \ln \hat{x}_i + (1 - x_i)\ln(1 - \hat{x}_i)], \quad (7.4.11)$$

$$Loss_{\text{KL}} = \frac{1}{2}\sum_{j=1}^{N}[1 + \log(\sigma_j^{(i)})^2 - (m_j^i)^2 - (\sigma_j^{(i)})^2], \quad (7.4.12)$$

where x_i represents the original input feature for the ith sample, while $\hat{x}_i$ denotes the ith dimension of the reproduced original input feature. The loss function consists of two parts: (1) the cross-entropy loss function, which is used to measure the degree of difference between the reproduced features and the original input features; (2) the relative entropy loss function, i.e., the KL (Kullback–Leibler) divergence, which is used to measure the degree of difference between the standard multivariate Gaussian distributions of the $p(z|x)$.

7.4.2 Stacked variational autoencoder

7.4.2.1 *Structure of the stacked variational autoencoder*

The Stacked Variational Autoencoder (SVAE) is a deep network architecture formed by stacking multiple VAEs, which progressively reduce the dimensionality of input features and extract high-level representations. The training process of the entire model is divided into two stages: unsupervised pretraining and supervised fine-tuning. The architecture of the SVAE and logistic evaluation model is illustrated in Figure 7.8 [146]. In Figure 7.8, the circle dots in the input layer represent neuron $z^{(k)}$, which is the high-level feature extracted by the kth VAE.

In the pretraining stage, the evaluation model starts training from the lowest level VAE. After fully completing learning the features of this layer, the higher-order features outputted from this layer of the VAE will be used as the input of its previous layer, and the training of its previous layer will be continued until all the VAEs have been

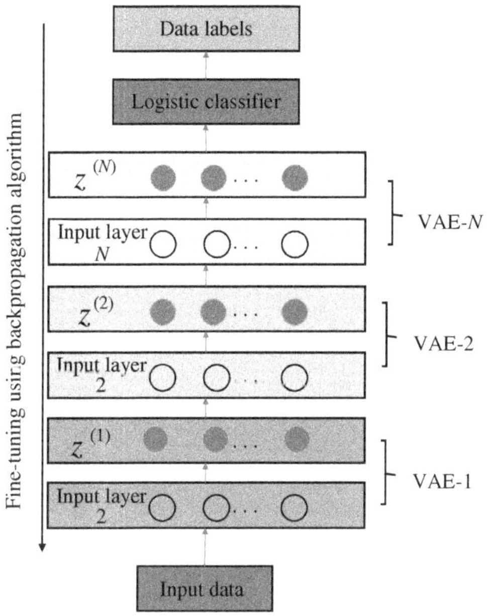

Figure 7.8. SVAE+Logistic evaluation model.

fully trained. In this way, the pretraining of the discriminative model is completed and the extraction of the original high-dimensional features is realized. The whole pretraining process does not require the participation of the labeled data and is an unsupervised learning process. The pretraining enables the parameters of the VAE to converge to a better local optimal solution and, at the same time, can reduce the effect of gradient dispersion when using the BackPropagation (BP) algorithm to perform fine-tuning. The SVAEs have a strong generalization ability, and the extracted higher-order features are noise-resistant through the learning of the distribution of the features and the incorporation of Gaussian noise in the training process. Moreover, compared with directly extracting features using a single VAE, the SVAE, due to its deep network structure, can extract more abstract high-order features and has better fitting ability for high-dimensional nonlinear systems, making it more suitable for complex classification tasks. The higher-order features extracted by the SVAE are input into the logistic classifier, and the parameters of the whole network are supervised and fine-tuned using the BP algorithm, and the optimal parameters $\boldsymbol{\theta}_{\text{opt}}$ of the evaluation model are obtained according to Equation (7.4.6) as follows:

$$\boldsymbol{\theta}_{\text{opt}} = \arg\min_{\boldsymbol{\theta}}[-\boldsymbol{x}\ln\hat{\boldsymbol{x}} - (1-\boldsymbol{x})\ln(1-\hat{\boldsymbol{x}})],$$

where the function argmin(·) is the value of $\boldsymbol{\theta}$ when the function (·) is minimized, $\boldsymbol{\theta}$ is the model's parameter matrix, $\boldsymbol{x}$ is the expected label value of the training samples, and $\hat{\boldsymbol{x}}$ is the predicted label value of the training samples.

7.4.2.2 *L2 Regularization*

To enhance the generalization capability of the discriminative model, L2 regularization technique is introduced.

The loss function $Loss_{\text{Sparse}}$ with L2 regularization is defined as

$$Loss_{\text{Sparse}} = Loss_{\text{VAE}} + \lambda \sum_{q} w_q^2, \quad (7.4.13)$$

where $Loss_{\text{VAE}}$ is the original objective function, w_q represents the weight parameters of the neurons, and λ is the regularization coefficient. By adding the L2 penalty term to the loss function, the discriminative model is encouraged to use smaller weight values during training, which to some extent reduces the overfitting of the model and enhances its generalization ability.

7.4.3 Deep convolutional variational autoencoder

7.4.3.1 *Architecture of deep convolutional variational autoencoder*

Convolutional Neural Networks (CNNs) usually consist of an input layer, a convolutional layer, a pooling layer, an activation function, a fully connected layer, and an output layer. The convolutional layer consists of multiple feature surfaces, each of which is made up of multiple neurons, the input of the neurons in the current layer is connected to the local region of the previous layer of feature surfaces through the convolution kernel, the convolution operation is performed by using connection weights and bias, the activation function is activated to get the input value of the neurons in the current layer, and the size of the connection weights is determined by the

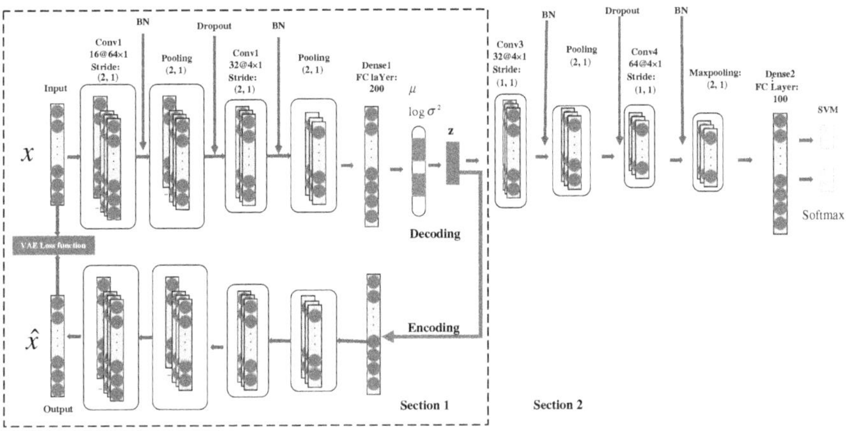

Figure 7.9. Structure of the deep convolutional variational autoencoder.

size of the convolution kernel. The pooling layer is generally located after the convolutional layer, similar to the downsampling operation, and plays a role in secondary feature extraction. Each neuron in the fully connected layer is fully connected to all neurons in the previous layer, which can integrate local information with class differentiation in the convolutional or pooling layer.

In a standard VAE, the neural network resembles a Multi-Layer Perceptron (MLP) with fully connected layers. The CNN was proposed to construct the VAE [147] to reduce the network's complexity. This results in the Deep Convolutional Variational Autoencoder (DCVAE) is illustrated in Figure 7.9.

The DCVAE model consists of two main components: Section 1 within the dashed box represents the encoding and decoding process of the Variational AutoEncoder (VAE), while section 2 except for section 1 in Figure 7.9 comprises a multi-layer CNN. The training of the DCVAE model is accomplished through unsupervised learning in the VAE and supervised learning in the CNN. In the VAE encoding phase, the input layer is followed by the first convolutional layer (Conv1), where "16 @ 64 × 1" denotes 16 feature maps with convolutional kernels of size (64, 1). The stride is set to (2, 1), meaning that the kernel slides with a vertical step of 2 and no horizontal movement. The output of Conv1 undergoes batch normalization (BN) method and serves as the input to the pooling layer. The ReLU is used as the activation function. The BN method normalizes the inputs over mini-batches to reduce input distribution shift, which facilitates faster convergence and enhances generalization performance. Maximum pooling with a stride of 2 is employed. The output of the first pooling layer then goes through a dropout operation, which introduces noise and serves as the input to the second convolutional layer (Conv2). Conv2 is also followed by BN normalization and maximum pooling. The result is passed to a fully connected layer with 200 neurons, and the latent mean vector m and the logarithmic variance $\log \sigma^2$ are outputted. Using the reparameterization sampling, the latent feature z can be obtained. Due to the VAE being unsupervised learning, it is necessary to reconstruct the input data through the decoding process to complete the training. The decoding process is the reverse operation of the encoding process, replacing the convolution operation with deconvolution. Once the VAE is trained, the latent variable z is fed into Conv3 in the CNN. Conv3 has 32 feature maps with a kernel size of (4, 1) and a stride of 1. Its output is batch-normalized, max-pooled, and passed through a dropout layer. Conv4 consists of 64 feature maps, also with a kernel of size (4, 1) and a stride of 1. After Conv4, the output is normalized and max-pooled again. The pooled output is connected to a fully connected layer with 100 neurons and subsequently fed into a Softmax classifier. The loss function for training the multi-layer CNN is constructed using cross-entropy. The network parameters are updated through back-propagation and fine-tuning. After multiple training iterations, the DCVAE model is fully optimized.

7.4.3.2 *Dropout method*

The dropout method is an effective approach to preventing overfitting in deep neural networks. In a given layer of the network, a subset of neurons is randomly deactivated—meaning that they are not activated and do not connect to neurons in the preceding or subsequent layers. As a result, the connections between neurons in the network become sparse, which compels the network to learn robust and useful connections among randomly selected subsets of neurons, even in the absence of some weight connections. Although dropout and regularization both serve to reduce network complexity by limiting weight dependencies, their mechanisms differ. Dropout technique does not modify the loss function; instead, it alters the network architecture directly during training. A hyperparameter p denotes the probability that a neuron is activated in the current layer, while $(1-p)$ represents the probability that a neuron is dropped.

Dropout technique can alter the connection type within a neural network and effectively train the network with different configurations.

The hyperparameter p adopts fixed values between (0,1), usually is set to 0.5, and the use of dropout technique in which layer of the network is also based on experience. By disrupting connections between neurons, dropout functions are similar to adding noise to the network input. A higher p corresponds to lower noise intensity, while a lower p results in greater noise. In this section, a variable dropout technique is adopted, as illustrated in Equation (7.4.14), where the value of p gradually decreases over time. Moreover, higher values of p occur more frequently than lower values. A larger p helps the network learn fine-grained features of the data, while a smaller p is profit to learn robust discriminative features; thereby it can reduce the model's sensitivity to small perturbations. As shown in Figure 7.18, dropout technique is applied after the first maximum pooling layer in the training phase of both the VAE and the multi-layer CNN within the DCVAE architecture:

$$p(i) = 1 - \frac{0.5}{\text{epochs} - \text{epoch}(i)}, \tag{7.4.14}$$

where *epochs* is the number of iterative training of the network, *epoch*(i) represents the ith iteration of training, and $p(i)$ denotes the value of hyperparameter p for the ith training.

7.4.3.3 *Update of learning rate*

The learning rate is a crucial hyperparameter that governs the speed at which weights are updated during backpropagation in a neural network. A larger learning rate allows for faster descent along the gradient direction but may cause the network to overshoot local optima. Conversely, the smaller the learning rate, the slower the updating speed. The smaller the probability of missing local minima, the more the time spent on convergence. To accelerate convergence speed, a relatively large learning rate is used at the beginning of training, and a smaller one is adopted as the training approaches the maximum number of iterations. In this study, the Stochastic Gradient Descent (SGD) method is employed for updating the network parameters. The learning rate η is adjusted according to

$$\eta_i = \begin{cases} 1 \times 10^{-2}, & acc\ (i+1) \geq acc(i) \\ 1 \times 10^{-4}, & acc\ (i+1) < acc(i)\ or\ epoch(i) \geq epochs^{-3}, \end{cases} \tag{7.4.15}$$

where the initial learning rate is set to 1×10^{-2}, $acc(i)$ denotes the classification accuracy at the ith training evaluation, and $epoch(i)$ represents the number of iterations at the ith training epoch.

7.5 Autoencoder-Based Echo State Network

To construct a process dynamic model with moving characteristics, dynamic neural networks [148] and Recurrent Neural Networks (RNNs) [149, 150] can be adopted, as they offer strong capabilities in sequential data processing. The Echo State Network (ESN) is a novel type of RNN distinguished by the use of a Dynamic Neurons Reservoir (DNR), which consists of a large number of sparsely connected neurons (typically 1%–5% connectivity) and possesses temporal memory functionality. The generalized inverse method is used to obtain the output weights, the global optimal solution can be obtained, and the learning speed is fast. However, when dealing with high-dimensional or complex process data, the ESNs require a large number of reservoir nodes. As sample data are mapped through these numerous nodes, the resulting output of the reservoir becomes high-dimensional and complex; it results in increasing computational cost and potentially reducing the model's precision. Moreover, the output weights are typically computed using multiple methods such as generalized inverse method or least square one, which are less accurate in the presence of high dimensionality or collinearity among features. The large-scale mappings in the DNR can lead to collinear output matrices and further degrade the accuracy of the ESN model. To address this, dimensionality reduction can be applied to the reservoir output. Autoencoding neural networks serve as an effective nonlinear feature extraction technique. Similar in structure and functionality to auto-associative neural networks [151, 152], autoencoders have a symmetric architecture in which the input and output layers are identical, and the hidden (bottleneck) layer has fewer nodes than the input/output layers. Through this bottleneck structure, the autoencoder can compress the input data while removing noise. Therefore, the extracted features achieve both dimensionality reduction and denoising. Importantly, due to the use of nonlinear activation functions in the hidden layers, the extracted features are free from collinearity.

Accordingly, Ref. [153] proposed a novel model, i.e., the Features Extracted from Auto-Encoder based Echo State Network (FEAE-ESN), which combined the strengths of both the ESN and autoencoding networks.

7.5.1 Echo state network

The Echo State Network (ESN) is a special type of RNN, as illustrated in Figure 7.10.

Figure 7.10 shows the structure of an ESN consisting of N input nodes, a reservoir with J neurons, and M output nodes. The basic

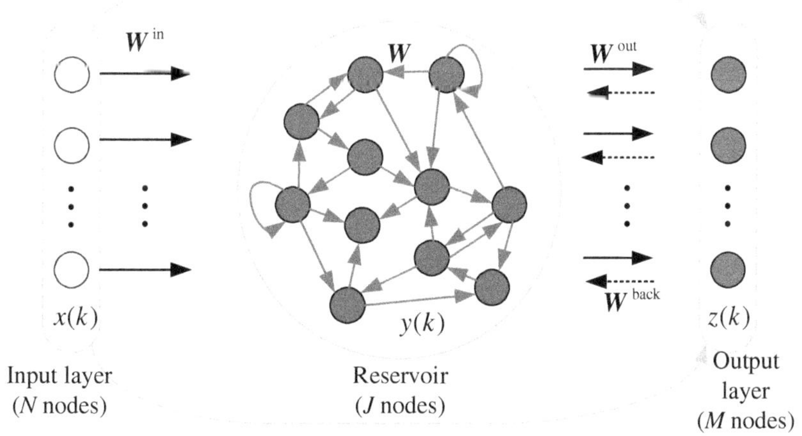

Figure 7.10. Structure of the ESN.

equations of the ESN are as follows:

$$y(k+1) = f(W^{in}x(k+1) + Wy(k) + W^{back}z(k)), \quad (7.5.1)$$

$$z(k+1) = f^{out}(W^{out}(x(k+1), y(k+1), z(k))), \quad (7.5.2)$$

where x, y, and z represent the input, state, and output variables of the ESN, respectively. Matrices W^{in}, W, and W^{back} denote the input weight matrix, reservoir weight matrix, and feedback weight matrix, respectively. W^{out} is the output connection weight matrix. f is the internal activation function, which is set as the hyperbolic tangent function (Tanh), and f^{out} is the activation function of the output layer, set as the identity function.

In practical applications, in order to ensure the performance and stability of the ESN, the spectral radius of the reservoir weight matrix must be less than 1. The output weights can be obtained using the Moore–Penrose pseudo-inverse; however, when the number of neurons in the reservoir is too large, the output dimension (i.e., the connection from the reservoir to the output layer) becomes excessively high, and multicollinearity may occur among the data, which affects the solution of the output weights.

To address these issues, an autoencoding network is used for dimensionality reduction and removal of multicollinearity, thereby improving the precision of the ESN. The bottleneck layer in the autoencoder can effectively reduce the dimensionality of the input data and remove noise among the data, which can in turn reduce the dimensionality of the reservoir's output, eliminate multicollinearity among its components, and enhance the modeling accuracy of the ESN.

7.5.2 Autoencoder-based echo state network

The FEAE-ESN model is constructed through the following steps:

Step 1: Obtain training data samples

Training data samples with P group $S_p(x_{pi}, y_{pj})$ are obtained, which include N inputs and M outputs, where $p = 1, 2, \ldots, P$; $i = 1, 2, \ldots, I$; $j = 1, 2, \ldots, J$; $x_{pi} \in R^J, z_{pi} \in R^I$.

Step 2: Get the new state variable

The sample S_h passes through the reservoir of the ESN to get the new state variable as

$$y(k+1) = f(\boldsymbol{w}^{\text{in}}\boldsymbol{x}(k+1) + \boldsymbol{w}y(k) + \boldsymbol{w}^{\text{back}}z(k)), \tag{7.5.3}$$

where k represents the data sampling time point, y is the state variable of the ESN, and $\boldsymbol{w}^{\text{in}}$, $\boldsymbol{w}$, and $\boldsymbol{w}^{\text{back}}$ are generated by randomization.

Step 3: Process data

After the state variables have been updated by S_p, the data matrix of the connecting output nodes is represented as

$$\boldsymbol{A} = [\boldsymbol{x}(k+1), \boldsymbol{y}(k+1), \boldsymbol{z}(k)]. \tag{7.5.4}$$

For data samples with P sets of N inputs and M outputs, the data matrix connecting the output nodes is denoted as

$$\boldsymbol{H} = \begin{bmatrix} \boldsymbol{x}_1(k+1), \boldsymbol{y}_1(k+1), \boldsymbol{z}_1(k) \\ \boldsymbol{x}_2(k+1), \boldsymbol{y}_2(k+1), \boldsymbol{z}_2(k) \\ \vdots \\ \boldsymbol{x}_P(k+1), \boldsymbol{y}_P(k+1), \boldsymbol{z}_P(k) \end{bmatrix}. \tag{7.5.5}$$

The matrix data in Equation (7.5.5) is reduced in dimensionality, de-linearized, and fed into an autoencoding network, where the number of autoencoding input nodes and the number of output nodes are the N_{col} columns of the matrix $\boldsymbol{H}$, the number of nodes in the mapping layer is N_{map}, and the number of neurons in the bottleneck layer is set to N_{bot}. After the autoencoding network has been trained by the BP algorithm of the error, the output of the mapping layer is given by $\boldsymbol{F} = \boldsymbol{f}(\boldsymbol{H}\boldsymbol{w} + \boldsymbol{b})$, the output of the bottleneck layer is defined as $\boldsymbol{G} = \boldsymbol{f}(\boldsymbol{F}\boldsymbol{w}' + \boldsymbol{b})$, $\boldsymbol{F}$ is the output matrix of the mapping layer in the autoencoding neural network when the input is matrix $\boldsymbol{H}$, $\boldsymbol{G}$ is the output matrix of the bottleneck layer in the autoencoding network, and $\tilde{\boldsymbol{w}} = [\tilde{\boldsymbol{w}}_{ji}]_{N_{\text{col}} \times N_{\text{map}}}$ is used to represent the matrix of the connection weights between the neurons of both the input layer and the mapping layer. $\tilde{\boldsymbol{w}}_{ji}$ denotes the connection weights between the ith input neuron in the input layer and the jth neuron in the mapping layer, $\boldsymbol{b} = [b_1, b_2, \ldots, b_{N_{\text{map}}}]$ is the threshold vector of the mapping layer, $\boldsymbol{w}' = [w'_{mj}]_{N_{\text{map}} \times N_{\text{bot}}}$ is the connection weight matrix

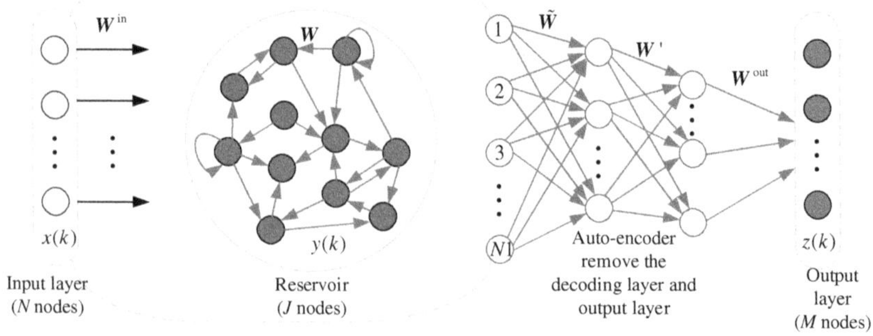

Figure 7.11. Structure of FEAE-ESN model.

between the mapping layer and bottleneck layer, w'_{mj} is the connection weight between the jth mapping neuron and the neuron of the mth bottleneck layer, $\boldsymbol{b}' = [b'_1, b'_2, \ldots, b'_{N_{\text{bot}}}]$ is the threshold vector of the bottleneck layer, and $f(\cdot)$ is the activation function and selected as

$$f(x) = \frac{1}{1+e^{-x}}. \tag{7.5.6}$$

Step 4: Output connection weight matrix

The output nodes of the ESN are linearly summed by

$$\boldsymbol{w}^{out} = \boldsymbol{G}^{-1}\boldsymbol{Y}, \tag{7.5.7}$$

where $\boldsymbol{w}^{out}$ is the output connection weight matrix, $\boldsymbol{Y}$ is the desired output value, and $\boldsymbol{G}^{-1}$ is the generalized inverse of the matrix $\boldsymbol{G}$.

Step 5: Verify performance

The properties of the ESN are verified using testing samples s_t.

In summary, the structure of the FEAE-ESN model is shown in Figure 7.11, and its establishment steps are shown in Figure 7.12.

7.6 Deep Canonical Correlation Sparse Autocoder

Canonical Correlation Analysis (CCA) is a generalized method for combining two or more data types [215]. The CCA modulus finds components that are correlated with each other by finding the maximum correlation between linear combinations of variants from two types of data. When the data is characterized by high dimensionality but small samples, the CCA modulus will face the problem of

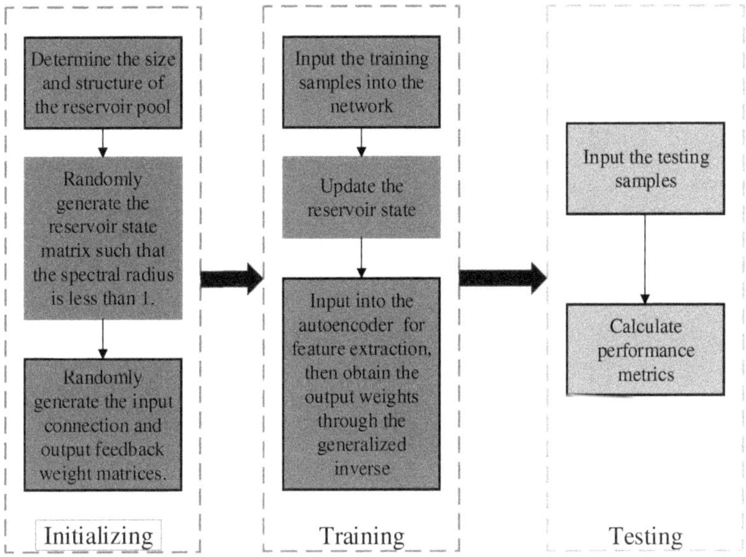

Figure 7.12. Modeling process of the FEAE-ESN.

overfitting. To solve this problem, the CCA modulus adds sparse penalty terms such as L1-norm or a combination of L1-norm and L2-norm (elastic network) to the traditional CCA modulus to select a small number of features [154–156]. More effective penalty terms have also been added, such as group sparse CCA [157], structured sparse CCA [158], joint sparse CCA [159] and adaptive sparse CCA [160]. However, complex nonlinear relationships among multiple data types are not considered in linear CCA modulus, so Deep CCA (DCCA) modulus has been proposed [161]. Correlation autoencoders are used to optimize the combination of typical correlations between two datasets. The DCCA and correlation autoencoders focus on nonlinear information transformations but neglect effective nonlinear dimensionality reduction. Therefore, a challenge issue is how to achieve effective nonlinear information transformation and nonlinear dimensionality reduction at the same time.

In order to find the complex nonlinear relationship between the two types of data and to solve the small-sample problem, a Deep Canonically Correlated Sparse Autoencoder (DCCSAE) is proposed in Ref. [162].

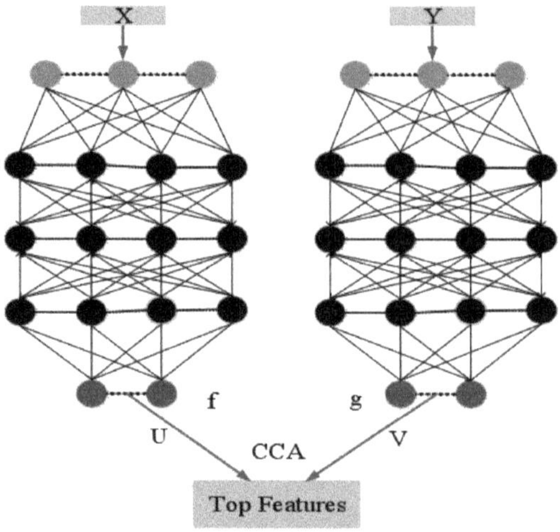

Figure 7.13. Modulus structure of deep canonical correlation analysis.

7.6.1 Deep typical correlation analysis

Andrew *et al.* [161] designed a Deep Neural Network (DNN) for typical correlation analysis called the DCCA, as shown in Figure 7.13. The DCCA modulus can overcome the CCA modulus limitation of not being able to detect complex nonlinear correlations. In the DCCA modulus, two DNNs can learn the nonlinear representation of each dataset. The DCCA modulus is obtained by maximizing the typical correlation between the outputs of two DNNs, namely

$$(\boldsymbol{U}', \boldsymbol{V}') = \max_{\boldsymbol{W}_f, \boldsymbol{W}_g, \boldsymbol{U}, \boldsymbol{V}} \frac{1}{N} \mathrm{tr}(\boldsymbol{U}^T \boldsymbol{f}(X) \boldsymbol{g}(Y)^T \boldsymbol{V}), \qquad (7.6.1)$$

where

$$\boldsymbol{U}^{\mathrm{T}} \left(\frac{1}{N} \boldsymbol{f}(X) \boldsymbol{f}(X)^{\mathrm{T}} + r\boldsymbol{I} \right) \boldsymbol{U} = \boldsymbol{I},$$

$$\boldsymbol{V}^{\mathrm{T}} \left(\frac{1}{N} \boldsymbol{g}(\boldsymbol{y}) \boldsymbol{g}^{\mathrm{T}}(\boldsymbol{y}) + s\boldsymbol{I} \right) \boldsymbol{V} = \boldsymbol{I},$$

$$\boldsymbol{u}_i^{\mathrm{T}} \boldsymbol{f}(X) \boldsymbol{g}(Y)^{\mathrm{T}} \boldsymbol{v}_j = 0, \quad i \neq j,$$

where N denotes the total number of data, $\boldsymbol{X}$ and $\boldsymbol{Y}$ denote the input matrices of the two datasets, $\boldsymbol{I}$ is the unit matrix, $\boldsymbol{f}(\mathrm{X})$ and $\boldsymbol{g}(\mathrm{Y})$ denote the nonlinear representations of the two DNNs with

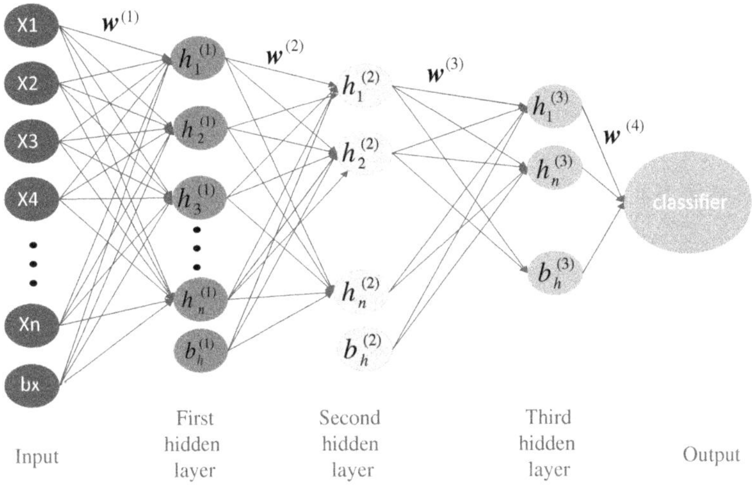

Figure 7.14. Structure of the SSAE.

parameters W_f and W_g, respectively, and $\boldsymbol{U}$ and $\boldsymbol{V}$ are the projection vectors of the final output.

7.6.2 Stacked sparse autoencoder

Autoencoders can be further stacked together to obtain more information. By using multiple autoencoders, the Stacked Sparse AutoEncoder (SSAE) [163] was obtained, as shown in Figure 7.14, where the encoding layer is connected to the input layer of the next Sparse AutoEncoder (SAE) for better feature extraction.

7.6.3 DCCSAE

The DCCA modulus works well when exploring nonlinear mappings between two types of data. The SAE has had great success in seeking nonlinear representations of individual data. However, the DCCA modulus is unable to achieve effective nonlinear dimensionality reduction and the SAE is unable to explore correlations between cross-modal data. Therefore, it is necessary to combine the DCCA modulus with the SAE to form the Deep Canonical Correlation Sparse AutoEncoder (DCCSAE), as shown in Figure 7.15. The DCCSAE seeks a deep network representation of the two datasets

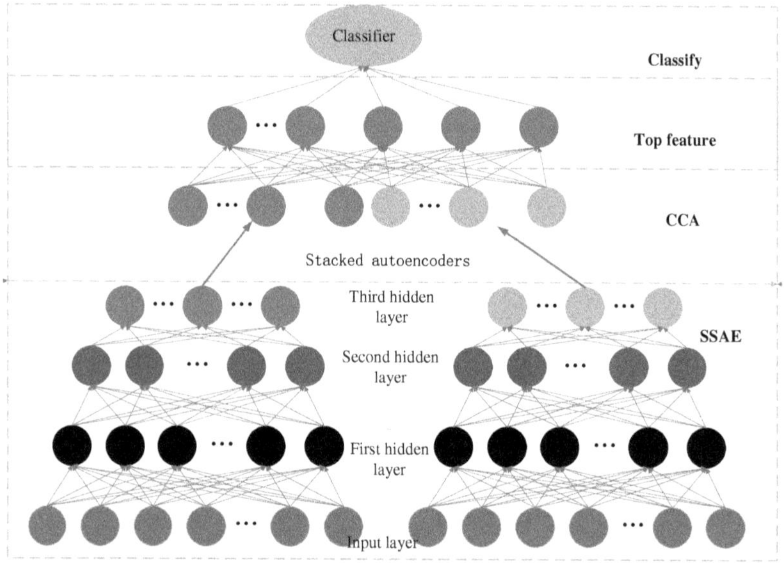

Figure 7.15. Structure of the DCCSAE.

that maximizes the typical correlation between the two datasets while minimizing the reconstruction error of the SAE. The cost function of the DCCSAE is defined as

$$
\begin{aligned}
J_{\mathrm{DCCSAE}} &= \min_{W_f W_g U V} -\frac{1}{N}\mathrm{tr}(\boldsymbol{U}^{\mathrm{T}}\boldsymbol{f}(X)\boldsymbol{g}(Y)^{\mathrm{T}}\boldsymbol{V}) + \\
&= \frac{\lambda}{N}||\hat{\boldsymbol{x}}_i - \boldsymbol{x}_i||^2 + ||\hat{\boldsymbol{y}}_i - \boldsymbol{y}_i||^2 + \alpha J_{\mathrm{KL}}(\rho||\hat{\rho}_j) \\
&\quad + \beta J_{\mathrm{KL}}(\sigma||\hat{\sigma}_k),
\end{aligned} \tag{7.6.2}
$$

where f and g are used to extract nonlinear features of the DNNs for each dataset while encoding each input. $\boldsymbol{U} = [u_1, u_2, \ldots, u_L]$ and $\boldsymbol{V} = [v_1, v_2, \ldots, v_L]$ are the projections of the DNN's outputs to the CCA directions of the top layers with$L$ units, and $\hat{\boldsymbol{x}}_i$ and $\hat{\boldsymbol{y}}_i$ denote the reconstructions of the input $\boldsymbol{x}_i$ and input $\boldsymbol{y}_i$, respectively. $J_{\mathrm{KL}}(\cdot)$ is defined as

$$
J_{\mathrm{KL}}(\rho||\hat{\rho}) = \sum_{j=1}^{N_h}\left[\rho\log\frac{\rho}{\hat{\rho}_j} + (1-\rho)\log\frac{1-\rho}{1-\hat{\rho}_j}\right], \tag{7.6.3}
$$

where $\hat{\rho} = \frac{1}{N}\sum_{i=1}^{N} h_j(x_i)$ denotes the average of all node activations of the neuron j in the hidden layer or all nodes in the input layer, and the same applies to $J_{\text{KL}}(\sigma \parallel \hat{\sigma}_j)$.

7.7 Conditional Dual Adversarial Autoencoding Network

Currently, the biggest problem of infrared target recognition based on deep learning network is that it is difficult to obtain enough labeled real target datasets, which leads to insufficient training of the model and affects the overall performance of the system. Reference [164] first proposed an infrared target simulation method based on Generative Adversarial Network (GAN), but due to some problems such as unstable GAN training, the generation of target images is not ideal. To address the above problems, a conditional dual adversarial autoencoding modeling method was proposed [164]. By combining GAN and Variational AutoEncoding (VAE), the proposed model can generate different classes of infrared targets under the premise of ensuring image authenticity and network diversity, which can provide richer samples for the training of target recognition algorithms and improve the accuracy of target recognition.

7.7.1 Conditional dual adversarial autoencoding model

Compared to the GAN, the VAE training is usually more stable and able to cover all samples in the dataset. However, the VAE actually optimizes the lower bound of the log-likelihood function rather than the likelihood function itself when calculating the KL divergence, and thus, the actual generated images are usually blurry. Based on combining the respective advantages and disadvantages of the GAN and the VAE, Ref. [164] proposed a Conditional Dual Adversarial AutoEncoder (CDAAE) for infrared target modeling, which can generate multiple categories of infrared target images based on the VAE by adding the discriminator and target category information. The modeling framework of the CDAAE consists of four networks, as shown in Figure 7.16, namely encoder E, decoder D_g, latent space discriminator D_1, and sample discriminator D_s.

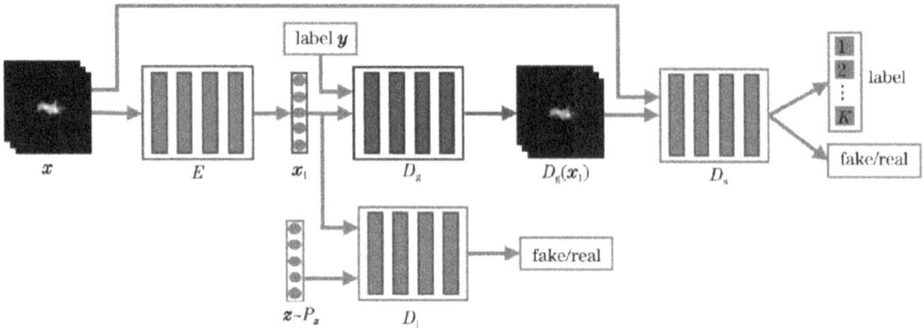

Figure 7.16. Structure of CDAAE model.

In the CDAAE model, in order to make the data $\boldsymbol{x}_1$ mapped to the latent space satisfy a specific data distribution f_z, instead of using the approximate computation of KL divergence in the VAE, a latent space discriminator D_1 is designed to learn the data distribution through a neural network and forms the first heavy confrontation in the model. The latent space data $\boldsymbol{x}_1$ outputted from the encoder and combined with the category label $\boldsymbol{y}$ passes through the decoder D_g to generate the reconstructed sample $D_g(\boldsymbol{x}_1)$. In addition to the conventional way of optimizing the decoder by calculating the Mean Square Error (MSE) of the reconstructed sample $D_g(\boldsymbol{x}_1)$ and the real sample $\boldsymbol{x}$, the Mean Average Error (MAE) criterion is used to design the sample discriminator D_s, in order to logically discriminate and classify the reconstructed sample $D_g(\boldsymbol{x}_1)$ and the real sample $\boldsymbol{x}$, and the decoder is further constrained and the effect of the sample generation is effectively improved to form the second heavy confrontation in the model. Through the training of the whole network, only the category labels and the random sampling data conforming to the distribution f_z are input into the decoder to obtain the realistic and rich infrared target images.

7.7.2 Principle of the algorithm

In order to efficiently generate infrared target images that meet the requirements, Literature [144] established an infrared modeling method based on the CDAAE, and Literature [164] examined the pattern recognition capabilities of Error Backpropagation Algorithm (EBPA), the Deep Autoencoder Learning Algorithm based on

Cross-Entropy Function (CDAA), and so on. The whole algorithm is divided into two stages: model training and target generation. In the model training phase, the potential data distribution of the samples is learned through neural networks, and the learned network parameters are utilized in the target generation phase to generate infrared targets of the desired categories.

7.7.2.1 *Model training*

The first heavy confrontation is between the encoder E and the latent space discriminator D_1. The network structure of both E and D_1 is given in Figure 7.17. The encoder E contains two convolutional layers and two fully connected layers, with four middle layers, in which all the convolutional layers have a convolutional kernel size of 4×4 and a step size of 2. The latent space discriminator D_1 contains three fully connected layers and uses them as the middle layers. In order to improve the training efficiency and stability, all interlayers are processed using the BN method after completing the nonlinear transformation of the data, and the LReLU function is used as the activation function.

At the beginning of the training phase, M images are first randomly selected from the dataset to form a batch sample set $\boldsymbol{x}$, which is used as the input to the encoder E. After $\boldsymbol{x}$ is processed through two layers of convolution (Conv1, Conv2), its dimension is reduced from four dimensions to two dimensions, and then the data are compressed through two fully connected layers (FC1, FC2), the latent space data $\boldsymbol{x}_1$ with the dimension of $(M \times N_1)$ are outputted; thus, the mapping from the original data to the latent space data is completed. In order to make the mapped data satisfy a specific data distribution f_z, the latent space data $\boldsymbol{x}_1$ and the randomly sampled data z from f_z are input into the latent space discriminator D_1, and after being processed by the three fully connected layers (FC1, FC2, FC3), the discriminant probability is outputted through the Sigmoid function. Finally, the entropy loss functions $Loss_E$ of the encoder and $Loss_{D_1}$ of the latent space discriminator are calculated using the discriminative probabilities, respectively, and the parameters $\boldsymbol{\theta}_E$ and $\boldsymbol{\theta}_{D1}$ of the two networks are updated by the BP algorithm to achieve the goal of minimizing the loss functions. The $Loss_E$ and $Loss_{D_1}$ are

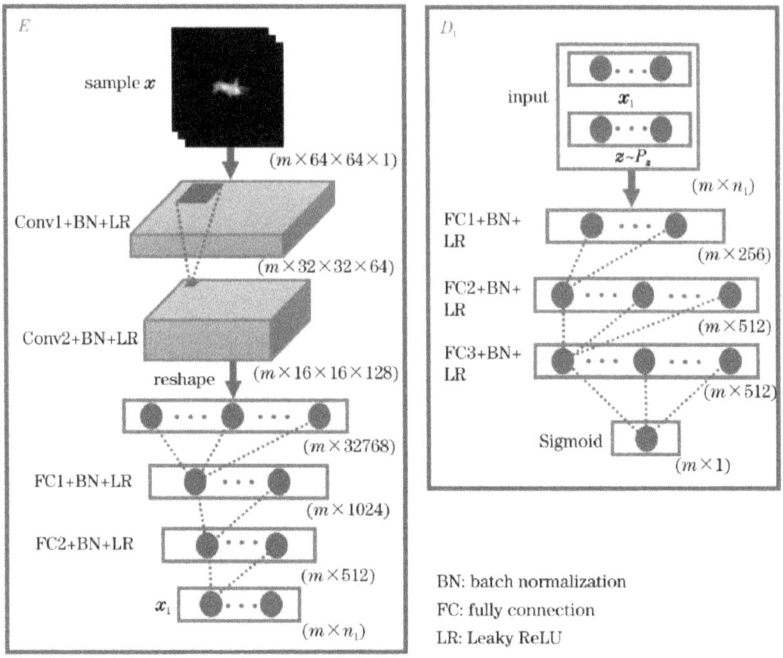

Figure 7.17. Structure of encoder and latent space discriminator.

expressed as

$$Loss_E = -E_{\boldsymbol{x}\sim f_x}\{\log D_1[E(\boldsymbol{x})]\}, \tag{7.7.1}$$

$$Loss_{D_1} = -E_{\boldsymbol{x}\sim f_x}\{1 - \log D_1[E(\boldsymbol{x})]\} - E_{z\sim f_z}\{\log[D_1(\boldsymbol{z})]\}. \tag{7.7.2}$$

After completing the training of the first heavy confrontation, the second heavy confrontation is carried out. The second heavy confrontation is between the decoder D_g and the sample discriminator D_s. Figure 7.18 shows the network structure of the decoder D_g and the sample discriminator D_s. Since decoding is actually the inverse of encoding, decoder D_g adopts a symmetric network structure, which consists of three fully connected layers and two deconvolutional layers to realize the mapping of latent space data to the reconstructed samples. The sample discriminator D_s is a network structure with a similar middle layer, including two convolutional layers and two fully connected layers, because the input data are the

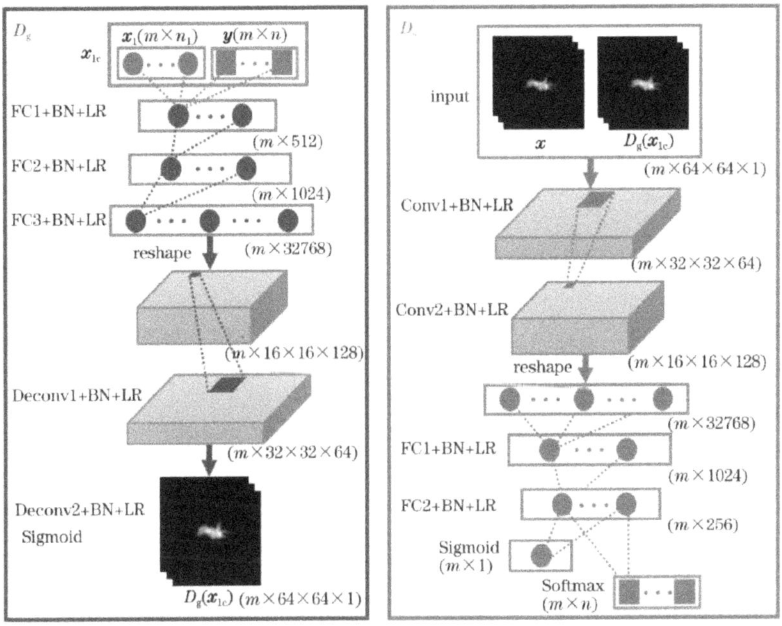

Figure 7.18. Structure of decoder and sample discriminator.

same as the encoder. All the convolutional and inverse convolutional layers of decoder D_g and sample discriminator D_s have a convolutional kernel size of 4×4 and a step size of 2. The BN processing is performed at the end of all the middle layers and the LReLU function is used as the activation function. Figure 7.18 shows that the sample discriminator D_s will eventually not only identify the authenticity of the samples but also output the class labels of the samples; thus, the network also has the ability to label the unclassified samples.

The second heavy confrontation process begins with concatenating the output $\boldsymbol{x}_1$ of the encoder E from the first heavy confrontation process with the corresponding class labels $\boldsymbol{y}$ of the sample set, which serves as the input $\boldsymbol{x}_{lc}$ to the decoder D_g, where the class label $\boldsymbol{y}$ is an encoded label matrix with a dimension of $m \times n$ and obtained by applying one-hot encoding to the original categorical labels. The combined representation $\boldsymbol{x}_{lc}$ is processed through three fully connected layers (FC1, FC2, FC3) and followed by two deconvolutional layers (Deconv1, Deconv2) to perform upsampling; the sample $D_g(\boldsymbol{x}_{lc})$ is ultimately reconstructed. On the basis of making use of the reconstructed sample $D_g(\boldsymbol{x}_{lc})$, the reconstructed loss function $Loss_{D_g-recon}$ combining the MSE with the MAE is used as part of the decoder's total loss function $Loss_{D_g}$ and expressed as

$$Loss_{D_{g-recon}} = \beta \times ||D_g(\boldsymbol{x}_{lc}) - \boldsymbol{x}||_2^2 + (1-\beta) \times ||D_g(\boldsymbol{x}_{lc}) - \boldsymbol{x}||_1, \tag{7.7.3}$$

where $||D_g(\boldsymbol{x}_{lc}) - \boldsymbol{x}||_2^2$ is the MSE of the reconstructed samples and original samples, $||D_g(\boldsymbol{x}_{lc}) - \boldsymbol{x}||_1$ is the MAE of the reconstructed samples and original samples, and β is the weight coefficient of the MSE and MAE, with a value range of (0, 1).

The generated reconstructed samples $D_g(\boldsymbol{x}_{lc})$ and real samples $\boldsymbol{x}$ are fed into the sample discriminator D_s. After a series of processing in the convolutional layers (Convl, Conv2) and fully connected layers (FC1, FC2), the final outputs are the discriminative probability transformed by the Sigmoid function and the category probability transformed by the Softmax function. The entropy loss functions $Loss_{D_{\mathrm{g}}-\mathrm{entropy}}$ of the decoder and $Loss_{D_{\mathrm{s}}}$ of the sample discriminator can be calculated by, respectively,

$$\begin{aligned} Loss_{D_{g-entropy}} &= -\mathbb{E}_{\boldsymbol{x}_{lc}\sim f_{\boldsymbol{x}_{lc}}}\{\log f_{D_s}(1|D_g(\boldsymbol{x}_{lc}))\} \\ &\quad - \mathbb{E}_{\boldsymbol{x}_{lc}\sim f_{\boldsymbol{x}_{lc}}}\{\log f_{D_s}(y|D_g(\boldsymbol{x}_{lc}))\}, \end{aligned} \tag{7.7.4}$$

where $f_{D_s}(1|D_g(\boldsymbol{x}_{lc}))$ is the probability that the sample discriminator determines the reconstructed sample to be true when the input is the reconstructed sample $D_g(\boldsymbol{x}_{lc})$ and $f_{D_s}(\boldsymbol{y}|D_g(\boldsymbol{x}_{lc}))$ is the probability that the sample discriminator estimates the reconstructed sample category to be y. And

$$\begin{aligned} Loss_{D_s} &= -\mathbb{E}_{\boldsymbol{x}\sim f_{\boldsymbol{x}}}\{1 - \log\{f_{D_s}(0|D_g(\boldsymbol{x}_{lc}))\}\} \\ &\quad - \mathbb{E}_{\boldsymbol{x}\sim f_{\boldsymbol{x}}}\{\log[f_{D_s}(1|\boldsymbol{x})]\} - \mathbb{E}_{\boldsymbol{x}\sim f_{\boldsymbol{x}}}\{\log[f_{D_s}(\boldsymbol{y}|\boldsymbol{x})]\}, \end{aligned} \tag{7.7.5}$$

where when the input is a reconstructed sample $D_g(\boldsymbol{x}_{lc})$, $f_{D_s}(0|D_g(\boldsymbol{x}_{lc}))$ is the probability that the sample discriminator judges the reconstructed sample to be false. When the input is a real sample $\boldsymbol{x}$, $f_{D_s}(1|D_g(\boldsymbol{x}_{lc}))$ is the probability that the sample discriminator judges the real sample to be true, and $f_{D_s}(\boldsymbol{y}|D_g(\boldsymbol{x}_{lc}))$ is the probability that the sample discriminator estimates the category of the real sample to be $\boldsymbol{y}$.

Combining Equations (7.7.4) and (7.7.5), the total loss function of decoder D is defined as

$$Loss_{D_g} = Loss_{D_g-entropy} + Loss_{D_g-recon}$$

At the end of the second heavy confrontation, the $Loss_{D_g}$ and $Loss_{D_s}$ are used to update the network parameters $\boldsymbol{\theta}_{D_g}$ of the decoder and $\boldsymbol{\theta}_{D_s}$ of the sample discriminator in backpropagation direction, respectively. A whole training is thus completed. Through continuous iterations, the model will eventually converge or locally converge.

7.7.2.2 *Target generation*

At the end of model training, a specific distribution f_z is randomly sampled in the latent space according to the category labels and the number of samples of the targets. The randomly sampled data $\boldsymbol{z}$ and the corresponding category labels are fed into the decoder D_g to randomly generate diverse targets.

7.8 Application Model of Auto-Encoding Model

7.8.1 Soft measurement model based on sparse autoencoding of mutual information

Soft measurement techniques can be broadly classified into two categories: one is the white-box model based on mechanistic analysis, which requires a deep understanding of the process mechanism of the process object and is more difficult to establish a model. The other category is based on a data-driven black-box model, which requires only the use of historical data of the process to build a model without in-depth analysis of the process mechanism, and the difficulty in modeling is relatively low. Commonly used data-driven soft-measurement-based modeling methods include principal component regression, partial least squares, artificial neural networks, and support vector machines.

When the amount of data is large, the dimensionality of variables is high, and the possible correlation between variables is strong, it is important to extract the intrinsic characteristics of the data before modeling. The commonly used Principal Component Analysis (PCA) method and Kernel Principal Component Analysis (KPCA) method can be used for nonlinear dimensionality reduction. Compared with the PCA method, the KPCA method maps the data to a high-dimensional space by introducing a kernel function and then performs dimensionality reduction based on the PCA method on the mapped data [165, 166]. However, it may be difficult to extract effective data features for complex, strongly nonlinear, and uncertain industrial processes. Therefore, Literature [167] proposed the use of sparse autoencoders for feature extraction, soft measurement modeling, and prediction of complex industrial process data. Sparse autoencoders are able to obtain a more meaningful feature representation of the input data by introducing a sparsity limit in the nodes of the hidden layer.

The reconstruction error of the sparse autoencoder during feature extraction treats the reconstruction accuracy of each dimension of the input samples as equal, but in fact that the correlation between the individual input variables (each dimension of the samples) of the soft measurement model and the output ones is constantly changing. In order to extract stronger correlation features with the output variables, the reconstruction accuracy of the dimension corresponding to the samples should be higher for input variables of high correlation with the output variables. Therefore, Ref. [167] has introduced mutual information to calculate the correlation between input variables and output ones to weight the reconstruction error term of sparse autoencoders and further improve the accuracy of soft sensing models. Mutual information is a useful measure of information in information theory, which can be seen as the amount of information contained in one random variable about another random variable. It is a measure of the statistical correlation between two random variables and can provide the degree of linear and nonlinear correlation between two variables. A least square support vector machine soft sensing modeling method based on mutual information sparse autoencoder can be obtained by calculating the correlation between input variables and output ones using mutual information.

7.8.1.1 *Mutual information*

Mutual information is a measure of statistical correlation between two random variables that gives the degree of linear and nonlinear correlation between the two random variables. Let $I(\boldsymbol{x}, \boldsymbol{y})$ denote the mutual information between two random variables $\boldsymbol{x}$ and $\boldsymbol{y}$. It is non-negative and can be expressed as

$$I(\boldsymbol{x},\boldsymbol{y}) = \sum f(\boldsymbol{x},\boldsymbol{y}) \log \frac{f(\boldsymbol{x},\boldsymbol{y})}{f(\boldsymbol{x})f(\boldsymbol{y})}, \tag{7.8.1}$$

where $f(\boldsymbol{x},\boldsymbol{y})$ denotes the joint probability density between $\boldsymbol{x}$ and $\boldsymbol{y}$, and $f(\boldsymbol{x})$ and $f(\boldsymbol{y})$ denote their marginal probability densities, respectively. The solution of mutual information is mainly based on the estimation of the probability densities $f(\boldsymbol{x},\boldsymbol{y})$, $f(\boldsymbol{x})$, and $f(\boldsymbol{y})$. However, since there is no *a priori* knowledge of the data distribution, it is necessary to fit the probability density distribution from the data itself alone, so the density estimation is more complicated, and the non-parametric kernel density estimation method is used for the calculation.

Assuming that the data sample set is the mutual information between the input and output variables in each dimension, it can be expressed as $\boldsymbol{x} = [x_1, x_2, \ldots, x_N]^T$, and $\boldsymbol{y} = [y_1, y_2, \ldots, y_N]^T$, and the mutual information between the input and output variables in each dimension can be expressed as

$$c_d = I\{\boldsymbol{x}_d, \boldsymbol{y}\} = \sum f\{\boldsymbol{x}_d, \boldsymbol{y}\} \log \frac{f\{\boldsymbol{x}_d, \boldsymbol{y}\}}{f(\boldsymbol{x}_d)f(\boldsymbol{y})}, \tag{7.8.2}$$

where $\boldsymbol{x}_d = [x_{1d}, x_{2d}, \ldots, x_{ND}]^{\mathrm{T}}$ denotes the N sampled values of the dth-dimensional input variable, and $d = 1, 2, \ldots, D$.

The derived mutual information is then used to compute the weights corresponding to each dimension of the input variables, i.e.,

$$\lambda_d = \frac{c_d}{\sum_{d=1}^{D} c_d}, \quad \sum_{d=1}^{D} \lambda_d = 1. \tag{7.8.3}$$

7.8.1.2 *Mutual information sparse autoencoder*

The derived weights are used to weight the reconstruction error term at these points; the loss function of weighted AE can be defined as

$$J_{w\text{AE}} = Loss(\boldsymbol{x}, \hat{\boldsymbol{x}}) = \frac{1}{2N}\sum_{i=1}^{N}\sum_{d=1}^{D}\lambda_d[\boldsymbol{x}_{id} - \hat{\boldsymbol{x}}_{id}]^2.$$

Therefore, the final cost function on the mutual information sparse autoencoder is defined as

$$J_{\text{cost}} = J_{w\text{AE}} + \gamma J_{\text{weight}} + \beta J_{\text{sparse}}, \tag{7.8.4}$$

where β denotes the coefficient of the sparse term, which is generally given empirically or experimentally. Minimization of J_{cost} using the proposed Newton method (Limited-memory BFGS, LBFGS) yields the required parameter $\boldsymbol{\theta}$ of the encoder. The LBFGS method is a modification of Newton's method with the following steps:

Step 1: Initialization

Given the initial symmetric positive definite matrix $\boldsymbol{H}_0$, the error limit $\varepsilon > 0$, the initial value of the parameter matrix $\boldsymbol{\theta}_0 \in R^{N\times 1}$, and the number of iterations $k = 0$.

Step 2: Calculate the gradient

The gradient of the cost function $J_{\text{cost}}(\boldsymbol{\theta}(k))$ in Equation (7.8.4) is recorded as $g(k)$ and calculated by $g(k) = \nabla f(\boldsymbol{\theta}(k)) = \nabla J_{\text{cost}}(\boldsymbol{\theta}(k))$. If $||g(k)|| \leq \varepsilon$, the algorithm terminates. Otherwise, the algorithm makes $\boldsymbol{d}(k) = -\boldsymbol{H}(k)g(k)$ and goes to Step 3.

Step 3: Determine the step size

The step size $\alpha(k)$ is determined (generally set the initial step size to 1) by the linear searching, which needs to satisfy the Wolfe–Powell condition, i.e.,

$$\begin{aligned}
&J_{\cos t}(\boldsymbol{\theta}(k) + \alpha(k)\boldsymbol{d}(k)) \leq J_{\cos t}(\boldsymbol{\theta}(k)) + c_1\alpha(k)\boldsymbol{g}^T(k)\boldsymbol{d}(k)\\
&\boldsymbol{g}^T(k+1)\boldsymbol{d}(k) \geq c_2\boldsymbol{g}^T(k)\boldsymbol{d}(k)\\
&\boldsymbol{g}(k+1) = \nabla J_{\cos t}(\boldsymbol{\theta}(k) + \alpha(k)\boldsymbol{d}(k)),
\end{aligned}$$

where $0 < c_1 < \frac{1}{2}$, $c_1 < c_2 < 1$, then $\boldsymbol{\theta}(k+1) = \boldsymbol{\theta}(k) + \boldsymbol{\alpha}(k)\boldsymbol{d}(k)$.

Step 4: Calculate variables

The following variables are updated or calculated by

$$\begin{aligned} \boldsymbol{s}(i) &= \boldsymbol{\theta}(i+1) - \boldsymbol{\theta}(i) \\ \boldsymbol{t}(i) &= g(i+1) - g \\ u(i) &= \frac{1}{\boldsymbol{t}^T(i)\boldsymbol{s}(i)} \\ \boldsymbol{V}(i) &= \boldsymbol{I} - \boldsymbol{u}(i)\boldsymbol{t}(i)\boldsymbol{s}^T(i), \end{aligned}$$

where $i = k - l, \ldots, k$, $l = \min\{k, r-1\}, r$ is a non-negative integer to control the number of storage in the vector set $\{\boldsymbol{s}(i), \boldsymbol{t}(i)\}_{i=k-l}^{k}$, such that $\boldsymbol{H}^0(k) = \frac{\boldsymbol{s}^T(k)\boldsymbol{t}(k)}{\boldsymbol{t}^T(k)\boldsymbol{t}(k)}\boldsymbol{I}_o$.

Step 5: Calculate $\boldsymbol{H}(k+1)$

$\boldsymbol{H}(k+1)$ can be calculated by

$$\begin{aligned} \boldsymbol{H}(k+1) = {} & (\boldsymbol{V}^T(k) \cdots \boldsymbol{V}^T(k-1))\boldsymbol{H}^0(k)(\boldsymbol{V}(k-l) \cdots \boldsymbol{V}(k)) \\ & + u(k-l)(\boldsymbol{V}^T(k) \cdots \boldsymbol{V}^T(k-l+1)\boldsymbol{s}(k-l)\boldsymbol{s}^T(k-l) \\ & \times (\boldsymbol{V}(k-l+1) \cdots \boldsymbol{V}(k)) \\ & + u(k-l+1)(\boldsymbol{V}^T(k) \cdots \boldsymbol{V}^T(k-l+2))\boldsymbol{s}(k-l+1)\boldsymbol{s}^T \\ & \times (k-l+1)(\boldsymbol{V}(k-l+2) \cdots \boldsymbol{V}(k)) \\ & + \cdots + u(k)\boldsymbol{s}(k)\boldsymbol{s}^T(k), \end{aligned}$$

Step 6: Continue to execute the conditions

Let $k = k + 1$ and go to Step 2.

7.8.1.3 *Least square support vector machines for soft measurement modeling*

The basic idea of the Support Vector Machine (SVM) model is to map the input vectors to a high-dimensional feature space and then construct the optimal decision function. Least Square SVM (LSSVM) model is an extension of the SVM, which can solve only linear equations and hence has faster convergence speed. A set of nonlinear mappings is used to map the samples from the original space to the high-dimensional feature space, i.e., which can be denoted as $\boldsymbol{R}(x) = \{\varphi(\boldsymbol{x}_1), \varphi(\boldsymbol{x}_2), \ldots, \varphi(\boldsymbol{x}_N)\}$, then the optimal decision function is constructed in the high-dimensional space and can be recorded as $y(x) = \boldsymbol{w}^T\varphi(\boldsymbol{x}) + \boldsymbol{b}$. Based on the principle of structural risk minimization, the objective function of the LSSVM model is described as

$$\min_{\boldsymbol{w},e} J(\boldsymbol{w}, e) = \frac{1}{2}\boldsymbol{w}^T\boldsymbol{w} + \frac{\lambda}{2}\sum_{i=1}^{n} e_i^2$$

$$s.t : y_i = \boldsymbol{w}^T\varphi(x_i) + \boldsymbol{b} + e_i \quad i = 1, 2, \ldots, n,$$

where λ is the regularization parameter, the Lagrangian method is used to solve the following equation:

$$Loss(\boldsymbol{w}, \boldsymbol{b}, \boldsymbol{e}, \boldsymbol{\alpha}) = J(\boldsymbol{w}, e) - \sum_{i=1}^{N} \alpha_i\{\boldsymbol{w}^T\varphi(\boldsymbol{x}_i) + \boldsymbol{b} + e_i - \boldsymbol{y}_i\},$$

where α_i is the Lagrange multiplier, which can be obtained by the KKT condition:

$$\frac{\partial L}{\partial \boldsymbol{w}} = 0, \ \frac{\partial L}{\partial \boldsymbol{b}} = 0, \ \frac{\partial L}{\partial \boldsymbol{e}} = 0, \ \frac{\partial L}{\partial \boldsymbol{\alpha}} = 0.$$

Obtainable

$$\boldsymbol{w} = \sum_{i=1}^{n} \alpha_i \boldsymbol{\varphi}\left(x_i\right)$$

$$\sum_{i=1}^{n} \alpha_i = 0,$$

$$\alpha_i = ce_i,$$

$$\boldsymbol{w}^T\varphi(x_i) + \boldsymbol{b} + e_i - y_i = 0.$$

In the above equations, we eliminate $\boldsymbol{w}$, e_i, a system of matrix linear equations is represented as

$$\begin{bmatrix} 0 & \boldsymbol{A}^T \\ \boldsymbol{A} & \boldsymbol{B} + \boldsymbol{c}^{-1}\boldsymbol{I} \end{bmatrix} \begin{bmatrix} \boldsymbol{b} \\ \boldsymbol{\alpha} \end{bmatrix} = \begin{bmatrix} \boldsymbol{0} \\ \boldsymbol{y} \end{bmatrix},$$

where $\boldsymbol{A} = [1, 1, \ldots, 1]^T$, $\boldsymbol{y} = [y_1, y_2, \ldots, y_N]^T$, $\boldsymbol{\alpha} = [\alpha_1, \alpha_2, \ldots, \alpha_N]^T$, $\boldsymbol{B} = \{K_{ij} = k(\boldsymbol{x}_i, \boldsymbol{x}_j)\}$, $k(\boldsymbol{x}_i, \boldsymbol{x}_j)$ is the kernel function, and $i, j = 1, 2, \ldots, N$. Different SVMs can be constructed by selecting different kernel functions. Since the radial basis function has good tracking performance, it is more suitable to be selected for modeling soft measurements in the form of

$$k(\boldsymbol{x}_i, \boldsymbol{x}) = \exp\left(\frac{-||\boldsymbol{x} - \boldsymbol{x}_i||^2}{2\sigma^2}\right),$$

where σ is the bandwidth of the kernel function.

The final system of equations is solved to obtain the regression function, i.e.,

$$f(\boldsymbol{x}) = \sum_{i=1}^{N} \alpha_i k(\boldsymbol{x}_i, \boldsymbol{x}) + b.$$

The hyperparameters c and σ of the LSSVM model can be obtained by minimizing the Root Mean Square Error (RMSE) optimization by genetic algorithm (GA).

The combination of a mutual information sparse auto-encoder and LSSVM-based soft measurement modeling is performed as follows:

Step 1: Divide the datasets

The datasets $\{X, Y\}$ are divided into training samples and testing samples. In this section, $\boldsymbol{x}_{\text{train}}$ and $\boldsymbol{y}_{\text{train}}$ are used to denote the inputs and outputs of the training samples, and $\boldsymbol{x}_{\text{test}}$ and $\boldsymbol{y}_{\text{test}}$ represent the inputs and outputs of the testing samples, respectively.

Step 2: Normalize the data

The data are normalized by placing them between 0 and 1.

Step 3: Calculate the mutual information

The mutual information between each dimensional input variable and the output variable is calculated and estimated using Equation (7.8.2) and then the corresponding weights are calculated using Equation (7.8.3).

Step 4: Train the proposed mutual information sparse autoencoder

The proposed mutual information sparse autoencoder with the inputs of the training samples $\boldsymbol{x}_{\text{train}}$ is trained. After the parameter vector $\boldsymbol{\theta} = \{\boldsymbol{w}_1, \boldsymbol{w}_2, \boldsymbol{b}_1, \boldsymbol{b}_2\}$ of the autoencoder is calculated by minimizing the cost function J_{cost} (Equation (7.8.4)), we can obtain the features $\boldsymbol{h}_{\text{train}}$ of $\boldsymbol{x}_{\text{train}}$.

Step 5: Optimize parameters

$\boldsymbol{h}_{\text{train}}$ is used as the input data and $\boldsymbol{y}_{\text{train}}$ is used to denote the output data. The LSSVM model is trained by optimizing the parameters through a genetic algorithm.

Step 6: Estimate model

The trained model is applied to the testing samples, i.e., the hidden layer feature vector $\boldsymbol{h}_{\text{train}}$ of $\boldsymbol{x}_{\text{train}}$ is extracted, and it is input into the trained LSSVM model to obtain its model estimate $\hat{\boldsymbol{y}}_{\text{train}}$.

Step 7: Establish soft measurement modeling and predict process

Block diagram of the specific soft measurement modeling and prediction process is shown in Figure 7.19.

7.8.2 Fuzzy inference model based on deep autoencoding network

The network fuzzy inference system with deep autoencoder is an approach that combines a deep autoencoding network with a fuzzy inference system. In this case, the fuzzy inference system uses the

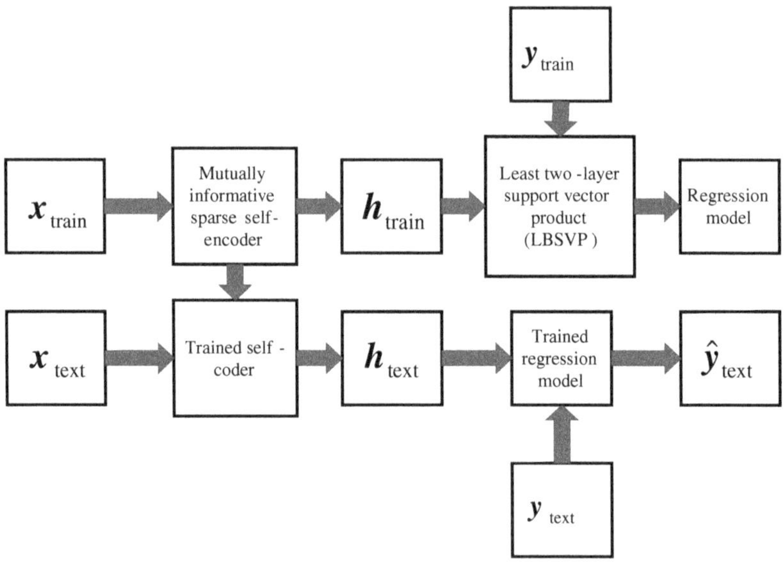

Figure 7.19. Soft measurement modeling and prediction process [167].

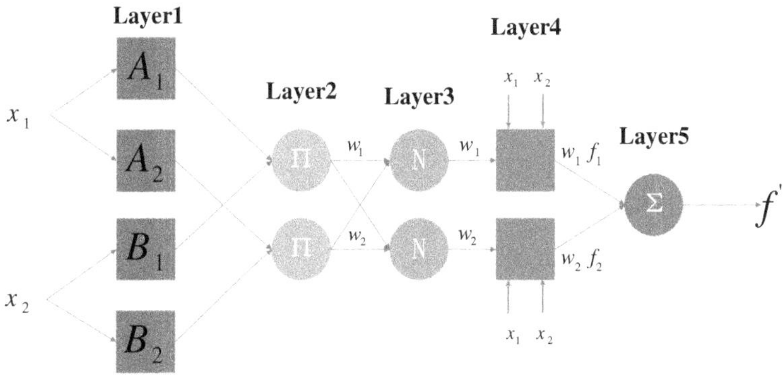

Figure 7.20. Deep autoencoding fuzzy inference system.

Sugeno fuzzy model [168], which is more capable of describing the dynamic properties of nonlinear systems compared to the traditional if-then inference model.

7.8.2.1 *Fuzzy inference systems*

Without loss of generality, the fuzzy if-then rule for the two-input first-order Sugeno model is only illustrated as follows:

$$\text{Rule 1: if}(x_1 \ is \ A_1) \text{ and } (x_2 \ is \ B_1) \text{ then } (f_1 = p_1x_1 + q_1x_2 + r_1) \tag{7.8.5}$$

$$\text{Rule 2: if}\,(x_1 \text{ is } A_2) \text{ and } (x_2 \text{ is } B_2) \text{ then}\,(f_2 = p_2x_1 + q_2x_2 + r_2) \tag{7.8.6}$$

where x_1 and x_2 are the input values, (A_1, A_2) and (B_1, B_2) are the fuzzy sets, f_1 and f_2 are the weighted outputs of each rule under the corresponding fuzzy rule, and (p_1, p_2), (q_1, q_2), and (r_1, r_2) are the trained weighting coefficients. The fuzzy inference system is shown in Figure 7.20.

The first layer is the fuzzification processing layer, where all nodes are adaptive nodes and their output is the fuzzy affiliation of the input, i.e.,

$$O_i^{(1)} = \mu_{A_i}(x_1) \quad i = 1, 2, \tag{7.8.7}$$

$$O_i^{(1)} = \mu_{B_i}(x_2) \quad i = 3, 4, \tag{7.8.8}$$

where $\mu_{A_i}(x_1)$ and $\mu_{B_i}(x_2)$ can be any fuzzy affiliation function and the bell-shaped affiliation function is generally used to carry out the fuzzification of the input signal, i.e.,

$$\mu_{A_i}(x) = \frac{1}{1 + \left\{\left(\frac{x-c_i}{a_i}\right)^2\right\}^{b_i}}, \tag{7.8.9}$$

where a_i, b_i, and c_i are the parameters of the affiliation function and μ_{A_i} is the output of the first layer.

The second layer is the confidence processing layer, i.e., it is multiplied by the output signals of the first layer to obtain the confidence under each rule, and the output of this layer is given by

$$O_i^{(2)} = w_i = \mu_{A_i}(x_1)\mu_{B_i}(x_2) \quad i = 1, 2. \tag{7.8.10}$$

The third layer is the normalization processing layer to obtain the normalized confidence metrics.

The fourth layer is the rule output layer, i.e., the weighted output under the corresponding output rule for the input data is calculated.

The fifth layer is the final result layer, i.e., it outputs the final reasoning result, i.e.,

$$f = (\bar{w}_1 x_1)p_1 + (\bar{w}_1 x_2)q_1 + (\bar{w}_1)r_1 + (\bar{w}_2 x_1)p_2 + (\bar{w}_2 x_2)q_2 + \bar{w}_2 r_2. \tag{7.8.11}$$

The improvement of inference accuracy of a fuzzy inference system, i.e., the optimal weighting coefficients (p_1, p_2), (q_1, q_2), and (r_1, r_2), can be determined by combining the gradient descent method with the least square method.

7.8.2.2 *Deep autoencoding network-fuzzy inference system [233]*

When the number of input features is large, the traditional fuzzy inference system will have the disadvantage of excessive computational complexity, i.e., the training time is too long, so it can be improved by using a deep autoencoding network. For the input layer to the confidence processing layer, encoders are employed for improvement, while the autoencoding network is used to improve the rule output layer to the final inference output layer. It is worthwhile to set the input data as N-dimension, the plausibility processing layer as J-dimension, the rule output layer as M-dimension, and the final inference result layer as 1-dimensional. Thus, the mapping of the input layer to the confidence processing layer and the mapping of the rule output layer to the inference output layer are expressed as follows:

$$\boldsymbol{H} = \boldsymbol{w}\boldsymbol{\mu}(\boldsymbol{x}), \tag{7.8.12}$$

$$\boldsymbol{Y} = \bar{\boldsymbol{w}}\boldsymbol{X} + \boldsymbol{r}, \tag{7.8.13}$$

where $\boldsymbol{\mu} = \{\boldsymbol{\mu}_{A_1}, \boldsymbol{\mu}_{A_2}, \boldsymbol{\mu}_{B_1}, \boldsymbol{\mu}_{B_2}, \ldots\}$ is the affiliation function, $\boldsymbol{x} = \{x_1, x_2, \ldots, x_N\}$ is the input feature vector, and $\boldsymbol{w}$ and $\bar{\boldsymbol{w}}$ are the weight matrices of the corresponding dimensions, respectively.

Let the set of input samples be $\boldsymbol{S} = [s_1, s_2, \ldots, s_{N_s}]$, each of which contains N feature measures, i.e., $\boldsymbol{s}_p = [s_{p1}, s_{p2}, \ldots, s_{pN_s}]$, and the loss function of the deep autoencoding network is defined as

$$Loss(\boldsymbol{w}, \bar{\boldsymbol{w}}) = -\frac{1}{N_S}\sum_{i=1}^{N_S}\sum_{p=1}^{N}[s_{ip}\ln(y_i) + (1 - s_{ip})\ln(1 - y_i)]. \tag{7.8.14}$$

The loss function in Equaion (7.8.14) defines the error between the input feature vector and the final output result after passing through the proposed deep autoencoding fuzzy system. Therefore, the loss function can be minimized to train the weight matrices $\boldsymbol{w}$ and $\bar{\boldsymbol{w}}$, i.e., the minimization problem of the objective function can be described as

$$\arg\min_{\boldsymbol{w}, \bar{\boldsymbol{w}}} Loss(\boldsymbol{w}, \bar{\boldsymbol{w}}). \tag{7.8.15}$$

The actual solution can also be solved by the gradient descent method with the following steps:

Step 1: Calculate the gradient of $Loss(\boldsymbol{w}, \bar{\boldsymbol{w}})$

The partial derivatives of $Loss(\boldsymbol{w}, \bar{\boldsymbol{w}})$ with respect to the weight matrix $\boldsymbol{w}$ and $\bar{\boldsymbol{w}}$ are calculated. Since the number of columns of $\boldsymbol{b}$ is the same, it is useful to let $\boldsymbol{w}' = (\boldsymbol{w}^{\boldsymbol{T}}, \bar{\boldsymbol{w}}^{\mathrm{T}})^{\mathrm{T}}$. The partial derivative is then represented as

$$\frac{\partial}{\partial \boldsymbol{w}'} Loss(\boldsymbol{w}'). \tag{7.8.16}$$

Step 2: Continuously update w'

For each iteration of the training samples, the adjustment of their weight matrices is reduced along the direction that minimizes the gradient, which is mathematically expressed as

$$\boldsymbol{w}'(k+1) = \boldsymbol{w}'(k) - \eta \frac{\partial}{\partial \boldsymbol{w}'(k)} Loss(\boldsymbol{w}'(k)), \tag{7.8.17}$$

where η denotes the step size of the algorithm.

7.8.3 Fast sparse autoencoding network based on feature clustering

After Bengio [169] proposed the idea of initializing each layer of the neural network with an autoencoder, Bengio and Ranzato *et al.* proposed the Sparse AutoEncoding (SAE) network to achieve deeper mining of data. Theory research has shown that the more the number of nodes in hidden layer, the better the processing effect of the network, but the increase in the number of nodes in the hidden layer will lead to an increase in the number of parameters of the network which in turn increases the training time of the network dramatically and even leads to the lack of storage space in the hardware and other problems, an important research topic is how to train the network quickly. Literature [169] proposed a fast sparse autoencoding network based on feature clustering. First, K-mean clustering is used to cluster the existing features to reduce feature redundancy, and the optimal number of classifications for clustering is selected as the number of essential features of the network, and the essential features of the network can be obtained via retraining network, as well as they are rotated and elastically twisted to expand the feature diversity and ensure the processing effect of the network.

7.8.3.1 *Features of clustering feature*

The high similarity of some of the features calculated in the visualized network features indicates a large feature redundancy in general network training. Therefore, Literature [169] performed K-mean clustering of existing features to reduce the large feature redundancy and select the number of best classified ones as the number of essential

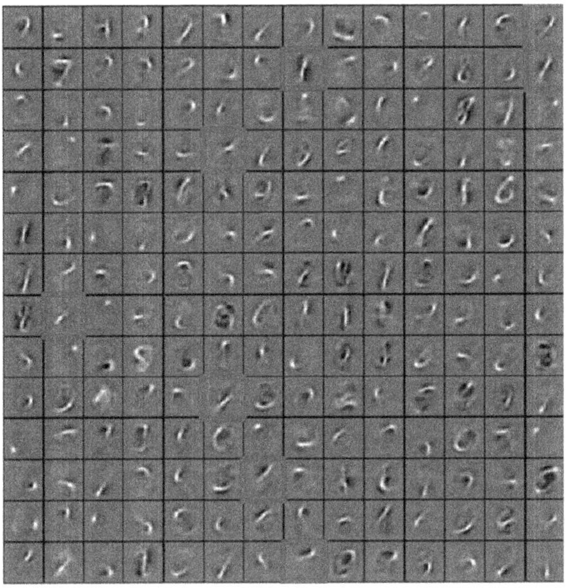

Figure 7.21. Visualization of primary features of a classical sparse autoencoding network [169].

features for the network and performed rotation and elastic twist operations on the essential features to expand their diversity.

7.8.3.1.1 *Characteristic repetition phenomenon*

In order to explore the information learned by the network, the features learned by the hidden layer of the network can be visualized according to mode maximization, whose main principle is to solve the input mode that maximizes the response of the nodes in the hidden layer and visualize it. The results of visualizing 196 primary features of handwritten digital image data processed by the classical sparse autoencoding network are shown in Figure 7.21. Figure 7.21 shows that the primary features of the network are the edge information of the input data, and several of them have high similarity, and a class of features with high similarity is boxed. A large number of experiments have confirmed that this kind of special positive redundancy phenomenon prevails, and the redundancy of features can be reduced, and the training time shortened by appropriately reducing the number of nodes in the hidden layer of the network.

When adjusting the network parameters or repeating the experiment many times, it is necessary to continuously initialize the network for repeated training. If the number of essential features in the network is determined by the K-mean clustering method, the number of neurons in the hidden layer of the network can be initialized directly using the number of essential features, which is much less than the number of neurons in the hidden layer of the general network, so the number of features to be trained is reduced dramatically, the time-consuming training of the network also is reduced.

7.8.3.1.2 *Feature clustering*

To reduce the redundancy of features, K-mean clustering can be performed on existing features to obtain the optimal number of categories, also known as the number of essential features in the network. Due to the strong aggregation of data within the same category and the large distance between data within different categories after being best classified, the optimal number of classifications can be determined based on the principle of maximizing the ratio of inter-class distance to intra-class distance.

The method for finding the optimal number of clusters is to set the maximum number of categories to be $C_{\max}$ and then madke the number of categories traverse from 1 to $C_{\max}$, and for each number of categories, there are the following calculation formulas:

$$
\begin{aligned}
a(i) &= \frac{1}{N_i - 1} \sum_{j=1}^{N_i} \|\boldsymbol{x}_i - \boldsymbol{x}_j\| \\
b(i) &= \min_{\substack{1 \le p \le \mathbb{N}_c \\ c \ne c_i}} \|\boldsymbol{x}_i - \boldsymbol{x}_c\| \qquad (7.8.18) \\
S_c &= \sum_{i=1}^{N} \frac{b(i) - a(i)}{\max(a(i), b(i))},
\end{aligned}
$$

where $a(i)$ is the average distance between sample $\boldsymbol{x}_i$ and other samples $\boldsymbol{x}_j$ in the same category, N_i is the total number of samples in the category where sample $\boldsymbol{x}_i$ is located, $b(i)$ is the minimum distance between sample x_i and the center sample of other categories $\boldsymbol{x}_j$, c_i indicates that the sample $\boldsymbol{x}_i$ is located in the category, $\boldsymbol{x}_c$ is the center sample of the cth category of data, c is the number of categories of the samples, S_c is the aggregation effect indicator for dividing the samples into N_c categories, N is the total number of sample, and the bigger the S_c, the better thr aggregation of classification results. After the test traverses the number of categories, the number of categories N_c corresponding to the largest value of S_c is obtained, and then the number of categories with the highest number of occurrences in the result is selected as the optimal number of classifications through multiple repetitions of the test, after obtaining the number of intrinsic features, the number of nodes in the hidden layer of the autoencoding network can be initialized to the number of intrinsic features, and the intrinsic features of the data obtained by the training of the auto-encoding network can be reconfigured. The feature redundancy of the network can be reduced when using essential features, but it can also reduce the classification accuracy of the network slightly, so it is also necessary to increase the diversity of features by rotating and twisting operations.

7.8.3.2 *Improvement of feature diversity*

Figure 7.21 shows that the subtle differences between similar features constitute the feature diversity. Generally, expansion of image feature diversity can be done by rotating operations to enrich the directionality of the feature images, or by distorting the image data to increase the diversity of the images [170, 171].

The feature images are rotated at different angles to increase the directionality of the feature images, while elastic twisting of some of the rotated features can further increase the image diversity of the features. The first step of the elastic warping operation is to perform a random mapping of the images by using the bilinear difference method to calculate the pixel values at coordinates $(x + c_1, y + c_2)$, and then the pixel values at coordinates (x, y) are calculated by

$$f^*(x, y) = f(x + \alpha \cdot c_1, y + \alpha \cdot c_2), \tag{7.8.19}$$

where α is the elastic deformation parameter that controls the degree of elastic distortion, $f^*(x, y)$ denotes the new pixel value of the images at the coordinate point (x, y), and c_1 and c_2 are random numbers of size between -1 and 1. The second step of the elastic distortion operation is to perform Gaussian blurring on the results obtained in the first step. Figure 7.22 shows a comparison

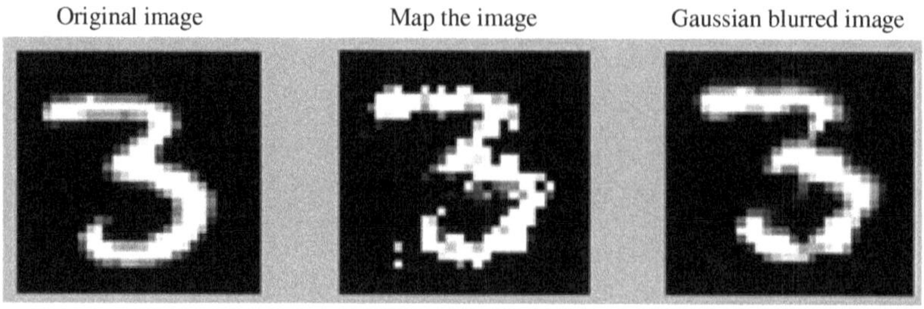

Figure 7.22. Graphs of image distortion results.

of the results of the two steps of elastic distortion for an ordinary digital images, which shows that if only the images are mapped randomly, the digital edges of the images are jagged, and the difference between the original images is too large, so the second step of the distortion operation can make the edges of the data tend to be smoothed.

The diversity of the essential features of the network will be greatly optimized after the rotating and twisting operation, and the rich diversity of the network features can prevent the network from overfitting so that the network features can add new information to the top of the original feature images; thus, a higher classification accuracy can be obtained.

7.8.3.3 *Fast sparse autoencoding network based on feature clustering*

The architecture of a fast sparse autoencoding network based on feature clustering is as follows [172]:

Step 1: Initialization and method determination

A portion of the data is randomly selected as training data to be input into the autoencoding network, and hyperparameters such as the number of layers in the network and the number of nodes in each layer are set to train the network using the LBFCS gradient descent method.

Step 2: Perform K-mean clustering

K-mean clustering on the first trained feature of autoencoding network is performed to obtain the optimal number of clusters N_c.

Step 3: Construct a new autoencoding network and extract the features of the trained hidden layer

A new autoencoding network is constructed, the number of nodes in the hidden layer is set to N_h, and the features of the trained hidden layer are extracted.

Step 4: Expand the features

The features are expanded by twisting and rotating the N_h features.

Step 5: Construct and train the second autoencoding network

The second autoencoding network is constructed to train the learning network using the features obtained in Step 4 as inputs.

Step 6: Termination of running or not

Fast sparse autoencoding network repeats from Step 2 to Step 5 until the set initial number of layers in the network is reached, and the fine-tuning of the network is completed using all the data.

7.8.4 Extreme learning machine based on stacked denoising sparse autoencoder

Extreme Learning Machine (ELM) [173] is a simple and efficient Single Hidden Layer Feedforward Neural Network (SLFN). The input weights and hidden layer's bias in the ELM are randomly generated, and its output weights are obtained by solving the minimized squared loss function, which can effectively solve the problems of slow convergence speed and easily produce local optimal solutions of traditional SLFN. Numerous studies have shown the following: (1) The ELM randomly determines the input weights and hidden layer's bias, which can improve the learning speed of the whole network, but the random assignment of parameters in the hidden layer makes the ELM require more nodes in the hidden layer than traditional tuning-based learning algorithms. However, too many nodes of the hidden layer tend to weaken the sparsity of the hidden layer, generate redundant nodes, increase the complexity of the network structure, and reduce the classification accuracy. In this regard, the researcher adopts the swarm intelligence optimization algorithm to optimize the network structure in order to improve the overall performance. The network structure, improved by using swarm intelligence algorithm to optimize the parameters of nodes in the hidden layer, can improve the generalization ability and stability of the ELM, but it also increases the computational complexity and has poor performance when dealing with large-scale high-dimensional datasets. (2) The ELM can be used to reduce the computational complexity and training time of the deep network when dealing with large-scale high-dimensional data. For example, the ELM-AE has good feature representation and classification performance [174]. Discriminative Graph regularized Extreme Learning Machine AutoEncoder (GELM-AE) can extract more abstract high-level features and improve the overall performance of the network model [175]. The integrated CNN2ELM learning framework combining the CNN and the ELM can improve the robustness of the recognition of the age of the face image [176].

The stacked Denoising Sparse Auto-Encoder (sDSAE) incorporates sparsity constraints to optimize the network structure, which can better extract the deeper features of the data, while the denoising process reduces the noise interference and enhances the robustness of the algorithm. In Literature [177], the sDSAE-ELM, which has combined the sDSAE and the ELM, utilizes the sDSAE to generate input weights and hidden layer bias of the ELM to solve the problem of complex network structure and weak robustness caused by random assignment of the weights and hidden layer bias in the ELM while retaining the advantage of fast training speed of the ELM.

7.8.4.1 *Extreme learning machine*

The Extreme Learning Machine (ELM) is an SLFN algorithm with fast learning capability; its network structure is shown in Figure 7.23.

In the ELM, there are N nodes in the input layer, J nodes in the hidden layer, and M nodes in the output layer. And the connection weights between the nodes in the input and the nodes in the hidden layers and the bias of the nodes in the hidden layer are randomly generated. Assuming that there are N samples $(\boldsymbol{x}_i, \boldsymbol{y}_{Ti})$, where $\boldsymbol{x}_i = [x_{i1}, x_{i2}, \ldots, x_{in}]^T \in R^n$ and $\boldsymbol{y}_{Ti} = [y_{Ti1}, y_{Ti2}, \ldots, y_{Tim}]^{\mathrm{T}} \in R^m$, the output of the network is defined as

$$f(\boldsymbol{x}_i) = \sum^{J} \boldsymbol{w}_{mj} g(\boldsymbol{w}_{ji}, \boldsymbol{b}_j, \boldsymbol{x}_i), \tag{7.8.20}$$

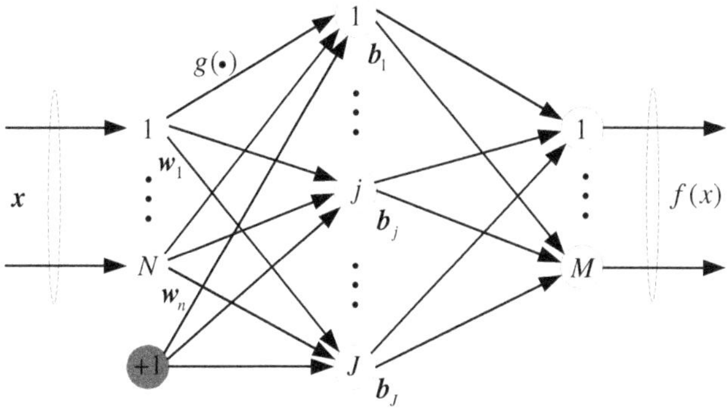

Figure 7.23. Structure of ELM model.

where $j = 1, 2, \ldots, J$, $\boldsymbol{w}_j = [w_{j1}, w_{j2}, \ldots, w_{jN}]^T$ is the input weight vectors between nodes in the Nth input layer and the nodes of the jth hidden layer. b_j is the bias value of the node of the jth hidden layer, $g(\cdot)$ is the activation function of node in the hidden layer, and $\boldsymbol{w}_{mj} = [w_{m1}, \beta_{m2}, \ldots, \beta_{MJ}]^T$ is the output weight vector between the node of the jth hidden layer and the node of the Mth output layer. Let $\boldsymbol{h}(x) = [g_1(x), g_2(x), \ldots, g_J(x)]$ denote the output of the hidden layer corresponding to the input data $\boldsymbol{x}$. The output matrix of the data samples in the hidden layer is denoted as $\boldsymbol{H} = [\boldsymbol{h}^{\mathrm{T}}(x_1), \boldsymbol{h}^{\mathrm{T}}(x_2), \ldots, \boldsymbol{h}^{\mathrm{T}}(x_N)]^T$, i.e.,

$$\boldsymbol{H} = \begin{bmatrix} g(\boldsymbol{w}_1, \boldsymbol{b}_1, \boldsymbol{x}_1) & g(\boldsymbol{w}_2, \boldsymbol{b}_2, \boldsymbol{x}_1) & \cdots & g(\boldsymbol{w}_l, \boldsymbol{b}_l, \boldsymbol{x}_1) \\ g(\boldsymbol{w}_1, \boldsymbol{b}_1, \boldsymbol{x}_2) & g(\boldsymbol{w}_2, \boldsymbol{b}_2, \boldsymbol{x}_2) & \cdots & g(\boldsymbol{w}_l, \boldsymbol{b}_l, \boldsymbol{x}_2) \\ \vdots & \vdots & & \vdots \\ g(\boldsymbol{w}_1, \boldsymbol{b}_1, \boldsymbol{x}_N) & g(\boldsymbol{w}_2, \boldsymbol{b}_2, \boldsymbol{x}_N) & \cdots & g(\boldsymbol{w}_l, \boldsymbol{b}_l, \boldsymbol{x}_N) \end{bmatrix}. \tag{7.8.21}$$

Let $\hat{\boldsymbol{x}} = [\hat{\boldsymbol{x}}_1, \hat{\boldsymbol{x}}_2, \ldots, \hat{\boldsymbol{x}}_N]^{\mathrm{T}}$ denote the target output of the samples, then the matrix expression for this system is represented as

$$\boldsymbol{H}\boldsymbol{w} = \hat{\boldsymbol{x}}. \tag{7.8.22}$$

The training process of the network is equivalent to solving the least square solution $\hat{\boldsymbol{w}}$ of Equation (7.8.22), which yields

$$\|\boldsymbol{H}\hat{\boldsymbol{w}} - \hat{\boldsymbol{x}}\| = \min \|\boldsymbol{H}\boldsymbol{w} - \hat{\boldsymbol{x}}\|. \tag{7.8.23}$$

Typically, the number of nodes l in the hidden layer is less than the number of training samples N. Therefore, $\boldsymbol{w}$ is solved to yield

$$\boldsymbol{w} = \boldsymbol{H}^{\dagger}\hat{\boldsymbol{x}}, \tag{7.8.24}$$

where $\boldsymbol{H}^{\dagger}$ denotes the Moore–Penrose (MP) generalized inverse of $\boldsymbol{H}$. This solution is unique and makes the minimum of the training errors of the network. The expression of $\boldsymbol{w}$ is written as

$$\boldsymbol{w} = \begin{cases} \left(\boldsymbol{H}^{\mathrm{T}}\boldsymbol{H}\right)^{-1}\boldsymbol{H}^{\mathrm{T}}\hat{\boldsymbol{x}}, \boldsymbol{H}^{\mathrm{T}}\boldsymbol{H} & \text{exotic} \\ \boldsymbol{H}^{\mathrm{T}}\left(\boldsymbol{H}\boldsymbol{H}^{\mathrm{T}}\right)^{-1}\hat{\boldsymbol{x}}, \boldsymbol{H}\boldsymbol{H}^{T} & \text{non-exotic} \end{cases}. \tag{7.8.25}$$

To obtain better learning ability, the orthogonal projection method is used to calculate the output weights by adding a positive real matrix $\frac{1}{C}$ to the diagonal line of $\boldsymbol{H}^{T}\boldsymbol{H}$. Equation (7.8.25) is transformed into

$$\boldsymbol{w} = \begin{cases} \left(\frac{\boldsymbol{I}}{\boldsymbol{C}} + \boldsymbol{H}^{\mathrm{T}}\boldsymbol{H}\right)^{-1}\boldsymbol{H}^{\mathrm{T}}\boldsymbol{x}, \boldsymbol{H}^{\mathrm{T}}\boldsymbol{H} & \text{exotic} \\ \boldsymbol{H}^{\mathrm{T}}\left(\frac{\boldsymbol{I}}{\boldsymbol{C}} + \boldsymbol{H}\boldsymbol{H}^{\mathrm{T}}\right)^{-1}\boldsymbol{x}, \boldsymbol{H}\boldsymbol{H}^{\mathrm{T}} & \text{non-exotic} \end{cases}, \tag{7.8.26}$$

where $\boldsymbol{I}$ is the unit matrix and $\boldsymbol{C}$ is the regularization factor.

The learning process of the ELM is shown in Algorithm 7.1.

Algorithm 7.1. ELM

Input training set $\{\boldsymbol{x}_i, \hat{\boldsymbol{x}}_i\}$, $(\boldsymbol{x}_i \in R^N,\ \hat{\boldsymbol{x}}_i \in R^M, i = 1, 2, \ldots, N)$, activation function $g(\cdot)$, number of nodes J in the hidden layer, output weights $\boldsymbol{w}$.

Step 1: Generate input weights $\boldsymbol{w}_j$ and hidden layer bias $\boldsymbol{b}_j$ randomly.

Step 2: Calculate output matrix $\boldsymbol{H}$ of hidden layer according to Equation (7.8.21).

Step 3: Calculate the output weights $\boldsymbol{w}$ according to Equation (7.8.26).

7.8.4.2 *Noise-canceling sparse autoencoder*

The traditional AutoEncoder (AE) has limited ability to reconstruct the input data and poor ability to extract data features [178]. On

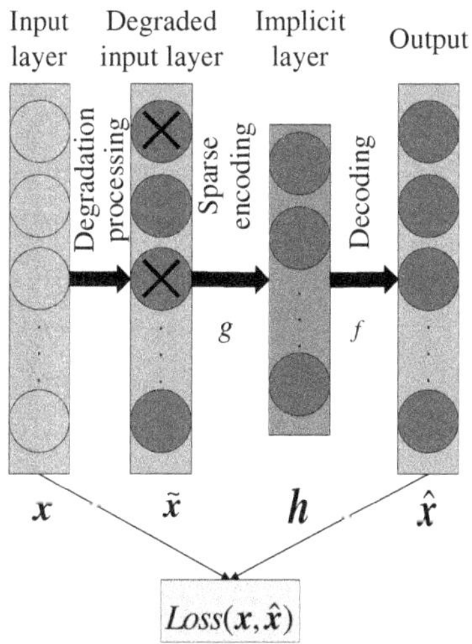

Figure 7.24. Structure of the DSAE.

the basis of the AE, sparsity constraints are added to obtain the Sparse AutoEncoder (SAE) with good regulating ability, which is conducive to the extraction of more representative features and the improvement in classification accuracy of the algorithm. Denoising Sparse Auto-Encoder (DSAE) is based on the SAE and degradation of the original sample data, which aims to exclude noise interference, better reconstruct the original input, and enhance the robustness of the algorithm. The structure of the DSAE is shown in Figure 7.24.

The training process of the DSAE includes three stages: degradation, sparse encoding, and decoding. First, the original input data $\boldsymbol{x}$ is set to 0 according to the pre-set degradation rate v to get the degraded data. Then, the degraded data $\tilde{\boldsymbol{x}}$ is sparsely encoded to get the encoding data $\boldsymbol{h}$. Finally, the encoded data $\boldsymbol{h}$ is used to get the reconstruction data $\hat{\boldsymbol{x}}$. On this basis, the parameters of each layer are adjusted to minimize the reconstruction error, which is represented by the loss function $Loss(\boldsymbol{x}, \hat{\boldsymbol{x}})$ to obtain the optimal representation of the input features.

The formulas for the sparse encoding and decoding processes, respectively, are written as

$$h = g(\tilde{\boldsymbol{x}}) = s_g(\tilde{\boldsymbol{w}\boldsymbol{x}} + b), \tag{7.8.27}$$

$$\hat{\boldsymbol{x}} = f(\boldsymbol{h}) = s_f(\boldsymbol{w}\boldsymbol{h} + \boldsymbol{b}'), \tag{7.8.28}$$

where $s_g(\boldsymbol{x})$ is the activation function, generally taken as the Sigmoid function, $\boldsymbol{w}$ and $\boldsymbol{b}$ are the weight matrix and bias vector for sparse encoding, $\boldsymbol{w}'$ and $\boldsymbol{b}'$ are the weight matrix and bias vector for decoding, respectively, and $\boldsymbol{w}' = \boldsymbol{w}^{\mathrm{T}}$. Assuming that the training set is recorded as $D = \{x_i\}_{i=1}^{N}$, the overall loss function of the DSAE is defined as

$$Loss(\boldsymbol{x}, \hat{\boldsymbol{x}}) = \frac{1}{N}\sum_{D} J(\boldsymbol{x}, \hat{\boldsymbol{x}}) + \frac{\lambda}{2}||\boldsymbol{w}||_F^2 + \beta\sum_{j=1}^{N_h} J_{KL}(\rho \parallel \hat{\rho}_j), \tag{7.8.29}$$

where $J(\boldsymbol{x}, \hat{\boldsymbol{x}}) = \frac{1}{2} \|\boldsymbol{x} - \hat{\boldsymbol{x}}\|_2^2$ in the first term of the right of the equal sign is the squared error term, the second part is the weight decay term (also known as the regularization term), which aims to reduce the size of the weights to prevent overfitting, λ is the weight decay parameter, the third part is the sparse penalty term, β is the sparse penalty weight, and N_h is the number of nodes in the hidden layer.

7.8.4.3 *Stack denoising sparse autoencoder*

The DSAE belongs to a shallow network with limited learning ability, while the sDSAE is stacked by multiple DSAEs, which are trained layer by layer by taking the output of the previous hidden layer as the input of the subsequent hidden layer, and its overall performance is better than that of the shallow network when dealing with high-dimensional large datasets. However, the performance of the sDSAE depends on the number of network layers and the number of nodes. The more the number of network layers, the better its performance is not, and too many layers are prone to cause gradient dispersion phenomenon and also training overfitting [179]. Therefore, a two-layer sDSAE is set up in this section.

7.8.4.4 *sDSAE-ELM algorithm*

The ELM randomly generates input weights and hidden layer bias during the training process; in order to get the ideal classification effect, it often needs to generate a large number of nodes in the hidden layer, and too many nodes in the hidden layer will lead to a complex structure of the network and affect the overall learning performance. In order to avoid too many random redundant nodes in the ELM, the sDSAE is used to obtain the feature expression of the input data, and the sparsity restriction is added so that the network can learn the better structural features in the input data; thus, it can better describe the input data, provide the ELM with the required input weights and hidden layer bias for the ELM, and be more conducive to the ELM classification.

Theoretically, the sDSAE-ELM is able to obtain better input weights and hidden layer biases than the ELM. On the one hand, the sDSAE-ELM utilizes the structure of sDSAE with sparsification to learn and train the original input data, and the obtained input weights and biases of the hidden layer are used as the input weights and biases of the hidden layer in the sDSAE-ELM, respectively. It contains the relevant feature information of the input data, which is conducive to discovering more essential high-level abstract features and promotes the data reconstruction and the overall performance, whereas the input weights and biases of hidden layer in the ELM are randomly assigned, and are not related to the input data, and no promoting effect on the data reconstruction and the performance of the overall algorithm. On the other hand, the sDSAE-ELM generates the input weights and biases of the hidden layer in the ELM through the sDSAE, which overcomes the drawbacks of the ELM that generates redundant nodes due to the random assignment of hidden layer parameters and reduces the classification accuracy of the algorithm. In addition, the sDSAE-ELM optimizes the network's structure, as shown in Figure 7.25, which degrades the original input data, thus to effectively eliminate the interference of noise and enhance robustness.

The training architecture of the sDSAE-ELM is as follows:

Step 1: Preprocess the original input x

Based on the above discussion, the DSAE1 is trained using the gradient descent method to obtain the output $\boldsymbol{h}_1$ and network parameter $\boldsymbol{w}_1$, $\boldsymbol{b}_1$ of the first hidden layer. $\boldsymbol{h}_1$ is a highly abstracted result of the original input data and network parameters, which better reflects the essential characteristics of the input data, and makes the algorithm more robust due to the degradation of the original inputs and the addition of sparsity constraints to the network, as well as reduces

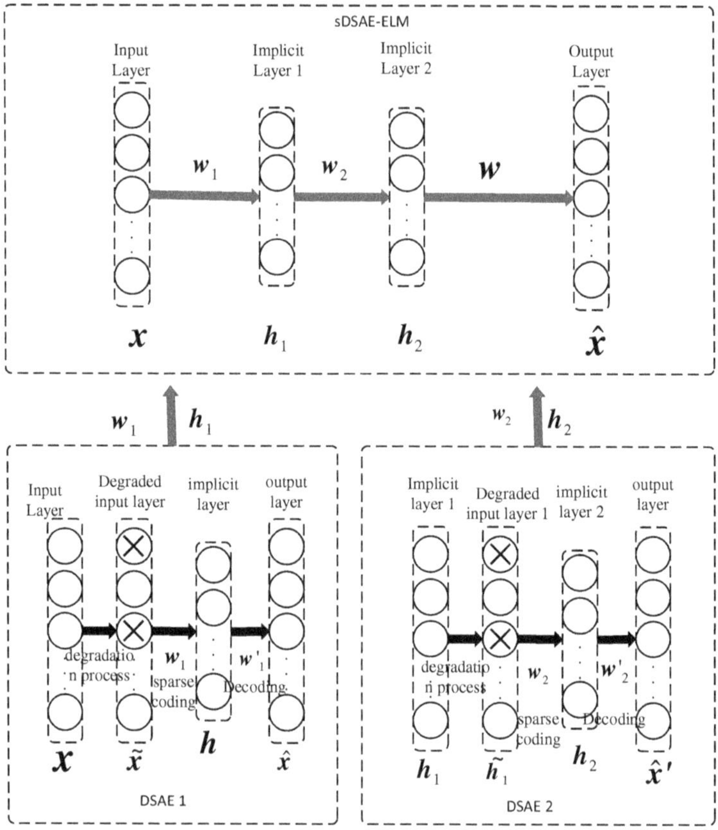

Figure 7.25. Structure of sDSAE-ELM model.

the dimensionality of the data when the dimensionality of the original inputs is high.

Step 2: Train DSAE2 using gradient descent method to determine the parameters of the ELM

Compared with traditional learning algorithms, the ELM not only learns faster but also has better classification performance. However, compared with tuning-based learning algorithms, the ELM requires more nodes of the hidden layer due to the randomness of the generation of its input weights and bias of the hidden layer. This process is the same as Step 1, which obtains the output of the second hidden layer $\boldsymbol{h}_2$ and the network parameters $\boldsymbol{w}_2, \boldsymbol{b}_2$, where $\boldsymbol{w}_2$ is used as the input weights of the ELM, $\boldsymbol{b}_2$ is used as the bias of the hidden layer in the ELM, and the output matrix is $\boldsymbol{h}_2$. This step is capable of overcoming the problem of the random generation of parameters of the hidden layer in the ELM, optimizing the structure of the network, and improving the stability of the ELM.

Step 3: Calculate output weights

Classification is performed using the ELM with input data as $\boldsymbol{h}_1$, input weights and biases of the hidden layer as $\boldsymbol{w}_2$ and $\boldsymbol{b}_2$, the output matrix of the hidden layer as $\boldsymbol{h}_2$, respectively, and output weights are obtained according to Equation (7.8.21).

The learning process of sDSAE-ELM is shown in **Algorithm 7.2**.

Algorithm 7.2. sDSAE-ELM

Input: Training set $\{\boldsymbol{x}_i, \hat{\boldsymbol{x}}_i\}, (\boldsymbol{x}_i \in \mathrm{R}^N, \hat{\boldsymbol{x}}_i \in \mathrm{R}^M, i = 1, 2, \ldots, N)$, activation function of each DSAE, degradation rate v, and sparsity parameter β.

Output: Weight $\boldsymbol{w}$

Step 1: Preprocess the original input $\boldsymbol{x}$

The original input $\boldsymbol{x}$ is preprocessed to train the DSAE1 to obtain the output $\boldsymbol{h}_1$ of the first hidden layer and the network parameters $\boldsymbol{w}_1, \boldsymbol{b}_1$.

Step 2: Input $\boldsymbol{h}_1$

$\boldsymbol{h}_1$ is input to the DSAE2, after training the DSAE2, the output $\boldsymbol{h}_2$ of the second hidden layer, and the optimal network parameters $\boldsymbol{w}_2$, b_2 are obtained.

Step 3: Calculate weight $\boldsymbol{w}$

Parameters $\boldsymbol{h}_1$,$\boldsymbol{w}_2$, and b_2 are taken as the inputs, input weights, and hidden layer's bias of the ELM, respectively, and the output of the hidden layer in the ELM is $\boldsymbol{h}_2$. $\boldsymbol{w}$ is obtained by calculating according to Equation (7.8.26).

7.9 Example 7: Modulation Recognition Algorithm Based on Improved LDA and Autoencoder

Automatic Modulation Recognition (AMR) is a complex and difficult technique before demodulation at the receiver side, and it has a wide range of applications in both military and civilian fields.

Therefore, it is of great significance to study automatic modulation signal recognition.

The flow of the AMR algorithm based on pattern recognition is as follows: signal preprocessing, extraction of signal features, and classification algorithm. Signal features include instantaneous amplitude, phase and frequency, higher-order cumulations, cyclic spectrum, and so on. Classification algorithms include Support Vector Machine (SVM), K-Nearest Neighbor (KNN), and Simple Bayes. However, these algorithms are basically suitable for Gaussian white noise channels or cases with less interference.

In this section, a modulation identification algorithm based on the LDA method and the SSDAE is investigated for Gaussian white noise channels and Rayleigh fading channels [179–181].

7.9.1 Feature extraction

7.9.1.1 *Signal modeling*

Under a complex channel, the signal received by the receiver is represented as

$$\boldsymbol{r}(k) = \boldsymbol{x}(k) \otimes \boldsymbol{h}(k) + \boldsymbol{n}(k), \tag{7.9.1}$$

where $\boldsymbol{r}(k)$ is the received signal vector, $\boldsymbol{x}(k)$ is the transmitted signal vector, $\boldsymbol{h}(k)$ is the channel effect under the complex channel, such as Rayleigh fading, Doppler shift, and frequency and phase errors. And $\boldsymbol{n}(k)$ is Gaussian white noise with zero mean.

7.9.1.2 *Higher-order cumulants*

The higher-order moments or complex random process of the modulated signal $X(k)$ are represented as

$$M_{pq} = \mathrm{E}[X(k)^{p-q} X^*(k)^q], \tag{7.9.2}$$

where * denotes the conjugate and cumulants of $X(k)$ from the second order to eighth order are calculated by

$$C_{20} = M_{20}, \tag{7.9.3}$$

$$C_{21} = M_{21}, \tag{7.9.4}$$

$$C_{40} = M_{40} - 3M_{20}^2, \tag{7.9.5}$$

$$C_{41} = M_{41} - 3M_{21}M_{20}, \tag{7.9.6}$$

$$C_{42} = M_{42} - |M_{20}|^2 - 2M_{21}^2, \tag{7.9.7}$$

$$C_{60} = M_{60} - 15M_{40}M_{20} + 30M_{20}^3, \tag{7.9.8}$$

$$C_{61} = M_{61} - 5M_{40}M_{21} - 10M_{20}M_{41} + 30M_{20}^3M_{20}^2, \tag{7.9.9}$$

$$C_{63} = M_{63} - 6M_{20}M_{41} - 9M_{42}M_{21} + 18M_{20}^2M_{21} + 12M_{21}^3, \tag{7.9.10}$$

$$C_{80} = M_{80} - 28M_{60}M_{20} - 35M_{40}^2 + 420M_{40}M_{20}^2 - 630M_{20}^4. \tag{7.9.11}$$

The various types of modulated signals are substituted from Equation (7.9.3) to Equation (7.9.11) to calculate the cumulative amount of each order, i.e., the feature parameters to be extracted.

7.9.2 Algorithm design

7.9.2.1 *A-ALDA algorithm*

The LDA is a dimensionality reduction feature extraction algorithm similar to the PCA method, which ensures that the samples have the maximum interclass distance and the minimum intraclass distance in the new subspace, i.e., the projected samples have the best separability. By combining this method with the anti-alias principle, we can obtain the anti-alias linear discriminant analysis (A-ALDA) method.

Suppose that there are a total of N samples and M classes. That is, the set of both samples and labels is denoted as $\boldsymbol{D} = \{(\boldsymbol{x}_1, y_1), (\boldsymbol{x}_2, y_2), \ldots, (\boldsymbol{x}_N, y_N)\}$, $\boldsymbol{x}_i$ are features, and $y_i \in \{Y_1, Y_2, \ldots, Y_M\}$ are labels, where Y_i is a category. In addition, $N_j(j = 1, 2, \ldots, M)$ is defined as the number of samples of category j, $\boldsymbol{X}_j$ represents the set of samples of category j, $\boldsymbol{m}_j$ indicates the mean vector of the samples of category j, and $\sum_j$ is the covariance matrix of the samples of category j, and the intra-class scatter matrix $\boldsymbol{S}_w$ is calculated by

$$\boldsymbol{S}_w = \sum_{j=1}^{M} \boldsymbol{S}_{wj} = \sum_{j=1}^{M} \sum_{\boldsymbol{x} \in \boldsymbol{X}_j} (\boldsymbol{x} - \boldsymbol{m}_j)(\boldsymbol{x} - \boldsymbol{m}_j)^{\mathrm{T}}. \tag{7.9.12}$$

The interclass scatter matrix $\boldsymbol{S}_b$ is calculated by

$$S_b = \sum_{j=1}^{n} (\boldsymbol{m}_j - \boldsymbol{m})(\boldsymbol{m}_j - \boldsymbol{m})^T, \tag{7.9.13}$$

where $\boldsymbol{m}$ is the mean vector of all samples.

The multi-class optimization objective function of the LDA method is described as

$$\underbrace{\arg\max}_{w} J(w) = \frac{\prod\limits_{diag} \boldsymbol{w}^{\mathrm{T}} \boldsymbol{S}_b \boldsymbol{w}}{\prod\limits_{diag} \boldsymbol{w}^{\mathrm{T}} \boldsymbol{S}_w \boldsymbol{w}} = \prod_{i=1}^{d} \frac{\boldsymbol{w}_i^{\mathrm{T}} \boldsymbol{S}_b \boldsymbol{w}_i}{\boldsymbol{w}_i^{\mathrm{T}} \boldsymbol{S}_w \boldsymbol{w}_i}. \tag{7.9.14}$$

The rightmost side of Equation (7.9.14) is the form of generalized Rayleigh entropy, whose maximum value is the largest eigenvalue of $\boldsymbol{S}_w^{-1}\boldsymbol{S}_b$. The projection matrix is the tensor matrix of the eigenvector corresponding to the largest eigenvalue. At this time, the projection matrix $\boldsymbol{w}$ is the matrix tented by the eigenvector corresponding to d maximum eigenvalues. Based on the LDA projection, the eigen parameters are written as follows:

$$\boldsymbol{y} = \boldsymbol{w}^{\mathrm{T}} \boldsymbol{x}. \tag{7.9.15}$$

In Equation (7.9.13), the inter-class scatter matrix is the covariance matrix of the mean of the samples in each class with the total mean of the samples in all classes, and the effect of the projection is that the samples in each class are far from the total mean of the samples. In order to improve the differentiation of each class of samples from the confused samples of the surrounding aliases, the interclass scatter matrix $\boldsymbol{S}_{Ab}$ of the A-ALDA is calculated by

$$\boldsymbol{S}_{Ab} = \sum_{j=1}^{n} w_j^2 \left(\boldsymbol{m}_j - \frac{\boldsymbol{m} + \boldsymbol{m}_{Aj}}{2}\right)\left(\boldsymbol{m}_j - \frac{\boldsymbol{m} + \boldsymbol{m}_{Aj}}{2}\right)^{\mathrm{T}}, \tag{7.9.16}$$

where w_j^2 is the weight coefficient for each category, which is the square of the inverse Euclidean distance from the mean of the samples in this category to the confused samples in this category, and $w_j^2 = \left(1/\sqrt{(m_j - m_{Aj})^2}\right)^2$, i.e., the closer a category is to the confused

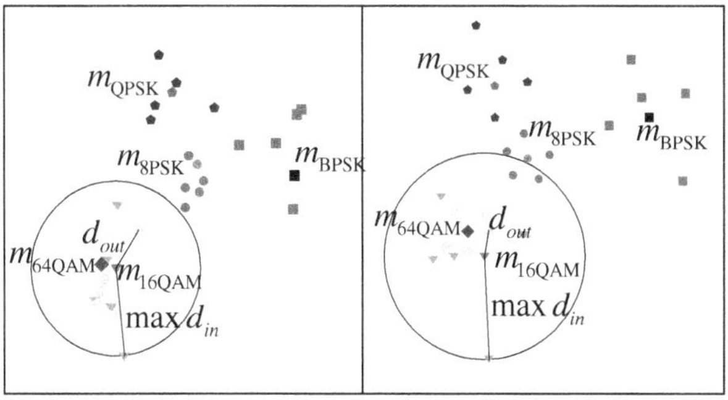

(a) (b)

Figure 7.26. Selection of n_{Aj} values.

samples, the greater the weight. m_{Aj} is the mean of the confused samples around m_j, defined as

$$m_{Aj} = \frac{1}{N_{Aj}} \sum_{i=1}^{N_{Aj}} x_i^{Aj}, \tag{7.9.17}$$

where N_{Aj} is the number of the confused samples around the samples of the class j and x_i^{As} is the confused samples. The choice of the value of N_{Aj} directly affects the performance of the A-ALDA, and the process of taking the value of N_{As} is shown in Figure 7.26.

In Figure 7.26(a), we can know that the five classes of samples and the corresponding sample means of each class are marked with the same shape and different colors, m_{16QAM} is the sample mean point of the 16QAM, and the means of other signal samples are marked in the same way. $\max d_{in}$ is the maximum Euclidean distance between the samples within the class and the mean of the class, and d_{out} is the Euclidean distance between the samples outside the class and the mean of the class. The number of confusing samples n_{Aj} is the number of all out-of-class samples that satisfy $d_{out} < \max d_{in}$, i.e., the number of samples that does not belong to 16QAM samples within the circle. However, for individual signals that are heavily affected by noise, the case of Figure 7.26(b) occurs. Due to the presence of samples that are heavily affected by noise, the circle with radius $\max d_{in}$ contains other classes of signals that are not originally confused samples, in which case $\max d_{in}$ is defined as

$$\max d_{in} = \frac{2\sum_{s=(N_j+q*N_j)/2}^{N_j} d_s}{(1-q)N_j}, \tag{7.9.18}$$

where N_j is the number of samples in the class j, $\max d_{in}$ is the mean of the furthest sample points $\frac{1}{2}N_j(1-q)$ in the class j, q is a constant close to 1, and $\max d_{in}$ is the maximum Euclidean distance from the mean for samples in class j when q is equal to 1. d_s is the Euclidean distance from the mean of the class j in all samples, in increasing order of distance. As a result, the maximum Euclidean distance of the samples within a class from the mean of that class is the average of the distances of some of the more distant samples and thus can avoid the effect of heavy noise points.

7.9.2.2 *Sparse denoising autoencoder*

Traditional single-layer sparse autoencoders tend to lose the deep feature information of the samples when learning features, which makes the features lack robustness and greatly reduces the classification accuracy. The network designed in this section is shown in Figure 7.27.

In Figure 7.27, $\boldsymbol{x}_i$ is the original cumulant feature, and $\boldsymbol{x}$ is the combination of the cumulant feature and the new feature after the A-ALDA projection. The SSDAE in this section adopts a three-layer autoencoder structure, its each layer all adds noise reduction to the

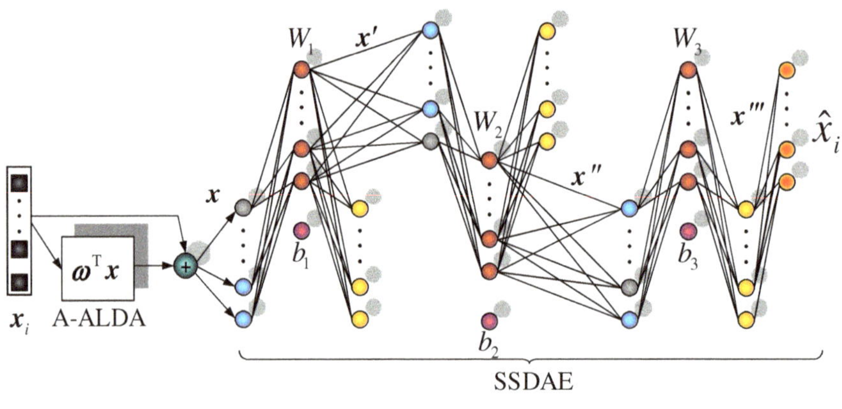

Figure 7.27. Structure of A-ALDA and SSDAE.

input to randomly make its input neurons be set to zero via using a corruption probability, and the gray neurons in the input layer are set to zeros. The feature output of the autoencoder is the output of the hidden layer, which is the output of the red neuron in Figure 7.27. $\boldsymbol{w}_1$, $\boldsymbol{w}_2$, and $\boldsymbol{w}_3$ denote the weights of the SSDAE with three layers, and $\boldsymbol{b}_1$, $\boldsymbol{b}_2$, and $\boldsymbol{b}_3$ denote the bias terms of the SSDAE with three layers, respectively. The learned feature representations of the AE with three layers are, respectively, given by

$$\boldsymbol{x}' = f(\boldsymbol{w}_1\boldsymbol{x} + \boldsymbol{b}_1), \tag{7.9.19}$$

$$\boldsymbol{x}'' = f(\boldsymbol{w}_2\boldsymbol{x}' + \boldsymbol{b}_2), \tag{7.9.20}$$

$$\boldsymbol{x}''' = f(\boldsymbol{w}_3\boldsymbol{x}'' + \boldsymbol{b}_3), \tag{7.9.21}$$

where $f(g)$ is the sigmoid function, i.e., $f(z) = 1/(1 + e^{-z})$. The reconstruction error cost function for updating the weight parameters of network during training is defined as

$$J(\boldsymbol{w}, \boldsymbol{b}) = \frac{1}{N}\sum_{n=1}^{N}\frac{1}{2}\left\|x_n' - x_n\right\|^2. \tag{7.9.22}$$

In order to improve the learning ability of the autoencoding network, a sparsity restriction is imposed on the autoencoder, i.e., only some of the neurons in the hidden layer are "activated" at the same time. As a result, Kullback–Leibler (KL) divergence is introduced to measure the similarity between the average activation of neurons in a hidden layer and a given sparsity parameter. KL divergence is defined as

$$J_{KL}(\rho\|\hat{\rho}_j) = \sum_{j=1}^{N_h}\left[\rho\log\frac{\rho}{\hat{\rho}_j} + (1-\rho)\log\frac{1-\rho}{1-\hat{\rho}_j}\right], \tag{7.9.23}$$

where ρ is the sparsity parameter and N_h is the number of neural units in the hidden layer. $\hat{\rho}_j$ is the average activation of the neural units in the hidden layer, defined as $\hat{\rho}_j = \frac{1}{m}\sum_{n=1}^{m} h_j(x_n)$.

Therefore, the cost function of the SSDAE is defined as

$$J_{\text{SAE}} = J(\boldsymbol{w}, \boldsymbol{b}) + \beta J_{KL}(\rho\|\hat{\rho}_j), \tag{7.9.24}$$

where β is the sparse term penalty factor.

7.9.3 Simulation experiments and result analyses

7.9.3.1 *Signal generation*

The simulation experiment system was Ubuntu 16.04, and the Python programming language was used for all simulations. In this section, 100,000 data samples of each signal were simulated under different signal lengths. Among them, 50,000 samples were used as a training set and 50,000 samples were used as a testing set. The simulation conditions were set as follows: (1) generating binary Bernoulli distributed with a coefficient of 0.5; (2) randomly modulating the binary sequences into five types of modulated signals, BPSK, QPSK, 8PSK, 16QAM, and 64QAM, and keeping the number of each type uniform; (3) interpolating zeros between the data symbols and reducing the inter-encoding interference with a rooted ascending cosine filter; (4) simulating the Rayleigh fading channel and the Doppler shift; (5) simulating Gaussian white noise channels with Signal-to-Noise Ratios (SNRs) ranging from 0 dB to 20 dB in steps of 2 dB.

7.9.3.2 *Selection of confusion points*

For the selection of the value of N_{Aj}, N_{Aj} was the number of out-of-class samples that satisfy all the classes of $d_{out} < \max d_{in}$, where d_{out} was a deterministic value, then the parameter that directly affects the value taken for N_{Aj} was $\max d_{in}$. $\max d_{in}$ was determined by the parameter q. Table 7.1 gives the effect of different values of q on the recognition accuracy when the signal length was 512 and the SNRs were 0 dB, 10 dB, and 20 dB, respectively.

Table 7.1 shows that the value of q is taken too small so that the value of $\max d_{in}$ is smaller, fewer samples satisfy $d_{out} < \max d_{in}$, and more confusing samples are excluded, so the selection of confusing points is irrational and recognition accuracy is also lower. The highest

Table 7.1. Effect of the value of q on recognition accuracy.

SNR \ q	0.90	0.92	0.94	0.96	0.98	1.00
0	0.764	0.768	0.778	0.786	0.794	0.785
10	0.862	0.868	0.872	0.875	0.878	0.877
20	0.872	0.875	0.879	0.882	0.886	0.886

recognition accuracy at different SNRs is achieved when the value of q is taken as 0.98, and n_{Aj} determined in this way is the most appropriate.

7.9.3.3 *Simulation experiments under finite signal length*

The variation of the recognition accuracy of the algorithm with the SNR under finite signal length was discussed. Two metrics, silhouette coefficient [182] and Calinski–Harabasz index [183], were used to evaluate the feature separability; the larger the value of the metrics, the better the feature separability. The evaluation scores of original features, reconstructed features of the LDA, and reconstructed features of the A-ALDA are shown in Table 7.2.

Table 7.2 shows that the scores of both metrics of the A-ALDA are higher than the scores of original features and reconstructed features of the LDA method, and the anti-confusion LDA method can achieve better feature reconstruction results.

In this section, five existing modulation recognition algorithms were compared. The recognition accuracies of the five algorithms are shown in Figure 7.28.

Figure 7.28 shows that under complex channels, the proposed algorithm in this section has a recognition rate of 79.4% at an SNR of 0 dB, which has an improvement of about 5% comparison with the average recognition rate of the SAE and ANC algorithms without the feature selection and reconstruction process, and has an improvement of about 3.5% comparison with the average recognition rate of the GP-KNN and PCA-SVM algorithms. At a high SNR of 20 dB, the recognition rate of the proposed algorithms in this section is

Table 7.2. Evaluation scores of original features and reconstructed features using LDA and A-ALDA.

	Silhouette coefficients	Calinski–Harabasz index
Primitive	0.112	15.845
LDA	0.299	20.662
A-ALDA	0.320	40.272

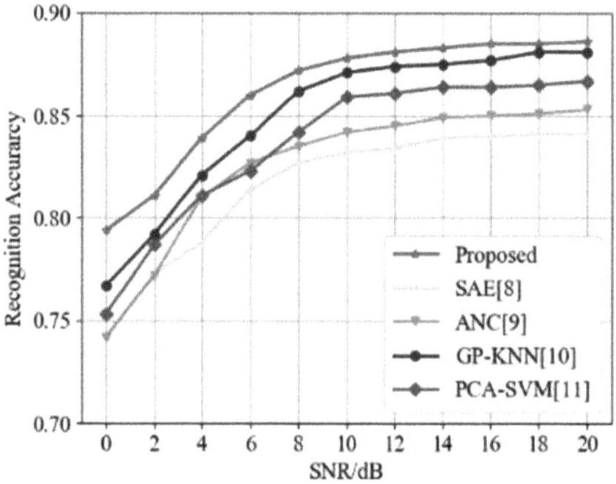

Figure 7.28. Modulation recognition accuracy of five algorithms under complex channels.

Table 7.3. Effect of LDA and A-ALDA algorithms on recognition rate.

Algorithms	SNR	512	1024	2048
LDA	0	0.77 ± 0.02	0.81 ± 0.02	0.82 ± 0.02
	4	0.86 ± 0.02	0.86 ± 0.02	0.88 ± 0.01
	8	0.86 ± 0.02	0.87 ± 0.01	0.90 ± 0.01
	12	0.87 ± 0.01	0.87 ± 0.01	0.91 ± 0.01
	16	0.88 ± 0.01	0.89 ± 0.01	0.92 ± 0.01
A-ALDA	0	0.78 ± 0.02	0.81 ± 0.02	0.83 ± 0.01
	4	0.86 ± 0.02	0.88 ± 0.01	0.89 ± 0.01
	8	0.87 ± 0.01	0.88 ± 0.01	0.93 ± 0.01
	12	0.88 ± 0.01	0.89 ± 0.01	0.93 ± 0.01
	16	0.88 ± 0.01	0.91 ± 0.01	0.94 ± 0.01

about 88.6%, which has an improvement of about 0.5% comparison with that of the GP-KNN and has an improvement of about 2% comparison with that of the PCA-SVM. Table 7.3 gives the effect of the two feature reconstruction algorithms, i.e., the LDA method and A-ALDA, on the recognition rate for different signal lengths and different SNRs.

Table 7.3 shows that the recognition rate of the A-ALDA is slightly improved over LDA at a signal length of 512, and the average improvement of the recognition rate of the proposed algorithm over

LDA is about 1.4% at a signal length of 1024. At a signal length of 2048, the average improvement is about 2%. It can be seen that the improved A-ALDA algorithm can improve the recognition accuracy, and the longer the signal length, the more obvious the recognition rate improvement.

7.9.3.4 *Simulation experiments under phase and frequency errors*

The simulation conditions were set to SNR = 10dB, and the signal length was 2048. Here, phase and frequency errors were considered separately, and first, only the carrier phase error was considered to exist, and the carrier frequency was known at the receiver. In this case, the expression for the signal is written as

$$\boldsymbol{r}(k) = \mathrm{e}^{\mathrm{j}\theta_0}\boldsymbol{s}(k) \otimes \boldsymbol{h}(k) + \boldsymbol{n}(k). \tag{7.9.25}$$

The phase error testing range was set to [−10°,10°], the step size was set to 2°, and 20,000 samples were generated for each signal. The recognition accuracies of the five algorithms with the effect of phase error are shown in Figure 7.29(a).

The effect of frequency error on the recognition accuracies of the two algorithms was then considered. When the carrier phase is known at the receiver, the expression of the signal is then written as

$$r(k) = \mathrm{e}^{\mathrm{j}\frac{2\pi\cdot k\cdot f_0}{f}}\boldsymbol{s}(k) \otimes \boldsymbol{h}(k) + \boldsymbol{n}(k), \tag{7.9.26}$$

where f_0 is the frequency error and f_0/f is the relative frequency. The frequency error test range was set to 1×10^{-4}–2×10^{-4} with a step size of 0.2×10^{-4}. Twenty thousand samples were still generated for each signal. The recognition accuracies of the five algorithms with the effect of frequency error are shown in Figure 7.29(b).

Figure 7.29 shows that the recognition accuracy of the proposed algorithm is higher than the other algorithms when the phase error is from −4° to 4° but improves lower in other ranges. The reason is that when the phase offset is large, there is more noise in the cumulant features of each signal, and the number of confused sample points for feature reconstruction increases, and it is difficult for the A-ALDA to make a better projection transformation. For the same reason, the average recognition accuracy of the proposed algorithm

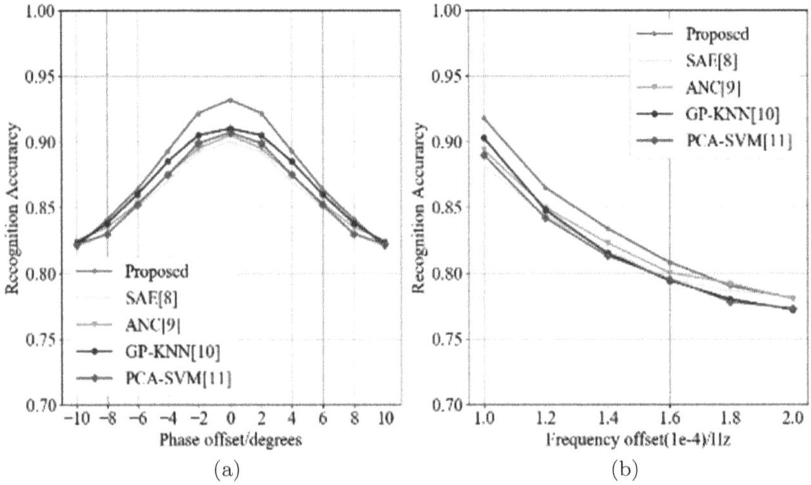

Figure 7.29. Recognition accuracy with phase and frequency error. (a) Presence of phase error. (b) Presence of frequency error.

in this section is improved by 2.3% compared to other algorithms on the case of frequency errors from 1×10^{-4} to 1.5×10^{-4}, but the recognition rate is lower in other frequency deviation bands.

In summary, the proposed A-ALDA algorithm is utilized to construct the new features by taking into account the specific value of the modulated signal accumulation so that the new features have good separability. The inclusion of the SSDAE, which is a deep neural network with sparsity, can learn the main components of the samples, and the addition of noise reduction prevents the network from overfitting, reduces the noise learned by the neural units, and improves the robustness of the algorithm, as well as facilitates the learning and classification of the signal sample features. The proposed algorithm has improved the recognition rate compared to existing modulated signal recognition algorithms under complex channels and also outperforms other algorithms under conditions of limited signal length and interferences from the phase and frequency errors.

Chapter 8

Recurrent Neural Networks

Summary

This chapter starts with an analysis of the structure, principles, and computational architecture of Recurrent Neural Networks (RNNs), the loss function of RNNs is defined, the BPTT algorithm is derived, and the problems of gradient vanishing and exploding are analyzed. The first-order Stochastic Gradient Descent (SGD) method is used, and the RNN algorithm based on SGD optimization is discussed. Detailed analysis is conducted on the principle of the Recursive Least Squares (RLS) algorithm, and an RLS optimized RNNs algorithm (RLS-RNNs) is studied. Improvements were made to RLS-RNNs from two aspects: adaptive adjustment of forgetting factor and prevention of overfitting. Non-invasive load identification is taken as an example; the deep learning RNN model is extended to a correlated RNN model and applied to the identification of non-invasive power loads. Based on establishing intrinsic connections between inputs and outputs, reliable identification of load feature labels is achieved. By selecting the activation function and cross-entropy loss function reasonably, the problem of non-invasive load identification problem is effectively solved. Image compression is taken as another example; the combination of the DTCWT with RNN encoders improves the compression ratio for low pixel values by increasing the number of network iterations and modifying wavelet components, while the higher the pixel value, the higher the compression ratio.

8.1 Introduction

Recurrent Neural Network (RNN), as an effective deep learning model, when short-term memory dependencies on data over time are introduced, makes its parameter training more difficult than feed-forward neural networks. One of the key issues for the effective utilization of the RNNs is whether they can be efficiently trained or optimized. At present, the optimization algorithms of mainstream RNNs mainly include first-order Stochastic Gradient Descent (SGD) method, adaptive learning rate algorithm, and second-order SGD method. Among them, the first-order SGD method, called first-order optimization algorithm, includes SGD method, and the Momentum SGD (MSGD) algorithm, Nesterov momentum algorithm, adaptive learning rate algorithm, AdaGrad algorithm, RMSProp algorithm, AdaDelta algorithm, Adam algorithm, etc. These methods or algorithms perform better than the SGD method. Whereas the second-order gradient descent algorithm, called second-order optimization algorithm, uses the second-order gradient information of the objective function to optimize parameters. Compared with the first-order optimization algorithm, the second-order optimization algorithms still have excessive computational complexity; it is unsuitable to make them process large-scale data. Moreover, the obtained high-precision solutions have limited improvement on the model's generalization ability and sometimes even affect generalization. Therefore, it is very difficult to currently make wide use of the second-order optimization algorithms to train the RNNs.

Various neural networks based on Recursive Least Squares (RLS) algorithm belong to another category [184, 185]. Among them, a multi-layer perceptron based on the RLS algorithm, the linear input layer parameter matrix solving algorithm based on RLS algorithm, and multi-layer RNNs based on the RLS algorithm require storing a covariance matrix for each neuron, which incurs significant spatiotemporal overhead. As for optimization RNN based on extended Kalman filter, it is necessary to calculate the Jacobian matrix to achieve linearization, which also incurs significant spatiotemporal overhead. However, the RLS solution of parameters for Echo State Network (ESN) can be achieved by approximating nonlinear systems as linear systems, but it is limited to solving the parameters of the output layer in the ESN and is not suitable for general RNN training optimization.

In this chapter, on the basis of introducing the RNN model, the RNN based on RLS optimization (RLS-RNN) algorithm is introduced.

8.2 RNN Model

8.2.1 RNN principle

Recurrent Neural Networks (RNNs) are mainly used to process sequential data, and their principle is shown in Figure 8.1.

In Figure 8.1, the RNNs contain input units, with the input set marked as $\{x(k), k = 0, 1, 2, \ldots\}$. The output set of output units is recorded as $\{y(k), k = 0, 1, 2, \ldots\}$. The output set of hidden units is represented as $\{s(k), k = 1, 2, \ldots\}$.

In Figure 8.1(a), there is a unidirectional flow of information from the input unit to the hidden unit, while another unidirectional flow of information is from the hidden unit to the output unit. In some cases, the RNNs can break the limitations of the latter by guiding information from the output unit to the hidden unit, which is called "Back Propagation", and the input of the hidden layer also includes the state of the previous hidden layer, i.e., the nodes within the hidden layers can be self-connected or interconnected.

The RNN is expanded into a fully neural network, as shown in Figure 8.1(b). The calculation architecture for this network is as follows:

Step 1: $\boldsymbol{x}(k)$ indicates the input at the time $k(k = 1, 2, 3 \ldots)$.

Step 2: $\boldsymbol{s}(k)$ denotes the state vector of the hidden layer at the moment k, which is the memory unit of the network and can be calculated based on the output of the current input layer and the state vector of the previous hidden layer by

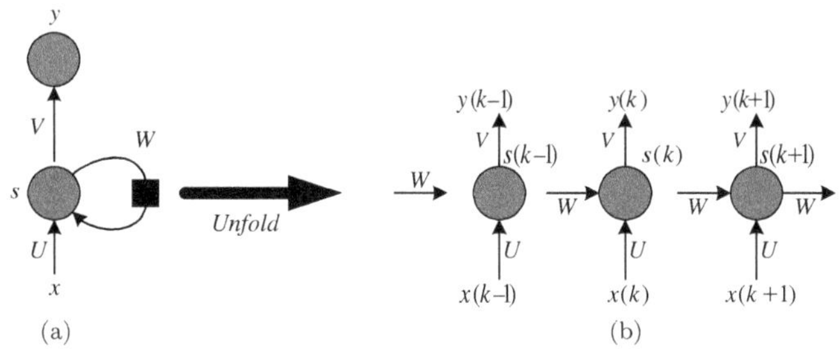

Figure 8.1. Recurrent neural networks.

$\boldsymbol{s}(k) = \sigma(\boldsymbol{U}\boldsymbol{x}(k) + \boldsymbol{W}\boldsymbol{s}(k-1))$, where σ is usually a nonlinear activation function, such as tanh or ReLU function. However, when the state vector $\boldsymbol{s}(0)$ of the first hidden layer is calculated, it is necessary to use $\boldsymbol{s}(-1)$, but it doesn't exist. So, it is generally set as $\boldsymbol{0}$ vectors in implementation. $\boldsymbol{U}$ and $\boldsymbol{W}$ are vectors of sharing parameters.

Step 3: $y(k)$ is the output of the kth step and is calculated by $y(k) = \text{Softmax}(Vs(k))$.

Note the following points:

(1) The hidden layer state $s(k)$ is the memory unit of the network, which contains the hidden layer state of all previous steps. The output $y(k)$ of the output layer is only related to the current state $s(k)$. In practice, in order to reduce the complexity of the network, $s(k)$ often only contains the hidden layer states of the previous few steps rather than all steps.
(2) In the RNNs, for each input step, each layer shares its own parameters **U**, **V**, and **W**. This indicates that each step is doing the same thing only under different input conditions; thus, it greatly reduces the number of parameters that need to be learned in the network.
(3) Each step in Figure 8.1 will have an output, but it is not necessary. We only need to care about the final output, without knowing the output of each layer. Similarly, it is also not necessary to have input for each step. The key to the RNNs lies in the hidden layer, which can capture the information of the sequence.

After receiving the input $x(k)$ at time $k, s(k)$ is the state of the hidden layer and $o(k)$ is the output value. The key point is that the value of $s(k)$ depends on not only $x(k)$ but also $s(k-1)$. The calculation formula for the RNNs is given by

$$y(k) = \sigma(\boldsymbol{V}\boldsymbol{s}(k)), \tag{8.2.1}$$

$$s(k) = g(\boldsymbol{U}\boldsymbol{x}(k) + \boldsymbol{W}\boldsymbol{s}(k-1)). \tag{8.2.2}$$

Equation (8.2.1) is the calculation formula for the output layer, which is a fully connected layer, meaning that its each node is connected to each node of the hidden layer. $\boldsymbol{V}$ is the weight matrix of the output layer, and $\sigma(\cdot)$ is the activation function. Equation (8.2.2) is the calculation formula for the hidden layer, which is the loop layer.

$\boldsymbol{U}$ is the weight matrix of input $\boldsymbol{x}$, $\boldsymbol{W}$ is the previous value and serves as the weight matrix of this input $s(k-1)$, and $g(\cdot)$ is the activation function.

Equations (8.2.1) and (8.2.2) indicate that the difference between a recurrent layer and a fully connected layer is that the recurrent layer has an additional weight matrix $\boldsymbol{W}$.

If Equation (8.2.2) is repeatedly substituted into Equation (8.2.1), we get

$$\begin{aligned} y(k) &= \sigma(\boldsymbol{V}\boldsymbol{s}(k)) \\ &= \sigma(k-1)(\boldsymbol{V}g(\boldsymbol{U}\boldsymbol{x}(k)+\boldsymbol{W}\boldsymbol{s}(k-1))) \\ &= \sigma(k-1)(\boldsymbol{V}g(\boldsymbol{U}\boldsymbol{x}(k)+\boldsymbol{W}(g(\boldsymbol{U}\boldsymbol{x}(k-1)+\boldsymbol{W}\boldsymbol{s}(k-2)))) \\ &= \sigma(k-1)(\boldsymbol{V}g(\boldsymbol{U}\boldsymbol{x}(k)+\boldsymbol{W}(g(\boldsymbol{U}\boldsymbol{x}(k-1)+\boldsymbol{W}(g(\boldsymbol{U}\boldsymbol{x}(k-2) \\ &\quad +\boldsymbol{W}\boldsymbol{s}(k-3))))) \\ &= \sigma(k-1)(\boldsymbol{V}g(\boldsymbol{U}\boldsymbol{x}(k)+\boldsymbol{W}(g(\boldsymbol{U}\boldsymbol{x}(k-1)+\boldsymbol{W}(g(\boldsymbol{U}\boldsymbol{x}(k-2) \\ &\quad +\boldsymbol{W}(g(\boldsymbol{U}\boldsymbol{x}(k-3)+\ldots))))). \end{aligned} \tag{8.2.3}$$

Equation (8.2.3) indicates that the output $y(k)$ of the RNN is influenced by previous inputs $x(k), x(k-1), x(k-2), x(k-3), \ldots,$ which is why the RNNs can look ahead with any number of input values.

8.2.2 Loss function of the RNN

According to Equations (8.2.1) and (8.2.2), the loss function of the RNN is defined as

$$E = \sum_{k=1}^{T} \text{Loss}(k), \tag{8.2.4}$$

where Loss(k) is defined as the cross-loss function, i.e.,

$$\text{Loss}(k) = -\boldsymbol{y}^{\mathrm{T}}(k) \log \hat{y}(k). \tag{8.2.5}$$

8.2.3 BPTT algorithm

If the RNNs are expanded, the parameters $\boldsymbol{W}, \boldsymbol{U},$ and $\boldsymbol{V}$ of the networks are shared, and in the use of gradient descent methods, the output of each step depends on not only the network at the current time but also the state of the network at several previous times. For example, when $k = 4$, three more steps need to be passed backwards, and various gradients need to be added to the following three steps. This learning algorithm is called as Backpropagation Through Time (BPTT) algorithm, which is a training algorithm for loop layers. Its basic principle is the same as that of the BP algorithm, and it also includes the same four steps [186]:

Step 1: Forward calculation

The output value of each neuron is calculated in the forward direction.

Step 2: Reverse calculation

The error term e_j of each neuron is the partial derivative of the error function e with respect to the weighted input net_j of neuron j.

Step 3: Gradient calculation

The gradient of each weight is calculated.

Step 4: Update weights

The weights are updated using the stochastic gradient descent method.

Note that in vanilla RNN training, the BPTT algorithm cannot solve the long-term dependency problem (i.e., the current output is related to a long sequence, usually beyond ten steps) because the BPTT algorithm can bring about the so-called gradient vanishing/exploding problem.

8.2.3.1 *Forward calculation*

Assuming the dimension of the input vector $\boldsymbol{x}$ is N and the dimension of the output vector $\boldsymbol{s}$ is M, then the dimension of matrix $\boldsymbol{U}$ is $N \times M$, and the dimension of matrix $\boldsymbol{W}$ is $M \times M$. The expansion of Equation (8.2.2) is written as

$$\begin{bmatrix} s_1(k) \\ s_2(k) \\ \vdots \\ s_M(k) \end{bmatrix} = f\left(\begin{bmatrix} u_{11}u_{12}\cdots u_{1N} \\ u_{21}u_{22}\cdots u_{2N} \\ \vdots \\ u_{M1}u_{M2}\cdots u_{MN} \end{bmatrix}\begin{bmatrix} x_1 \\ x_2 \\ \vdots \\ x_M \end{bmatrix} + \begin{bmatrix} w_{11}w_{12}\cdots w_{1M} \\ w_{21}w_{22}\cdots w_{2M} \\ \vdots \\ w_{M1}w_{M2}\cdots w_{MM} \end{bmatrix}\begin{bmatrix} s_1(k-1) \\ s_2(k-1) \\ \vdots \\ s_M(k-1) \end{bmatrix}\right), \tag{8.2.6}$$

where $s_j(k)$ represents the value of the jth element of the vector $\boldsymbol{s}$ at time k. u_{ji} represents the weight between the ith neuron of the input layer to the jth neuron of the loop layer. w_{ji} represents the ith weight of the jth neuron at time $k-1$ of the loop layer.

8.2.3.2 *Error calculation*

The BTPP algorithm propagates the error term $e^l(k)$ at time k in the lth layer along two directions: One direction is to pass it to the previous layer to obtain $e^{l-1}(k)$, and this part is only related to the weight matrix $\boldsymbol{U}$. Another direction is to propagate it along the timeline to the initial moment k_0 to get $e^l(0)$, which is only related to the weight matrix $\boldsymbol{W}$.

The current vector $\boldsymbol{net}(k)$ is used to represent the weighted input of the neuron at time k, and

$$\boldsymbol{net}(k) = \boldsymbol{U}\boldsymbol{x}(k) + \boldsymbol{W}\boldsymbol{s}(k-1), \tag{8.2.7}$$

$$\boldsymbol{s}(k-1) = g(\boldsymbol{net}(k-1)). \tag{8.2.8}$$

Thereby,

$$\frac{\partial \boldsymbol{net}(k)}{\partial \boldsymbol{net}(k-1)} = \frac{\partial \boldsymbol{net}(k)}{\partial \boldsymbol{s}(k-1)}\frac{\partial \boldsymbol{s}(k-1)}{\partial \boldsymbol{net}(k-1)}. \tag{8.2.9}$$

Let $\boldsymbol{a}$ and $\boldsymbol{a}^T$ represent column and row vectors, respectively. The first term of Equation (8.2.6) is the derivative of a vector function on a vector, i.e., a Jacobian matrix, i.e.,

$$\frac{\partial \boldsymbol{net}(k)}{\partial \boldsymbol{s}(k-1)} = \begin{bmatrix} \frac{\partial net_1(k-1)}{\partial s_1(k-1)} & \frac{\partial net_1(k)}{\partial s_2(k-1)} & \cdots & \frac{\partial net_1(k)}{\partial s_M(k-1)} \\ \frac{\partial net_2(k)}{\partial s_1(k-1)} & \frac{\partial net_2(k)}{\partial s_2(k-1)} & \cdots & \frac{\partial net_2(k)}{\partial s_M(k-1)} \\ \vdots & & & \\ \frac{\partial net_M(k)}{\partial s_1(k-1)} & \frac{\partial net_M(k)}{\partial s_2(k-1)} & \cdots & \frac{\partial net_M(k)}{\partial s_M(k-1)} \end{bmatrix}$$

$$= \begin{bmatrix} w_{11} & w_{12} & \cdots & w_{1M} \\ w_{21} & w_{22} & \cdots & w_{2M} \\ \vdots & & & \\ w_{M1} & w_{M2} & \cdots & w_{MM} \end{bmatrix} = \boldsymbol{W}. \tag{8.2.10}$$

Similarly, the second term of Equation(8.2.6) is also a Jacobian matrix, i.e.,

$$\frac{\partial \boldsymbol{s}(k-1)}{\partial \boldsymbol{net}(k-1)} = \begin{bmatrix} \frac{\partial s_1(k-1)}{\partial net_1(k-1)} & \frac{\partial s_1(k-1)}{\partial net_2(k-1)} & \cdots & \frac{\partial s_1(k-1)}{\partial net_M(k-1)} \\ \frac{\partial s_2(k-1)}{\partial net_1(k-1)} & \frac{\partial s_2(k-1)}{\partial net_2(k-1)} & \cdots & \frac{\partial s_2(k-1)}{\partial net_M(k-1)} \\ \vdots & & & \\ \frac{\partial s_M(k-1)}{\partial net_1(k-1)} & \frac{\partial s_M(k-1)}{\partial net_2(k-1)} & \cdots & \frac{\partial s_M(k-1)}{\partial net_M(k-1)} \end{bmatrix}$$

$$= \begin{bmatrix} g'(net_1(k-1)) & 0 & \cdots & 0 \\ 0 & g'(net_2(k-1)) & \cdots & 0 \\ \vdots & & & \\ 0 & 0 & \cdots & g'(net_M(k-1)) \end{bmatrix}$$

$$= \mathrm{diag}[g'(net_1(k-1))], \tag{8.2.11}$$

where diag$[\boldsymbol{a}]$ represents a diagonal matrix created by a vector $\boldsymbol{a}$, i.e.,

$$\mathrm{diag}[\boldsymbol{a}] = \begin{bmatrix} a_1 & 0 & \cdots & 0 \\ 0 & a_2 & \cdots & 0 \\ \vdots & & & \\ 0 & 0 & \cdots & a_n \end{bmatrix}.$$

Finally, the two items are combined; we obtain

$$\frac{\partial \boldsymbol{net}(k)}{\partial \boldsymbol{net}(k-1)} = \frac{\partial \boldsymbol{net}(k)}{\partial \boldsymbol{s}(k-1)} \frac{\partial \boldsymbol{s}(k-1)}{\partial \boldsymbol{net}(k-1)}$$
$$= \boldsymbol{W}\text{diag}[g'(net_1(k-1))]$$
$$= \begin{bmatrix} w_{11}g'(net_1(k-1)) & w_{12}g'(net_2(k-1)) & \cdots & w_{1M}g'(net_M(k-1)) \\ w_{21}g'(net_1^{t-1}(k-1)) & w_{22}g'(net_2(k-1)) & \cdots & w_{2M}g'(net_2(k-1)) \\ \vdots & & & \\ w_{M1}g'(net_1(k-1)) & w_{M2}g'(net_2(k-1)) & \cdots & w_{MM}g'(net_2(k-1)) \end{bmatrix}. \tag{8.2.12}$$

Equation (8.2.12) describes the law of passing error e forward in time by one moment. Based on this law, the error term $e(k)$ at any time k can be obtained by

$$e^T(k) = \frac{\partial E}{\partial \boldsymbol{net}(k)}$$
$$= \frac{\partial L}{\partial \boldsymbol{net}(n)} \frac{\partial \boldsymbol{net}(n)}{\partial \boldsymbol{net}(k)}$$
$$= \frac{\partial L}{\partial \boldsymbol{net}(n)} \frac{\partial \boldsymbol{net}(n)}{\partial \boldsymbol{net}(n-1)} \frac{\partial \boldsymbol{net}(n-1)}{\partial \boldsymbol{net}(n-2)} \cdots \frac{\partial \boldsymbol{net}(k+1)}{\partial \boldsymbol{net}(k)}$$
$$= \boldsymbol{W}\text{diag}[g'(\boldsymbol{net}(t-1))]\boldsymbol{W}\text{diag}[g'(\boldsymbol{net}(t-2))] \cdots$$
$$\times W\text{diag}[g'(\boldsymbol{net}(k))]\mathbf{e}^l(n)$$
$$= e^T(n) \prod_{i=k}^{n-1} \boldsymbol{W}\text{diag}[g'(\boldsymbol{net}(i))]. \tag{8.2.13}$$

Equation (8.2.13) is an algorithm that backpropagates the error term over time.

The loop layer backpropagates the error term to the previous layer of the network, which is exactly the same as a general fully connected layer, and will only be briefly described as follows.

The relationship between the weighted input $\boldsymbol{net}^l$ of the loop layer and the weighted input $\boldsymbol{net}^{l-1}$ of the previous layer is represented as

$$\boldsymbol{net}^l(k) = \boldsymbol{U}\boldsymbol{a}^{l-1}(k) + \boldsymbol{W}\boldsymbol{s}(k-1), \tag{8.2.14}$$

$$\boldsymbol{a}^{l-1}(k) = g^{l-1}(\boldsymbol{net}^{l-1}(k)), \tag{8.2.15}$$

where $\boldsymbol{net}^l(k)$ is the weighted input of the neuron of the lth layer (assuming that the lth layer is a recurrent layer), $\boldsymbol{net}^{l-1}(k)$ is the weighted input of the neurons of the $l-1$th layer, and $\boldsymbol{a}^{l-1}(k)$ is the output of the neurons of the $l-1$ layer. g^{l-1} is the activation function of the l-1th layer:

$$\begin{aligned}\frac{\partial \boldsymbol{net}^l(k)}{\partial \boldsymbol{net}^{l-1}(k)} &= \frac{\partial \boldsymbol{net}^l(k)}{\partial \boldsymbol{a}^{l-1}(k)} \frac{\partial \boldsymbol{a}^{l-1}(k)}{\partial \boldsymbol{net}^{l-1}(k)} \\ &= \boldsymbol{U}\text{diag}[g'^{l-1}(\boldsymbol{net}^l(k))]. \end{aligned} \tag{8.2.16}$$

So,

$$\begin{aligned}(e^{l-1}(k))^T &= \frac{\partial E}{\partial \boldsymbol{net}^{l-1}(k)} \\ &= \frac{\partial L}{\partial \boldsymbol{net}^l(k)} \frac{\partial \boldsymbol{net}^l(k)}{\partial \boldsymbol{net}^{l-1}(k)} \\ &= (e^l(k))^T \boldsymbol{U}\text{diag}[g'^{l-1}(\boldsymbol{net}^l(k))]. \end{aligned} \tag{8.2.17}$$

Equation (8.2.7) is called as the algorithm that passes the error term to the upper layer algorithm.

8.2.3.3 Calculation of weight gradient

Now, we have finally reached the final step of the BPTT algorithm: Calculate the gradient of each weight.

First, the gradient $\frac{\partial E}{\partial \boldsymbol{W}}$ of the error function E on the weight matrix $\boldsymbol{W}$ can be calculated by the following:

Figure 8.2 shows the quantities calculated in the first two steps so far, including the output $\boldsymbol{s}(k)$ of the loop layer at each time step k and the error term $e(k)$.

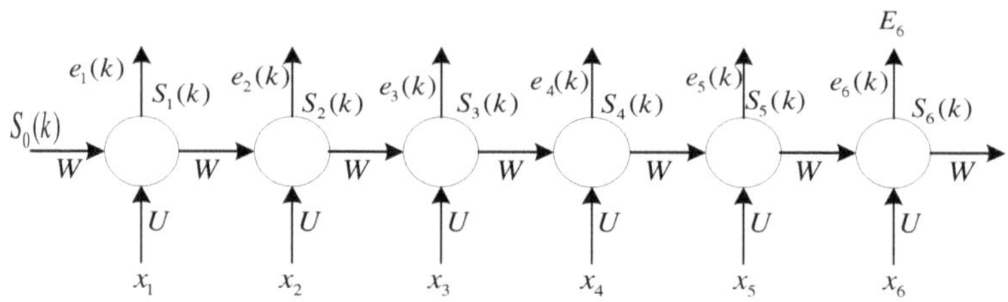

Figure 8.2. Gradient calculation procedure of weight matrix W.

Weight gradient calculation algorithm for fully connected networks: As long as the error term $e(k)$ at any given time k and the output $\boldsymbol{s}(k-1)$ of the loop layer at the previous time $k-1$ are known, the gradient $\nabla_{\boldsymbol{W}(k)}E$ can be calculated by

$$\nabla_{\boldsymbol{W}(k)}E = \begin{bmatrix} e_1(k)s_1(\boldsymbol{k}-1) & e_1(k)s_2(\boldsymbol{k}-1) & \cdots & e_1(k)s_M(\boldsymbol{k}-1) \\ e_2(k)s_1(\boldsymbol{k}-1) & e_2(k)s_2(\boldsymbol{k}-1) & \cdots & e_2(k)s_M(\boldsymbol{k}-1) \\ \vdots & & & \\ e_M(k)s_1(\boldsymbol{k}-1) & e_M(k)s_2(\boldsymbol{k}-1) & \cdots & e_M(k)s_M(\boldsymbol{k}-1) \end{bmatrix}, \tag{8.2.18}$$

where $e_i(k)$ represents the ith component of the error term vector at time k and $s_i(k-1)$ denotes the output value of the ith neuron of the loop layer at the time $k-1$.

Since

$$\boldsymbol{net}(k) = \boldsymbol{U}\boldsymbol{x}(k) + \boldsymbol{W}\boldsymbol{s}(k-1), \tag{8.2.19}$$

$$\begin{bmatrix} net_1(k) \\ net_2(k) \\ \vdots \\ net_M(k) \end{bmatrix} = \boldsymbol{U}\boldsymbol{x}(k) + \begin{bmatrix} w_{11} & w_{12} & \cdots & w_{1M} \\ w_{21} & w_{22} & \cdots & w_{2M} \\ \vdots & & & \\ w_{M1} & w_{M2} & \cdots & w_{MM} \end{bmatrix} \begin{bmatrix} s_1(k-1) \\ s_2(k-1) \\ \vdots \\ s_M(k-1) \end{bmatrix}$$

$$= \boldsymbol{U}\boldsymbol{x}(k) + \begin{bmatrix} w_{11}s_1(k-1) & w_{12}s_2(k-1) & \cdots & w_{1M}s_M(k-1) \\ w_{21}s_1(k-1) & w_{22}s_2(k-1) & \cdots & w_{2M}s_M(k-1) \\ \vdots & & & \\ w_{M1}s_1(k-1) & w_{M2}s_2(k-1) & \cdots & w_{MM}s_M(k-1) \end{bmatrix}. \tag{8.2.20}$$

Since taking the derivative of $\boldsymbol{W}$ is not related to $\boldsymbol{U}\boldsymbol{x}(k)$, it will no longer be considered. Now, we only consider taking the derivative of the weight term w_{ji}. Equation (8.2.20) indicates that w_{ji} is only related to $net_j(k)$; therefore,

$$\frac{\partial E}{\partial w_{ji}} = \frac{\partial E}{\partial net_j(k)} \frac{\partial net_j(k)}{\partial w_{ji}} = e_j(k)s_i(k-1). \tag{8.2.21}$$

According to the above rule, the matrix in Equation (8.2.16) can be generated.

After the gradient $\nabla_{\boldsymbol{W}(k)}E$ of the weight matrix $\boldsymbol{W}$ at time k has been obtained, the final gradient $\nabla_{\boldsymbol{W}}E$ is the sum of the gradients at each time, i.e.,

$$\begin{aligned}\nabla_{\boldsymbol{W}(k)}E &= \sum_{i=1}^{k}\nabla_{\boldsymbol{W}(k)}\text{loss}(k)\\ &= \begin{bmatrix} e_1(k)s_1(k-1) & e_1(k)s_2(k-1) & \cdots & e_1(k)s_M(k-1)\\ e_2(k)s_1(k-1) & e_2(k)s_2(k-1) & \cdots & e_2(k)s_M(k-1)\\ \vdots & & & \\ e_M(k)s_1(k-1) & e_M(k)s_2(k-1) & \cdots & e_M(k)s_M(k-1)\end{bmatrix} + \cdots\\ &\quad + \begin{bmatrix} e_1(k)s_1(0) & e_1(k)s_2(0) & \cdots & e_1(k)s_M(0)\\ e_2(k)s_1(0) & e_2(k)s_2(0) & \cdots & e_2(k)s_M(0)\\ \vdots & & & \\ e_M(k)s_1(0) & e_M(k)s_2(0) & \cdots & e_M(k)s_M(0)\end{bmatrix}.\end{aligned} \tag{8.2.22}$$

Equation (8.2.22) is the formula for calculating the gradient of weight matrix $\boldsymbol{W}$ of the loop layer. Now, we explain why the final gradient is the sum of gradients at each time step because

$$\boldsymbol{net}(k) = \boldsymbol{Ux}(k) + \boldsymbol{W}g(\boldsymbol{net}(k-1)), \tag{8.2.23}$$

where $\boldsymbol{Ux}(k)$ is completely independent of $\boldsymbol{W}$; it can be considered as a constant. Now, we consider the part to the right of the plus sign in the first equation because both $\boldsymbol{W}$ and $g(net(k-1))$ are functions of $\mathbf{W}$, so

$$(uv)' = u'v + uv',$$

$$\frac{\partial \boldsymbol{net}(k)}{\partial \boldsymbol{W}} = \frac{\partial \boldsymbol{W}}{\partial \boldsymbol{W}}g(\boldsymbol{net}(k-1)) + \boldsymbol{W}\frac{\partial g(\boldsymbol{net}(k-1))}{\partial \boldsymbol{W}}.$$

The final $\nabla_{\boldsymbol{W}}E$ that needs to be calculated is given by

$$\begin{aligned}\nabla_{\boldsymbol{W}}E &= \frac{\partial E}{\partial \boldsymbol{W}} = \frac{\partial E}{\partial \boldsymbol{net}}\frac{\partial \boldsymbol{net}}{\partial \boldsymbol{W}}\\ &= \boldsymbol{e}^T(k)\frac{\partial \boldsymbol{W}}{\partial \boldsymbol{W}}g(\boldsymbol{net}(k-1)) + \boldsymbol{e}^T(k)\boldsymbol{W}\frac{\partial g(\boldsymbol{net}(k-1))}{\partial \boldsymbol{W}}.\end{aligned} \tag{8.2.24}$$

The part to the left of the plus sign in Equation (8.2.24) is first calculated. $\frac{\partial \boldsymbol{W}}{\partial \boldsymbol{W}}$ is the derivative of a matrix, and its result is a four-dimensional tensor, i.e.,

$$\frac{\partial \boldsymbol{W}}{\partial \boldsymbol{W}} = \begin{bmatrix} \frac{\partial w_{11}}{\partial \boldsymbol{W}} & \frac{\partial w_{12}}{\partial \boldsymbol{W}} & \cdots & \frac{\partial w_{1M}}{\partial \boldsymbol{W}} \\ \frac{\partial w_{21}}{\partial \boldsymbol{W}} & \frac{\partial w_{22}}{\partial \boldsymbol{W}} & \cdots & \frac{\partial w_{2M}}{\partial \boldsymbol{W}} \\ \vdots & & & \\ \frac{\partial w_{M1}}{\partial \boldsymbol{W}} & \frac{\partial w_{M2}}{\partial \boldsymbol{W}} & \cdots & \frac{\partial w_{MM}}{\partial \boldsymbol{W}} \end{bmatrix}$$

$$= \begin{bmatrix} \begin{bmatrix} \frac{\partial w_{11}}{\partial w_{11}} & \frac{\partial w_{11}}{\partial w_{12}} & \cdots & \frac{\partial w_{11}}{\partial w_{1M}} \\ \frac{\partial w_{11}}{\partial w_{21}} & \frac{\partial w_{11}}{\partial w_{22}} & \cdots & \frac{\partial w_{11}}{\partial w_{2M}} \\ \vdots & & & \\ \frac{\partial w_{11}}{\partial w_{M1}} & \frac{\partial w_{11}}{\partial w_{M2}} & \cdots & \frac{\partial w_{11}}{\partial w_{MM}} \end{bmatrix} \begin{bmatrix} \frac{\partial w_{12}}{\partial w_{11}} & \frac{\partial w_{12}}{\partial w_{12}} & \cdots & \frac{\partial w_{12}}{\partial w_{1M}} \\ \frac{\partial w_{12}}{\partial w_{21}} & \frac{\partial w_{12}}{\partial w_{22}} & \cdots & \frac{\partial w_{12}}{\partial w_{2M}} \\ \vdots & & & \\ \frac{\partial w_{12}}{\partial w_{M1}} & \frac{\partial w_{12}}{\partial w_{M2}} & \cdots & \frac{\partial w_{12}}{\partial w_{MM}} \end{bmatrix} \cdots \end{bmatrix}$$

$$= \begin{bmatrix} \begin{bmatrix} 1 & 0 & \cdots & 0 \\ 0 & 0 & \cdots & 0 \\ \vdots & & & \\ 0 & 0 & \cdots & 0 \end{bmatrix} \begin{bmatrix} 0 & 1 & \cdots & 0 \\ 0 & 0 & \cdots & 0 \\ \vdots & & & \\ 0 & 0 & \cdots & 0 \end{bmatrix} \cdots \end{bmatrix}. \tag{8.2.25}$$

Since $\boldsymbol{s}(k-1) = g(\boldsymbol{net}(k-1))$ is a column vector, the four-dimensional tensor shown in Equation (8.2.25) is multiplied by this vector $\boldsymbol{s}(k-1)$ to yield a three-dimensional tensor, which is then left multiplied by the row vector $\boldsymbol{e}^T(k)$ to obtain the final matrix:

$$\boldsymbol{e}^T(k)\frac{\partial \boldsymbol{W}}{\partial \boldsymbol{W}} g(\boldsymbol{net}(k-1)) = \boldsymbol{e}^T(k)\frac{\partial \boldsymbol{W}}{\partial \boldsymbol{W}} \boldsymbol{s}(k-1)$$

$$= \boldsymbol{e}^T(k) \begin{bmatrix} \begin{bmatrix} 1 & 0 & \cdots & 0 \\ 0 & 0 & \cdots & 0 \\ \vdots & & & \\ 0 & 0 & \cdots & 0 \end{bmatrix} \begin{bmatrix} 0 & 1 & \cdots & 0 \\ 0 & 0 & \cdots & 0 \\ \vdots & & & \\ 0 & 0 & \cdots & 0 \end{bmatrix} \cdots \end{bmatrix} \begin{bmatrix} s_1(k-1) \\ s_2(k-1) \\ \vdots \\ s_M(k-1) \end{bmatrix}$$

$$= e^T(k) \begin{bmatrix} \begin{bmatrix} s_1(k-1) \\ 0 \\ \vdots \\ 0 \end{bmatrix} & \begin{bmatrix} s_2(k-1) \\ 0 \\ \vdots \\ 0 \end{bmatrix} & \cdots \end{bmatrix}$$

$$= \begin{bmatrix} e_1(k)s_1(k-1) & e_1(k)s_2(k-1) & \cdots & e_1(k)s_M(k-1) \\ e_2(k)s_1(k-1) & e_2(k)s_2(k-1) & \cdots & e_2(k)s_M(k-1) \\ \vdots & & & \\ e_M(k)s_1(k-1) & e_M(k)s_2(k-1) & \cdots & e_M(k)s_M(k-1) \end{bmatrix}$$

$$= \nabla_{\boldsymbol{W}(k)} E. \tag{8.2.26}$$

The part to the right of the plus sign in Equation (8.2.24) is calculated again, i.e.,

$$\begin{aligned} \boldsymbol{e}^T(k)\boldsymbol{W}\frac{\partial f(\boldsymbol{net}(k-1))}{\partial \boldsymbol{W}} &= \boldsymbol{e}^T(k)\boldsymbol{W}\frac{\partial g(\boldsymbol{net}(k-1))}{\partial \boldsymbol{net}(k-1)}\frac{\partial \boldsymbol{net}(k-1)}{\partial \boldsymbol{W}} \\ &= \boldsymbol{e}^T(k)\boldsymbol{W}g'(\boldsymbol{net}(k-1))\frac{\partial \boldsymbol{net}(k-1)}{\partial \boldsymbol{W}} \\ &= \boldsymbol{e}^T(k)\frac{\partial \boldsymbol{net}(k)}{\partial \boldsymbol{net}(k-1)}\frac{\partial \boldsymbol{net}(k-1)}{\partial \boldsymbol{W}} \\ &= \boldsymbol{e}^T(k-1)\frac{\partial \boldsymbol{net}(k-1)}{\partial \boldsymbol{W}}. \end{aligned} \tag{8.2.27}$$

So, the recursive formula is written as

$$\begin{aligned} \nabla_{\boldsymbol{W}} E = \frac{\partial E}{\partial \boldsymbol{W}} &= \frac{\partial E}{\partial \boldsymbol{net}(k)}\frac{\partial \boldsymbol{net}(k)}{\partial \boldsymbol{W}} \\ &= \nabla_{\boldsymbol{W}(k)} E + \boldsymbol{e}^T(k-1)\frac{\partial \boldsymbol{net}(k-1)}{\partial \boldsymbol{W}} \\ &= \nabla_{\boldsymbol{W}(k)} E + \nabla_{\boldsymbol{W}(k-1)} E + \boldsymbol{e}^T(k-2)\frac{\partial \boldsymbol{net}(k-2)}{\partial \boldsymbol{W}} \\ &= \nabla_{\boldsymbol{W}(k)} E + \nabla_{\boldsymbol{W}(k-1)} E + \cdots + \nabla_{\boldsymbol{W}(1)} E \\ &= \sum_{n=1}^{k} \nabla_{\boldsymbol{W}(k)} E. \end{aligned} \tag{8.2.28}$$

The above procedure has proved that the final gradient $\nabla_{\boldsymbol{W}} E$ is the sum of the gradients at each moment.

Similar to the weight matrix $\boldsymbol{W}$, the calculation method for the weight matrix $\boldsymbol{U}$ can be obtained by

$$\nabla_{U(k)} E = \begin{bmatrix} e_1(k)x_1(k) & e_1(k)x_2(k) & \cdots & e_1(k)x_N(k) \\ e_2(k)x_1(k) & e_2(k)x_2(k) & \cdots & e_2(k)x_N(k) \\ \vdots & & & \\ e_M(k)x_1(k) & e_M(k)x_2(k) & \cdots & e_M(k)x_N(k) \end{bmatrix}. \tag{8.2.29}$$

Equation (8.2.29) is the gradient of the error function at time k to the weight matrix $\boldsymbol{U}$. Like the weight matrix $\boldsymbol{W}$, the final gradient is also the sum of gradients at each moment, i.e.,

$$\nabla_{\boldsymbol{U}} E = \sum_{i=1}^{k} \nabla_{\boldsymbol{U}(t)} E. \tag{8.2.30}$$

8.2.3.4 *Gradient explosion and disappearance problems*

The RNN cannot handle longer sequences well. One main reason is that the RNNs are prone to gradient explosion and vanishing during training, which results in gradients not being able to propagate continuously over longer sequences during training, making it difficult for the RNNs to be used to capture long-distance effects.

Why do the RNNs experience gradient explosion and vanishing problems? Now, we conduct a detailed analysis. According to Equation (8.2.11), we get

$$\boldsymbol{e}^T(k) = \boldsymbol{e}^T(n) \prod_{i=k}^{n-1} \boldsymbol{W} \mathrm{diag}[g'(\boldsymbol{net}(i))], \tag{8.2.31}$$

$$||\boldsymbol{e}^T(k)|| \leq ||\boldsymbol{e}^T(n)|| \prod_{i=k}^{n-1} ||\boldsymbol{W}|| ||\mathrm{diag}[g'(\boldsymbol{net}(i))]|| \leq ||\boldsymbol{e}^T(n)|| (\beta_W \beta_f)^{n-k}, \tag{8.2.32}$$

where β is the upper bound of the modulus of the matrix. Since Equation (8.2.32) is an exponential function, if $n - k$ is large (i.e., looking far ahead), it will cause the corresponding error term to grow or shrink very quickly and result in gradient explosion and vanishing problems (depending on whether β is greater than or less than 1).

Generally speaking, gradient explosions are easier to handle. When the gradient explodes, the program will converge to a NaN error. You can also set a gradient threshold, which can be directly truncated when the gradient exceeds this threshold.

Gradient vanishing is more difficult to detect and also more challenging to handle. Overall, there are three methods to address the problem of gradient vanishing:

(1) Reasonable initialization weight values. Initialization of weights can ensure that each neuron does not take maximum or minimum values as much as possible to avoid areas where gradients disappear.
(2) The ReLU function is used instead of sigmoid and Tanh as activation functions.
(3) The most popular approach is to use the RNNs with other structures, such as Long Short-Term Memory Networks (LTSMs) and Gated Recurrent Units (GRUs).

8.3 RNNs Based on SGD Optimization

The model structure of the RNNs for processing temporal data, as shown in Figure 8.3, typically consists of an input layer, a hidden layer (also known as a recurrent layer), and an output layer.

In Figure 8.3, $\boldsymbol{X}^l(k) \in \mathbb{R}^{m\times a}$, $\boldsymbol{S}^l(k) \in \mathbb{R}^{m\times h}$, and $\boldsymbol{y}^l(k) \in \mathbb{R}^{m\times d}$ represent the input values of input layer, output values of the hidden layer, and output layer values for the sth batch of training sample data at time k, where m is the small batch size, a is the dimension of training sample data, h is the number of neurons in the hidden layer, and d is the number of neurons in output layer. $\boldsymbol{U}^{l-1} \in \mathbb{R}^{a\times h}$, $\boldsymbol{W}^{l-1} \in \mathbb{R}^{h\times h}$, and $\boldsymbol{V}^{l-1} \in \mathbb{R}^{h\times d}$ are the parameter matrices from the input layer to the hidden layer, the hidden layer at the adjacent moment, and the hidden layer to the output layer during the training of the sth batch of data, respectively. $(\boldsymbol{b}^{l-1})^H \in \mathbb{R}^{1\times h}$ and $\boldsymbol{b}^{O(l-1)} \in \mathbb{R}^{1\times d}$ are the bias parameter matrices for the hidden layer and output

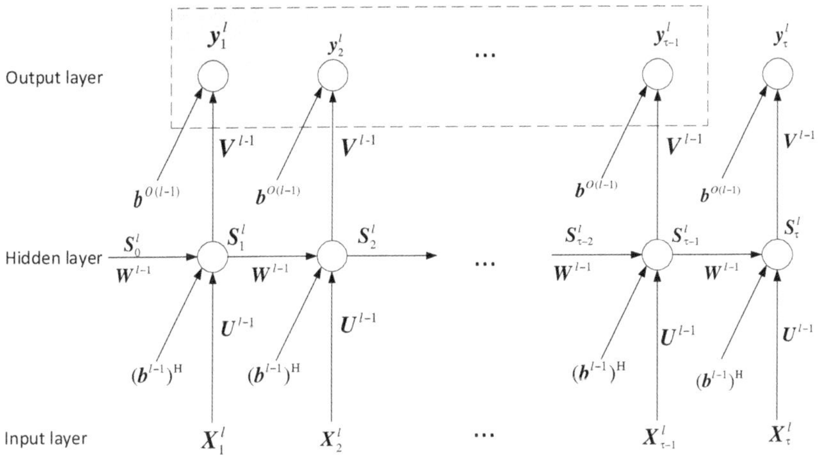

Figure 8.3. Structure of the RNN.

layer, respectively; τ indicates that the current sequence data has a common time step. The core idea of the RNNs is to share parameters at different time steps of the model, where the values of the hidden layer at each time step are weighted and input into the calculation of the next time step so that the weight parameters can learn the correlation features between different time steps of the sequence data and generalize them. The output layer selects which time steps to output based on actual problems, including common classification and prediction problems for sequential data. For prediction problems of sequence data, the output layer has outputs at every time step. For the classification problem of sequence data, only the last time step of the output layer is outputted, i.e., the time step in the dashed box in Figure 8.3 is not outputted. The RNNs can obtain actual output through forward propagation, and the calculation process is as follows:

$$\boldsymbol{S}^l(k) = g(\boldsymbol{x}^l(k)\boldsymbol{U}^{l-1} + \boldsymbol{S}^l(k-1)\boldsymbol{W}^{l-1} + \mathbf{1} \times (\boldsymbol{b}^{l-1})^H), \tag{8.3.1}$$

$$\boldsymbol{y}^l(k) = \sigma(\boldsymbol{S}^l(k)\boldsymbol{V}^{l-1} + \mathbf{1} \times \boldsymbol{b}^{\boldsymbol{O}(l-1)}), \tag{8.3.2}$$

where l is the column vector with m rows and its element value is 1 and $g(\cdot)$ and $\sigma(\cdot)$ are the activation functions for the hidden layer and output layer, respectively. Popular activation functions include the sigmoid function and tanh function. In order to facilitate subsequent derivation and simplify the expression, the above two equations can be further expressed as augmented matrices, i.e.,

$$\boldsymbol{S}^l(k) = g((\boldsymbol{R}^l(k))^H(\boldsymbol{\Theta}^{l-1})^H), \tag{8.3.3}$$

$$\boldsymbol{y}^l(k) = \sigma(\boldsymbol{R}^{\boldsymbol{O}(l)}(k)\boldsymbol{\Theta}^{\boldsymbol{O}(l-1)}), \tag{8.3.4}$$

where $(\boldsymbol{R}^l(k))^H \in \boldsymbol{R}^{m\times(a+h+1)}$ and $\boldsymbol{R}^{O(l)}(k) \in \boldsymbol{R}^{m\times(h+1)}$ are the input augmentation matrices of the hidden layer and the output layer, respectively; $(\boldsymbol{\Theta}^{l-1})^H \in \boldsymbol{R}^{(a+h+1)\times h}$ and $\boldsymbol{\Theta}^{O(l-1)} \in \boldsymbol{R}^{(h+1)\times d}$ are the weight augmentation matrices for the hidden layer and input layer, respectively:

$$(\boldsymbol{R}^l(k))^H = [\boldsymbol{x}^l(k) \quad \boldsymbol{S}^l(k-1) \quad \mathbf{1}], \tag{8.3.5}$$

$$\boldsymbol{R}^{O(l)}(k) = [\boldsymbol{S}^l(k-1) \quad \mathbf{1}], \tag{8.3.6}$$

$$(\boldsymbol{\Theta}^{l-1})^H = [(\boldsymbol{U}^{l-1})^T \quad (\boldsymbol{W}^{l-1})^T \quad (\boldsymbol{b}^{l-1})^H], \tag{8.3.7}$$

$$\boldsymbol{\Theta}^{O(l-1)} = [(\boldsymbol{V}^{l-1})^H \quad \boldsymbol{b}^{O(l-1)}]^{\mathrm{T}}. \tag{8.3.8}$$

The updating method of parameters in the RNNs is closely related to the optimization algorithm used. The optimization of the RNN model based on the SGD method usually relies on minimizing the objective function through backpropagation [187]. Common objective functions include cross-entropy function, Mean Square Error (MSE) function, and logistic function. We now adopt the objective function based on the MSE criterion, which is calculated by

$$\hat{J}(\boldsymbol{\Theta}^{l-1}) = \frac{\sum_{t=t_0}^{\tau} \left\|\hat{\boldsymbol{y}}^l(k) - \boldsymbol{y}^l(k)\right\|_{\mathrm{F}}^2}{2m(\tau - t_0 + 1)}, \tag{8.3.9}$$

where $\hat{\boldsymbol{y}}^l(k) \in \boldsymbol{R}^{m\times d}$ is the expected output corresponding to $\boldsymbol{x}^l(k)$. $\boldsymbol{\Theta}^H$ denotes the matrix of all parameters in the network and t_0 represents the starting output time step of the output layer. If it is a classification problem, then $t_0 = \tau$. If it is a sequence prediction problem, then $t_0 = 1$.

Let $\hat{\boldsymbol{\nabla}}^{O(l)} = \partial\hat{J}(\boldsymbol{\Theta}^{l-1})/\partial\boldsymbol{\Theta}^{O(l-1)}$, according to Equation (8.3.9) and the *Chain Rule*, $\hat{\nabla}_s^o$ is calculated by

$$\hat{\boldsymbol{\nabla}}^{O(l)} = \sum_{k=t_0}^{\tau} (\boldsymbol{R}^{O(l)}(k))^{\mathrm{T}} \hat{\boldsymbol{\Delta}}^{O(l)}(k), \tag{8.3.10}$$

where $\hat{\Delta}^{o(l)}(k) = \partial\hat{J}(\boldsymbol{\Theta}^{l-1})/\partial\boldsymbol{Z}^{O(l)}(k)$, i.e.,

$$\hat{\Delta}^{O(l)}(k) = \begin{cases} \frac{(\boldsymbol{y}^l(k)-\hat{\boldsymbol{y}}^l(k))\odot\sigma'(Z^{O(l)}(k))}{m(\tau-t_0+1)}, & k \geq t_0 \\ 0, & k < t_0 \end{cases}, \tag{8.3.11}$$

where $\odot$ is the Hadamard product, $Z^{O(l)}(k)$ is the non-activated linear output of the output layer, i.e.,

$$\boldsymbol{Z}^{O(l)}(k) = \boldsymbol{R}^{O(l)}\boldsymbol{\Theta}^{O(l-1)}. \tag{8.3.12}$$

The updating formula of parameters for this layer is given by

$$\Theta^{O(l)} = \Theta^{O(l-1)} - \alpha\hat{\nabla}^{O(l)}, \tag{8.3.13}$$

where α is the learning step size.

Let $\hat{\nabla}^{lH} = \partial J(\boldsymbol{\Theta}^{l-1})/\partial(\boldsymbol{\Theta}^{l-1})^H$, according to the BPTT algorithm and Equation (8.3.13) and the *Chain Rule*, we can obtain

$$\hat{\nabla}^{l\mathrm{H}} = \sum_{t=1}^{\tau} ((\boldsymbol{R}^l(k))^{\mathrm{H}})^{\mathrm{T}}(\hat{\boldsymbol{\Delta}}^l(k))^{\mathrm{H}}, \tag{8.3.14}$$

where $(\ddot{\boldsymbol{\Delta}}^l(k))^{\mathrm{H}} = \partial\hat{J}(\boldsymbol{\Theta})/\partial(\boldsymbol{Z}^l(k))^{\mathrm{H}}$ is the gradient of the objective function for the inactive linear output of the hidden layer, i.e.,

$$(\hat{\boldsymbol{\Delta}}^l(k))^{\mathrm{H}} = \begin{cases} ((\tilde{\boldsymbol{\Delta}}^l(k))^{\mathrm{H}}((\tilde{\boldsymbol{\Theta}}^{l-1})^{\mathrm{H}})^{\mathrm{T}}) \odot g'((\boldsymbol{Z}^l(k))^{\mathrm{H}}) & k < \tau \\ (\hat{\boldsymbol{\Delta}}^{O(l)}(k)(\boldsymbol{V}^{l-1})^T) \odot g'(\boldsymbol{Z}^l(k))^{\mathrm{H}}) & k = \tau \end{cases}, \tag{8.3.15}$$

where $(\tilde{\boldsymbol{\Delta}}^l(k))^{\mathrm{H}} = [\hat{\boldsymbol{\Delta}}^{O(l)}(k), (\hat{\boldsymbol{\Delta}}^l(k+1))^{\mathrm{H}}]$, $(\tilde{\boldsymbol{\Theta}}^{l-1})^{\mathrm{H}} = [\boldsymbol{V}^{l-1}, \boldsymbol{W}^{l-1}]$, $(\boldsymbol{Z}^l(k))^{\boldsymbol{H}}$ is the inactive linear output of the hidden layer, i.e,

$$(\boldsymbol{Z}^l(k))^{\mathrm{H}} = (\boldsymbol{R}^l(k))^{\mathrm{H}}(\boldsymbol{\Theta}^{l-1})^{\mathrm{H}}. \tag{8.3.16}$$

The updating formula of parameters for this layer is given by

$$(\boldsymbol{\Theta}^l)^{\mathrm{H}} = (\boldsymbol{\Theta}^{l-1})^{\mathrm{H}} - \alpha(\hat{\nabla}^l)^{\mathrm{H}}. \tag{8.3.17}$$

8.4 RNNs Based on RLS Optimization

8.4.1 RLS algorithm

The RLS is a recursive algorithm of least squares optimization, which not only has a fast convergence speed but also is suitable for online learning. Assuming that the current training sample input $\boldsymbol{x} = \{x_1, x_2, \ldots, x_N\}$ corresponds to the expected output set $\boldsymbol{y} = \{y_1, y_2, \ldots, y_N\}$. The objective function is usually defined as

$$J(\boldsymbol{w}) = \frac{1}{2}\sum_{n=1}^{N} \lambda^{N-n}(y_n - \boldsymbol{w}^{\mathrm{T}}\boldsymbol{x}_n)^2, \tag{8.4.1}$$

where $\boldsymbol{w}$ is the weight vector and $\lambda \in (0, 1]$ is the forgetting factor.

Let $\nabla_{\boldsymbol{w}} J(\boldsymbol{w}) = 0$, we get

$$\boldsymbol{w}_{\mathrm{opt}} = \sum_{n=1}^{N} \lambda^{N-n}(\boldsymbol{x}_n\boldsymbol{x}_n^{\mathrm{T}})^{-1}\boldsymbol{x}_n y_n. \tag{8.4.2}$$

After organizing, we can obtain

$$\boldsymbol{w} = \boldsymbol{w}_{\mathrm{opt}} = \mathbf{A}^{-1}\boldsymbol{b}, \tag{8.4.3}$$

where

$$\boldsymbol{A} = \sum_{n=1}^{N} \lambda^{N-n}\boldsymbol{x}_n\boldsymbol{x}_n^{\mathrm{T}}, \tag{8.4.4}$$

$$\boldsymbol{b} = \sum_{n=1}^{N} \lambda^{N-n}\boldsymbol{x}_n y_n. \tag{8.4.5}$$

In order to avoid expensive matrix inversion operations and be suitable for online learning, let

$$\boldsymbol{P}(k) = \boldsymbol{A}^{-1}(k). \tag{8.4.6}$$

The recursive formulas for Equations (8.4.5) and (8.4.6) are written as

$$\boldsymbol{A}(k) = \lambda\boldsymbol{A}(k-1) + \boldsymbol{x}(k)\boldsymbol{x}^{\mathrm{T}}(k), \tag{8.4.7}$$

$$\boldsymbol{b}(k) = \lambda\boldsymbol{b}(k-1) + \boldsymbol{x}(k)y(k). \tag{8.4.8}$$

According to the Sherman Morrison Woodbury formula, we have

$$\boldsymbol{P}(k) = \frac{1}{\lambda}\boldsymbol{P}(k-1) - \frac{1}{\lambda}\boldsymbol{G}(k)\boldsymbol{u}^{\mathrm{T}}(k), \tag{8.4.9}$$

where

$$\boldsymbol{u}(k) = \boldsymbol{P}(k-1)\boldsymbol{x}(k), \tag{8.4.10}$$

$$\boldsymbol{G}(k) = \frac{\boldsymbol{u}(k)}{\lambda + \boldsymbol{u}^{\mathrm{T}}(k)\boldsymbol{x}(k)}, \tag{8.4.11}$$

where $\boldsymbol{G}(k)$ is the gain vector. Further substituting Equations (8.4.7), (8.4.8), and (8.4.9) into Equation (8.4.4) to yield

$$\boldsymbol{w}(k) = \boldsymbol{w}(k-1) - \boldsymbol{G}(k)e(k), \tag{8.4.12}$$

where

$$e(k) = (\boldsymbol{w}^{\mathrm{T}}(k-1)\boldsymbol{x}(k) - y(k)). \tag{8.4.13}$$

8.4.2 RNNs based on RLS optimization

Although the RLS algorithm has a fast learning speed, it is only applicable to linear systems. Note that in the RNNs, if the activation function is not considered, the output calculations of the hidden layer and output layer are still linear. In this section, we propose a new mini-batch RLS optimization algorithm based on this characteristic. Assuming that the activation function $\sigma(\cdot)$ of the output layer has an inverse function $\sigma^{-1}(\cdot)$, and following the RLS algorithm, the objective function of the output layer is defined as

$$J(\boldsymbol{\Theta}) = \sum_{n=1}^{l}\sum_{k=t_0}^{\tau} \frac{\lambda^{l-n}\left\|\hat{\boldsymbol{Z}}^{O(n)}(k) - \boldsymbol{Z}^{O(n)}(k)\right\|_{\mathrm{F}}^{2}}{2m(\tau - t_0 + 1)}, \tag{8.4.14}$$

where l denotes the total number of batches of training samples. $\hat{\boldsymbol{Z}}^{O(n)}(k)$ is the non-activated linear expected value of the output layer, i.e.,

$$\hat{\boldsymbol{Z}}^{O(n)}(k) = \sigma^{-1}(\hat{\boldsymbol{y}}^{(n)}(k)). \tag{8.4.15}$$

Therefore, the parameter optimization problem in the RNNs can be described as

$$\boldsymbol{\Theta}^l = \arg\min_{\boldsymbol{\Theta}} J(\boldsymbol{\Theta}). \tag{8.4.16}$$

Since the forward propagation of the RNNs does not involve updates of weight parameters, the forward propagation calculation of our proposed algorithm is basically the same as that of the SGD-RNN, and $\boldsymbol{S}^l(k)$ uses Equation (8.3.3) for calculation. Their only difference is that it does not need to calculate $\boldsymbol{y}^l(k)$ but needs to use Equation (8.3.12) to calculate $\boldsymbol{Z}^{O(l)}(k)$. This section will only consider the derivation of parameter updates for the output and hidden layers of the RLS-RNN.

8.4.2.1 *Parameter update of RLS-RNN output layer*

Let $\nabla_{\boldsymbol{\Theta}^O} = \partial J(\boldsymbol{\Theta})/\partial \boldsymbol{\Theta}^O$; based on Equation (8.4.14) and the *Chain Rule*, it can be concluded that

$$\nabla_{\boldsymbol{\Theta}^O} = \sum_{n=1}^{l}\sum_{k=t_0}^{\tau} (\boldsymbol{R}^{O(l)}(k))^T \Delta^{O(n)}(k), \tag{8.4.17}$$

where $\boldsymbol{\Delta}^{O(n)}(k) = \partial J(\boldsymbol{\Theta})/\partial \boldsymbol{Z}^{O(n)}(k)$, and

$$\Delta^{O(n)}(k) = \begin{cases} \frac{\lambda^{s-n}(\boldsymbol{Z}^{O(n)}(k)-\hat{\boldsymbol{Z}}^{O(n)}(k))}{m(\tau-t_0+1)}, & k \geq t_0 \\ 0, & k < t_0 \end{cases}. \tag{8.4.18}$$

In order to obtain the optimal parameters $\hat{\boldsymbol{\Theta}}^O$, further let $\nabla_{\boldsymbol{\Theta}^O} = 0$, i.e.,

$$\sum_{n=1}^{l}\sum_{k=t_0}^{\tau} (\boldsymbol{R}^{O(n)}(k))^{\mathrm{T}} \Delta^{O(n)}(k) = 0. \tag{8.4.19}$$

By substituting (8.4.18) into Equation (8.4.19), we obtain

$$\sum_{n=1}^{l}\sum_{k=t_0}^{\tau} \frac{\lambda^{l-n}(\boldsymbol{R}^{O(n)}(k))^{\mathrm{T}}(\boldsymbol{R}^{O(n)}(k)\boldsymbol{\Theta}^O - \hat{\boldsymbol{Z}}^{O(n)}(k))}{m(\tau-t_0+1)} = 0. \tag{8.4.20}$$

The least squares solution of $\Theta^{O(l)}$ obtained by organizing is given by

$$\boldsymbol{\Theta}^{O(l)} = \hat{\boldsymbol{\Theta}}^{O} = (\boldsymbol{A}^{O(l)})^{-1}\boldsymbol{B}^{O(l)}, \tag{8.4.21}$$

where

$$\boldsymbol{A}^{O(l)} = \sum_{n=1}^{l}\sum_{k=t_0}^{\tau} \frac{\lambda^{l-n}(\boldsymbol{R}^{O(n)}(k))^{\mathrm{T}}\boldsymbol{R}^{O(n)}(k)}{m(\tau - t_0 + 1)}, \tag{8.4.22}$$

$$\boldsymbol{B}^{O(l)} = \sum_{n=1}^{l}\sum_{k=t_0}^{\tau} \frac{\lambda^{l-n}(\boldsymbol{R}^{O(n)}(k))^{\mathrm{T}}\boldsymbol{Z}^{O*(n)}(k)}{m(\tau - t_0 + 1)}. \tag{8.4.23}$$

Similar to the derivation of the RLS algorithm, the recursive forms of Equations (8.4.22) and (8.4.23) are written as

$$\boldsymbol{A}^{O(l)} = \lambda\boldsymbol{A}^{O(l-1)} + \sum_{k=t_0}^{\tau}\sum_{n=1}^{m} \frac{\boldsymbol{R}^{O(l)}(k,n)(\boldsymbol{R}^{O(l)}(k,n))^{\mathrm{T}}}{m(\tau - t_0 + 1)}, \tag{8.4.24}$$

$$\boldsymbol{B}^{O(l)} = \lambda\boldsymbol{B}^{O(l-1)} + \sum_{k=t_0}^{\tau}\sum_{n=1}^{m} \frac{\boldsymbol{R}^{O(l)}(k,n)((\boldsymbol{Z}^{O(l)}(k,n))^{*})^{\mathrm{T}}}{m(\tau - t_0 + 1)}, \tag{8.4.25}$$

where $\boldsymbol{R}^{O(l)}(k,n) \in \mathbb{R}^{h+1}$ is the nth column vector of $(\boldsymbol{R}^{O(l)}(k))^{\mathrm{T}}$, and $(\boldsymbol{Z}^{O(l)}(k,n))^{*} \in \mathbb{R}^{d}$ is the kth column vector of $((\boldsymbol{Z}^{l}(k))^{*})^{\mathrm{T}}$. However, due to the fact that the RNNs here are trained in minibatches, Equation (8.4.24) cannot directly use the Sherman Morrison Woodbury formula to obtain inverse of $\boldsymbol{A}^{O(l)}$. Considering that the samples $\boldsymbol{\Theta}^{O(l-1)}$, $\boldsymbol{A}^{O(l-1)}$, and $\boldsymbol{B}^{O(l-1)}$ in the same batch are the same, we adopt the idea of using SGD method to calculate the average gradient of minibatches and use the average approximation method to deal with this problem. Equations (8.4.24) and (8.4.25) can be rewritten as

$$\boldsymbol{A}^{O(l)} = \sum_{k=t_0}^{\tau}\sum_{n=1}^{m} \frac{\boldsymbol{A}^{O(l)}(k,n)}{m(\tau - t_0 + 1)}, \tag{8.4.26}$$

$$\boldsymbol{B}^{O(l)} = \sum_{k=t_0}^{\tau}\sum_{n=1}^{m} \frac{\boldsymbol{B}^{O(l)}(k,n)}{m(\tau - t_0 + 1)}, \tag{8.4.27}$$

where

$$\boldsymbol{A}^{O(l)}(k,n) = \lambda \boldsymbol{A}^{O(l-1)} + \boldsymbol{R}^{O(l)}(k,n)(\boldsymbol{R}^{O(l)}(k,n))^{\mathrm{T}}, \quad (8.4.28)$$

$$\boldsymbol{B}^{O(l)}(k,n) = \lambda \boldsymbol{B}^{O(l-1)} + \boldsymbol{R}^{O(l)}(k,n)((\boldsymbol{Z}^{O(l)}(k,n))^{*})^{\mathrm{T}}. \quad (8.4.29)$$

Then, the approximate formula for $(\boldsymbol{A}^{O(l)})^{-1}$ and $\boldsymbol{\Theta}^{O(l)}$ can be written as

$$(\boldsymbol{A}^{O(l)})^{-1} \approx \sum_{k=t_0}^{\tau} \sum_{n=1}^{m} \frac{(\boldsymbol{A}^{O(l)}(k,n))^{-1}}{m(\tau - t_0 + 1)}, \quad (8.4.30)$$

$$\boldsymbol{\Theta}^{O(l)} \approx \sum_{k=t_0}^{\tau} \sum_{n=1}^{m} \frac{(\boldsymbol{A}^{O(l)}(k,n))^{-1} \boldsymbol{B}^{O(l)}(k,n)}{m(\tau - t_0 + 1)}. \quad (8.4.31)$$

Let $\boldsymbol{P}^{O(l)} = (\boldsymbol{A}^{O(l)})^{-1}$, according to Equations (8.4.30) and (8.4.31), and the Sherman Morrison Woodbury formula, the updating formulas after sorting are written as

$$\boldsymbol{P}^{O(l)} \approx \frac{1}{\lambda} \boldsymbol{P}^{O(l-1)} - \sum_{k=t_0}^{\tau} \sum_{n=1}^{m} \frac{\boldsymbol{G}^{O(l)}(k,n)(\boldsymbol{\Lambda}^{O(l)}(k,n))^{\mathrm{T}}}{\lambda m(\tau - t_0 + 1)}, \quad (8.4.32)$$

$$\boldsymbol{\Theta}^{O(l)} \approx \boldsymbol{\Theta}^{O(l-1)} - \sum_{t=t_0}^{\tau} \sum_{n=1}^{m} \boldsymbol{G}^{O(l)}(k,n)(\Delta^{O(l)}(k,n))^{\mathrm{T}}, \quad (8.4.33)$$

where $\Delta^{O(l)}(k,n) \in \mathbb{R}^d$ is the nth column vector of $(\Delta^{O(l)}(k))^{\mathrm{T}}$, and

$$\boldsymbol{\Lambda}^{O(l)}(k,n) = \boldsymbol{P}^{O(l-1)} \boldsymbol{R}^{O(l)}(k,n), \quad (8.4.34)$$

$$\boldsymbol{G}^{O(l)}(k,n) = \frac{\boldsymbol{\Lambda}^{O(l)}(k,n)}{\lambda + (\boldsymbol{\Lambda}^{O(l)}(k,n))^{\mathrm{T}} \boldsymbol{R}^{O(l)}(k,n)}. \quad (8.4.35)$$

8.4.2.2 *Update of parameters of hidden layer in RLS-RNN*

Let $\nabla_{\boldsymbol{\Theta}^H} = \partial J(\boldsymbol{\Theta})/\partial \boldsymbol{\Theta}^H$; based on Equation (8.4.14) and the *Chain Rule*, it can be concluded that

$$\nabla_{\boldsymbol{\Theta}^H} = \sum_{n=1}^{l} \sum_{k=1}^{\tau} (\boldsymbol{R}^{H(n)}(k))^{\mathrm{T}} (\boldsymbol{\Delta}^{(n)}(k))^{H}, \quad (8.4.36)$$

where $(\boldsymbol{\Delta}^{(n)}(k))^H = \partial J(\boldsymbol{\Theta})/\partial(\boldsymbol{Z}^{(n)}(k))^{\mathrm{H}}$; its specific formula based on the BPTT algorithm is written as

$$(\boldsymbol{\Delta}^{(n)}(k))^{\mathrm{H}} = \begin{cases} ((\hat{\boldsymbol{\Delta}}^{(n)}(k))^{\mathrm{H}}((\tilde{\boldsymbol{\Theta}}^{(n-1)})^{\mathrm{H}})^{\mathrm{T}}) \odot g'((\boldsymbol{Z}^{(n)}(k))^{\mathrm{H}}), & k < \tau \\ (\boldsymbol{\Delta}^{O(n)}(k)(\boldsymbol{V}^{(n-1)})^{\mathrm{T}} \odot g'(\boldsymbol{Z}^{\mathrm{H}}(k)), & k = \tau \end{cases}, \tag{8.4.37}$$

where $(\hat{\boldsymbol{\Delta}}^{(n)}(k))^{\mathrm{H}} = [\boldsymbol{\Delta}^{O(n)}(k), (\boldsymbol{\Delta}^{(n)}(k+1))^{\mathrm{H}}]$, further let $\nabla_{\boldsymbol{\Theta}^H} = 0$, we obtain

$$\sum_{n=1}^{l}\sum_{k=1}^{\tau}((\boldsymbol{R}^{(n)}(k))^{\mathrm{H}})^{\mathrm{T}}(\Delta^{(n)}(k))^{\mathrm{H}} = 0. \tag{8.4.38}$$

However, Equation (8.4.37) is very complex and $g'(\boldsymbol{Z}^{(l)\mathrm{H}})$ is generally nonlinear, so Equation (8.4.37) cannot be substituted into Equation (8.4.38) to obtain the least squares solution of parameter $\boldsymbol{\Theta}^{\mathrm{H}}$ in the hidden layer. Next, a new method is proposed to derive the equivalent form of $\Delta_n^{\mathrm{H}}(k)$; in order to obtain the least square solution of $\boldsymbol{\Theta}^{\mathrm{H}}$, a new objective function $J^{\mathrm{H}}(\boldsymbol{\Theta}^{\mathrm{H}})$ of the hidden layer is defined as

$$J^H(\boldsymbol{\Theta}^H) = \sum_{n=1}^{l}\sum_{k=1}^{\tau}\frac{\lambda^{l-n}}{2m\tau}\left\|\hat{\boldsymbol{Z}}^{(n)\mathrm{H}}(k) - \boldsymbol{Z}^{(n)\mathrm{H}}(k)\right\|_{\mathrm{F}}^2, \tag{8.4.39}$$

where $\hat{\boldsymbol{Z}}^{(n)\mathrm{H}}(k)$ is the expected value of the non-activated linear output of the layer. Obviously, if $J(\boldsymbol{\Theta}) \to 0$, then $J^{\mathrm{H}}(\boldsymbol{\Theta}^{\mathrm{H}}) \to 0$, i.e.,

$$\boldsymbol{\Theta}_{\mathrm{opt}}^{(l)\mathrm{H}} = \arg\min_{H} J^{\mathrm{H}}(\boldsymbol{\Theta}^{\mathrm{H}}). \tag{8.4.40}$$

Let $\partial J^{\mathrm{H}}(\boldsymbol{\Theta}^{\mathrm{H}})/\partial\boldsymbol{\Theta}^{\mathrm{H}} = 0$; we obtain

$$\sum_{n=1}^{l}\sum_{k=1}^{\tau}\frac{\lambda^{l-n}}{m\tau}(\boldsymbol{R}^{(n)\mathrm{H}}(k))^{\mathrm{T}}(\boldsymbol{Z}^{(n)\mathrm{H}}(k) - \hat{\mathbf{Z}}^{(n)\mathrm{H}}(k)) = 0. \tag{8.4.41}$$

Comparing Equation (8.4.38) with Equation (8.4.41), another equivalent definition of $\boldsymbol{\Delta}^{(n)\mathrm{H}}(k)$ is given by

$$\boldsymbol{\Delta}^{(n)\mathrm{H}}(k) = \frac{\eta\lambda^{l-n}}{m\tau}(\boldsymbol{Z}^{(n)\mathrm{H}}(k) - \hat{\mathbf{Z}}^{(n)\mathrm{H}}(k)), \tag{8.4.42}$$

where η is the scaling factor. In theory, there should be some differences in η corresponding to different mini-batch data. However, considering that each batch of minibatch data is randomly selected from the entire training set, this difference may be ignored. From Equation (8.3.12), we can know $\boldsymbol{Z}^{(n)\mathrm{H}}(k) = \boldsymbol{R}^{(n)\mathrm{H}}(k)\boldsymbol{\Theta}^{\mathrm{H}}$. After Equation (8.4.42) is substituted into Equation (8.4.38), we obtain

$$\sum_{n=1}^{l}\sum_{l=1}^{\tau}\frac{\eta\lambda^{l-n}}{m\tau}(\boldsymbol{R}^{(n)\mathrm{H}}(k))^{\mathrm{T}}(\boldsymbol{R}^{(n)\mathrm{H}}(k)\boldsymbol{\Theta}^{\mathrm{H}} - \hat{\boldsymbol{Z}}^{(n)\mathrm{H}}(k)) = 0. \quad (8.4.43)$$

Further sorting, the least square solution of $\boldsymbol{\Theta}^{(l)\mathrm{H}}$ is obtained by

$$\boldsymbol{\Theta}^{(l)\mathrm{H}} = \boldsymbol{\Theta}^{(l)\mathrm{H}} = (\boldsymbol{A}^{(l)\mathrm{H}})^{-1}\boldsymbol{B}^{(l)\mathrm{H}}, \quad (8.4.44)$$

where

$$\boldsymbol{A}^{(l)\mathrm{H}} = \sum_{n=1}^{l}\sum_{k=1}^{\tau}\frac{\eta\lambda^{L-n}}{m\tau}(\boldsymbol{R}^{(n)\mathrm{H}}(k))^{\mathrm{T}}\boldsymbol{R}^{(n)\mathrm{H}}(k), \quad (8.4.45)$$

$$\boldsymbol{B}^{(l)\mathrm{H}} = \sum_{n=1}^{l}\sum_{k=1}^{\tau}\frac{\eta\lambda^{l-n}}{m\tau}(\boldsymbol{R}^{(n)\mathrm{H}}(k))^{\mathrm{T}}\boldsymbol{Z}^{(n)H}(k). \quad (8.4.46)$$

The derivation process of the recursive least square solution for Equation (8.4.44) is similar to that of parameter update in the output layer. Let $\boldsymbol{P}^{(l)\mathrm{H}} = (\boldsymbol{A}^{(l)\mathrm{H}})^{-1}$, and based on the approximate average solution method mentioned earlier, we obtain

$$\boldsymbol{P}^{(l)\mathrm{H}} \approx \frac{1}{\lambda}(\boldsymbol{P}^{l-1})^{\mathrm{H}} - \frac{1}{\lambda m\tau}\sum_{k=1}^{\tau}\sum_{n=1}^{m}\boldsymbol{G}^{(l)\mathrm{H}}(k,n)(\boldsymbol{\Lambda}^{(l)\mathrm{H}}(k,n))^{\mathrm{T}}, \quad (8.4.47)$$

$$\boldsymbol{\Theta}^{(l)\mathrm{H}} \approx (\boldsymbol{\Theta}^{(l-1)})^{\mathrm{H}} - \frac{1}{\eta}\sum_{k=1}^{\tau}\sum_{n=1}^{m}\boldsymbol{G}^{(l)\mathrm{H}}(k,n)(\Delta^{(l)\mathrm{H}}(k,n))^{\mathrm{T}}, \quad (8.4.48)$$

where $\boldsymbol{\Delta}^{(l)\mathrm{H}}(k,n) \in \mathbb{R}^{h}$ is the nth column vector of $(\boldsymbol{\Delta}^{(l)\mathrm{H}}(k))^{\mathrm{T}}$, and

$$\boldsymbol{\Lambda}^{(l)\mathrm{H}}(k,n) = (\boldsymbol{P}^{(l-1)})^{\mathrm{H}}\boldsymbol{R}^{(l)\mathrm{H}}(k,n), \quad (8.4.49)$$

$$\boldsymbol{G}^{(l)\mathrm{H}}(k,n) = \frac{\boldsymbol{\Lambda}^{(l)\mathrm{H}}(k,n)}{\frac{\lambda}{\eta} + (\boldsymbol{\Lambda}^{(l)\mathrm{H}}(k,n))^{\mathrm{T}}\boldsymbol{R}^{(l)\mathrm{H}}(k,n)}. \quad (8.4.50)$$

It should be noted that since the expected output $\hat{\boldsymbol{Z}}^{(l)\mathrm{H}}(k)$ of the hidden layer is not known, $\boldsymbol{\Delta}^{(l)\mathrm{H}}(k)$ cannot actually be obtained through Equation (8.4.42). Fortunately, Equation (8.3.14) is equivalent to Equation (8.4.41), so in the specific implementation of the algorithm, Equation (8.4.37) will be used to replace Equation (8.4.42).

Overall, the RLS-RNN architecture may be shown in Figure 8.4.

8.4.3 Improvement of RLS-RNN

8.4.3.1 *λ adaptive adjustment*

Numerous studies have shown that the value of the forgetting factor λ has a significant impact on the performance of the RLS algorithm, especially when dealing with time-varying tasks in the RLS algorithm. Due to the fact that the RLS-RNN is built on the basis of traditional RLS algorithm, its convergence quality is also easily affected by the value of λ. Here, an adaptive adjustment method of λ is directly presented. For the lth mini-batch sample, the forgetting factor used by the RLS-RNN is defined as

$$\lambda^{(l)} = \begin{cases} \lambda_{\max}, & \text{if } \sigma^{(l)e} \leq \kappa\sigma^{(l)v} \\ \min\left(\frac{q^{(l)}\sigma^{(l)v}}{\xi+\left|\sigma^{(l)c}-\sigma^{(l)v}\right|}, \lambda_{\max}\right), & \text{otherwise,} \end{cases} \tag{8.4.51}$$

where $\lambda_{\max}$ is close to 1 and $\kappa > 1$ is used to control update of $\lambda^{(l)}$ and generally recommended to take 2. Usually, the smaller the κ value, the more frequent the $\lambda^{(l)}$ updates; ξ is a very small constant to prevent the denominator from being zero during calculation; $q^{(l)}$, $\sigma^{e(l)}$, and $\sigma^{(l)v}$ are calculated by

$$q^{(l)} = \mu_0 q^{(l-1)} + \frac{(1-\mu_0)\sum\limits_{k=t_0}^{\tau}\sum\limits_{n=1}^{m}(\mathbf{\Lambda}^{O(l)}(k,n))^{\mathrm{T}}\boldsymbol{R}^{O(l)}(k,n)}{m(\tau-t_0+1)},$$

$$\sigma^{e(l)} = \mu_1\sigma^{e(l-1)} + \frac{(1-\mu_1)\sum\limits_{k=t_0}^{\tau}\sum\limits_{n=1}^{m}\left\|\mathbf{\Delta}^{O(l)}(k,n)\right\|_2^2}{\mathrm{md}(\tau-t_0+1)},$$

$$\sigma^{v(l)} = \mu_2\sigma^{v(l-1)} + \frac{(1-\mu_2)\sum\limits_{k=t_0}^{\tau}\sum\limits_{n=1}^{m}\left\|\mathbf{\Delta}^{O(l)}(k,n)\right\|_2^2}{\mathrm{md}(\tau-t_0+1)},$$

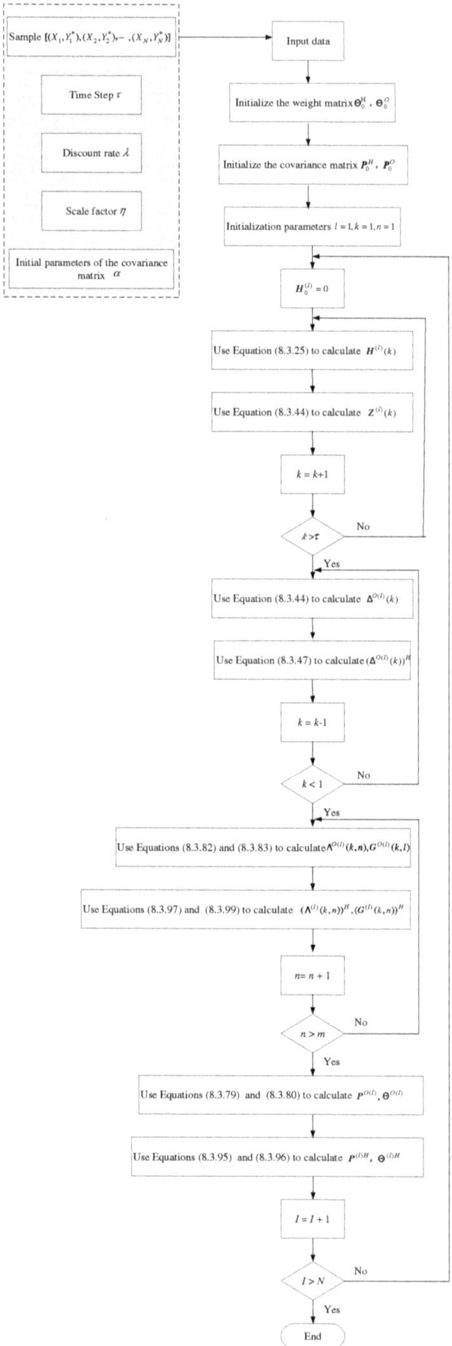

Figure 8.4. Flow of R LS-RNN.

where it is recommended to take μ_0 as 7/8. Usually, $\zeta_1 \geq 2$, $\mu_1 = 1 - 1/(\zeta_1 m)$, and $\zeta_2 > \zeta_1$. Of course, the above method, used to update $\lambda^{(l)}$, will introduce new hyperparameters, which will bring certain difficulties to the debugging of the RLS-RNN. A fixed λ for training is also a feasible approach.

8.4.3.2 *Prevention of overfitting*

Although traditional RLS algorithms converge quickly, they often face the risk of overfitting, and the RLS-RNN also faces this risk. Similarly, some research results on this issue in the field of the RLS algorithm can be utilized to improve the RLS-RNN.

A traditional RLS algorithm proposed by Eksioglu may be added a regularization term during parameter updates. Based on this idea, Equations (8.4.33) and (8.4.48) can be slightly improved by redefining them as

$$\boldsymbol{\Theta}^{O(l)} \approx \boldsymbol{\Theta}^{O(l-1)} - \sum_{k=1}^{\tau} \boldsymbol{G}^{O(l)}(k)\boldsymbol{\Delta}^{O(l)}(k) - \gamma \boldsymbol{P}^{O(l-1)}\mathrm{sgn}(\boldsymbol{\Theta}^{O(l-1)}),$$

$$\boldsymbol{\Theta}^{(l)H} \approx (\boldsymbol{\Theta}^{l-1})^{H} - \frac{1}{\eta}\sum_{k=t_0}^{\tau} \boldsymbol{G}^{(l)H}(k)\Delta^{(l)H}(k) - \gamma(\boldsymbol{P}^{(l-1)})^{H}\mathrm{sgn}((\boldsymbol{\Theta}^{(l-1)})^{H}),$$

where γ is the regularization factor. $\boldsymbol{G}^{O(l)}(k) = [\boldsymbol{G}^{O(l)}(k,1), \boldsymbol{G}^{O(l)}(k,2), \ldots, \boldsymbol{G}^{O(l)}(k,m)]$ and $\boldsymbol{G}^{(l)\mathrm{H}}(k) = [\boldsymbol{G}^{(l)\mathrm{H}}(k,1), \boldsymbol{G}^{(l)\mathrm{H}}(k,2), \ldots, \boldsymbol{G}^{(l)\mathrm{H}}(k,m)]$.

In fact, besides this method, readers can also use other regularization methods to further improve the RLS-RNN.

8.5 Example 8: A Non-Invasive Load Identification Method Based on Correlated RNN Model

Traditional load monitoring usually involves installing an intelligent sensor at each load component or electrical connection to monitor its working status, which is not user-friendly and costly. However, non-invasive electric load monitoring (NILM) is an economical and convenient monitoring method that replaces sensor networks with software programs and uses "soft computing" at load checkpoints instead of "hard measurement" near devices. It only requires the installation of monitoring equipment at the distribution line to monitor load types, electricity consumption details, and other information. It is an inevitable trend for deepening the reform of smart meters. This not only provides users with detailed electricity usage information but also provides data support for power companies to accurately predict the proportion of load composition, which is of great significance for promoting interaction between power companies and users.

In recent years, various load identification algorithms have emerged in research on the NILM technology, such as Karhunen–Loéve transform algorithm, group search machine learning algorithm, cepstral smoothing load decomposition method, heuristic NILM method, Bayesian NILM method, load identification algorithm based on improved Hidden Markov Model (HMM), and ant colony optimization algorithm based on improved time–frequency analysis NILM system. Among these algorithms, the research on supervised learning algorithms is still the focus on designing load identification algorithm. This section introduces the application of the RNN model to solve the problem of non-invasive power load identification [188]. Through its characteristics of remembering historical input feature quantities, an intrinsic correlation between input and output is established to achieve load identification of time series inputs.

8.5.1 Load identification method for associated RNN models

8.5.1.1 *Load identification model*

In non-invasive load monitoring, as the power data monitored at the main distribution line are also a type of time series data, the correlation RNN model can effectively establish intrinsic correlations based on the characteristics of power load data to achieve accurate identification and decomposition of power loads.

Figure 8.5 shows the entire load identification process, in which load event detection is an important step in the pre-processing of the RNN model identification.

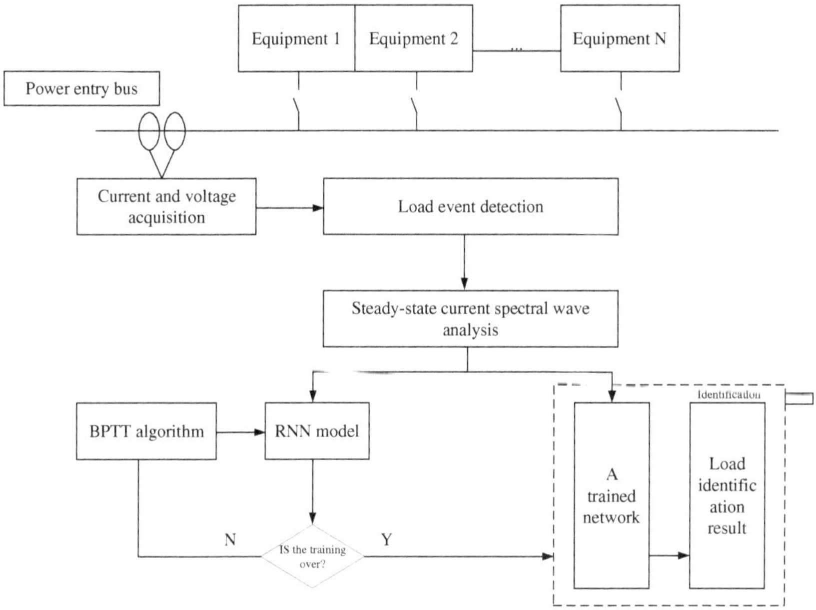

Figure 8.5. Load identification architecture associated with RNN model.

In Figure 8.5, the RNN model adopts the model described in Section 8.1 and uses gradient descent method to train the learning ability of the network at each level.

8.5.1.2 *Load event detection*

Load event detection is an indicator that reflects load switching and is a crucial step in load identification. An adaptive changing point optimization algorithm for load event detection is adopted, whose basic idea is to divide a time window of a certain length into two categories based on a certain point in time. The previous point in time is classified as one category, and the subsequent point in time is considered as one category, so as to minimize the sum of variances within the same category and maximize the sum of variances between different categories. At this point, if the average distance between the two types exceeds a certain threshold, it is considered a changing point and there is a load event.

For ease of description, assuming that at a certain time point k, the L data samples $\{x_i\}$ within the time window are divided into two categories, the probabilities of C_0 categories $\{x_1, x_2, \ldots, x_k\}$ and C_1 categories $\{x_{k+1}, x_{k+2}, \ldots, x_L\}$ are defined as, respectively,

$$p_r(C_0) = \sum_{i=1}^{k} p_i = p_{rk}, \tag{8.5.1}$$

$$p_r(C_1) = \sum_{i=k+1}^{L} p_i = 1 - p_{rk}. \tag{8.5.2}$$

Their mean values are defined as

$$m_0 = \sum_{i=1}^{k} x_i p_r(x_i|C_0), \tag{8.5.3}$$

$$m_1 = \sum_{i=k+1}^{L} x_i p_r(x_i|C_1), \tag{8.5.4}$$

where p_i is the probability of x_i appearing in dataset $\{x_i\}$ and p_i satisfies $\sum_{i=1}^{L} p_i = 1$.

According to the classification rules, the probabilities of the samples belonging to the class C_0 and the class C_1 can be calculated, respectively, as follows:

$$p_r(x_i|C_0) = p_i \bigg/ \sum_{i=1}^{k} p_i = p_i/p_{rk}, \tag{8.5.5}$$

$$p_r(x_i|C_1) = p_i \bigg/ \sum_{i=k+1}^{L} p_i = p_i/(1 - p_{rk}). \tag{8.5.6}$$

Therefore, when the objective function reaches its minimum, as shown in Equation (8.5.7), potential changing points can be obtained. Combined with the difference between the means of two types, when the difference exceeds a certain threshold, it is considered that there is a load event, and this point is also called a changing point. Generally speaking, this threshold is determined by the load with the minimum power:

$$\min\left[\sum_{i=1}^{k}(x_i - m_0)^2 p_r(x_i|C_0) + \sum_{i=k+1}^{L}(x_i - m_1)^2 p_r(x_i|C_1)\right]. \tag{8.5.7}$$

8.5.1.3 *Feature extraction*

Accurate and reliable non-invasive load identification methods are usually based on the selection of load characteristics of electrical equipment. Usually, the steady-state current of electrical equipment during normal operation has certain statistical characteristics, which are manifested in the harmonic components of the current.

For ease of description, assuming that there are N electrical devices in the user's household, the established equations during the switching process of the electrical devices are given by

$$\begin{cases} I_1 = \sum_{i=1}^{K}\sum_{n=1}^{M} I_{1n}\cos(i\omega t + \theta_{1i}) \\ I_2 = \sum_{i=1}^{K}\sum_{n=1}^{M} I_{2n}\cos(i\omega t + \theta_{2i}) \\ \vdots \\ I_N = \sum_{i=1}^{K}\sum_{n=1}^{M} I_{Nn}\cos(i\omega t + \theta_{Ni}) \end{cases}, \tag{8.5.8}$$

where I_{ij} is the amplitude of the current I_i of the ith branch decomposed into j harmonics. Additionally, each item $\cos(k\omega t + \theta_k)$ can be written as a combination of $\{\cos(k\omega t), \sin(k\omega t)\}$, $L = 1, 2, \ldots, K$, i.e.,

$$\cos(k\omega t + \theta_k) = s_{ik}\cos(k\omega t) + t_{ik}\sin(k\omega t), \tag{8.5.9}$$

where ω is the fundamental angular frequency, k represents the harmonic order, and s_{ik} and t_{ik} represent coefficients, respectively. In practical engineering projects, due to sampling frequency constraints, only the first k harmonic components are usually taken.

According to *Kirchhoff's law*, the current at an inlet node is obtained by linearly superimposing the currents flowing through each branch. Therefore, in the process of load identification, the change information of harmonic components before and after the changing point is an important characteristic of load switching.

8.5.1.4 *RNN model-based load identification architecture*

After load event detection and feature extraction, the RNN for non-invasive load decomposition model may be designed based on the time series characteristics of load data. Its main architecture is as follows:

Step 1: Prepare training and testing sample data

When preparing training samples, the sampling device obtains harmonic data at intervals of 2 seconds, and the switch of the equipment is in the state corresponding to Table 8.1, where N represents the type of load equipment. In different combinations, the order of devices should not be considered in principle. Second, there is no situation where two devices are turned on at the same time, i.e., after one device is turned on, if the next device needs to be turned on, it is necessary to wait for 1 minute or wait for it to run to steady state, and the data is saved in the MySQL database.

Step 2: Determine the number of neurons in each layer

When the RNN is applied to load identification, the quality of the identification results is determined by the number of neurons and layers of network. The number of nodes in input layer is determined by the dimension of the input load feature vector, while the dimension of the corresponding expected load label vector determines the number of output nodes. The selection of the number of neurons in the hidden layer is usually related to engineering practice. Generally, the fewer neurons there are, the worse the identification performance

Table 8.1. Meaning of output layer variables.

	Meaning of output layer variables						
Work state combination	Electric heating system	Refrigerator	TV	Air conditioner	Rice cooker	Disinfection cabinet	Hot water kettle
S_1	0	0	0	0	0	0	1
S_2	0	0	0	0	0	1	0
S_3	0	0	0	0	0	1	1
$\vdots$	$\vdots$	$\vdots$	$\vdots$	$\vdots$	$\vdots$	$\vdots$	$\vdots$
S_{2^N-1}	1	1	1	1	1	1	1

under mixed features. As the number of neurons increases, the recognition effect will gradually improve, but if it exceeds a certain number, the recognition effect will tend to stabilize. Based on practical experience, it is generally recommended to choose twice the number of input neurons or more.

Based on the combination of electrical working states, 16 neurons of the input layer are selected and input temporal variables are denoted as $\{x_1, x_2, \ldots, x_T\}$, where the input time series of neurons at the sampling time t is recorded as $x_t = [I_L^t(1), I_L^t(2), \ldots, I_L^t(16)]$.

Step 3: Define output layer variables

Experiments on 7 representative types of electrical appliances were conducted, 7 neurons in the output layer were selected, and the output variables were defined according to Table 8.1. Among them, "0" indicates that the appliance is in a stopped state, and "1" indicates that the appliance is in a running state.

Step 4: Determine the number of network's layers

For each load feature input signals corresponding to each load label, the input data at each sampling time point corresponds to a layer of the network. The hidden layer has the function of recognizing details in the learning process of neural networks, which can better distinguish categories. Therefore, relatively speaking, the more layers there are, the higher the recognition accuracy, but the training process is also more complex. Usually, the number of hidden layers is related to the input feature category, and in the case of a single feature, one hidden layer is generally set.

Step 5: Determine the activation functions for each layer

In neural networks, the activation function represents the mapping relationship between nodes in the network that produce corresponding outputs for a given input. Due to the highly unstable convergence characteristics of network models constructed by binary and linear functions, standardized shape activation functions such as the Softsign function are commonly used to solve these issues.

The Softsign function is defined as

$$f(x) = \frac{1}{1 + \exp\{-\lambda(x - b)\}}, \tag{8.5.10}$$

where the parameters $\lambda = 1$ and $b = 0$.

Step 6: Determine the loss function

The loss function is directly related to the network's ability to correct weights, thereby it determines the learning capability of the RNN. To avoid problems on $g'(\cdot)$ and $f'(\cdot)$ such as gradient vanishing, the cross-entropy loss function is chosen to minimize the deviation between the output of the training sample set and the expected value, i.e.,

$$L(y_t, \hat{y}_t) = -\frac{1}{N}\sum_{t=1}^{T}\sum_{j=1}^{N}[y_t(j) \ln y_t(j) + (1 - \hat{y}_t(j) \ln (1 - y_t(j)))]. \tag{8.5.11}$$

8.5.2 Simulation experiments and result analyses

8.5.2.1 *Experiments on load event detection*

The voltage and current signals collected at a certain energy research center were used as experimental data. Figure 8.6 shows the changes in effective current values over a certain period of time, including the switching of electrical equipment. Figure 8.6 shows that there

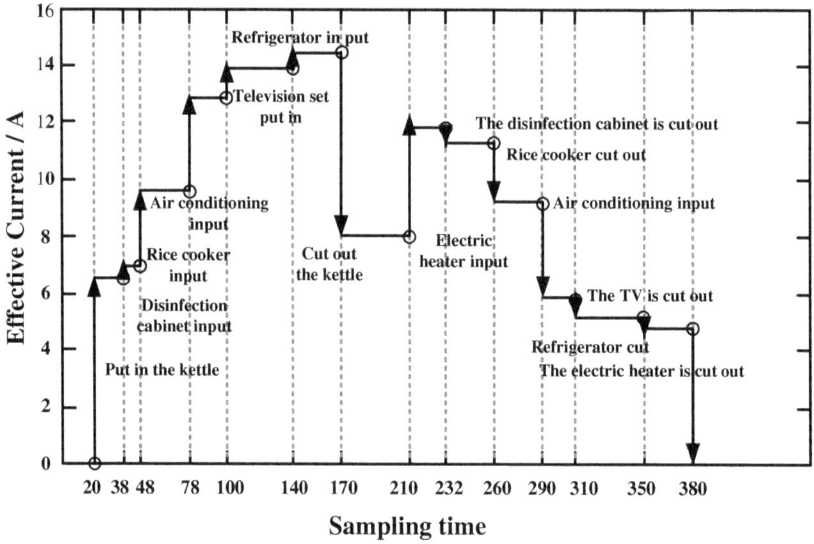

Figure 8.6. Time series of effective current values.

Table 8.2. Detection results of changing points.

Window size	Detection results	Missed or incorrectly detected
4~28	20, 38, 48, 78, 100, 140, 170, 210, 232,260, 260, 310, 350, 380	0
26~50	20, 48, 78, 100, 140, 170, 210, 232,260, 260, 310, 350, 380	1
51~52	20, 48, 78, 100, 140, 170, 210, 260,260, 310, 350, 380	2
53~70	20, 48, 78, 100, 140, 170, 210, 260,260, 310, 380	3

are 14 changing points in total, including 20, 38, 48, 78, 100, 140, 170, 210, 232, 260, 290, 310, 350, and 380, when actual load events occur. Table 8.2 presents the detection results obtained at different time windows and indicates that the proposed method in this section can effectively capture the occurrence points of load events in shorter time windows. Usually, the window size is set to 10.

When a mutation point is detected, it indicates that the electrical equipment has undergone input or disconnection operations. At this point, the load characteristics of the steady-state section are extracted and the trained RNN model is used to identify the load. The entire process includes two stages: training and testing.

8.5.2.2 *Offline training*

A total of 27 groups of data were collected for 7 electrical devices, with sampling time intervals of 2 seconds, and recorded in the MySQL database. Each group corresponds to the working state combination in Table 8.1, and 40 time points from each group were selected as training samples. An example of the training model is shown in Figure 8.7.

This training model characterized by harmonic components makes each group of training samples relatively independent, so only the samples of this category need to be considered during training, thereby improving learning speed.

The specific training steps of the entire RNN model are as follows:

Step 1: Initialize parameters

The learning step size was initialized to 0.01, the input dimension was 16, the number of hidden layer nodes was set to 32, and the

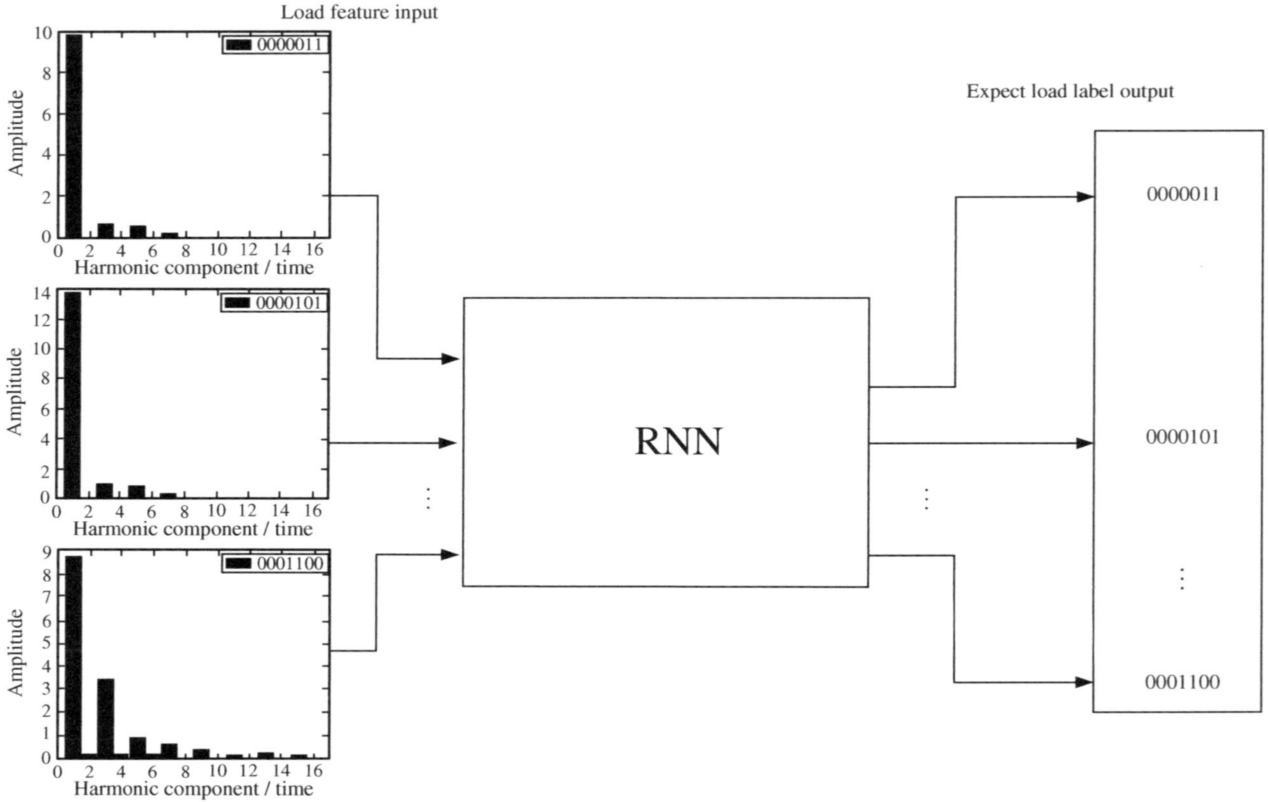

Figure 8.7. Training framework based on features of harmonic components.

number of output layer nodes was equal to 7. All elements of the weight parameter matrix were initialized to the value of $(-1, 1)$, and the error limit *err*_gate was set to 1×10^{-4}.

Step 2: Training

The RNN training process was evaluated based on the variance between the actual and expected values outputted by the network:

$$\text{error} = \sum_{t}^{T} \sum_{j}^{N} (y_t(j) - y_t(j))^2. \quad (8.5.12)$$

When the maximum number of iterations was set to 1000, the average error eventually reached 2.3308×10^{-4}, and the training process ends.

Figure 8.8 shows the variation curve of the loss function with the number of iterations during the training process. Figure 8.8 indicates that when the sigmoid function is used as the activation function, the output quickly saturates, making the RNN appear to have gradient vanishing and a slower convergence speed.

Figure 8.9 shows the convergence process of the RNN when the Softsign function is used as the activation function and indicates that the RNN has a faster convergence speed and does not encounter gradient vanishing problems. Specifically, the average error curve when

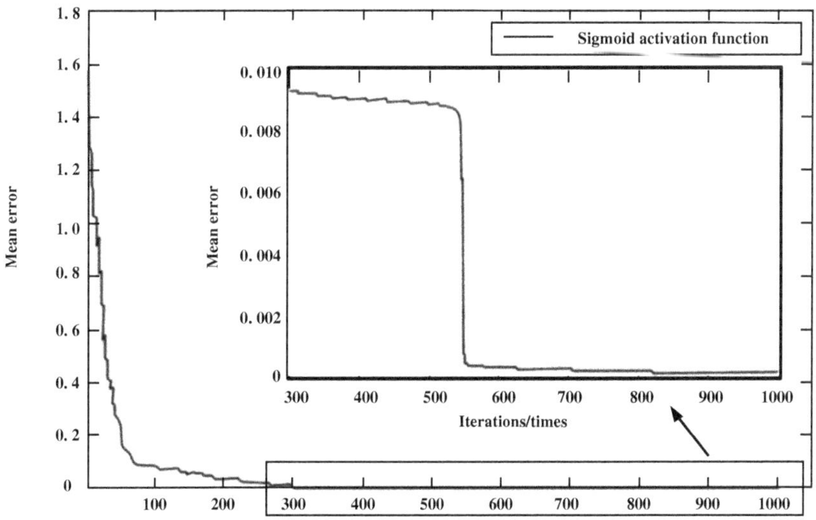

Figure 8.8. Average error curve.

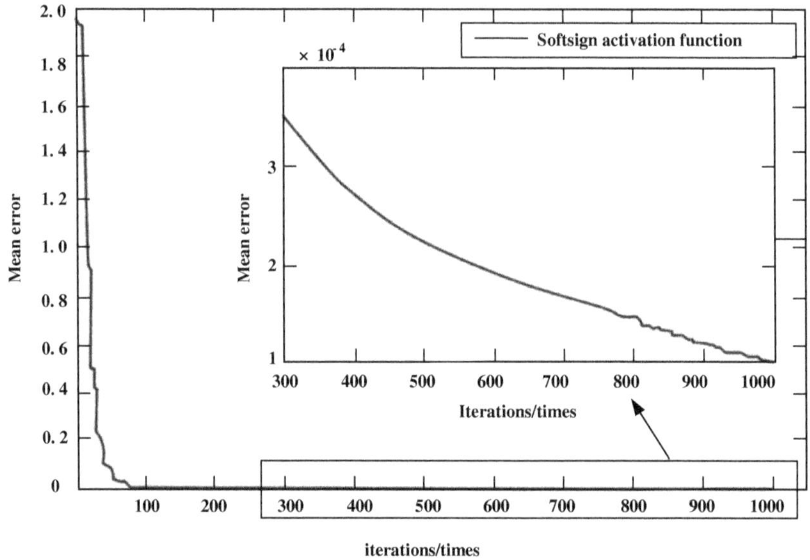

Figure 8.9. Average error curves with softsign.

using Cross-Entropy Error (CEE) as the loss function is shown in Figure 8.10. Figure 8.10 shows that the CEE loss function can effectively solve the "gradient vanishing" problem that occurs during the RNN training and converges quickly.

8.5.2.3 *Online identification*

In order to achieve better statistical identification accuracy, during the actual load identification testing process, the load data was first recorded in the MySQL database. The sampling time points were spaced every 2 seconds, and the electrical devices with a status combination of 1 were turned on at 1-minute intervals in sequence to ensure that load events were detected. When all electrical devices with a status combination of 1 were turned on for more than 20 seconds, they should be turned off sequentially. The data were obtained from 10 time points when the electrical equipment was fully open and identified. The final identification accuracy of the testing sample set is shown in Table 8.3. Table 8.3 shows that the accuracy is higher when there are fewer combinations of electrical equipment. When there are many devices, some identification errors may occur.

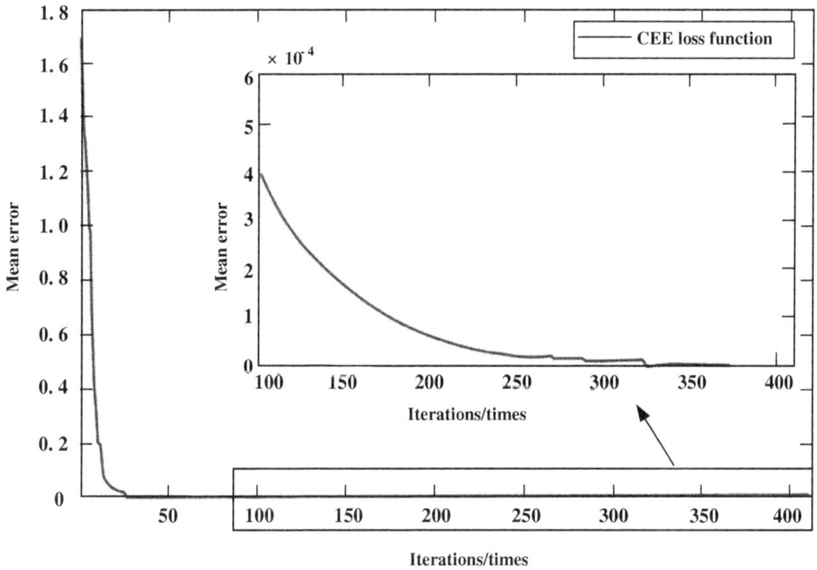

Figure 8.10. Average error curves with loss function.

Table 8.3. Identification accuracy ($H = 32$).

Equipment status combination	Sample capacity	Test sample	Correct identification sample	Identification accuracy %
1 device	$C_7^1 = 7$	70	70	100
2 devices	$C_7^2 = 21$	210	210	100
3 devices	$C_7^3 = 35$	350	328	63.14
4 devices	$C_7^4 = 35$	350	251	71.71
5 devices	$C_7^5 = 21$	210	146	70.65
6 devices	$C_7^6 = 7$	70	53	75.71
7 devices	$C_7^7 = 1$	10	10	100
Total		1270	1071	84.33

Its main reason may be the similarity of harmonic characteristics of electrical equipment after superposition, which causes feature aliasing and leads to identification errors in the RNN model.

For this purpose, load identification was conducted in Ref. [188] by increasing the number of nodes in different hidden layers. Figure 8.11 shows the identification results for nodes with

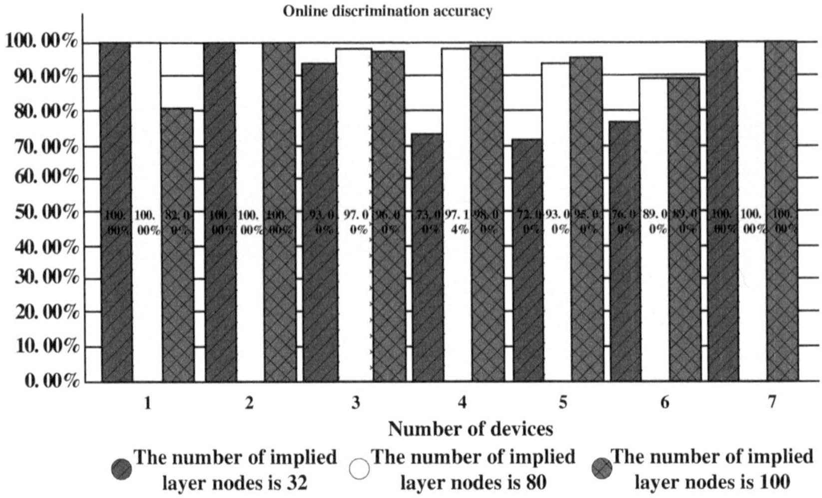

Figure 8.11. Comparison of the number of nodes in different hidden layers.

32, 80, and 100, respectively. Figure 8.11 shows that as the number of hidden layer nodes increases, the accuracy of load identification will improve to a certain extent, but when the number of nodes increases to a certain extent, saturation occurs, i.e., the accuracy no longer increases significantly. Therefore, in practical situations, it is necessary to further adjust the number of hidden layer nodes based on experiments.

In addition, the addition of the hidden layers can enhance the load identification capability of the RNN models to some extent. Table 8.4 presents the accuracy of load identification for different sample sizes under two hidden layers and 32 nodes. Compared with a single-layer hidden layer, effective discrimination can be obtained in the case of multiple device mixing, which improves the accuracy of load identification by 10%. In theory, multi-layer hidden layers can learn the detailed information of samples; thus, high ability to identify load events can be achieved.

Table 8.5 shows the actual output of the RNN model corresponding to Figure 8.7. Table 8.5 shows that the proposed method in Ref. [188] can effectively identify the working state combination of electrical equipment at the changing point. Therefore, in the subsequent steady-state operation stage, the identified load category will

Table 8.4. Identification accuracy ($H_1 = 32, H_2 = 32$).

Equipment status combination	Sample capacity	Test sample	Correct identification sample	Identification accuracy %
1 device	$C_7^1 = 7$	70	70	100
2 devices	$C_7^2 = 21$	210	210	100
3 devices	$C_7^3 = 35$	350	336	66
4 devices	$C_7^4 = 35$	350	304	86.86
5 devices	$C_7^5 = 21$	210	164	62.38
6 devices	$C_7^6 = 7$	70	62	88.57
7 devices	$C_7^7 = 1$	10	10	100
Total		1270	1186	63.63

Table 8.5. Identification results of the example.

Sample time		Electric heating system	Refrigerator	TV	Air-conditioning	Rice cooker	Disinfection cabinet	Kettle
20	$\hat{y}$	0	0	0	0	0	0	1
	y	0.001205	0.000307	0.005265	0.001126	0.004223	0.001206	0.664531
38	$\hat{y}$	0	0	0	0	0	1	1
	y	0.001416	0.000253	0.001616	0.001267	0.003271	0.66445	0.661556
48	$\hat{y}$	0	0	0	0	1	1	1
	y	0.001632	0.000477	0.005411	0.001586	0.666882	0.664633	0.670042
78	$\hat{y}$	0	0	0	1	1	1	1
	y	7e-05	0.00031	0.00388	0.666877	0.667634	0.666264	0.658413
100	$\hat{y}$	0	0	1	1	1	1	1
	y	0.000164	8.15e-05	0.666652	0.67776	0.664674	0.661156	0.641676
140	$\hat{y}$	0	1	1	1	1	1	1
	y	7.66e-06	0.666668	0.666027	0.667537	0.668723	0.665403	0.003218
170	$\hat{y}$	0	1	1	1	1	1	1
	y	1.08e-05	0.666201	0.666158	0.668681	0.664686	0.662415	0.002015
210	$\hat{y}$	1	1	1	1	1	1	1
	y	0.668472	0.667506	0.668666	0.683888	0.662722	0.664078	0.000243
232	$\hat{y}$	1	1	1	1	1	1	1
	y	0.668268	0.667474	0.66863	0.682728	0.661827	0.007128	0.000276
260	$\hat{y}$	1	1	1	1	1	1	1
	y	0.666035	0.666102	0.667474	0.666623	0.007883	0.006447	0.000718
260	$\hat{y}$	1	1	1	1	1	1	1
	y	0.66666	0.66644	0.663365	0.001224	0.001135	0.006765	1.01e-05
310	$\hat{y}$	1	1	1	0	0	0	0
	y	0.666883	0.661761	0.001574	0.006276	0.002373	0.004667	0.000137
350	$\hat{y}$	1	0	0	0	0	0	0
	y	0.667846	0.0063	0.001874	0.008075	0.000651	0.006455	0.002766

be consistent with the real load switching category, which can potentially improve the accuracy of load identification.

To further validate the effectiveness of the method in Ref. [188], a comparison was made with the particle swarm algorithm, as shown in Table 8.6. Table 8.6 shows that the identification of multiple devices in an open state can easily result in a discrepancy between the number of identified devices and the actual number of identified devices, so as to generate low identification performance. In this section, the RNN training and learning method was adopted, which can avoid deviating from the actual load switching situation due to the constraint of the optimal solution of the objective function to some extent. Moreover, the introduction of load event detection methods can also reduce the inconsistency between load identification categories and the switching of load categories in real situations.

In deep learning, the RNN models were applied to identify non-invasive power loads, memorizing historical input feature quantities and establishing an intrinsic relationship between input and output to achieve reliable identification of load feature label quantities. In response to the phenomenon of "gradient vanishing" in the training process of the RNN models on large sample data, the problem was effectively solved by selecting the Softsign function as the activation function and the cross-entropy loss function for experiments. Moreover, it is beneficial to increase the number of hidden layers and nodes in the hidden layers to improve the load identification ability of the RNN model.

Table 8.6. Identification accuracy.

Equipment status combination	Sample capacity	Test sample	Correct identification sample	Identification accuracy %
1 device	$C_7^1 = 7$	70	70	100
2 devices	$C_7^2 = 21$	210	203	66.67
3 devices	$C_7^3 = 35$	350	268	85.14
4 devices	$C_7^4 = 35$	350	268	76.57
5 devices	$C_7^5 = 21$	210	126	61.42
6 devices	$C_7^6 = 7$	70	58	72.5
7 devices	$C_7^7 = 1$	10	10	100
Total		1270		63.36

8.6 Example 9: An Image Compression Method Based on DTCWT and RNN Encoder

The limitations of communication bandwidth and storage space have led to an increasing research on image compression. The basic requirement of image compression is to compress the image without causing any loss. Discrete Cosine Transform (DCT) can compress images by reducing the energy of low-frequency components. Genetic algorithms based on wavelet transform can also be used for image compression, but they suffer from issues such as aliasing, phase, and shift variance.

The lossy image compression method based on Deep Neural Networks (DNNs) outperforms traditional image encoding standards, such as JPEG and JPEG2000. Encoding and decoding images iteratively will take more time. Autoencoders can obtain the compressed binary codes for image restoration and extract dense visual representations from images through image dimensionality reduction. Variational autoencoders are used for compression, and non-variational RNNs are used for variable-rate encoding.

This section introduces the use of Dual Tree Complex Wavelet Transform (DTCWT) to decompose an image into 8 sub-bands and then encodes the low-frequency sub-bands using an RNN encoder to achieve better accuracy in image compression and transmission [189]. The performance of the method in this section is verified in terms of compression ratio and bit per pixel (bpp).

8.6.1 Mathematical model

8.6.1.1 *DTCWT*

Complex Wavelet Transformer (CWT) is used to approximate amplitude or phase coefficients; it has translational invariance and non-aliasing characteristics. The CWT utilizes the phase representation, invariance, and non-aliasing of analytic wavelets or orthogonal wavelets to ensure stable amplitude. The real-valued wavelet filter in the DTCWT generates parallel decomposition trees for the real and imaginary parts of the transform, which facilitates the use of mature wavelet transform methods. Its main advantage is that it decomposes the signal along the direction of greater energy, while traditional DWT is not feasible.

The analytic wavelet $\varphi_c(t)$ consists of two real wavelets, $r(t)$ and $\varphi_c(t)$, which can be represented as

$$\varphi_c(t) = \varphi_r(t) + j\varphi_i(t), \tag{8.6.1}$$

where

$$\varphi_i(t) = \text{Hilbert}[\varphi_r(t)] = \frac{1}{\pi}\int_{-\infty}^{+\infty} \frac{\varphi_r(t)}{t-\tau} d\tau = \varphi_i(t) \otimes \frac{1}{\pi t}, \tag{8.6.2}$$

where $\otimes$ represents the convolution operation. The relationship between the Fourier transform of $\varphi_i(t)$ and $\varphi_r(t)$ is defined as

$$H_i(\omega) = \text{FT}\{\text{Hilbert}[\varphi_r(t)]\} = -j \cdot \text{sgn}(\omega) \cdot H_r(\omega). \tag{8.6.3}$$

8.6.1.2 *Recurrent neural network*

8.6.1.2.1 *Encoder*

The conditional probability is calculated using the context vector $\boldsymbol{c}_o$ of each output y_1. The background vector obtained from the encoder scans the input sequence and generates continuous images. The weighted average of all hidden RNN representations is derived from the context vector $\boldsymbol{c}_o$. The context vector $\boldsymbol{c}_o$ is represented as

$$\boldsymbol{c}_o = \sum_t \alpha_{\text{ot}} \boldsymbol{h}_t, \tag{8.6.4}$$

where $\alpha_{\text{ot}} \in [0,1]$, $\sum_t \alpha_{\text{ot}} = 1$; $\boldsymbol{h}_t = [\overrightarrow{\boldsymbol{h}}_t^{\text{T}}; \overleftarrow{\boldsymbol{h}}_t^{\text{T}}]^{\text{T}}$, $(\overrightarrow{\boldsymbol{h}}_t, \overleftarrow{\boldsymbol{h}}_t)$ represent the inputs of forward and backward RNNs, respectively. The context vector $\boldsymbol{c}_o$ is global, for example $\boldsymbol{c}_o = \boldsymbol{h}_t$. The background matrix does not rely on the 0 prefix to encode all parts into a group matrix graph. When the dimensionality of the vector is quite large, this strategy achieves the best level of machine translation. However, for long input sequences, when the model's size is relatively small, it is better to use the dynamic context vector $\boldsymbol{c}_o$ in Equation (8.6.4). The weight α_{ot} is calculated through a learning alignment model, i.e.,

$$\alpha_{\text{ot}} = \frac{\exp(e_{\text{ot}})}{\sum_{t'} (\exp(e_{\text{ot}}))}, \tag{8.6.5}$$

where

$$e_{\rm ot} = a(\boldsymbol{S}_{0-1}, \boldsymbol{h}_t). \tag{8.6.6}$$

If $a(\cdot)$ is a feedforward neural network, it represents the importance degree of calculating the prior hidden state of the RNN decoder S_{o-1} for each hidden representation. The alignment mode h_t is a neural network with a hidden layer, i.e.,

$$a(\boldsymbol{S}_{0-1}, \boldsymbol{h}_t) = \boldsymbol{v}\mathrm{T}^{\tau} \tanh(\boldsymbol{W}\boldsymbol{S}_{o-1} + \boldsymbol{U}\boldsymbol{h}_t), \tag{8.6.7}$$

where $\boldsymbol{W}$ and $\boldsymbol{U}$ are the weight matrix and $\boldsymbol{v}$ is a vector such that the output of $a(\cdot)$ is a scalar.

8.6.1.2.2 *Model training*

For a set of input–output vectors, by maximizing the average conditional logarithmic similarity, the model can be trained on all training sets, and

$$\widehat{\mathrm{M}} = \arg\max_{M} \frac{1}{N} \sum_{n=1}^{N} \log P(y_1^n, \ldots, y_0^n | x_1^n, x_2^n, \ldots, x_T^n, M), \tag{8.6.8}$$

where $\widehat{\mathrm{M}}$ represents the model set and N represents the number of training expressions. Due to the different features of the encoder and decoder, the model can be trained using the Stochastic Gradient Descent (SGD) method. Here, the Adadelta algorithm is used to automatically estimate the learning rate rather than empirical adjustment. In the experiment, the exponential scheduling method used to analyze the performance of the Adadelta algorithm is good, but it is more difficult to adjust the hyperparameters.

8.6.1.3 *Compression framework*

The compression framework is shown in Figure 8.12. Input images with different pixel values are viewed according to the RGB standard,

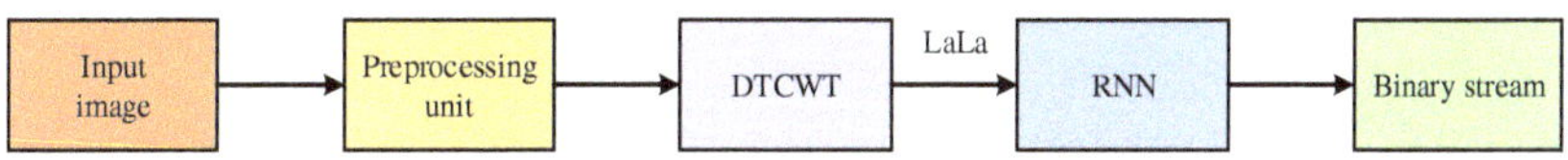

Figure 8.12. Compression scheme.

so each red, green, and blue element must provide three intensities; thus, it contains more information. For ease of understanding, the image is converted into a grayscale image, where grayscale is the only color, and all R, G, and B components have the same intensity and are given different intensity values. After the image matrix becomes a serial signal form, it is provided to the DTCWT, which uses two real DWTs. The first DWT provides a real transformation part, and the second DWT provides an imaginary part. These two real transformations use two different filters. These two filter sets are designed together to allow for overall conversion. In the implementation of the DTCWT, two filter banks, including analysis filter and synthesis filter, are used. LaHa and LaLa are employed for representing the upper filter band low-pass/high-pass filter pair, while the LbLb and the LbHb represent the lower filter/high-pass filter pair. For further processing, the LaLa component is considered as it has higher quality and accuracy compared to other components.

Neural networks provide real-time adaptive learning and implementation and are a recent scheduling method. Before using neural networks for encoding, neural networks will be trained first. After selecting the original weights uniformly, the method is learnt and trained. Here, a controlled training operation involving a system is used to provide the required output to the network by separately grading the system order or providing input for the desired product. The system uses weights in the hidden layer and processes to process records in the dataset one by one, and then compares the resulting output with the expected output. Then, the fault will be transmitted back through the system, which is triggered to adjust the weight of the input to the next record to be processed, and perform 1000 repetitions of neural network training.

The trained neural network is used to encode the LaLa component in the DTCWT for final encoding. The matrix structure of the LaLa components is combined with the trained autoencoder output for encoding, and the encoded output and trained output of the autoencoder are used together for the decoding process at the receiver.

8.6.1.4 *Implement the model*

The Gaussian-processed image is decomposed by the 11th and 17th tap symmetric filters of the DTCWT. The image is divided into three levels, and each level's decomposition contains both real and imaginary parts. The imaginary part has six directional bandwidths, which are ±150, ±450, and ±750.

As shown in Figure 8.13, the input image X is decomposed into eight sub-bands and several levels. First, row processing and column processing filters are used to process the images. A butterfly structure is designed, symbol inversion and second-order complement operations are utilized to reduce computational complexity, and division operations are replaced with threshold operations.

The three stages of wavelet transform are employed for decomposition to obtain 8 sub-bands. The six boundary directions of the input image are captured in six complex LH, HL, and HH sub-bands, respectively, where two complex LL sub-bands are taken as low-frequency components.

The RNN encoder receives the LaLa component output of the DTCWT. The implementation model of the RNN is shown in Figure 8.14. X1 and Y1 are input and output sequences. The definitions of feedforward function $a(\cdot)$, alignment vector generator, and contact vector generator are shown in Equations (8.6.4)–(8.6.8). The RNN encoder according to Equation (8.6.6) is trained.

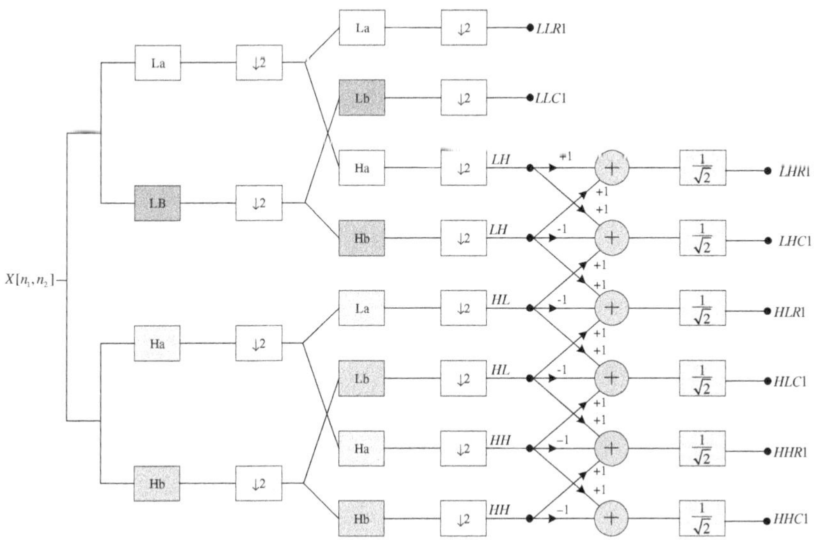

Figure 8.13. Calculation principle of the DTCWT.

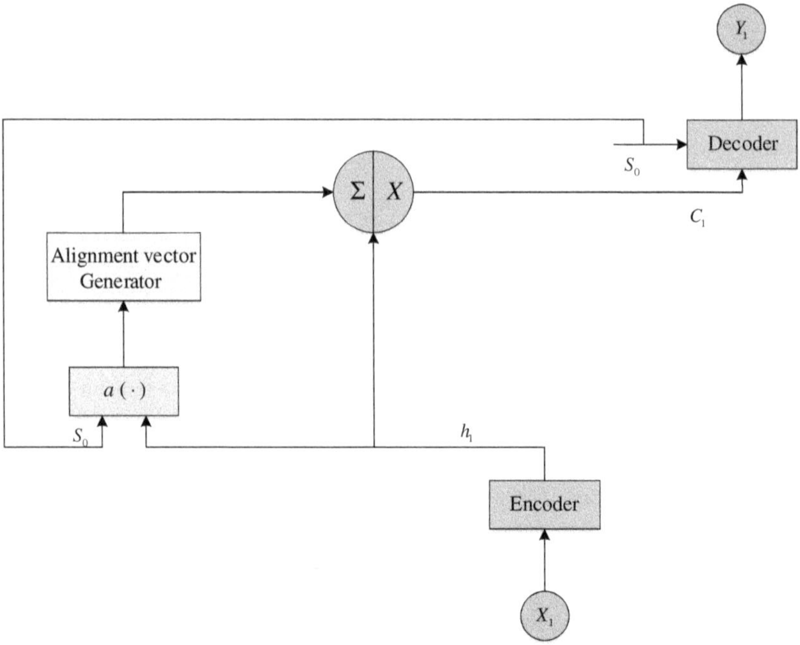

Figure 8.14. RNN model with single time step.

8.6.2 Simulation experiments and result analyses

This section presents the MATLAB simulation results of an image and provides the values of CR (%) and CR (BPP) for various images and their corresponding graphs. The DTCWT and neural encoding were used to compress the input image.

Figure 8.15(a) shows that the input image is RGB, so three intensities need to be specified for each pixel. Figure 8.15(b) shows a grayscale image stored as an 8-bit integer, and 256 possible different grayscale levels from black to white.

The trained images were encoded and decoded with different iterations. In the neural network, the LaLa components obtained through 1000 iterations of the DTCWT are shown in Figure 8.16.

Figure 8.17 shows the optimal training effect of using deep neural networks to train images by analyzing 1000 epochs. Figure 8.18 shows the CR (%) and CR (bpp) of the testing image shown in Figure 8.15.

The simulation of the proposed algorithm [189] is given in this section using the standard testing image, as shown in Figure 8.19.

(a) (b)

Figure 8.15. Image compression [189]. (a) Input image Barbara. (b) Grayscale image.

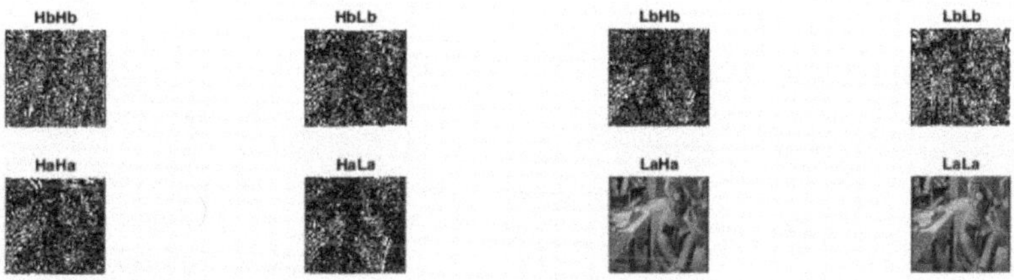

Figure 8.16. One-dimensional DTCWT for first-order decomposition [189].

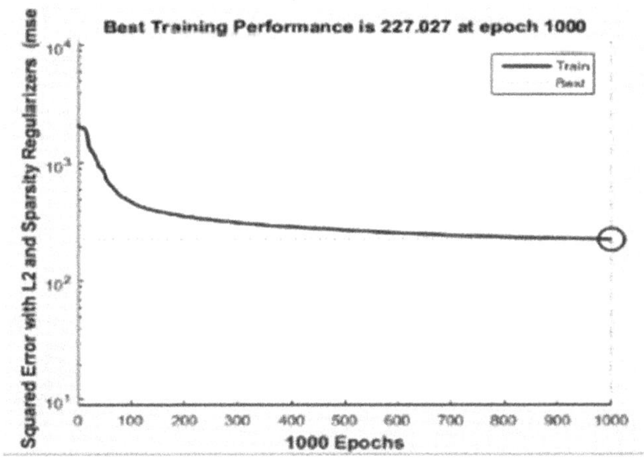

Figure 8.17. Performance after 1000 training sessions [189].

```
cr_bits =

   21.6964

Compressionratio in percentage=
   95.3909


cr_bpp =

    2.7121
```

Figure 8.18. Performance parameters in MATLAB [189].

Figure 8.19. Test image for analysis [189].

Figures 8.20 and 8.21 shows that the compression ratios of high pixel values in the images are better than those of low pixel values, and the compression ratios vary linearly from high pixel values to low pixel values.

In this section, an image compression method based on the DTCWT and RNN encoder[ ] is introduced. The simulation results of the introduced method are given. The simulation results have shown that compared to other traditional technologies, when the DTCWT

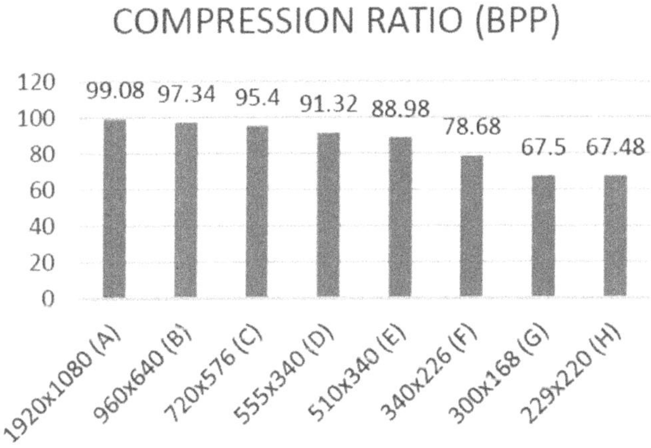

Figure 8.20. CR of test image [189].

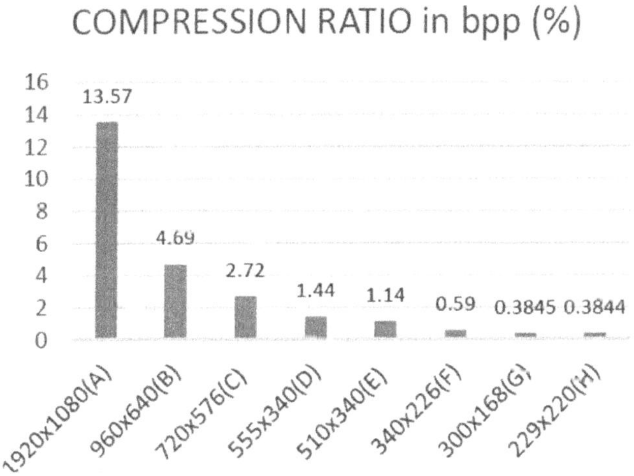

Figure 8.21. CR (bpp) analysis of various test images [189].

is used to compress images, the higher the pixel values, the higher the compression ratios, that when the pixel values of the image are high, the compression ratios can be up to 66.08%, and that for lower pixel values, the compression ratios can be improved by increasing the number of iterations of the neural network and modifying the wavelet components, and this new neural network model can also be used for video compression.

Chapter 9

Deep Recursive Cascaded Convolutional Neural Networks

Summary

Starting from the structure of Recursive Convolutional Neural Networks (RCNNs), in this chapter, the framework of Deep RCNNs (DRCNNs), including structural units and loss functions, is analyzed. The Bilinear Recurrent Neural Network (BRNN) and the particle swarm optimization algorithm-based BRNN are studied. On the basis of two-dimensional (2D) Convolutional Neural Networks (CNNs), three-dimensional (3D) CNNs (3D-CNNs) are introduced, and spatial features based on the 3D-CNN and Bi-directional RNN (BiRNN) are extracted. The structural principle of 3D-RCNNs is discussed. In the example of the image rain removal algorithm, attention mechanism and gated recurrent unit are applied to the RCNN to study the rain removal of a single image in different scenes, and good results are achieved. In the example of single-image super-resolution method, a cascaded recursive residual CNN is adopted to further expand the functionality of the RCNN, which can reconstruct more image details.

9.1 Introduction

Convolutional Neural Networks (CNNs) have the characteristics of weight sharing and local connections. In addition to the forward propagation process, the Backpropagation (BP) algorithm is utilized to calculate the reverse gradient and update the parameters of the network. The updating of network parameters is an important process in the training phase of the CNNs. Given a training set $\{x, y\}$, the parameters $\{(\boldsymbol{w}_1, \boldsymbol{b}_1), \ldots, (\boldsymbol{w}_l, \boldsymbol{b}_l), \ldots, (\boldsymbol{w}_L, \boldsymbol{b}_L)\}$ are updated by the BP algorithm to minimize the loss function $\mathrm{Loss}(x, y)$, i.e.,

$$\hat{\boldsymbol{\theta}} = \arg\min_{\boldsymbol{\theta}} \mathrm{Loss}(x, y), \tag{9.1.1}$$

where $\mathrm{Loss}(x, y)$ is the loss function determined by a specific task and x is the input for online testing. Once x is trained, it can be used for online testing of tasks, that is, for predicting the target output y.

9.2 Deep Recursive Convolutional Neural Network

9.2.1 Recursive convolutional neural network

In neural networks, the balance between the number of network parameters and the network's expressive ability is a challenge, yet the recursive structures of network, as shown in Figure 9.1, which includes multiple convolutional and activation layers, can effectively solve this problem and contains a recursive module repeated T times in the recursive structure. If a recursive structure is represented as a mapping function $f(\cdot)$, the recursion formula from the kth to the $k-1$th can be represented as follows:

$$s(k+1) = f(s(k)), \tag{9.2.1}$$

where $s(k)$ represents the feature map when the recursive structure is input for the kth time. The essence of recursive structure unfolded by repeating T times is to involve the sharing weights of the multiple network layers; thus, the number of network parameters can be

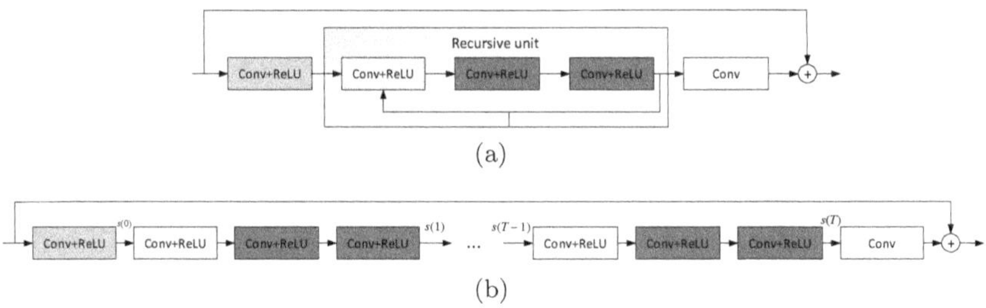

Figure 9.1. CNN with recursive structure. (a) CNN with recursive structure. (b) Unfolding form of Figure 9.1(a).

reduced while maintaining good expressive ability. The number of recursive structures T used repeatedly and the number of layers R contained in each recursive structure can be adjusted according to specific tasks, and T and R can be determined by experiments.

In Figure 9.1, the convolutional layers of the same color share weights.

9.2.2 Deep recursive cascaded convolutional neural network

The structure of a deep recursive cascaded convolutional neural network (DRCCN) [190] is shown in Figure 9.2, which is used to understand the mapping relationship between undersampled images and fully sampled images. The input of the network is undersampled K-space data with zero padding, which is transformed into 12-channel undersampled images through inverse Fourier transform. The output of the network involves the reconstructed 12-channel images from fully sampled K-space data. The real and imaginary parts of complex image data in the channel dimension are concatenated as input to the network. The entire network is composed of multiple cascaded modules together, which contains a Recursive Convolutional Neural Network unit (recursive CNN unit) and a Data Consistency (DC) unit. Each RCNN unit is a residual structure, as shown in Figure 9.2. Among the cascaded CNN units, the undersampled K-space data are used to update the output of each CNN unit to ensure the consistency of the data outputted from each unit. The convolution kernel size of each convolutional layer is 3×3. In each CNN unit, except for the last layer with 24 feature maps (where the first 12 feature maps are the real parts of the output complex image and the last 12 feature maps are the imaginary parts), the outputs of the other layers are all 64 feature maps. Except for the last layer, the

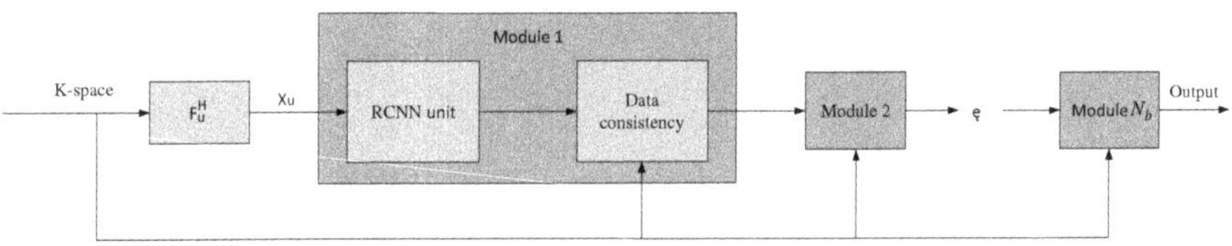

Figure 9.2. Structure of a recursive network.

Rectified Linear Unit (ReLU) is put after the other layers for non-linear activation. It is worth noting that the number of modules N_b in the entire network and the actual number of convolutional layers N_c in each CNN unit can be adjusted according to the actual task.

During the training process, the parameters in the network are calculated and updated by minimizing the loss function to obtain gradients. The absolute value of error between real data and the network's output is defined as the loss function and given by the following:

$$L(x, y) = \frac{1}{M} \sum_{m=1}^{M} \|y(x_m; \theta) - y_m\|, \tag{9.2.2}$$

where y is the output of the network, x_m represents the mth input in the training set (a 12-channel undersampled image obtained by transforming zero padding K-space data), y_m denotes the corresponding real data of the fully sampled images, M presents the number of training images in the training set, and $\boldsymbol{\theta}$ is regarded as the network parameters to be learned in the CNN.

The output of each cascaded CNN unit can gradually approach the real image, but for the positions already captured in K-space, the output of the CNN unit often becomes distorted. In order to ensure the consistency of data collected in K-space, it is necessary to add a data consistency layer after each CNN unit. Specifically, let f_l represent the K-space of the network output image, i.e., $f_{l,j} = Fy_{l,j} = Fy_{l,j}(x; \theta)$, where $f_l(k)$ represents the value of K-space at position k and j represents the plane coordinates of the image. To ensure data consistency, the K-space of the CNN output image needs to be updated with the K-space data that have already been collected, i.e.,

$$f_l^{\mathrm{rec}}(k) = \begin{cases} f_l(k) & \text{if } k \notin S \\ \frac{f_l(k)+\lambda f_0(k)}{1+\lambda} & \text{if } k \in S \end{cases}, \tag{9.2.3}$$

where $f_0(k)$ represents the collected K-space data, S represents the set of all collected positions, and λ is the weight parameter and is set to infinity, that is, if the corresponding position in K-space has already been sampled, the final updated K-space will use the sampled K-space values. Finally, the updated K-space is performed by inverse Fourier transform to output the reconstructed image of this module, which can be further optimized by the next module.

9.3 Bilinear Recurrent Neural Network Model

9.3.1 BRNN structure

The Bidirectional Recurrent Neural Network (BRNN) is a structure that combines bilinear polynomials and RNNs [191], as shown in Figure 9.3. It can approximate multiple nonlinear functions based on the characteristics of the training data without rapidly increasing network complexity and has advantages such as dynamic recursive calibration. Compared with higher-order neural networks, it is easier and faster to implement in the application process.

A bilinear polynomial is a simple periodic nonlinear system model. For a one-dimensional input/output situation, the input–output relationship of a linear polynomial is given by the following:

$$y(k) = \sum_{i=1}^{N-1}\sum_{j=1}^{N-1} b_i y(k-j)x(k-i) + \sum_{i=1}^{N-1} a_i y(k-i) + \sum_{i=1}^{N-1} c_i x(k-i), \tag{9.3.1}$$

where $x(i)$ is the input variable, $y(i)$ is the output variable, and N is the recursive number. When the BRNN model has N input neurons, M hidden neurons, and K-order polynomials, the input variable vector is defined as follows:

$$\boldsymbol{X}(k) = \{x_1(k), x_2(k), \ldots, x_N(k)\}^{\mathrm{T}}.$$

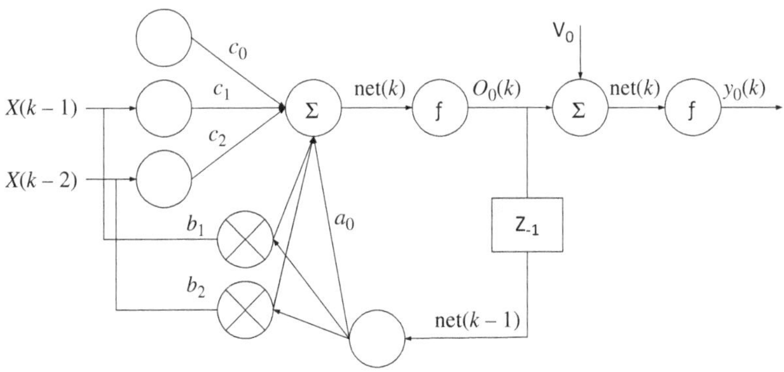

Figure 9.3. Structure of the BRNN.

The vector of the network connection weights from the input layer to the hidden layer is defined as follows:

$$\mathbf{net}(k) = \{\mathrm{net}_1(k), \mathrm{net}_2(k), \ldots, \mathrm{net}_N(k)\}^{\mathrm{T}}.$$

The $M \times K$-dimensional periodic vector is defined as follows:

$$\begin{aligned}\boldsymbol{Z}(k) = \{&\mathrm{net}_1(k-1), \ldots, \mathrm{net}_M(k-1), \ldots, \mathrm{net}_1(k-K), \ldots,\\ &\mathrm{net}_M(k-K)\}^{\mathrm{T}}.\end{aligned}$$

The periodic vector of the pth hidden layer neuron is defined as follows:

$$\mathbf{net}_p(k) = \boldsymbol{d}_p + \boldsymbol{A}_p^{\mathrm{T}}\boldsymbol{Z}(k) + \boldsymbol{Z}(k)\boldsymbol{B}_p\boldsymbol{X}(k) + \boldsymbol{C}_p^{\mathrm{T}}\boldsymbol{X}(k), \tag{9.3.2}$$

where $\boldsymbol{d}_p$ is the bias vector of the neurons, $\boldsymbol{A}_p$ is the periodic weight vector, $\boldsymbol{B}_p$ is the bilinear polynomial weight matrix, $\boldsymbol{C}_p$ is the feedforward weight vector, and T represents the transpose matrix. When the activation function of the hidden layer neurons is recorded as $g(\cdot)$, the output variable of the hidden layer neurons is defined as follows:

$$\boldsymbol{O}_p(k) = g(\mathbf{net}_p(k)). \tag{9.3.3}$$

The classic BRNN is a simple feedforward neural network from the hidden layer to the output layer. If $O_{lp}(k)$ is the output variable of the lth neuron in the pth hidden layer at time k, then the periodic vector and the output vector of the pth hidden layer nodes are defined, respectively, as follows:

$$\mathbf{net}_l(k) = \sum_{p=0}^{N_k} w_{lp}O_{lp}(k), \tag{9.3.4}$$

$$\boldsymbol{O}_l(k) = g(\mathbf{net}_l(k)), \tag{9.3.5}$$

where w_{lp} is the weight coefficient from the hidden layer to the output layer.

In summary, the output vector of the output layer is given by the following:

$$y_l(k) = \sigma\left(vl + \sum_{p=0}^{N_k-1} w_{lp}O_{lp}(k)\right). \tag{9.3.6}$$

9.3.2 Particle swarm optimization for BRNN

9.3.2.1 *BRNN optimization model*

The classical BRNN contains a large number of neurons and weight parameters, which result in a complex network structure and slower convergence speed. Therefore, in order to improve the convergence speed of the BRNN, the particle swarm algorithm is integrated with the BRNN to prune the redundant neurons and weights that exist in the classical BRNN model, reduce the complexity of the network, prevent it from falling into local optima, and improve the computational speed.

The structure of the BRNN consists of an input layer, a hidden layer, and an output layer. When the number of the output layer N_O is 1, the total number of multiplication terms N_c in Equation (9.3.1) is calculated by

$$N_c = (N_i + N_f + N_i \cdot N_f) \cdot N_h + N_h, \tag{9.3.7}$$

where N_i is the number of input layer neurons, N_f is the number of feedback nodes, and N_h is the number of the hidden layer neurons. In order to further improve the prediction accuracy of the BRNN model, it is necessary to extract a portion of the feedback input variables and add them to the linear feedback layer. Therefore, the optimized output variables of the BRNN model are converted from Equation (9.3.1) to the following:

$$y_p(k) = d_p + \sum_{i=0}^{N_i-1} \sum_{j=S}^{E} b_{pi} y_p(k-j) x(k-i) + \sum_{i=0}^{N_f-1} a_{pi} y_p(k-i)$$
$$+ \sum_{i=0}^{N_i-1} c_{pi} x(k-i), \tag{9.3.8}$$

where $E = p + (N_i - 1)/2$, $S = (N_i - 1)/2 - p$, and d_p denotes bias value.

Therefore, the computational complexity of the output variables N_c in the output layer of the BRNN model has changed into N_S, and N_S is calculated by the following:

$$N_S = N_i \cdot N_f + (2N_i + N_f) \cdot N_h + N_h. \tag{9.3.9}$$

The computational complexity of the output variables ΔN in the output layer of the optimized BRNN can be reduced and calculated by the following:

$$\Delta N = N_c - N_S = N_h \cdot N_f \cdot N_i - N_h \cdot N_i - N_f \cdot N_i. \quad (9.3.10)$$

In order to further reduce the computational load of the BRNN model, dynamically reduce weights and neurons in the network, and ensure the reliability of the output prediction results, the particle swarm algorithm is used to dynamically prune the BRNN model.

9.3.2.2 *BRNN optimization process*

The Particle Swarm Optimization (PSO) algorithm starts from the perspective of random solutions and iteratively seeks a global optimal solution. It has the advantages of high robustness, fast convergence, and simple implementation. So, the PSO algorithm is integrated into the BRNN optimization process to effectively improve the convergence speed and learning ability of the BRNN.

Based on the number of weight values and bias values in the BRNN, the searching space dimension of the particle swarm algorithm N_v is determined by the following:

$$N_v = N_i \cdot N_h + N_h \cdot N_f + N_i + N_h. \quad (9.3.11)$$

In an N_y-dimensional searching space and a population $\boldsymbol{X} = (x_1, x_2, \ldots, x_m)$ consisting of m particles, the qth particle is represented as an N_y-dimensional vector, that is, the spatial position of the qth particle in the N_v-dimensional space. The velocity of the qth particle is given by $\boldsymbol{v}_q = (v_{q1}, v_{q2}, \ldots, v_{qN_v})^{\mathrm{T}}$, and individual extremum and global extremum are denoted as $\boldsymbol{p}_q = (p_{q1}, p_{q2}, \ldots, p_{qN_v})^{\mathrm{T}}$ and $\boldsymbol{v}_q = (v_{q1}, v_{q2}, \ldots, v_{qN_v})^{\mathrm{T}}$, respectively. During each iteration of the particle swarm algorithm, the particles will update their iteration speed and position based on individual and global extremum values. The velocity and position of the qth particle are updated, respectively, by the following:

$$\begin{aligned} V_{qN_v}(k+1) = wv_{qN_v}(k) + c_1 r_1 (p_{qN_v}(k) - S_{qN_v}(k)) \\ + c_2 r_2 (p_{gN_v}(k) - S_{gN_v}(k)), \end{aligned} \quad (9.3.12)$$

$$x_{qN_v}(k+1) = x_{qN_v}(k) + v_{qN_v}(k+1), \quad (9.3.13)$$

where w is the inertia weight, c_1 and c_2 are the learning factors, and r_1 and r_2 are random number in $(0, 1)$.

The specific architecture for optimizing the BRNN is as follows:

Step 1: Initialize the network

The dimensions N_v, learning factors c_1 and c_2, and maximum number of iterations of the particle swarm algorithm are initialized, and c_1 and c_2 are usually set to be equal to the values ranging from 0 to 4. Among them, particle initialization mainly involves assigning random values to the initial velocity and spatial position of the particles.

Step 2: Determine the optimization model

The basic network topology structure of the BRNN optimization model is determined.

Step 3: Calculate the fitness function

A random population of particles is generated to optimize the training of the BRNN model and output fitness values that meet the accuracy requirements as follows:

$$\text{fit(net)} = \sum_{i=1}^{m} \sum_{j=1}^{N_o - 1} |y_{ij} - \hat{y}_{ij}|, \tag{9.3.14}$$

where y_{ij} and $\hat{y}_{ij}$ represent the true and predicted values of the testing set, respectively, and $\text{fit}(\cdot)$ indicates the fitness function.

Step 4: Start the particle swarm algorithm

The individual velocity and position of particles are updated according to Equations (9.3.12) and (9.3.13), the particle population with a certain probability is initialized, the particle population is updated, and the particle fitness is recalculated.

Step 5: Prune the operation

Comparing the fitness size before and after the update, if the updated fitness is smaller than the original fitness, the original particle population is deleted, the population is updated, and invalid hidden layer or input layer neurons and their corresponding weight values in the BRNN network are deleted.

Step 6: Repeat steps 4 and 5 in sequence

The minimum fitness and its corresponding particle population are searched within the maximum number of loop steps, and the calculation results are outputted.

Step 7: Set the connection weights and thresholds

Connection weights and thresholds are dynamically set for optimizing the BRNN based on the optimal particles. The BRNN optimized by the network prediction model outputs prediction results after training.

9.4 3D Recurrent Convolutional Neural Network

Hyperspectral images contain rich spectral and spatial information, and have different types of information with their own unique characteristics. The spectral resolution of spectral information in hyperspectral images is high, and each pixel has a continuous spectral curve, which means that the brightness values (digital numbers, DN) of pixels among a single band and adjacent bands have a certain correlation. As the distance between bands increases, this correlation will gradually weaken. In hyperspectral image classification, spatial information mainly refers to spatial contextual information, which is characterized by the fact that pixels that are closer in spatial position are more likely to belong to the same type of land cover than those that are farther away. Reasonable use of spatial information during the classification process can effectively improve classification accuracy and weaken salt-and-pepper phenomena. When classifying, it is necessary to consider the characteristics of both types of information and adopt appropriate feature extraction strategies to improve the classification performance of hyperspectral images.

9.4.1 Spatial feature extraction based on 3D-CNN

The CNN is initially applied to two-dimensional (2D) image data and has demonstrated excellent performance in extracting spatial features of images. The vast majority of the CNNs use 2D convolution kernels, but hyperspectral images have hundreds of bands (i.e., hundreds of 2D images). Therefore, when 2D convolution kernels are applied to the processing of hyperspectral images, a large number of parameters will be generated, which can easily lead to overfitting for hyperspectral images with limited standard parameter data.

A 3D convolution kernel can convolve in three directions simultaneously and output a third-order tensor. In the jth feature map in the ith convolutional layer of the 3D-CNN [192], the output of the position (x, y, z) is calculated by the following:

$$v_{ij}(x,y,z) = \sigma\left(b_{ij} + \sum_{k=1}^{m}\sum_{w=0}^{W_i-1}\sum_{h=0}^{H_i-1}\sum_{b=0}^{B_i-1} u_{ijk,hwb}\, v_{(i-1)k} \right.$$
$$\left. \times\ (x+w, y+h, z+b)\right), \tag{9.4.1}$$

where $\sigma(\cdot)$ is the activation function, b_{ij} is the bias, B_i is the size of the 3D convolution kernel in the spectral dimension, and W_i and H_i are the width and height of the 3D convolution kernel, respectively. $u_{ijk,hwb}$ is a convolution kernel connected to the kth feature map of the $(i-1)$th layer. 3D convolution involves fewer parameters and is more suitable for training tasks with limited samples compared to 2D convolution. In addition, the convolution results of 2D convolution kernels and 3D convolution kernels to hyperspectral images are compared and shown in Figure 9.4; the 2D convolution kernels used to extract features may result in the loss of 3D structural information of the hyperspectral images.

3D convolutional kernels can fully extract spatial features of hyperspectral image data while preserving the size of the data dimension in the spectral dimension. When the 3D-CNN is used to extract

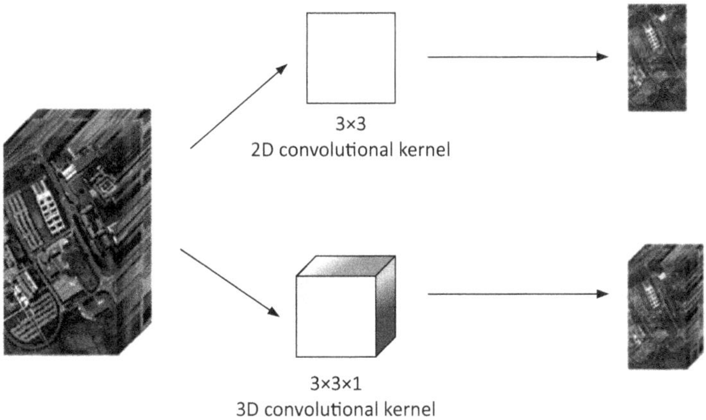

Figure 9.4. Comparison of two convolutional Kernel processing results.

spatial features, the size of the convolution kernel has a significant impact on the training performance and speed of the network. Research has shown that a convolution kernel of a size 3×3 can extract spatial features more effectively with fewer parameters. Therefore, the size of the 3D convolution kernel is set to $3 \times 3 \times 1$. The strategy for extracting spatial features using 3D-CNN is to extract data blocks of size $n \times n \times B$ (B refers to the number of spectral bands in the hyperspectral image) from the original hyperspectral images and use the 3D-CNN to convolve the data blocks with a kernel size of $3 \times 3 \times 1$. After doing several nonlinear transformations of convolutional layers, the spatial information of a certain range of neighborhoods around the central pixel is integrated into the central pixel to generate a vector of size $1 \times 1 \times B$, completing the extraction of spatial features.

9.4.2 BiRNN model

9.4.2.1 *Structural principle*

The spectral data of hyperspectral images is a type of sequential data with sequence correlation between bands. As the distance between bands increases, this correlation will gradually weaken. This phenomenon indicates that there is a stronger short-term dependence between each band in the spectral data, and it is dependent on both the preceding and following information. In order to reduce training costs, the loss of long-term information memory is not considered during training, but a standard RNN with fewer parameters may be contemplated. But standard RNNs only apply the previous information to the current output in processing sequence data and ignore the subsequent information. In some tasks, the output at the current time is not only related to past information but also to information at subsequent times. Therefore, a network layer that transmits information in reverse chronological order can be added to enhance the network's capabilities. So, there are the BiRNNs, which are some variants of the RNN [193]. Their basic idea is to train the same set of sequence data with forward RNNs and backward RNNs. The two RNNs are connected to the output layer at the same time, which provides contextual information for each output. The input sequences of two layers in the network are all X, but the direction of their

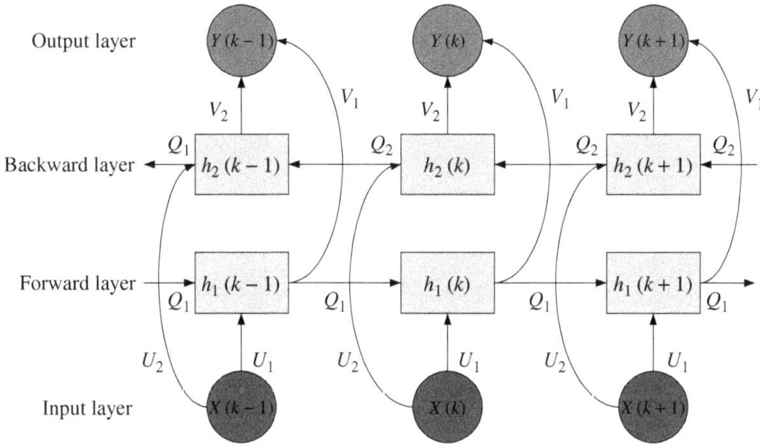

Figure 9.5. Schematic diagram of the BiRNN.

information transmission is opposite. Figure 9.5 shows the network structure of the BiRNN, which includes six weights: Weights (U_1, U_2) are from the input layer to forward and backward hidden layers, weights (Q_1, Q_2) are from the hidden layer to the hidden layer, and weights (V_1, V_2) are from forward and backward hidden layers to the output layer.

Assuming that the first layer transmits information in chronological order and the second layer transmits information in reverse chronological order, the hidden states of these two layers at time k are represented as follows:

$$\boldsymbol{h}^{(B)}(k) = g(\boldsymbol{U}^{(B)}\boldsymbol{h}^{(B)}(k-1) + \boldsymbol{W}^{(B)}\boldsymbol{X}(k-1) + \boldsymbol{b}^{(B)})$$
$$\boldsymbol{h}^{(F)}(k) = g(\boldsymbol{U}^{(F)}\boldsymbol{h}^{(F)}(k+1) + \boldsymbol{W}^{(F)}\boldsymbol{X}(k) + \boldsymbol{b}^{(F)})$$
$$y(k) = \boldsymbol{V}^{(F)}\boldsymbol{h}^{(F)}(k) + \boldsymbol{V}^{(B)}\boldsymbol{h}^{(B)}(k) + \boldsymbol{b}_y,$$

where $\boldsymbol{h}^{(B)}(k) = \boldsymbol{h}_1(k)$, and $\boldsymbol{h}^{(R)}(k) = \boldsymbol{h}_2(k)$. $\boldsymbol{U}^{(F)}$, $\boldsymbol{W}^{(F)}$, and $\boldsymbol{V}^{(F)}$ correspond to the weighted matrix at time k from the input layer to the hidden layer, from the hidden layer to the hidden layer, and from the hidden layer to the output layer in sequence, and $\boldsymbol{b}^{(F)}$, $\boldsymbol{b}^{(B)}$, and $\boldsymbol{b}_y$ denote bias at each step in sequence.

9.4.2.2 *Training architecture*

The training of the BiRNN adopts the BPTT algorithm, which is characterized by gradually calculating gradients from top to bottom, while forward propagation is the opposite of backpropagation. Forward propagation is a bottom-up algorithm that gradually calculates the final loss by inputting to calculate the output. Therefore, forward propagation and backward propagation are usually used to train the BiRNN to achieve the updating and optimize the parameters of the model.

9.4.2.2.1 *Forward propagation*

Step 1: Forward calculation once

The output of the forward hidden layer for each time step is calculated in the forward direction from time 1 to time T once.

Step 2: Reverse calculation once

The output of the backward hidden layer for each time step is obtained in reverse direction from time T to time 1.

Step 3: Output the final result

After calculating all the input moments in both forward and reverse directions, the final output is obtained by the forward and backward hidden layers based on each moment.

9.4.2.2.2 *Backward propagation*

Step 1: Calculate the terms of the output layer at all times.

Step 2: Update the previous layer

The BPTT algorithm is used to update the previous layer for the item $e(k)$ of all output layers.

Step 3: Update the subsequent layer

The BPTT algorithm is used to update the subsequent layer for the item $e(k)$ of all output layers.

9.4.3 3D-CRNN structure

The 3D-CRNN model consists of two parts: spatial feature extraction and spectral feature extraction. Among the two of them, the spatial feature extraction part is mainly composed of 3D-CNN [192], with convolution kernel sizes of $3 \times 3 \times 1$ and a stride of 1. There is no pooling layer between convolutional layers to preserve the feature information of small targets. In terms of the number of convolution kernels, according to the universal design ratio of the CNN, the number of convolution kernels in the later layer is twice that of the previous layer, and the number of convolution kernels in the initial layer is set to 4. The output of each convolutional layer is subjected to a batch normalization (BN) layer and ReLU activation function. The output of the last convolutional layer is discarded to avoid overfitting of the model due to dense sampling. In the process of preparing the dataset, it is first necessary to perform a certain amount of zero padding on the edges of the original images, and then select some pixel blocks with a size of $n \times n \times B$ as training and validation samples, with each pixel on the original image as the center point. Among others, $n \times n$ is the sampling size in the spatial dimension of the hyperspectral images, and B is the number of spectral bands. In order to meet the input format of the extracted spectral feature part, the size of the processed data must be $1 \times 1 \times B$. Therefore, for pixel blocks of different sizes input into the training network, the purpose of changing the output size can be achieved by changing the number of convolutional layers. For example, for a pixel block with a size of $5 \times 5 \times B$, the number of convolutional layers needs to be set to 2. For pixel blocks with a size of $7 \times 7 \times B$, the number of convolutional layers may be set to 3, and so on. The spectral feature extraction part is composed of the BiRNN, with a recursive layer and 32 hidden-layer features. The output of the BiRNN is used as input to the BN layer, the Tanh function is used as its activation function, and finally its output result is input into the fully connected (FC) layer, based on using Softmax as the loss function for training the classifier. The entire network structure is shown in Figure 9.6.

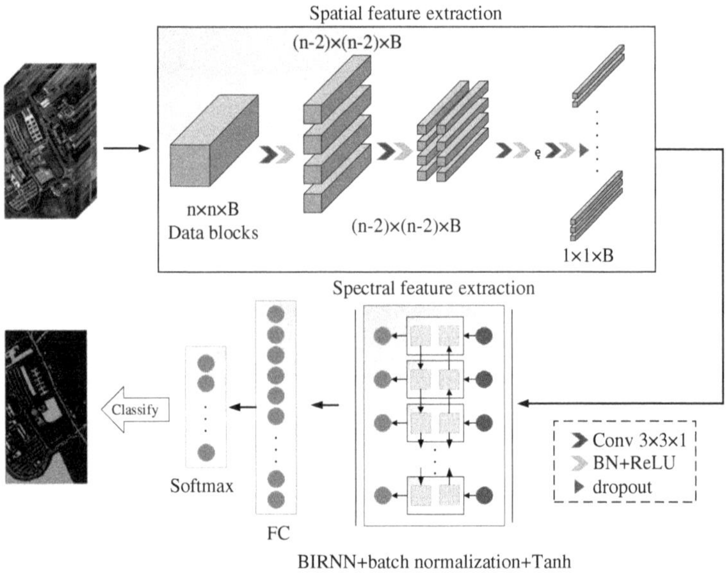

Figure 9.6. Structure diagram of 3D-RCNN.

9.5 Example 10: Image Rain Removal Method Based on Attention Mechanism and Gated Recurrent Unit

In this section, in response to the problem of stripe residual and blurred image details in processing images containing dense stripes using the GAN module, an AMGR-Net (Attention Mechanism and Gated Recurrent Network) that combined spatial attention mechanism, channel attention mechanism, and gated recurrent unit was proposed and validated through single-image rain removal simulation [194, 195]. The results indicate that the AMGR-Net method can better solve the problem of residual rain stripes and blurred details in the GAN module after processing images containing dense rain stripes by using the GAN.

9.5.1 Attention mechanism and gated recurrent model for image rain removal

9.5.1.1 *Network structure*

The network shown in Figure 9.7 is built on the basis of the Residual Generative Adversarial Network (ResGAN) and mainly includes gate

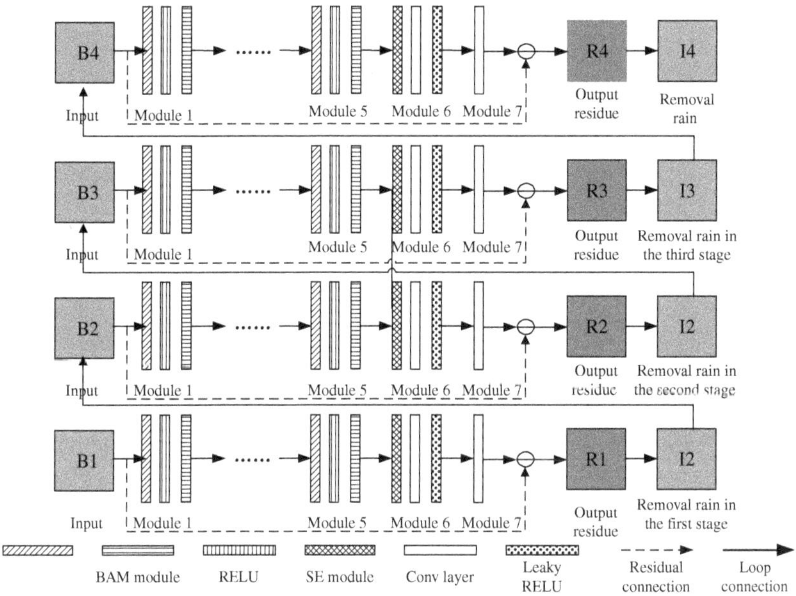

Figure 9.7. AMGR-Net.

control loop units, spatial attention modules, and channel attention modules.

9.5.1.1.1 *Gated recurrent unit*

To fully utilize the feature information from the previous stage, the RNN is used. To address the issue of weakening node memory performance in traditional RNNs, Gated Recurrent Units (GRUs) that excel at capturing long-term dependencies in data are chosen. The GRU includes an update gate and a reset gate.

As for the updating, it is used to control how many historical states are retained in the output state at the current time, as well as how many candidate states are retained at the current time. Given the feature mapping $x^j(k)$ of the jth layer in the kth time step, the mapped features $x^{j-1}(k)$ of the $j-1$th layer in the kth time step, and the hidden state $x^j(k-1)$ of the previous time step, the updating formula of the gate is written as follows:

$$z^j(k) = \sigma(W_z^j x^{j-1}(k) + U_z^j x^j(k-1) + b_z^j), \tag{9.5.1}$$

where W_z^i and U_z^j represent weight parameters and b_z^j represents deviation parameters. The Sigmoid function σ is used to transform the value of each element in the update gate to a value between 0 and 1.

For the reset gate, it is used to determine the degree of dependence of the candidate state at the current time on the network state at the previous time. The calculation formula of the reset gate is written as follows:

$$r^j(k) = \sigma(W_r^j x^{j-1}(k) + U_r^j x^j(k-1) + b_r^j), \qquad (9.5.2)$$

where W_r^j, U_r^j, represent weight parameters, and b_r^j denotes deviation parameters, respectively, and σ represents the Sigmoid function, which transforms the value of each element in the reset gate to a value within 0 and 1.

Candidate hidden states are calculated by the GRU to assist in the calculation of subsequent hidden states $x^j(k)$.

Firstly, the outputs of the reset gate $r^j(k)$ in the current time step and the hidden state $x^j(k-1)$ of the previous time step are calculated according to the Hadamard product operation. If the element values in the reset gate approach 0, it means resetting the corresponding hidden state value to 0, and discarding the hidden state of the previous time step. If the value of the element in the reset gate approaches 1, the hidden state of the previous time step is retained.

Secondly, $r^j(k) \odot x^j(k-1)$ and $x^{j-1}(k)$ are multiplied by the weight matrix W_n^j and U_n^j, respectively, and the results of these two parts are used as an independent variable for the Tanh activation function. The value range is compressed to $[-1, 1]$ to obtain the candidate hidden state $n^j(k)$. So, the candidate hidden state $n^j(k)$ in time step k is calculated by the following:

$$n^j(k) = \tanh(W_n^j x^{j-1}(k) + U_n^j(r^j(k) \odot x^j(k-1) + b_n^j)), \qquad (9.5.3)$$

where W_n^j, U_n^j, represent weight parameters, and b_n^j denotes deviation parameters, respectively.

Finally, the hidden state $x^j(k)$ in time step k is calculated. At this point, the output of the update gate $z^j(k)$ must be used to perform Hadamard product operations on the previous time step's state $x^j(k-1)$ and candidate hidden state $n^j(k)$, respectively. The final output of the network in the current time step is obtained by the following:

$$x^j(k) = z^j(k) \odot x^j(k-1) + (1 - z^j(k) \odot n^j(k)). \tag{9.5.4}$$

If the output of the update gate $z^j(k)$ approaches 0, it indicates that the information from the previous moment has been forgotten. Conversely, if the output of the updating gate $z^j(k)$ approaches 1, it shows that the current input information has been forgotten. If there are other values, it indicates different levels of importance. In summary, the GRU can effectively solve the forgetting problem, that is, the gradient vanishing problem in the backpropagation process.

9.5.1.1.2 *Spatial attention module*

We currently use spatial attention modules to enhance or suppress features at different spatial locations. In the problem of removing rain from a single image, it is particularly important to extract advanced features such as contextual information. Compared with the standard convolution operation, dilated convolution is beneficial for constructing more efficient spatial maps. Therefore, in the spatial attention module, the dilated convolution operation is utilized to expand the receptive field, and to make full use of contextual information. For the given input feature map $F \in \mathbb{R}^{C\times H\times W}$ and the calculated spatial attention map $M(F) \in \mathbb{R}^{C\times H\times W}$, the feature map F' is redefined as follows:

$$F' = F + F \otimes M(F), \tag{9.5.5}$$

where $\otimes$ represents the convolution operation.

The spatial attention module includes two hyperparameters: the expansion coefficient *rate* and the reduction rate r. Among the two of them, the expansion coefficient *rate* determines the size of the feature receptive field and promotes the aggregation of contextual information; the reduction rate r mainly controls the capacity and minimum value of the spatial attention module. Specifically, as shown in Figure 9.8, first, a 1×1 convolution is used to reduce the size of the features $F \in \mathbb{R}^{C\times H\times W}$ to $\mathbb{R}^{C/r\times H\times W}$ to achieve the goal of integrating and compressing feature maps in the channel dimension. After the feature is reduced, two 3×3 dilated convolutions are used to effectively utilize contextual information. Then, another 1×1 convolution

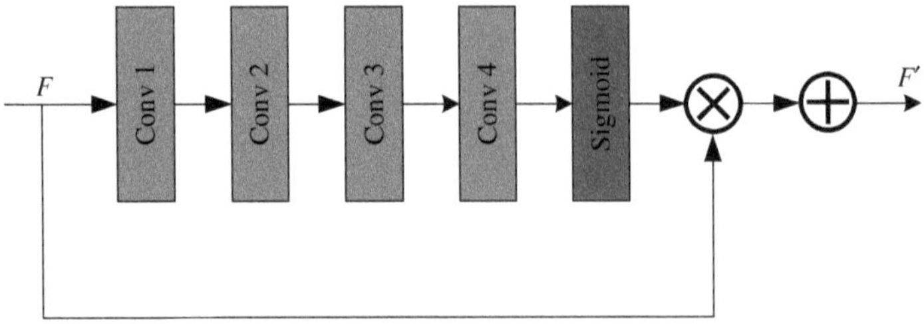

Figure 9.8. Structure of spatial attention module.

is used to simplify the features $\mathbb{R}^{1\times H\times W}$ to obtain a spatial attention map $M(F1)$. Second, the Sigmoid function is taken to obtain a spatial attention map $M(F)$ in the range of (0,1). Finally, it will be multiplied by the input feature map F and added to the original input feature map to obtain the refined feature map F'. The spatial attention $M(F)$ is calculated by the following:

$$M(F) = \sigma(BN(f_3^{1\times1}(f_2^{3\times3}(f_1^{3\times3}(f_0^{1\times1}(F)))))). \qquad (9.5.6)$$

9.5.1.1.3 *Channel attention module*

In module 6 of the AMGR-Net, the channel attention module is used to model the dependency relationships between channels in the feature map, thereby increasing the feature expression ability of this stripe removal network. As shown in Figure 9.9, the first step is to perform a squeezing operation. The input feature X (including feature maps) is subjected to a global average pooling operation. Each feature map is compressed to make it have a global receptive field; the C feature maps of feature X become a $1 \times 1 \times C$ sequence of real numbers. In order to utilize the aggregated information generated via the squeezing operation, it is necessary to perform an incentive operation by using parameters to generate weights for each feature channel, and to comprehensively capture the dependencies between channels. Two fully connected layers are introduced to not only limit the complexity of the model but also fully consider its generalization ability, with a dimensionality reduction ratio of 6. A simple threshold mechanism after two fully connected layers is adopted, and ReLU and Sigmoid activation functions are used, respectively. Finally, the output weights of the incentive operation are treated

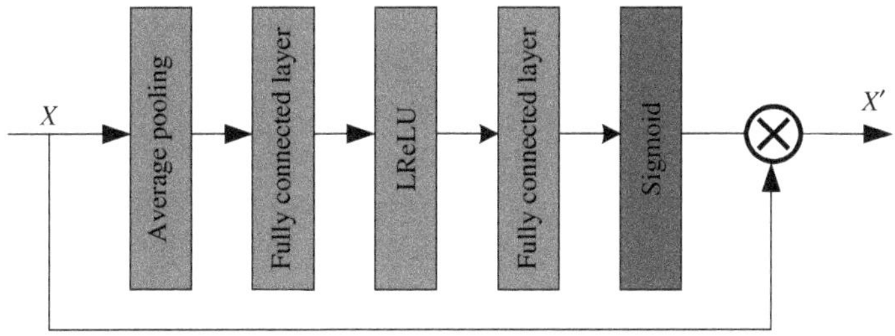

Figure 9.9. Structure of channel attention module.

as the weight information of each feature channel by resetting the weights; then, a multiplication operation is used to weight channel by channel onto the original input feature X to result in a new feature X'. This achieves recalibration of the original features in the channel dimension and increases the discriminative power of the features.

Due to the presence of different stripe layers in the input images, it is difficult to completely remove the stripes in the images through only one stage. Therefore, in this section, the RNN is used to decompose the removal stripe process into four stages, which treat the input images at different stages as time series containing stripe images of different levels. At the same time, in order to better utilize the feature information at the previous time step and guide the learning at subsequent time steps, the GRU loop units are used, which can be represented as follows:

$$O_1 = O(k), \tag{9.5.7}$$

$$R(k) = \sum_{i=1}^{n} \alpha_i \bar{R}^i(k) = f_{\mathrm{CNN+RNN}}(O(k), x(k-1)), \quad 1 \le k \le N_s, \tag{9.5.8}$$

$$O(k+1) = O(k) - R(k), \quad 1 \le k \le N_s, \tag{9.5.9}$$

$$R = \sum_{k=1}^{N_s} R(k), \tag{9.5.10}$$

where O represents the input image containing stripes, and N_s represents the number of stripe removal stages; $R(k)$ represents the output of the kth time step, $\bar{R}^i(k)$ is the ith stripe layer decomposed at the kth time step; $O(k)$ is the inputs at the kth time step; $x(k-1)$ is the hidden state at the $(k-1)$th time step, $O(k+1)$ is the stripe removal result outputted at the kth time step, and $R(k)$ is calculated by adding $\bar{R}^i(k)$ to different α values.

The AMGR-Net consists of 6 modules with 24 convolutional kernels. Among them, the first five modules include the gated loop units, spatial attention modules, and activation functions, and use a 3×3 convolutional kernel. The first module serves as an encoder to convert the image into a feature map. Based on considering that a large receptive field helps to obtain a large amount of contextual information, dilated convolutions with dilation factors of 1, 2, 4, and 8 are used from modules 2 to 5, respectively, with corresponding receptive field sizes of 5×5, 9×9, 17×17, and 33×33, as well as the sixth module, which includes convolutional layers, channel attention modules, and activation functions and uses a 3×3 convolution kernel with a receptive field size set to 35×35. After module 6, a 1×1 convolutional layer is used with a receptive field size set to 35×35 to act as a decoder for generating residual maps. In addition, the LReLU is used as the activation function to preserve stripe detail information with negative eigenvalues.

9.5.1.2 *Loss function*

Mean Square Error (MSE) is widely used to minimize loss function for training deep neural networks. A loss function based on the MSE may be called the MSE loss function. However, the MSE loss function, due to being penalized by the square, performs poorly on the edges of the images and often results in excessively smooth restoration results. To address these issues, the MSE and the SSIM are usually chosen as the loss functions to achieve a balance between improving the removal stripe effect of rain bands and better preserving detailed information. The loss function is defined as follows:

$$\text{Loss} = \sum_{k=1}^{N_s} a\text{Loss}_{\text{MSE}}(\hat{R}(k), R(k)) + (1-a)\text{Loss}_{\text{SSIM}}(\hat{R}(k), R(k)), \tag{9.5.11}$$

where Loss_{MSE} and $\text{Loss}_{\text{SSIM}}$ represent the MSE and SSIM loss functions, respectively, k represents the number of removal stripe stages, $\hat{R}(k)$ represents the predicted stripe information of the network, R represents the true stripe information, and α represents the parameter that balances the two losses.

9.5.2 Simulation experiments and result analyses

9.5.2.1 *Experimental data and training strategy*

9.5.2.1.1 *Single-image data*

The most difficult dataset, RainH1800, which contained five different directions and intensities of rain stripes, was chosen to train and test the AMGR-Net. In it, the training set contained 1800 pairs of synthetic rain maps and clear images. When training the network, 100 images with 64×64 blocks from each pair of the data in the training set were randomly generated. The testing dataset contained 100 dense rain stripe data (RainH100). Furthermore, the performance of the proposed method was tested using real rainy-day images.

9.5.2.1.2 *Training strategy*

In the experiments, the Pytorch framework was used to build the network, and it was implemented on a workstation with an NVIDIA GTX 1080 GPU. The Adam algorithm was used to optimize the network model. The batch size and initial learning rate for the training were set to 32 and 0.005, respectively. The number of training iterations was 20000. When the number of iterations was 15000 and 17500, the learning rate was divided by 10.

9.5.2.2 *Experimental process and result analyses*

A recurrent neural network was used to decompose the rain removal process into four stages. The AMGR-Net was trained on the synthetic dataset RainH1800. The performance of the AMGR-Net was evaluated by comparing the rain removal results output from each stage. Table 9.1 shows the average SSIM and PSNR metrics of the AMGR-Net model at different stages on the synthetic test dataset RainH100, which indicates that the rain removal results of the fourth stage (i.e., the final result graph of the AMGR-Net) are higher than those of the first three stages.

Table 9.1. Comparison of results at different stages of AMGR-Net.

Stages	PSNR	SSIM
1	25.63	0.845
2	28.05	0.880
3	28.82	0.890
4	**29.03**	**0.892**

Figure 9.10 shows the comparison of rain removal results of AMGR-Net in four stages. It can be seen from Figure 9.10 that the visual effect of the rain removal image in the first stage is poor, while in the second, third, and fourth stages, after removing rain stripes, the details and chromaticity information of the images are well preserved. The comparison of the PSNR with the SSIM for the four stages of rain removal results is shown in Table 9.2, which indicates that the PSNR and the SSIM of the fourth stage are higher than those of the first three stages. Through comparative experiments, it can be found that AMGR-Net is effective in removing rain stripes from images in the fourth stage.

9.5.2.3 *Comparison with four representative rain removal methods*

In order to verify the effectiveness of the AMGR-Net, the trained network model was tested on newly synthesized rain maps and real rain maps, and four representative rain removal methods were selected for comparison of the results. These four methods are DSC, GMM-LP, Residual Guidance Network (RGN), and Recurrent Squeeze and Excitation Context Aggregation Net (RESCAN).

9.5.2.3.1 *Synthetic rain map and result analyses*

The AMGR-Net was compared with the four representative rain removal methods on the synthetic rain map dataset, and the results are shown in Figure 9.11.

Figure 9.11 shows that there is a serious problem of residual rain stripes in the rain removal images processed by the DSC and GMM-LP methods. The main reason may be that the DSC and GMM-LP

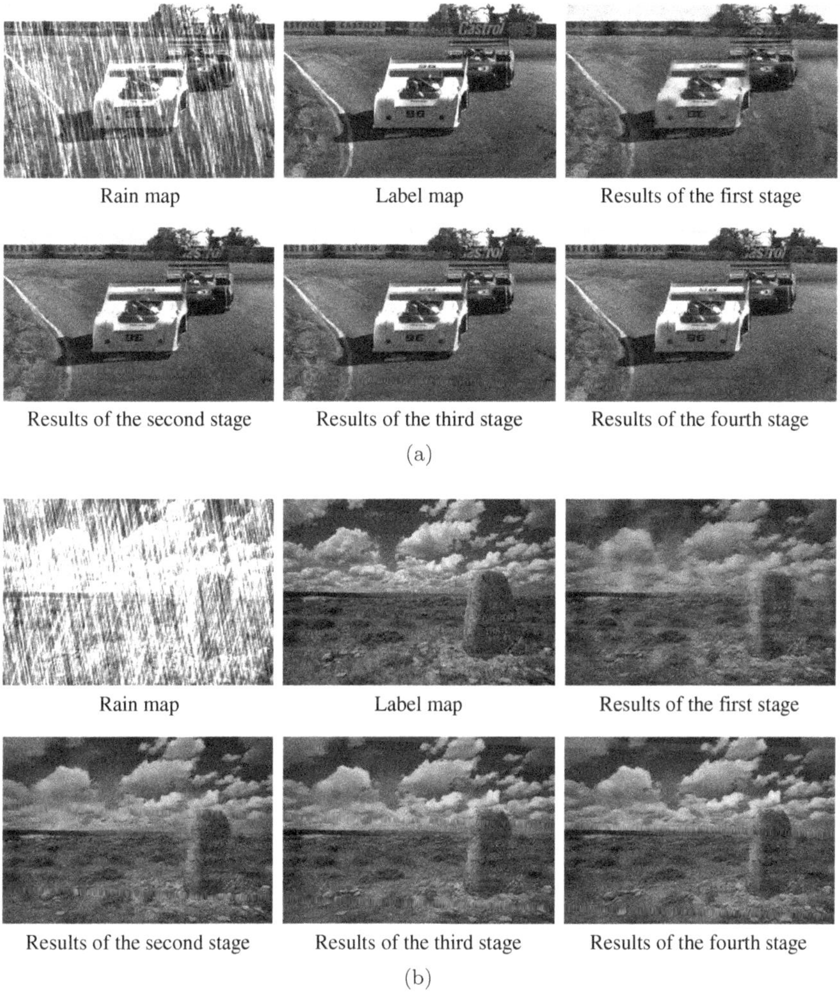

Figure 9.10. Precipitation results of AMGR-Net Model at different stages. (a) Racing car. (b) Stone tablet.

methods do not utilize high-level feature information in the original image. When the rain stripes in the image are severe, the rain removal effect of these two methods is the worst. The RGN and RESCAN methods can remove most of the rain stripes in the images, but they cannot effectively preserve the details of the images. In the AMGR-Net, spatial attention modules are used to enhance or suppress features at different spatial positions, and channel attention modules are

Table 9.2. Comparison of results at different stages in the AMGR-Net model.

Figure Number	Index	Stage 1	Stage 2	Stage 3	Stage 4
Figure 7.10	PSNR	26.43	28.99	29.64	**29.82**
	SSIM	0.857	0.900	0.903	**0.905**
Figure 7.11	PSNR	23.64	25.23	25.87	**26.14**
	SSIM	0.758	0.792	0.804	**0.808**

used to model the dependency relationships between channels in the feature map. These measures have increased the feature expression ability of this rain removal network. The AMGR-Net can accurately distinguish between non-rain stripes and rain stripes in the images by fully utilizing deep feature information, such as channel and spatial information. It achieves the effect of effectively removing rain stripes while preserving the original image details. In addition, recurrent neural networks can better utilize the feature information of the previous stage and guide the learning of subsequent stages. The rain removal result map of AMGR-Net has a better contrast effect, the restored image background is clearer, and a better visual effect of the images is achieved compared with the original rainless images.

Now, the PSNR and SSIM of the rain removal result maps are compared in Figure 9.11, and the PSNR and SSIM of the rain removal result maps on the dense rain stripe dataset RainH100 are further compared, as shown in Table 9.3. Table 9.3 shows that the AMGR-Net outperforms other methods in terms of the PSNR and the SSIM. The effectiveness of the AMGR-Net has been verified through these comparative experiments.

9.5.2.3.2 *Real rain map and result analyses*

In order to demonstrate that the AMGR-Net also has good rain removal effects on real rain maps, the experiments have been conducted on real rain maps using AMGR-Net, and the result maps of the four rain removal methods mentioned earlier are shown in Figure 9.12.

From Figure 9.12, it can be seen that the DSC and GMM-LP methods retain a significant amount of rain stripes. From a visual perspective, in the process of restoring real rain images, the rain

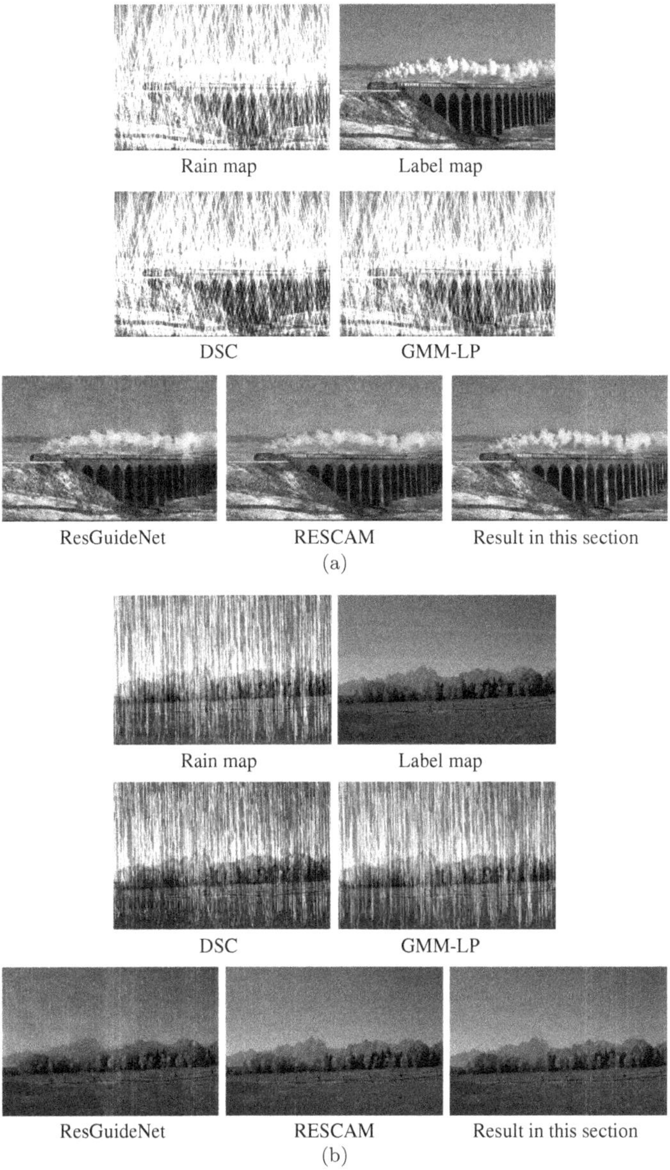

Figure 9.11. Comparison of experimental results on synthetic rain maps. (a) Dam. (b) Mountain scenery. (c) Elephants.

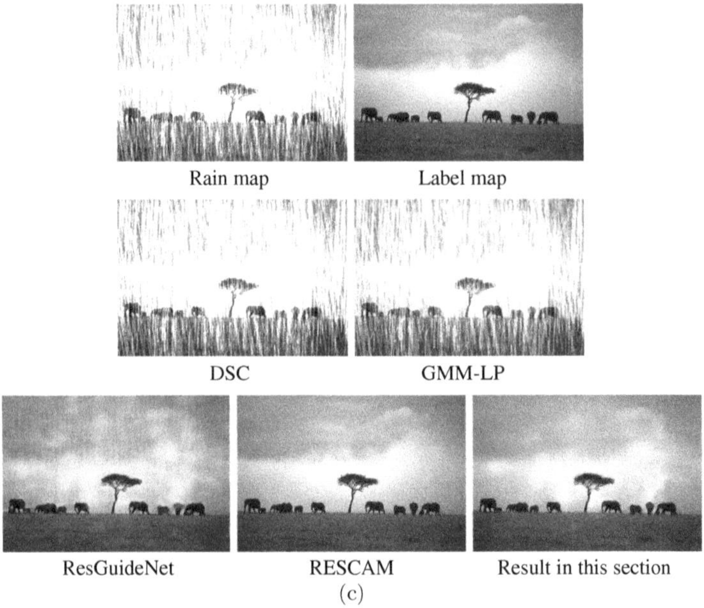

Figure 9.11. (*Continued*)

Table 9.3. Image restoration quality of different methods.

Image	Index	DSC	GMM-LP	RGN	RESCAN	Proposed AMGR-Net
a	PSNR	11.44	11.62	22.91	24.73	**25.71**
	SSIM	0.295	0.312	0.769	0.822	**0.883**
b	PSNR	15.63	14.97	27.71	32.29	**33.88**
	SSIM	0.237	0.249	0.879	0.886	**0.942**
c	PSNR	14.96	15.82	29.08	31.19	**31.83**
	SSIM	0.466	0.527	0.882	0.902	**0.945**
RainH100	PSNR	15.19	14.90	24.88	28.47	**29.03**
	SSIM	0.384	0.411	0.840	0.856	**0.892**

removal effect of the RGN and the RESCAN methods is not much different from that of the proposed AMGR-Net, but the proposed AMGR-Net can better preserve the details and chromaticity information of the images.

Figure 9.12. Comparison of real rain map results. (a) Natural landscape 1. (b) Natural landscape 2. (c) Natural landscape 3.

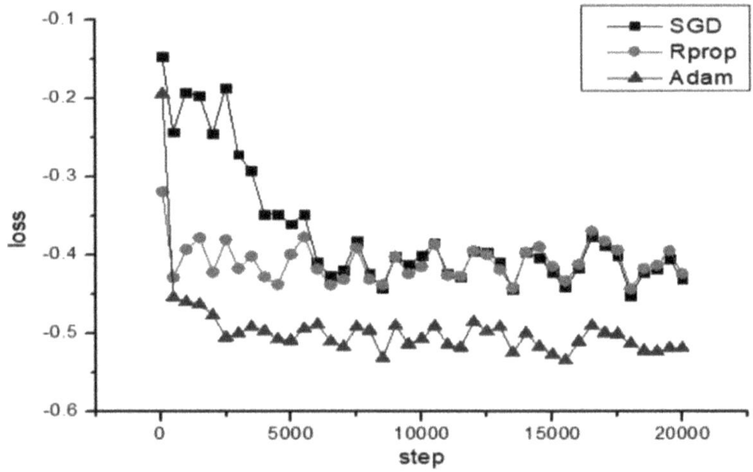

Figure 9.13. Convergence speeds of Rprop, SGD, and Adam algorithms.

9.5.2.3.3 *The influence of different optimization algorithms on the convergence speed*

A comparative experiment was conducted on the convergence speed of the SGD, Resilient Propagation (RProp) [196], and Adam algorithms. The results are shown in Figure 9.13, where graph lines with square, circle, and triangle curves represent the convergence tendency of the neural network when using SGD, RProp, and Adam algorithms, respectively. Figure 9.13 shows that the SGD and RProp algorithms have slower descent speeds and unstable convergence processes compared with the Adam algorithm, which verifies the effectiveness in the selection of the optimizer.

9.6 Example 11: Single-Image Super-Resolution Method Based on Cascaded Recursive Residual Convolutional Neural Network

The image super-resolution (SR) problem consists of two parts: single image and multi-image. The Single-Image Super-Resolution (SISR) method mainly includes interpolation, reconstruction, and learning-based methods [197]. The image super-resolution method aims to obtain High-Resolution (HR) images from Low-Resolution (LR) images. The emergence of algorithms and neighborhood embedding algorithms has slightly improved image quality. The earliest interpolation algorithms include linear interpolation, bicubic interpolation, and Lanczos. Other algorithms are neighborhood embedding and sparse coding, which achieve image super-resolution by learning the mapping function between low-resolution images and high-resolution images. Although these algorithms have a fast convergence speed, the quality of restoration images is not ideal and they are prone to producing flat textures, which cannot effectively restore high-frequency components. In recent years, with the emergence of CNNs, the performance of the SISR algorithm has significantly improved.

Super-Resolution CNN (SRCNN) is a pioneering research field of deep learning super-resolution reconstruction. Although there are only three layers in the network, it tends to map between the LR and the HR image spaces, and there are also some issues, such as loss of image details and long training times. So, the improved version of SRCNN (Fast SRCNN, FSRCNN) has the following three aspects:

(1) It uses deconvolution to enlarge the images, rather than the need for bicubic enlargement of LR images.
(2) The feature dimension is filled with smaller convolution kernels.
(3) Although the FSRCNN performs well, its shallow network structure limits its performance. The emergence of reset solves the problem of non-training when the network structure is deep.

In super-resolution using Very Deep Convolution Network (VDSR), since the low-frequency information carried by the input low-resolution image is largely similar to that of the high-resolution images, it spends a lot of time training this part. In fact, only the high-frequency residual part among high-resolution images and low-resolution images needs to be learned. Almost simultaneously with VDSR, effective super-resolution reconstruction algorithms include Deeply Recursive Convolution Network (DRCN) [198], Sub-Pixel Convolutional Neural Network (SPCNN), Laplacian SRN, and Super-Resolution Dense Network (SRDenseNet), but there are also areas that need improvement.

9.6.1 Network architecture

A Super-Resolution CNN (SRCNN) model is shown in Figure 9.14. The SRCNN model is divided into three parts: (1) color spot extraction and representation; (2) nonlinear mapping; and (3) reconstruction. In the SRCNN, low-resolution images enlarged by bicubic interpolation are used as input, but the high computational cost hinders its application. The FSRCNN has five parts: feature extraction layer, contraction layer, nonlinear mapping layer, expansion layer, and deconvolution layer.

Extremely deep CNNs are used to mix and train multiple different images, so that the trained model can solve the super-resolution problem of multiple different images. The VDSR takes the interpolated low-resolution image of the target size as the input of the network and then adds the image to the residual learned by the network to obtain the final output of the network. In the DRCN, recursive structure is first used to SISR convolutional neural networks. Meanwhile, residual learning, utilized to increase network depth and reduce training difficulty, has achieved good results. For the DRCN, its input is an interpolated image, which consists of three modules; among them, the first module uses embedded networks to extract feature maps, the second module is an inference network with the same nonlinear feature mapping function, and the last module is a reconstruction

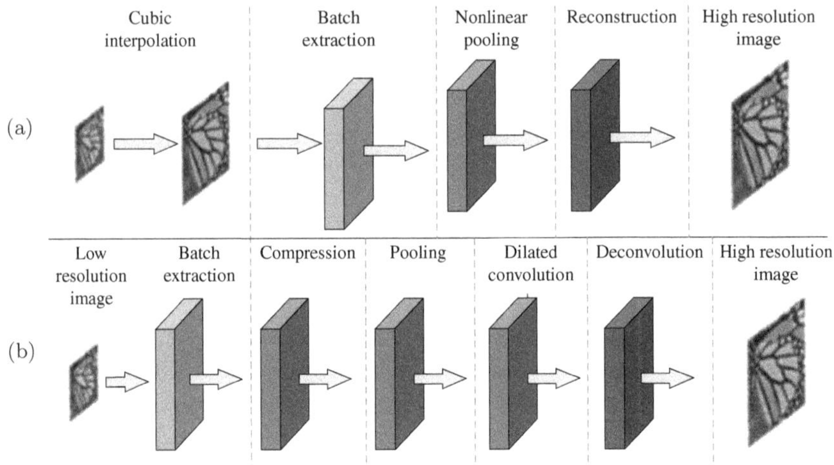

Figure 9.14. Structure of network. (a) SRCNN. (b) FSRCNN.

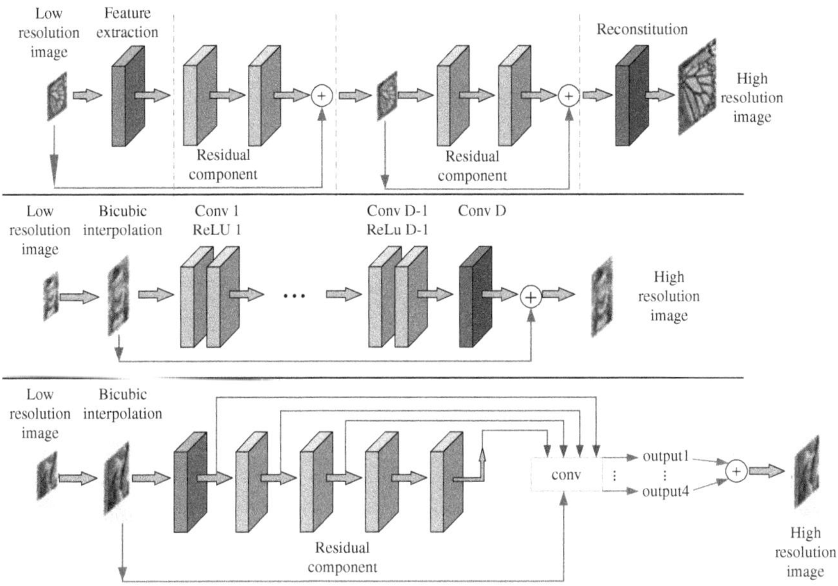

Figure 9.15. Three network architectures.

component that restores the final reconstruction result from the feature image. On the basis of the VDSR and DRCN, each remaining unit in the DRRN shares the same input, and each remaining unit in the Deep Recursive Residual Network (DRRN) uses the output of the first convolutional layer in the recursive block as input, which has a relatively deep structure to achieve performance gains. Figure 9.15 shows the operational framework of these methods. Inspired by the above-mentioned methods, researchers have proposed a Cascaded Recursive Residual Convolution Neural Network (CRRCNN).

9.6.2 Three-layer skip connection recursive residual network

The Cascaded Recursive Residual Convolutional Neural Network (CRRCNN) model is shown in Figure 9.16, which consists of three parts: color shift extraction, residual learning, and reconstruction. The input of the CRRCNN model is a low-resolution images without magnification, which cancels out the preprocessing of interpolating and enlarging the original low-resolution images. Due to the fact that

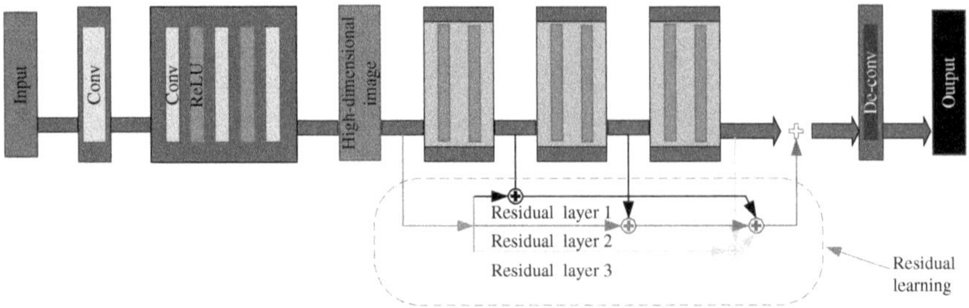

Figure 9.16. Structure of CRRCNN.

CRRCNN does not require expanding the image size outside the network, the last layer of the network is the deconvolution layer, which replaces the preprocessing step of bicubic enlargement and greatly improves the training speed. The network also adopts the recursive structure and residual learning, which has a significant effect on processing single-image super-resolution problems. Compared with other algorithms, the proposed method is simple and can produce better results.

9.6.2.1 *Spot extraction*

For single-image super-resolution, a popular strategy is to extract color spots through convolutional sets and then represent them using a set of pretrained methods such as the PCA, the DCT, and the Haar. The essence of the convolution operation is to calculate the pixels of an image using a filter. Currently, $\boldsymbol{x}$ represents low-resolution images in color spot extraction and P_1 represents an operation. The relationship between $\boldsymbol{x}$ and P_1 can be written as follows:

$$P_1 = \max(0, \boldsymbol{W} \otimes \boldsymbol{x} + b_1), \tag{9.6.1}$$

where b_1 represents the bias, $\boldsymbol{W}$ represents the m_1 filters with $C \times f_1 \times f_1$ dimensions, and C represents the number of channels in the input images. The features are extracted from the input images, and the ReLU function is used as the activation function.

9.6.2.2 *Cascaded recursive residual learning network*

In image restoration tasks, as the network layer deepens, many image details may be lost or damaged. The feature map obtained through a recursive residual learning network can provide more details. In addition, the output of each rolling basis used as a residual component helps to better restore the high-frequency components of the images. $\boldsymbol{x}$ represents a high-resolution image and $\boldsymbol{x}'$ is the result of residual layer 1. We have

$$\boldsymbol{x}' = \max(0, \boldsymbol{W}_1 \otimes \boldsymbol{x} + b_2), \tag{9.6.2}$$

$$\boldsymbol{x}^{\text{res1}} = \boldsymbol{x} + \boldsymbol{x}', \tag{9.6.3}$$

$$\boldsymbol{x}^{\text{res2}} = \text{ReLU}(F(\boldsymbol{x}', \boldsymbol{W}_2)) + \boldsymbol{x}, \tag{9.6.4}$$

$$\begin{aligned} \boldsymbol{x}^{\text{res3}} &= \text{ReLU}(F(F(\boldsymbol{x}', \boldsymbol{W}_2), \boldsymbol{W}_3)) + \boldsymbol{x} \\ &= \boldsymbol{x}'' + \boldsymbol{x}, \end{aligned} \tag{9.6.5}$$

where x^{res1}, x^{res2}, and x^{res3} represent the outputs of residual layer 1, residual layer 2, and residual layer 3, respectively, in Figure 9.16. $\text{ReLU}(x, w)$ denotes a nonlinear mapping equation of the ReLU function. $F(\boldsymbol{x}, \boldsymbol{w})$ represents a convolutional layer. The filter's sizes corresponding to $\boldsymbol{W}_2$ and $\boldsymbol{W}_3$ are $C \times f_3 \times f_3$ and $C \times f_4 \times f_4$.

9.6.2.3 *Reconstitution*

In previous methods, image interpolation was usually used in the preprocessing stage to perform enlargement operations. The last layer of the model in this section is the anti-roll base layer, which can reduce computational complexity and improve speed. The deconvolution layer can be represented as Decon $(-)$, where y represents the main residual component:

$$\boldsymbol{y} = \boldsymbol{x}^{\text{res1}} + \boldsymbol{x}^{\text{res2}} + \boldsymbol{x}^{\text{res3}} + \text{Decon}(n, m, \boldsymbol{x}''), \tag{9.6.6}$$

$$\boldsymbol{y}' = \min(||O - H(\boldsymbol{y}, \boldsymbol{w}^H)||^2), \tag{9.6.7}$$

$$\boldsymbol{y}'' = H(\boldsymbol{y}, \boldsymbol{w}^R), \tag{9.6.8}$$

where O represents the original image, $\boldsymbol{w}^H$ and $\boldsymbol{w}^R$ are the parameters of the filter in the convolutional layer, and $H(a, b)$ represents the transfer function. Final output $\boldsymbol{y}_{\text{out}}$ is obtained by the following:

$$\boldsymbol{y}_{\text{out}} = \boldsymbol{y} + \boldsymbol{y}''. \tag{9.6.9}$$

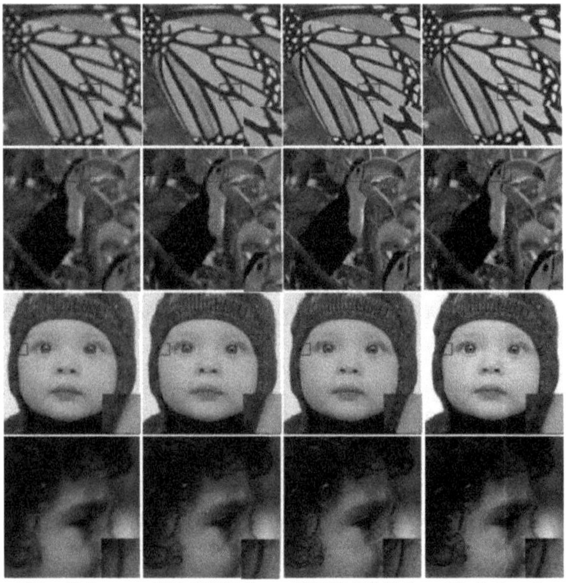

Figure 9.17. Recovering a 4-level super-resolution image from bicubic magnification.

9.6.3 Simulation experiments and result analyses

The experimental results of detecting single-channel images at 4× and 3× magnification are shown in Figures 9.17 and 9.18. The results in Figures 9.17 and 9.18 indicate that compared to the image quality obtained by other methods, the images processed by this proposed method are closer to the real images. The comparison between the corresponding peak signal-to-noise ratio (PSNR) is shown in Table 9.4. Table 9.4 indicates that the performance of the proposed method has a higher PSNR value than other methods. In addition, Figure 9.18 depicts the PSNR of different algorithms under triple magnification, and the results indicate that FSRCNN is competitive.

In summary, the image super-resolution method based on cascaded recursive residual CNN emphasizes the importance of the cascaded recursive residual learning network, which solves the problem of detail loss when the network layers are deepened. This proposed

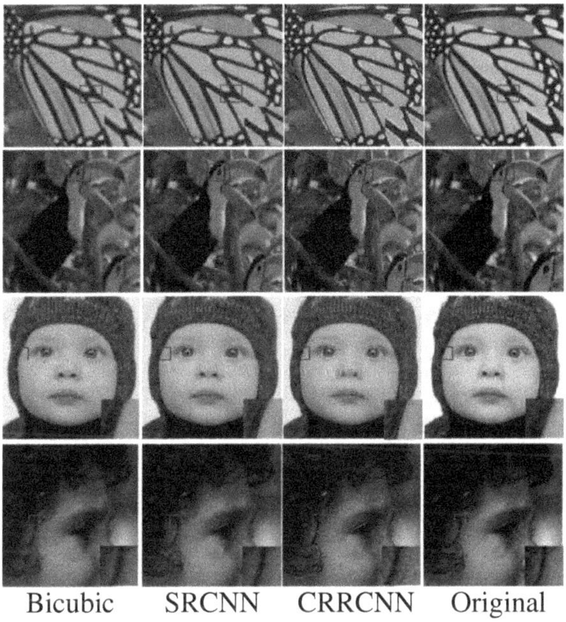

Figure 9.18. Recovering a 3-level super-resolution image from bicubic magnification.

Table 9.4. PSNR values for recovery ratios of 3 and 4.

Dataset	Magnification	Bicubic	SRCNN	VDSR	DRCN	CRRCNN
Dataset 5	3×	30.39	32.75	33.66	33.82	34.62
	4×	28.42	30.48	31.35	31.53	32.27
	5×	27.16	—	—	—	29.62
Dataset 14	3×	27.55	29.28	29.77	29.76	30.22
	4×	26.01	27.49	28.00	28.02	28.41
	5×	24.99	—	—	—	26.13

method can reconstruct more image details, make image edge information clearer, effectively suppress artifacts and outliers, has strong potential and universality, and can achieve accurate reconstruction of the single low-resolution images.

Chapter 10

Long Short-Term Memory Networks

Summary

Based on the relationship and differences between Long Short-Term Memory (LSTM) networks and Recurrent Neural Networks (RNNs), the state control gates are first introduced and the output of each gate is calculated. The BackPropagation Through Time (BPTT) algorithm of the LSTM network is derived in detail. On this basis, the structure and principle of the dual convolutional LSTM network are analyzed. Taking non-cooperative underwater acoustic communication modulation signal recognition as an example, a LSTM classifier based on instantaneous characteristics of communication signals and performance evaluation indicators is presented, and the classification performance is compared and analyzed. Taking fingerprint outdoor positioning as an example, a hybrid LSTM mechanism is introduced, and the accuracy of outdoor positioning is improved. Through these two examples, readers can understand how important it is to find effective solutions to problems.

10.1 Introduction

The topology structure of Recurrent Neural Network (RNN) has self-feedback mechanism and certain memory ability. Although the RNN can handle time-series data of any length, it is prone to gradient vanishing or exploding when training long sequences. To solve these problems, gate units of Long Short-Term Memory (LSTM) networks have been introduced to control the iteration speed of memory, and the efficiency and stability of prediction models have further been improved. The LSTM network has become the most popular RNN and has been successfully applied in many fields, such as speech recognition, image description, and natural language processing [198].

In this chapter, the basic principles and applications of the LSTM network are studied.

10.2 Long Short-Term Memory Network

The hidden layer of the original RNN only has one state h, which is very sensitive to short-term inputs. So, if another state c is added and used to preserve the long-term state, the state gates of the RNN and LSTM network are shown in Figure 10.1.

The newly added state c is called a cell state, and it is expanded according to the time dimension, as shown in Figure 10.2.

In Figure 10.2, there are three inputs to the LSTM network: the input of the network $\boldsymbol{x}(k)$ at time k, the output of the LSTM network $\boldsymbol{h}(k-1)$ at the previous time $k-1$, and the unit state $\boldsymbol{c}(k-1)$ at the previous time $k-1$. There are two outputs of the LSTM network: the output of the LSTM network $\boldsymbol{h}(k)$ at time k and the unit state $\boldsymbol{c}(k)$ at time k. Note that $\boldsymbol{x}(k)$, $\boldsymbol{c}(k)$, and $\boldsymbol{h}(k)$ all are vectors.

The key issue with the LSTM network is how to control the long-term unit state $\boldsymbol{c}$. Here, the idea of the LSTM network is to use

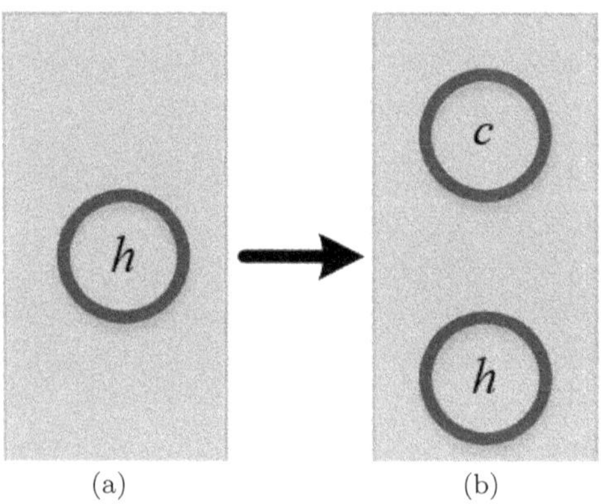

Figure 10.1. The state gates of RNN and LSTM. (a) Original RNN. (b) LSTM.

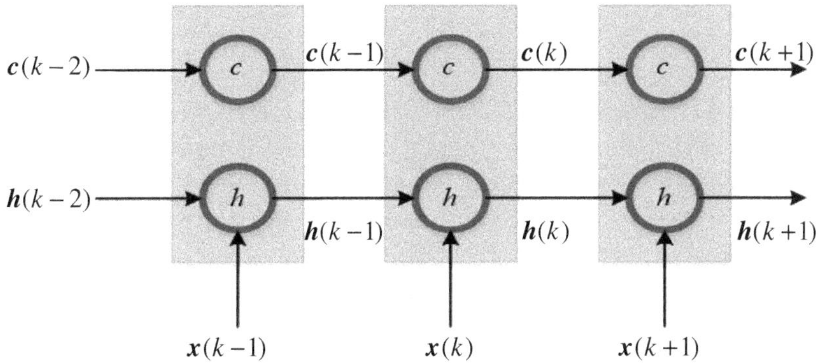

Figure 10.2. Status expansion.

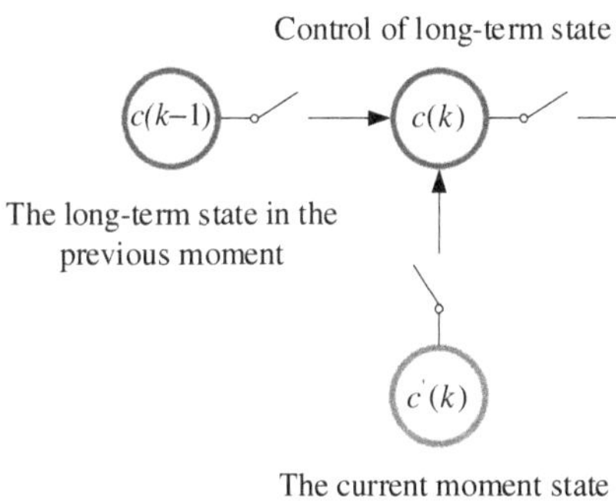

Figure 10.3. State control.

three control switches: the first switch of which is responsible for controlling the continued preservation of the long-term unit state c, the second switch of which is responsible for controlling the input of instant state to long-term state c, and the third switch of which is responsible for controlling whether to use the long-term unit state $\boldsymbol{c}$ as the output of the current LSTM network. The functions of the three switches are shown in Figure 10.3.

10.2.1 Forward calculation

How is the switch described earlier implemented in the LSTM network? It is necessary to introduce the concept of a gate. A gate is actually a fully connected layer, whose input is a vector and whose output is a real number vector between 0 and 1. Assume that $\boldsymbol{W}$ is the weight vector of the gate and $\boldsymbol{b}$ is the bias term, then the gate can be represented as

$$g(x) = \sigma(\boldsymbol{W}\boldsymbol{x} + \boldsymbol{b}). \tag{10.2.1}$$

The use of gates is to multiply the output vector of the gate by the element of the vector to be controlled. Since the output of the gate is a real vector between 0 and 1, when the gate's output is zero vector, anyone vector multiplied by it will turn into zero vector, which is equivalent to no information being able to pass through this gate. When the output is equal to one, anyone vector multiplied by it will not change, which is equivalent to any information being able to pass through this gate. Since the value range of the Sigmoid function σ is within (0,1), the states of the gates are always half open and half closed.

In the LSTM network, two gates are used to control the content of the unit state. Among them, one is the forget gate, which determines how much of the unit state $c(k-1)$ from the previous time $k-1$ is retained until the unit state $c(k)$ at the current time k, and the other is the input gate $x(k)$, which determines how much of the network's input is saved to the unit state $c(k)$ at the current time k, and the output gates are used to control how much unit state $c(k)$ is outputted as the current output value $h(k)$ of the LSTM network.

For the forgetting gate, there is the equation as follows:

$$f(k) = \sigma(\boldsymbol{W}_f[\boldsymbol{h}(t-1), \boldsymbol{x}(t)] + \boldsymbol{b}_f), \tag{10.2.2}$$

where $\boldsymbol{W}_f$ is the weight matrix of the forgetting gate, $[\boldsymbol{h}(t-1), \boldsymbol{x}(t)]$ represents the connection of two vectors into a longer vector, $\boldsymbol{b}_f$ denotes the bias term of the forgetting gate, and σ expresses the Sigmoid function. If the input dimension is D_x, the dimension of the hidden layer is D_h, and the dimension of the cell state is D_c (usually $D_c = D_{\mathrm{h}}$), then the dimension of the weight matrix $\boldsymbol{W}_f$ for the forgetting gate is $D_c \times (D_h + D_x)$. In fact, the weight matrix $\boldsymbol{W}_f$ is composed of two matrices concatenated together: One is $\boldsymbol{W}_{hf}$ corresponding to the input term $\boldsymbol{h}(k-1)$, and its dimension is $D_c \times D_h$. Another is $\boldsymbol{W}_{fx}$ corresponding to the input item $\boldsymbol{x}(k)$ with dimensions $D_c \times D_x$. The equation containing $\boldsymbol{W}_f$ can be written as

$$\begin{aligned}[\boldsymbol{W}_f]\begin{bmatrix} h(k-1) \\ x(k) \end{bmatrix} &= [\boldsymbol{W}_{fh}\ \boldsymbol{W}_{fx}]\begin{bmatrix} \boldsymbol{h}(k-1) \\ \boldsymbol{x}(k) \end{bmatrix} \\ &= \boldsymbol{W}_{fh}\boldsymbol{h}(k-1) + \boldsymbol{W}_{fx}x(k). \end{aligned} \tag{10.2.3}$$

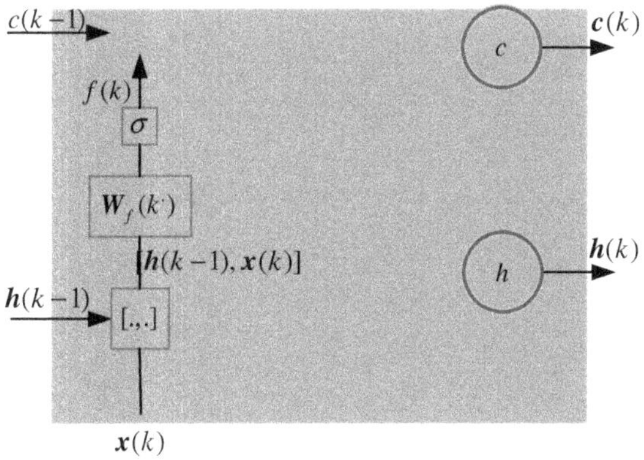

Figure 10.4. The calculation process of the forgetting gate.

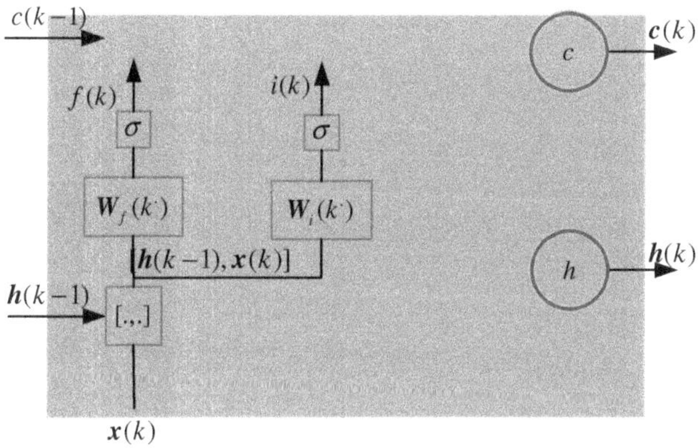

Figure 10.5. The calculation process of the input gate.

The calculation process of the forgetting gate is shown in Figure 10.4.

The forgetting gate is defined as

$$i(k) = \sigma(\boldsymbol{W}_i[\boldsymbol{h}(k-1), \boldsymbol{x}(k)] + \boldsymbol{b}_i), \tag{10.2.4}$$

where $\boldsymbol{W_i}$ is the weight matrix of the input gate and $\boldsymbol{b}_i$ is its bias term. The calculation process of the input gate is shown in Figure 10.5.

The current input unit state $\tilde{c}(k)$ is calculated to describe the current input based on the previous output and the current input.

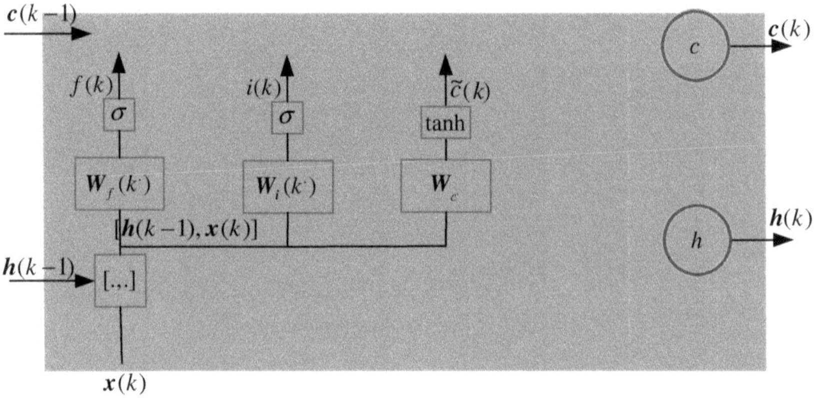

Figure 10.6. The calculation process of $\tilde{c}(k)$.

The current input unit state $\tilde{c}(k)$ is defined as

$$\tilde{c}(k) = \tanh(\boldsymbol{W}_c[\boldsymbol{h}(k-1), \boldsymbol{x}(k)] + \boldsymbol{b}_c). \tag{10.2.5}$$

The calculation process of $\tilde{c}(k)$ is shown in Figure 10.6.

Now, we calculate the unit state $\boldsymbol{c}(k)$ at the current time k, which is generated by multiplying the unit state $\boldsymbol{c}(k-1)$ at the previous time $k-1$ by the forgetting gate vector $\boldsymbol{f}(k)$ according to the element, then multiplying the current input unit state $\tilde{\boldsymbol{c}}(k)$ by the input gate vector $\boldsymbol{i}(k)$ according to the element, and finally adding up the two products. The calculation formula of $\boldsymbol{c}(k)$ is written as

$$\boldsymbol{c}(k) = \boldsymbol{f}(k) \odot \boldsymbol{c}(k-1) + \boldsymbol{i}(k) \odot \tilde{\boldsymbol{c}}(k), \tag{10.2.6}$$

where $\odot$ represents multiplying by elements. The calculation process of $\boldsymbol{c}(k)$ is shown in Figure 10.7.

In this way, in the LSTM network, the current memory state $\tilde{\boldsymbol{c}}(k)$ is combined with long-term memory state $\boldsymbol{c}(k-1)$ to form a new unit state $\boldsymbol{c}(k)$. Due to the control of the forgetting gate, it can store information from a long, long time ago. As a result of the control of the input gate, it can also prevent irrelevant content from entering memory. And the output gate controls the impact of long-term memory on the current output. The output of the output gate is defined as

$$o(k) = \sigma(\boldsymbol{W}_o[\boldsymbol{h}(k-1), \boldsymbol{x}(k)] + \boldsymbol{b}_o). \tag{10.2.7}$$

The calculation process of the output gate is shown in Figure 10.8.

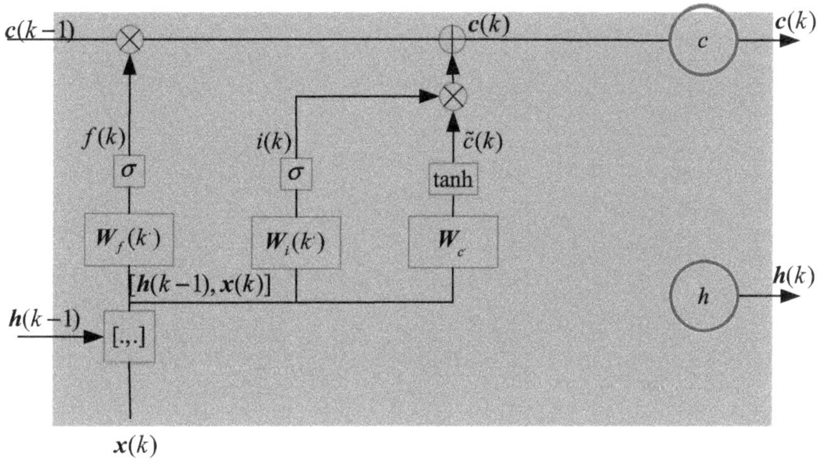

Figure 10.7. The calculation process of $\boldsymbol{c}(k)$.

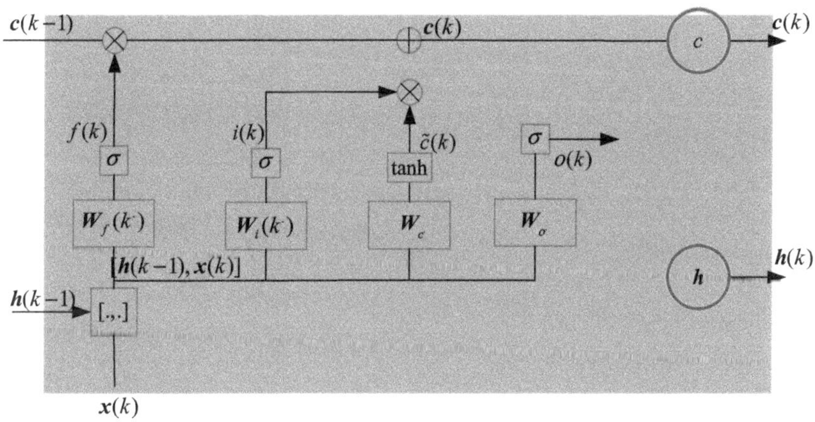

Figure 10.8. The calculation of the output gate.

The final output of the LSTM network is determined by both the output gate and the cell state, i.e.,

$$\boldsymbol{h}(k) = \boldsymbol{o}(k) \odot \tanh(\boldsymbol{c}(k)), \tag{10.2.8}$$

where $\odot$ represents that the elements at the corresponding positions of the two matrices are multiplied. When a diagonal matrix is right multiplied by a matrix, it is equivalent to multiplying the vector composed of the diagonal elements in the diagonal matrix by its

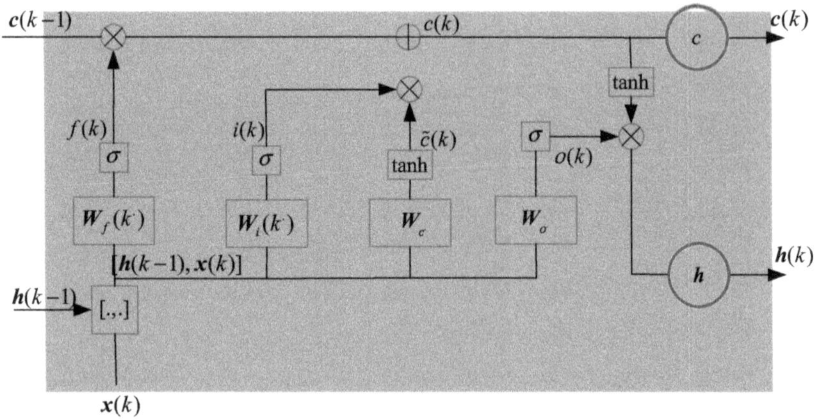

Figure 10.9. Output's calculation process of the LSTM network.

elements, i.e.,

$$\text{diag}[\boldsymbol{a}]\boldsymbol{X} = \boldsymbol{a} \odot \boldsymbol{X}.$$

When a row vector is right multiplied by a diagonal matrix, it is equivalent to multiplying the row vector composed of the diagonal elements in the diagonal matrix by its elements, i.e.,

$$\boldsymbol{a}^T \text{diag}[\boldsymbol{b}] = \boldsymbol{a} \odot \boldsymbol{b}.$$

The calculation process of the final output in the LSTM network is shown in Figure 10.9.

Equations (10.2.2) to (10.2.8) are all the formulas for the LSTM network forward calculation.

10.2.2 BPTT algorithm for LSTM network

The training algorithm for the LSTM network is still the backpropagation algorithm through BPTT. The main steps are as follows [199]:

Step 1: Forward calculation

The output value of each neuron is calculated along the forward propagation, which includes five vectors $\boldsymbol{f}(k), \boldsymbol{i}(k), \boldsymbol{c}(k), \boldsymbol{o}(k)$, and $\boldsymbol{h}(k)$ for the LSTM network.

Step 2: Backward calculation

The error term $\boldsymbol{e}(k)$ for each neuron is calculated along the backpropagation. Similar to recurrent neural networks, the backpropagation of the error terms in the LSTM network also includes two directions: One is backpropagation along time, which calculates the error term for each time step starting from the current moment. The other is to propagate the error term upwards.

Step 3: Gradient calculation

The gradient of each weight is calculated based on the corresponding error term.

Let the Sigmoid function be the activation function of the gate, and let the Tanh function be the activation function of the output. They and their derivatives are respectively obtained by

$$\sigma(x) = y = \frac{1}{1+e^{-x}}, \tag{10.2.9}$$

$$\sigma'(x) = y(1-y), \tag{10.2.10}$$

$$\tanh(x) = y = \frac{e^{x}-e^{-x}}{e^{x}+e^{-x}}, \tag{10.2.11}$$

$$\tanh'(x) = 1-y^{2} \tag{10.2.12}$$

From the above analyses, it can be seen that the derivatives of the Sigmoid fuction and Tanh function are both functions of the original function. In this way, once the value of the original function is calculated, it can be used to calculate the value of the derivative.

There are 8 sets of parameters that the LSTM network needs to learn, namely: the weight matrix $\boldsymbol{W}_f$ and bias term $\boldsymbol{b}_f$ of the forgetting gate, the weight matrix $\boldsymbol{W}_i$ and bias term $\boldsymbol{b}_i$ of the input gate, the weight matrix $\boldsymbol{W}_o$ and bias term $\boldsymbol{b}_o$ of the output gate, and the weight matrix $\boldsymbol{W}_c$ and bias term $\boldsymbol{b}_c$ of the computing unit state. Since two parts of each weight matrix use different formulas in BPTT algorithm, in subsequent derivations, the weight matrix $\boldsymbol{W}_f, \boldsymbol{W}_i, \boldsymbol{W}_c$, and $\boldsymbol{W}_o$ will all be divided into two separate matrices $\boldsymbol{W}_{fh}$ and $\boldsymbol{W}_{fx}$, $\boldsymbol{W}_{ih}$ and $\boldsymbol{W}_{ix}$, $\boldsymbol{W}_{ih}\boldsymbol{W}_{ix}$, $\boldsymbol{W}_{oh}$ and $\boldsymbol{W}_{ox}$, $\boldsymbol{W}_{ch}$ and $\boldsymbol{W}_{cx}$, respectively.

At time k, the output of the LSTM network is denoted as $\boldsymbol{h}(k)$. The error term $e(k)$ is defined as partial deriative of *Loss* with respect to $\boldsymbol{h}(k)$, i.e.,

$$e(k) = \frac{\partial \text{Loss}}{\partial \boldsymbol{h}(k)}. \tag{10.2.13}$$

Note that it is assumed here that the error term is the derivative of the loss function *Loss* with respect to the output $\boldsymbol{h}(k)$ rather than the derivative with respect to the weighted input $net^l(k)$. Since the LSTM network has four weighted inputs corresponding to $\boldsymbol{f}(k), \boldsymbol{i}(k), \boldsymbol{c}(k)$, and $\boldsymbol{o}(k)$, respectively, we now want to only pass one error term to the next layer instead of four error terms. But it is still necessary to define these four weighted inputs and their corresponding error terms. Four weighted inputs are represented as follows:

$$\begin{aligned} \boldsymbol{net}_f(k) &= \boldsymbol{W}_f\left[\boldsymbol{h}(k-1), \boldsymbol{x}(k)\right] + \boldsymbol{b}_f \\ &= \boldsymbol{W}_{fh}\boldsymbol{h}(k-1) + \boldsymbol{W}_{fx}\boldsymbol{x}(k) + \boldsymbol{b}_f, \end{aligned} \tag{10.2.14}$$

$$\begin{aligned} \boldsymbol{net}_i(k) &= \boldsymbol{W}_i\left[\boldsymbol{h}(k-1), \boldsymbol{x}(k)\right] + \boldsymbol{b}_i \\ &= \boldsymbol{W}_{ih}\boldsymbol{h}(k-1) + \boldsymbol{W}_{ix}\boldsymbol{x}(k) + \boldsymbol{b}_i, \end{aligned} \tag{10.2.15}$$

$$\begin{aligned} \boldsymbol{net}_{\tilde{c}}(k) &= \boldsymbol{W}_c\left[\boldsymbol{h}(k-1), \boldsymbol{x}(k)\right] + \boldsymbol{b}_c \\ &= \boldsymbol{W}_{ch}\boldsymbol{h}(k-1) + \boldsymbol{W}_{cx}\boldsymbol{x}(k) + \boldsymbol{b}_c, \end{aligned} \tag{10.2.16}$$

$$\begin{aligned} \boldsymbol{net}_o(k) &= \boldsymbol{W}_o\left[\boldsymbol{h}(k-1), \boldsymbol{x}(k)\right] + \boldsymbol{b}_o \\ &= \boldsymbol{W}_{oh}\boldsymbol{h}(k-1) + \boldsymbol{W}_{ox}\boldsymbol{x}(k) + \boldsymbol{b}_o. \end{aligned} \tag{10.2.17}$$

The four error terms corresponding four weighted inputs are defined as the partial deriative of loss function *Loss* with respect to four weighted inputs, respectively, i.e.,

$$e_f(k) \triangleq \frac{\partial \text{Loss}}{\partial \boldsymbol{net}_f(k)}, \tag{10.2.18}$$

$$e_i(k) \triangleq \frac{\partial \text{Loss}}{\partial \boldsymbol{net}_i(k)}, \tag{10.2.19}$$

$$e_{\tilde{c}}(k) \triangleq \frac{\partial \text{Loss}}{\partial \boldsymbol{net}_{\tilde{c}}(k)}, \tag{10.2.20}$$

$$e_o(k) \triangleq \frac{\partial \text{Loss}}{\partial \boldsymbol{net}_o(k)}. \tag{10.2.21}$$

10.2.3 Backpropagation of error term along time

To calculate the error term in backpropagation along time, the error term $e(k-1)$ at time $k-1$ is calculated by

$$\boldsymbol{e}^T(k-1) = \frac{\partial \text{Loss}}{\partial \boldsymbol{h}(k-1)} = \frac{\partial \text{Loss}}{\partial \boldsymbol{h}(k)} \frac{\partial \boldsymbol{h}(k)}{\partial \boldsymbol{h}(k-1)} = \boldsymbol{e}^T(k) \frac{\partial \boldsymbol{h}(k)}{\partial \boldsymbol{h}(k-1)}, \tag{10.2.22}$$

where $\frac{\partial \boldsymbol{h}(k)}{\partial \boldsymbol{h}(k-1)}$ is a Jacobian matrix. If the dimension of the hidden layer h is N_h, then the Jacobian matrix is a $N_h \times N_h$ matrix. Since $\boldsymbol{o}(k), \boldsymbol{f}(k), \boldsymbol{i}(k)$, and $\tilde{\boldsymbol{c}}(k)$ can be expressed in the form of a function of $\boldsymbol{h}(k-1)$, we obtain

$$\begin{aligned}
\boldsymbol{e}^{\mathrm{T}}(k) \frac{\partial \boldsymbol{h}(k)}{\partial \boldsymbol{h}(k-1)} &= \boldsymbol{e}^{\mathrm{T}}(k) \frac{\partial \boldsymbol{h}(k)}{\partial \boldsymbol{o}(k)} \frac{\partial \boldsymbol{o}(k)}{\partial \boldsymbol{net}_o(k)} \frac{\partial \boldsymbol{net}_o(k)}{\partial \boldsymbol{h}(k-1)} \\
&\quad + \boldsymbol{e}^{\mathrm{T}}(k) \frac{\partial \boldsymbol{h}(k)}{\partial \boldsymbol{c}(k)} \frac{\partial \boldsymbol{c}(k)}{\partial \boldsymbol{f}(k)} \frac{\partial \boldsymbol{f}(k)}{\partial \boldsymbol{net}_f(k)} \frac{\partial \boldsymbol{net}_f(k)}{\partial \boldsymbol{h}(k-1)} \\
&\quad + \boldsymbol{e}^{\mathrm{T}}(k) \frac{\partial \boldsymbol{h}(k)}{\partial \boldsymbol{c}(k)} \frac{\partial \boldsymbol{c}(k)}{\partial \boldsymbol{i}(k)} \frac{\partial \boldsymbol{i}(k)}{\partial \boldsymbol{net}_i(k)} \frac{\partial \boldsymbol{net}_i(k)}{\partial \boldsymbol{h}(k-1)} \\
&= \boldsymbol{e}_{\mathrm{o}}^{\mathrm{T}}(k) \frac{\partial \boldsymbol{net}_o(k)}{\partial \boldsymbol{h}(k-1)} + \boldsymbol{e}_f^{\mathrm{T}}(k) \frac{\partial \boldsymbol{net}_f(k)}{\partial \boldsymbol{h}(k-1)} \\
&\quad + \boldsymbol{e}_i^{\mathrm{T}}(k) \frac{\partial \boldsymbol{net}_i(k)}{\partial \boldsymbol{h}(k-1)} + \boldsymbol{e}_{\tilde{c}}^{\mathrm{T}}(k) \frac{\partial \boldsymbol{net}_{\tilde{c}}(k)}{\partial \boldsymbol{h}(k-1)}
\end{aligned} \tag{10.2.23}$$

After calculating each partial derivative in Equation (10.2.8), we can obtain

$$\frac{\partial \boldsymbol{h}(k)}{\partial \boldsymbol{o}(k)} = \text{diag}[\tanh(\boldsymbol{c}(k))], \tag{10.2.24}$$

$$\frac{\partial \boldsymbol{h}(k)}{\partial \boldsymbol{c}(k)} = \text{diag}[\boldsymbol{o}(k) \odot (1 - \tanh(\boldsymbol{c}(k))^2)]. \tag{10.2.25}$$

According to Equation (10.2.24), the following calculation formulas can be obtained:

$$\frac{\partial \boldsymbol{h}(k)}{\partial \boldsymbol{f}(k)} = \text{diag}[\boldsymbol{c}(k-1)], \tag{10.2.26}$$

$$\frac{\partial \boldsymbol{c}(k)}{\partial \boldsymbol{i}(k)} = \text{diag}[\tilde{\boldsymbol{c}}(k)], \tag{10.2.27}$$

$$\frac{\partial \boldsymbol{c}(k)}{\partial \tilde{\boldsymbol{c}}(k)} = \text{diag}[\boldsymbol{i}(k)]. \tag{10.2.28}$$

Since

$$\boldsymbol{o}(k) = \sigma(\boldsymbol{net}_o(k)), \tag{10.2.29}$$

$$\boldsymbol{net}_o(k) = \boldsymbol{W}_{oh}\boldsymbol{h}(k-1) + \boldsymbol{W}_{ox}\boldsymbol{x}(k) + \boldsymbol{b}_o, \tag{10.2.30}$$

$$\boldsymbol{f}(k) = \sigma(\boldsymbol{net}_f(k)), \tag{10.2.31}$$

$$\boldsymbol{net}_f(k) = \boldsymbol{W}_{fh}\boldsymbol{h}(k-1) + \boldsymbol{W}_{fx}\boldsymbol{x}(k) + \boldsymbol{b}_{\boldsymbol{f}}, \tag{10.2.32}$$

$$\boldsymbol{i}(k) = \sigma(\boldsymbol{net}_i(k)), \tag{10.2.33}$$

$$\boldsymbol{net}_i(k) = \boldsymbol{W}_{ih}\boldsymbol{h}(k-1) + \boldsymbol{W}_{ix}\boldsymbol{x}(k) + \boldsymbol{b}_i, \tag{10.2.34}$$

$$\tilde{c}(k) = \tanh(\boldsymbol{net}_{\tilde{c}}(k)), \tag{10.2.35}$$

$$\boldsymbol{net}_{\tilde{c}}(k) = \boldsymbol{W}_{ch}\boldsymbol{h}(k-1) + \boldsymbol{W}_{cx}\boldsymbol{x}(k) + \boldsymbol{b}_c. \tag{10.2.36}$$

So it is easy to obtain the following equations:

$$\frac{\partial \boldsymbol{o}(k)}{\partial \boldsymbol{net}_o(k)} = \text{diag}[\boldsymbol{o}(k) \odot (1 - \boldsymbol{o}(k))], \tag{10.2.37}$$

$$\frac{\partial \boldsymbol{net}_o(k)}{\partial \boldsymbol{h}(k-1)} = \boldsymbol{W}_{oh}, \tag{10.2.38}$$

$$\frac{\partial \boldsymbol{f}(k)}{\partial \boldsymbol{net}_f(k)} = \text{diag}[\boldsymbol{f}(k) \odot (1 - \boldsymbol{f}(k))], \tag{10.2.39}$$

$$\frac{\partial \boldsymbol{net}_f(k)}{\partial \boldsymbol{h}(k-1)} = \boldsymbol{W}_{fh}, \tag{10.2.40}$$

$$\frac{\partial \boldsymbol{i}(k)}{\partial \boldsymbol{net}_f(k)} = \text{diag}[\boldsymbol{i}(k) \odot (1 - \boldsymbol{i}(k))], \tag{10.2.41}$$

$$\frac{\partial \boldsymbol{net}_i(k)}{\partial \boldsymbol{h}(k-1)} = \boldsymbol{W}_{ih}, \tag{10.2.42}$$

$$\frac{\partial \tilde{\boldsymbol{c}}(k)}{\partial \boldsymbol{net}_f(k)} = \text{diag}[\tilde{\boldsymbol{c}}(k) \odot (1 - \tilde{\boldsymbol{c}}(k))], \tag{10.2.43}$$

$$\frac{\partial \boldsymbol{net}_{\tilde{c}}(k)}{\partial \boldsymbol{h}(k-1)} = \boldsymbol{W}_{ch}. \tag{10.2.44}$$

By substituting the above partial derivatives into Equation (10.2.23), we get:

$$\begin{aligned}\boldsymbol{e}(k-1) &= \boldsymbol{e}_o^{\mathrm{T}}(k)\frac{\partial \boldsymbol{net}_o(k)}{\partial \boldsymbol{h}(k-1)} + \boldsymbol{e}_f^{\mathrm{T}}(k)\frac{\partial \boldsymbol{net}_f(k)}{\partial \boldsymbol{h}(k-1)} \\ &\quad + \boldsymbol{e}_i^{\mathrm{T}}(k)\frac{\partial \boldsymbol{net}_i(k)}{\partial \boldsymbol{h}(k-1)} + \boldsymbol{e}_{\tilde{c}}^{\mathrm{T}}(k)\frac{\partial \boldsymbol{net}_{\tilde{c}}(k)}{\partial \boldsymbol{h}(k-1)} \\ &= \boldsymbol{e}_o^{\mathrm{T}}(k)\boldsymbol{W}_{oh} + \boldsymbol{e}_f^{\mathrm{T}}(k)\boldsymbol{W}_{fh} + \boldsymbol{e}_i^{\mathrm{T}}(k)\boldsymbol{W}_{ih} + \boldsymbol{e}_{\tilde{c}}^{\mathrm{T}}(k)\boldsymbol{W}_{ch}.\end{aligned} \tag{10.2.45}$$

According to the definition $\boldsymbol{e}_o(k), \boldsymbol{e}_f(k), \boldsymbol{e}_i(k)$, and $\boldsymbol{e}_{\tilde{c}}(k)$, we obtain

$$\boldsymbol{e}_o^{\mathrm{T}}(k) = \boldsymbol{e}^{\mathrm{T}}(k) \odot \tanh(c(k)) \odot \boldsymbol{o}(k) \odot (1 - \boldsymbol{o}(k)), \tag{10.2.46}$$

$$\begin{aligned}\boldsymbol{e}_f^{\mathrm{T}}(k) &= \boldsymbol{e}^{\mathrm{T}}(k) \odot \boldsymbol{o}(k) \odot (1 - \tanh(c(k))^2) \\ &\quad \odot \boldsymbol{c}(k-1) \odot \boldsymbol{f}(k) \odot (1 - f(k)),\end{aligned} \tag{10.2.47}$$

$$\begin{aligned}\boldsymbol{e}_i^{\mathrm{T}}(k) &= \boldsymbol{e}^{\mathrm{T}}(k) \odot \boldsymbol{o}(k) \odot (1 - \tanh(c(k))^2) \\ &\quad \odot \tilde{\boldsymbol{c}}(k) \odot \boldsymbol{i}(k) \odot (1 - i(k)),\end{aligned} \tag{10.2.48}$$

$$\begin{aligned}\boldsymbol{e}_{\tilde{c}}^{\mathrm{T}}(k) &= \boldsymbol{e}^{\mathrm{T}}(k) \odot \boldsymbol{o}(k) \odot (1 - \tanh(c(k))^2) \\ &\quad \odot \boldsymbol{i}(k) \odot (1 - \tilde{c}^2).\end{aligned} \tag{10.2.49}$$

The above formulas from Equations (10.2.45) to (10.2.49) are those propagating the error backwards along time for one moment. From this, it can be concluded that the formula for propagating the error term forward to any time k can be given by

$$\boldsymbol{e}^{\mathrm{T}}(n) = \prod_{j=n}^{k-1} \boldsymbol{e}_o^{\mathrm{T}}(j)\boldsymbol{W}_{oh} + \boldsymbol{e}_f^{\mathrm{T}}(j)\boldsymbol{W}_{fh} + \boldsymbol{e}_i^{\mathrm{T}}(j)\boldsymbol{W}_{ih} + \boldsymbol{e}_{\tilde{c}}^{\mathrm{T}}(j)\boldsymbol{W}_{ch}. \tag{10.2.50}$$

The error term will be transferred to the previous layer. Assuming that it is the lth layer at present, the error term for the $l-1$th layer is defined as the derivative of the error function with respect to the weighted input of the $l-1$th layer and is written as

$$\boldsymbol{e}^{l-1}(k) \triangleq \frac{\partial \mathrm{Loss}}{\boldsymbol{net}^{l-1}(k)}.$$

The calculation formula of the input $\boldsymbol{x}(k)$ for the LSTM network is defined as

$$x^l(k) = f^{l-1}(net^{l-1}(k)).$$

10.2.4 Calculation of weight gradient

The gradients of the loss function *Loss* on weight vector $\boldsymbol{W}_{fh}, \boldsymbol{W}_{ih}, \boldsymbol{W}_{ch}$, and $\boldsymbol{W}_{oh}$ are equal to the sum of the gradients at each time. Their gradients at time k are first calculated, and then their final gradients are calculated.

After the error terms $\boldsymbol{e}_o(k), \boldsymbol{e}_f(k), \boldsymbol{e}_i(k)$, and $\boldsymbol{e}_{\tilde{c}}(k)$ are obtained, it is easy to calculate the gradient of the loss function *Loss* on weight vector $\boldsymbol{W}_{oh}, \boldsymbol{W}_{ih}, \boldsymbol{W}_{fh}$, and $\boldsymbol{W}_{ch}$, i.e.,

$$\begin{aligned}\frac{\partial \text{Loss}}{\partial \boldsymbol{W}_{oh}(k)} &= \frac{\partial \text{Loss}}{\partial \boldsymbol{net}_o(k)} \frac{\partial \boldsymbol{net}_o(k)}{\partial \boldsymbol{W}_{oh}(k)} \\ &= \boldsymbol{e}_o(k)\boldsymbol{h}^T(k-1), \end{aligned} \tag{10.2.51}$$

$$\begin{aligned}\frac{\partial \text{Loss}}{\partial \boldsymbol{W}_{fh}(k)} &= \frac{\partial \text{Loss}}{\partial \boldsymbol{net}_f(k)} \frac{\partial \boldsymbol{net}_f(k)}{\partial \boldsymbol{W}_{fh}(k)} \\ &= \boldsymbol{e}_f(k)\boldsymbol{h}^T(k), \end{aligned} \tag{10.2.52}$$

$$\begin{aligned}\frac{\partial \text{Loss}}{\partial \boldsymbol{W}_{ih}(k)} &= \frac{\partial \text{Loss}}{\partial \boldsymbol{net}_i(k)} \frac{\partial \boldsymbol{net}_i(k)}{\partial \boldsymbol{W}_{ih}(k)} \\ &= \boldsymbol{e}_i(k)\boldsymbol{h}^T(k-1), \end{aligned} \tag{10.2.53}$$

$$\begin{aligned}\frac{\partial \text{Loss}}{\partial \boldsymbol{W}_{ch}(k)} &= \frac{\partial \text{Loss}}{\partial \boldsymbol{net}_{\tilde{c}}(k)} \frac{\partial \boldsymbol{net}_{\tilde{c}}(k)}{\partial \boldsymbol{W}_{ch}(k)} \\ &= \boldsymbol{e}_{\tilde{c}}(k)\boldsymbol{h}^T(k-1). \end{aligned} \tag{10.2.54}$$

The final gradients are obtained by adding the gradients at every now and then together as

$$\frac{\partial \text{Loss}}{\partial \boldsymbol{W}_{oh}} = \sum_{j=1}^{k} \boldsymbol{e}_o(j)\boldsymbol{h}^T(j-1), \tag{10.2.55}$$

$$\frac{\partial \text{Loss}}{\partial \boldsymbol{W}_{fh}} = \sum_{j=1}^{k} \boldsymbol{e}_f(j)\boldsymbol{h}^T(j-1), \tag{10.2.56}$$

$$\frac{\partial \text{Loss}}{\partial \boldsymbol{W}_{ih}} = \sum_{j=1}^{k} \boldsymbol{e}_i(j)\boldsymbol{h}^T(j-1), \tag{10.2.57}$$

$$\frac{\partial \text{Loss}}{\partial \boldsymbol{W}_{ch}} = \sum_{j=1}^{k} \boldsymbol{e}_{\tilde{c}}(j)\boldsymbol{h}^T(j-1). \tag{10.2.58}$$

The gradients of loss function *Loss* on the bias term vector $\boldsymbol{b}_f, \boldsymbol{b}_i, \boldsymbol{b}_c$, and $\boldsymbol{b}_o$ at time k are calculated by

$$\frac{\partial \text{Loss}}{\partial \boldsymbol{b}_o(k)} = \frac{\partial \text{Loss}}{\partial \boldsymbol{net}_o(k)} \frac{\partial \boldsymbol{net}_o(k)}{\partial \boldsymbol{b}_o(k)} = \boldsymbol{e}_o(k), \tag{10.2.59}$$

$$\frac{\partial \text{Loss}}{\partial \boldsymbol{b}_f(k)} = \frac{\partial \text{Loss}}{\partial \boldsymbol{net}_f(k)} \frac{\partial \boldsymbol{net}_f(k)}{\partial \boldsymbol{b}_f(k)} = \boldsymbol{e}_f(k), \tag{10.2.60}$$

$$\frac{\partial \text{Loss}}{\partial \boldsymbol{b}_i(k)} = \frac{\partial \text{Loss}}{\partial \boldsymbol{net}_i(k)} \frac{\partial \boldsymbol{net}_i(k)}{\partial \boldsymbol{b}_i(k)} = \boldsymbol{e}_i(k), \tag{10.2.61}$$

$$\frac{\partial \text{Loss}}{\partial \boldsymbol{b}_c(k)} = \frac{\partial \text{Loss}}{\partial \boldsymbol{net}_{\tilde{c}}(k)} \frac{\partial \boldsymbol{net}_{\tilde{c}}(k)}{\partial \boldsymbol{b}_c(k)} = \boldsymbol{e}_{\tilde{c}}(k). \tag{10.2.62}$$

The gradients of final bias, which are the sum of the bias gradients at each time step, are obtained by

$$\frac{\partial \text{Loss}}{\partial \boldsymbol{b}_0} = \sum_{j=1}^{k} \boldsymbol{e}_o(j), \tag{10.2.63}$$

$$\frac{\partial \text{Loss}}{\partial \boldsymbol{b}_i} = \sum_{j=1}^{k} \boldsymbol{e}_i(j), \tag{10.2.64}$$

$$\frac{\partial \text{Loss}}{\partial \boldsymbol{b}_f} = \sum_{j=1}^{k} \boldsymbol{e}_f(j), \tag{10.2.65}$$

$$\frac{\partial \text{Loss}}{\partial \boldsymbol{b}_c} = \sum_{j=1}^{k} \boldsymbol{e}_c(j), \tag{10.2.66}$$

For the gradients of loss function *Loss* on the weight vector $\boldsymbol{W}_{ox}, \boldsymbol{W}_{fx}, \boldsymbol{W}_{ix}$, and $\boldsymbol{W}_{cx}$, it only needs to be directly calculated based on the corresponding error term, i.e.,

$$\frac{\partial \text{Loss}}{\partial \boldsymbol{W}_{ox}} = \frac{\partial \text{Loss}}{\partial \boldsymbol{net}_o(k)} \frac{\partial \boldsymbol{net}_o(k)}{\partial \boldsymbol{W}_{ox}} = \boldsymbol{e}_o(k)\boldsymbol{x}^T(k), \tag{10.2.67}$$

$$\frac{\partial \text{Loss}}{\partial \boldsymbol{W}_{fx}} = \frac{\partial \text{Loss}}{\partial \boldsymbol{net}_f(k)} \frac{\partial \boldsymbol{net}_f(k)}{\partial \boldsymbol{W}_{fx}} = \boldsymbol{e}_f(k)\boldsymbol{x}^T(k), \tag{10.2.68}$$

$$\frac{\partial \text{Loss}}{\partial \boldsymbol{W}_{ix}} = \frac{\partial \text{Loss}}{\partial \boldsymbol{net}_i(k)} \frac{\partial \boldsymbol{net}_i(k)}{\partial \boldsymbol{W}_{ix}} = \boldsymbol{e}_i(k)\boldsymbol{x}^T(k), \tag{10.2.69}$$

$$\frac{\partial \text{Loss}}{\partial \boldsymbol{W}_{cx}} = \frac{\partial \text{Loss}}{\partial \boldsymbol{net}_{\tilde{c}}(k)} \frac{\partial \boldsymbol{net}_{\tilde{c}}(k)}{\partial \boldsymbol{W}_{cx}} = \boldsymbol{e}_{\tilde{c}}(k)\boldsymbol{x}^T(k). \tag{10.2.70}$$

All the above formulas are for the training algorithm of the LSTM network.

10.3 Dual Convolutional Long Short-Term Neural Networks

The specific structure of the Two-way Convolution LSTM (TCL) network is shown in Figure 10.10. It consists of five parts, including input layer, LSTM layer, CNN layer, fusion layer, and output layer. The dual neural networks are the LSTM network and CNN network, respectively. The amplitude and phase (A/P) components of the signals are used as inputs to the LSTM network, and the in-phase and quadrature (I/Q) components of the signals are regarded as inputs to the CNN.

10.3.1 Input layer

The inputs of the input layer are modulation signals in a complex form, and its output is a preprocessed two-dimensional matrix. For the RadioML2016.10a dataset, it is 2×128 dimensions. The I/Q components of the complex modulation signal are extracted and combined into a two-dimensional matrix being input to the CNN, while the A/P components are extracted and input to the LSTM network.

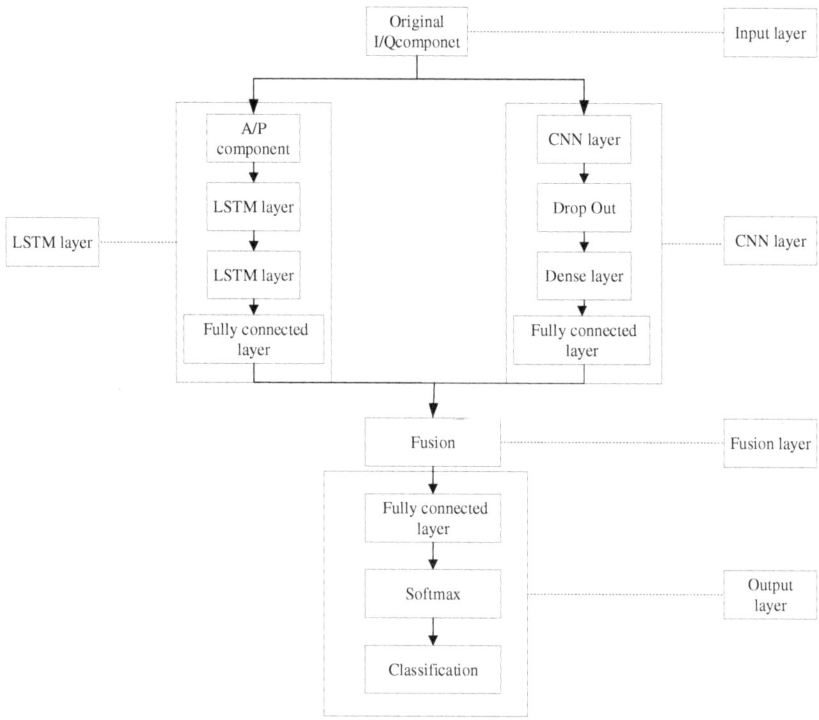

Figure 10.10. Two-way convolution LSTM.

10.3.2 CNN layer

Traditionally, a complete CNN consists of a feature extraction layer and a mapping layer. The CNN has two typical features: One is sparse connections, which belong to the convolutional kernels between two layers of neurons, and can improve computational efficiency, and reduce memory requirements. The other is weight parameter sharing, which refers to the convolution kernels sharing parameters at different input points; thus, the storage requirements of the model are significantly reduced. In the TCL model, the input of the CNN is the I/Q components, which mainly extract the spatial features of the signals. It only has a convolutional network layer, which is composed of convolutional filtering and nonlinear activation function ReLU. The filter size of the convolutional network layer is 1×3, with 256 filters. A small convolution kernel is utilized to improve computation speed and reduce parameters in the neural network. In addition, to prevent overfitting of the model, a dropout method may be used after the convolutional layer, where the dropout rate has been set to 0.5:

$$\mathrm{ReLU}(x) = \max(0, x). \tag{10.3.1}$$

10.3.3 LSTM layer

The LSTM network is an extension of Recurrent Neural Networks (RNNs), and its structure is shown in Figure 10.9. In the trained network, when there is no important information in the input sequence, the forgetting gate value of the LSTM network approaches 1 and the input gate value approaches 0. At this point, past memories are saved, thus long-term memory function can be achieved. When important information appears in the input sequence, the LSTM network should store it in memory, and then the value of the input gate will approach 1, while the value of the forgetting gate will approach 0, thus old memory information is forgotten and new important information is remembered. Through this design, the entire network is more likely to learn long-term dependencies between sequences. The number of layers in the LSTM network is set to 2, and each layer has 256 units to extract useful information from the modulation signals layer by layer. Meanwhile, to prevent overfitting, L2 regularization is applied to each layer in the LSTM network.

10.3.4 Fusion layer

The temporal features extracted by the LSTM network are fused with the spatial features extracted by the CNN, and these two features are connected by feature concatenation to form a new feature matrix, as shown in Equation (10.3.2), which is then passed to the output layer:

$$concat = f_{CNN}(\boldsymbol{x}^{I/Q}) \oplus f_{LSTM}(\boldsymbol{x}^{A/P}), \tag{10.3.2}$$

where *concat* represents the fused information, $f_{CNN}(\boldsymbol{x}^{I/Q})$ denotes the feature matrix of the I/Q components of the signal matrix $\boldsymbol{x}$ after passing through the CNN, $f_{LSTM}(\boldsymbol{x}^{A/P})$ presents the feature matrix of the A/P components of the signal matrix $\boldsymbol{x}$ after passing through the LSTM network, and $\oplus$ is the connection between the two feature matrices. The specific method is shown in Figure 10.11.

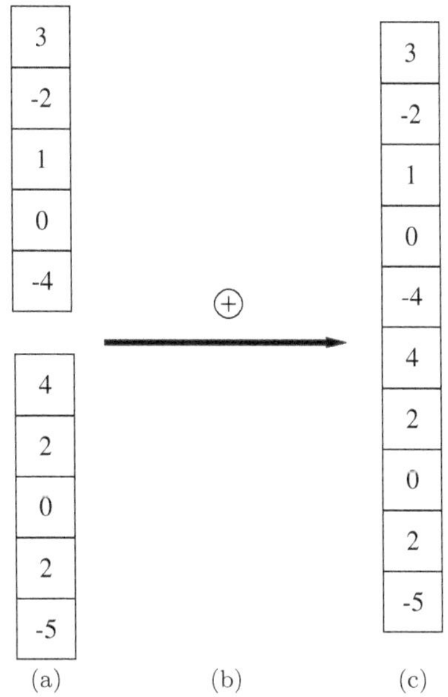

Figure 10.11. Feature concatenation. (a) Multivariate features. (b) Operation and process. (c) Fusion features.

Figure 10.11(a) shows the multivariate feature matrices obtained by two neural networks, and Figure 10.11(c) denotes the fusion feature matrix.

10.3.5 Output layer

The output layer adopts a fully connected layer and Softmax function. The fully connected layer maps the fused features to a sparse space, and the Softmax function maps the fully connected output to an interval $(0, 1)$. Its output is the probability of each modulation mode, and the node with the maximum output probability is selected as the prediction target. Finally, the modulation mode is determined. This prediction probability is calculated by

$$\hat{y}_i = \text{softmax}(z) = \frac{e^z}{\sum_{k=1}^{k} e^{z(k)}}, \tag{10.3.3}$$

where $\hat{y}_i$ represents the prediction probability of the ith modulation mode outputted by the predicted object, the *Softmax* is an activation function, z represents the input data, and k represents the number of modulation modes. In the TCL model, the cross-entropy loss function is used as the loss function. In addition, since cross entropy involves the calculation of probabilities for each mode, it almost always appears together with the *Softmax* function. Among them, the cross entropy loss function is defined as

$$\text{Loss} = -\frac{1}{N}\sum_i [y_i \log(\hat{y}_i) + (1 - y_i)\log(1 - \hat{y}_i)], \tag{10.3.4}$$

where $Loss$ represents the cross entropy loss function, N is the number of samples, y_i and $\hat{y}_i$ are the true probability and predicted probability of the ith modulation mode outputted by the predicted object, respectively. Backpropagation algorithm used to derive the loss function allows the network to learn better performance and to improve classification performance.

10.4 Example 12: Modulation Pattern Recognition Method for Non-Cooperative Underwater Acoustic Communication Signals Based on LSTM Network

In recent years, many modulation recognition methods for underwater acoustic signals have been proposed [200], such as support vector machine classification, maximum likelihood analysis based on the periodic characteristics of underwater acoustic signals, blind signal recognition algorithms, dual spectrum matrix feature recognition of underwater multicarrier modulation signals, and pattern recognition of underwater communication signals based on Support Vector Machine (SVM), etc. Although these methods have achieved some improvements in recognition rate, there is still room for improvement.

In this section, a Long Short-Term Memory (LSTM) classifier is introduced based on the instantaneous characteristics of underwater communication signals, which is used to identify and classify five commonly used modulation modes for underwater communication signals, and is validated using measured data from four sea areas, including the Songhua River.

10.4.1 LSTM classifier based on instantaneous characteristics of communication signals

10.4.1.1 *Instantaneous feature extraction of communication signals*

Real signal $x(t)$ can be represented by analytical signal $z(t)$ as

$$z(t) = x(t) + j\hat{y}(t), \tag{10.4.1}$$

where $\hat{y}(t)$ is the Hilbert transform of $z(t)$.

The instantaneous amplitude $a(t)$, instantaneous phase $\varphi(t)$, and instantaneous frequency $f(t)$ of the analytical signals are given by, respectively,

$$\begin{aligned} a(t) &= \sqrt{x^2(t) + \hat{y}^2(t)}, \\ \varphi(t) &= \arctan\left[\frac{\hat{y}(t)}{x(t)}\right] \\ f(t) &= \frac{1}{2\pi}\frac{\mathrm{d}\varphi(\mathrm{t})}{\mathrm{dt}}. \end{aligned} \tag{10.4.2}$$

The instantaneous characteristics of non-stationary signals are manifested as time-varying parameters, which are related to the average values that occur during the evolution process of the signals. The instantaneous frequency can also be described as the first-order conditional moment of the time–frequency distribution of the input signals. $f(t)$ in Equation (10.4.2) can also be expressed as

$$f(t) = \frac{\displaystyle\int_0^{\infty} fF(t,f)\mathrm{d}f}{\displaystyle\int_0^{\infty} F(t,f)\mathrm{d}f}, \tag{10.4.3}$$

where $F(t, f)$ is the time–frequency distribution.

The spectral entropy of signals is a measure of their spectral power distribution of signals. It is a concept based on Shannon entropy or information entropy in information theory. These signals are the normalized power distribution of a probability distribution in the frequency domain, and the spectral entropy equation can be derived from the signal power spectral equation and probability distribution equation. The instantaneous spectral entropy can well reflect the stability of the signals at various frequencies.

For a signal $x(k)$, its power spectrum is given by $G_X(m) = |X(m)|^2$, where $X(m)$ represents the discrete Fourier transform of $x(k)$. The probability distribution of the power spectrum is defined as

$$p(m) = \frac{G_x(m)}{\sum_i G_x(i)}. \tag{10.4.4}$$

Spectral entropy is calculated by

$$H = -\sum_{m=1}^{N} p(m) \log_2 p(m). \tag{10.4.5}$$

10.4.1.2 *LSTM classifier*

For problems highly correlated with time series, it is a great solution to use the LSTM network to solve these problems. Underwater acoustic communication signals are one-dimensional time signals, and there is temporal correlation between adjacent signal frames or long-distance frames. Therefore, the underwater acoustic communication signals can serve as inputs for the LSTM networks. In the LSTM units, storage units are specifically designed to preserve their historical information. The input gate, forgetting gate, and output gate are controlled to update the network's historical information. h represents the output, c denotes the value of the storage unit, and x serves as the input. The LSTM unit is shown in Figure 10.12.

The steps for updating the LSTM units are as follows:

Step 1: Calculate the state value of the candidate storage unit

The state value of the candidate storage unit at the current time k is calculated by

$$c(k) = \tanh(W_{xc}x(k) + W_{hc}h(k-1) + b_c), \tag{10.4.6}$$

where W_{xc} is the weight of the input data and W_{hc} denotes the connection weight outputted by the hidden layer unit at the previous moment $k-1$.

Step 2: Calculate the input gate

The input gate is calculated by

$$i(k) = \sigma(W_{xi}x(k) + W_{hi}h(k-1) + W_{ci}c(k-1) + b_i), \tag{10.4.7}$$

where σ is an activation function. The impact of the current input data on the state value of the storage unit is determined by the input

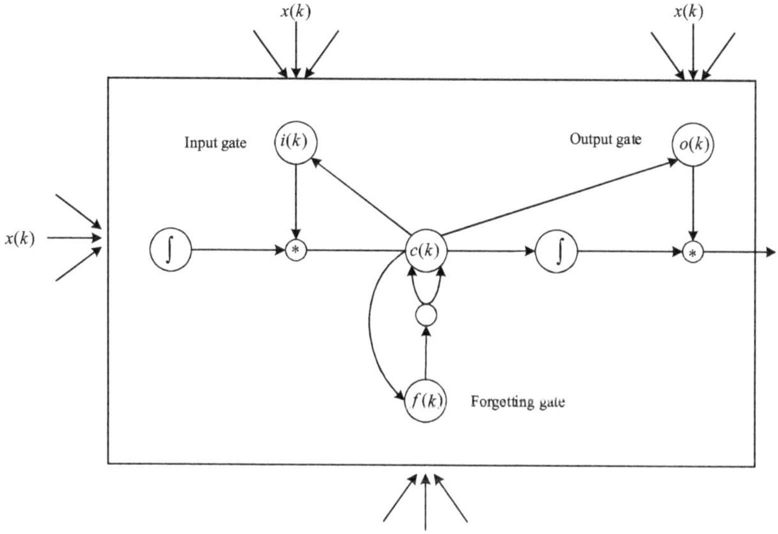

Figure 10.12. LSTM unit.

gate. The calculation of all gates is affected by not only the current input data but also the outputs of the hidden layer at the previous moment.

Step 3: Calculate forgetting gate

The threshold value for the forgetting gate is calculated by

$$f(k) = \sigma(W_{xf}x(k) + W_{hf}h(k-1) + W_{cf}c(k-1) + b_f). \quad (10.4.8)$$

Step 4: Calculate the current storage unit state

The state value of the current storage unit is calculated by

$$c(k) = f(k) \odot c(k-1) + i(k) \odot \tilde{c}(k), \quad (10.4.9)$$

where $\odot$ represents the dot product. $c(k-1)$ and the state values of the candidate storage unit $\tilde{c}(k)$ determine $c(k)$ whether to be updated at the current time k. $f(k)$ is combined with $i(k)$ to implement real-time adjustment of storage unit state value at the current time k.

Step 5: Calculate the output gate

The output gate is calculated by

$$o(k) = \sigma(W_{xo}x(k) + W_{ho}h(k-1) + W_{co}c(k-1) + b_o). \quad (10.4.10)$$

Step 6: Calculate the hidden layer output

The hidden layer output is calculated by

$$h(k) = o(k) \odot \tanh(c(k)). \tag{10.4.11}$$

10.4.2 Evaluation criteria

Assume that there are only two classification objectives in the prediction process: positive and negative. The number of correct samples classified as positive by the classifier is defined as *TP*, and the number of negative samples classified as positive by the classifier is defined as *TN*. The number of correct samples classified as negative by the classifier is defined as *FP*, and the number of negative samples classified as negative is defined as*FN*. The four terms of the confusion matrix are shown in Table 10.1. The total number of samples is defined as

$$N_{tot} = TP + FP + TN + FN. \tag{10.4.12}$$

Precision and recall are defined as

$$\begin{aligned} P &= \frac{TP}{TP + FP}, \\ R &= \frac{TP}{TP + FN}. \end{aligned} \tag{10.4.13}$$

The harmonic average of precision and recall is defined as

$$F1 = \frac{2 \times P \times R}{P + R} = \frac{2 \times TP}{N_{tot} + TP - TN}. \tag{10.4.14}$$

When comparing multiple confusion matrices and considering the classification performance comprehensively, the accuracy and

Table 10.1. The four terms of confusion matrix.

	Forecast results	
Real situation	positive	negative
positive	TP	FN
negative	FP	TN

recall, represented as $(P_1, R_1), (P_2, R_2), \ldots, (P_N, R_N)$, for each confusion matrix, are calculated separately. Macro precision (P_m), macro recall (R_m), and macro harmonic average $(F1_m)$ are calculated by, respectively,

$$\begin{aligned} P_m &= \frac{1}{N}\sum_{i=1}^{N} P_i, \\ R_m &= \frac{1}{N}\sum_{i=1}^{N} R_i, \\ F1_m &= \frac{2 \times P_m \times R_m}{P_m + R_m}. \end{aligned} \tag{10.4.15}$$

10.4.3 Simulation of anti-noise performance

The experiments on the anti-noise performance of the LSTM network based on the instantaneous characteristics of signals were carried out. A shallow sea with a negative sound velocity gradient was chosen and a water depth was set to 500 m. The depths of the transmitting transducer and the receiving transducer were 50 m and 100 m, respectively, with a distance of 5 km between them.

The recognition results based on the testing data under different Signal-to-Noise Ratios (SNRs) are shown in Table 10.2. Table 10.2 shows that the reliability of the LSTM network decreases with the decrease in the SNR. This phenomenon may be mainly influenced by the input characteristics. Under high SNR conditions, the signal features for the LSTM network are easier to learn and classify. However, at low SNRs, the signal's characteristics are affected by noises, which directly affects the learning and classification of the LSTM network.

Table 10.2. Simulation test results.

SNR	P_m	R_m	$F1_m$
−10 dB	91.76%	91.46%	91.49%
−5 dB	93.61%	93.26%	93.22%
0 dB	95.16%	95.02%	95.01%
5 dB	95.07%	95.02%	95.02%
10 dB	95.71%	95.28%	95.48%

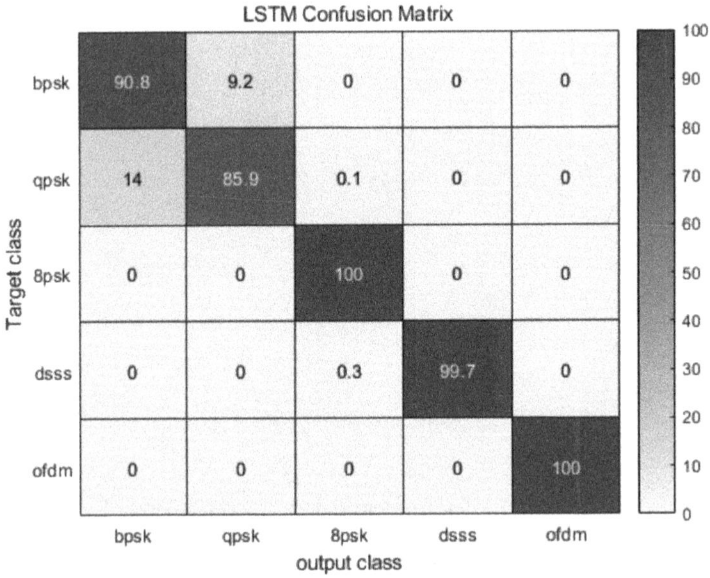

Figure 10.13. Confusion matrix of simulation testing results (SNR = 10 dB).

Figures 10.13–10.15 show the confusion matrix of testing data for the SNR of −10 dB, 5 dB, and −10 dB, respectively. Through confusion matrix analyses, the classifying quality of the LSTM network performs well on the OFDM signals and the DSSS signals. For MPSK signals, even at a high SNR of 10 dB, the recognition effect of the LSTM network is average, but the recognition effect is relatively poor under low SNR conditions. This indicates that there is no significant difference between instantaneous frequency features and spectral entropy features for the MPSK signals, and more time series features need to be extracted for the MPSK mode recognition. Under the SNR conditions of −10 dB and −5 dB, there is a high probability of misjudgment for the MPSK signals, and even for the DSSS signals.

10.4.4 Simulation experiments and result analyses

10.4.4.1 *Experimental datasets*

The experimental data came from fixed-point underwater acoustic communication in Songhua River, Bayuquan, Russian Island, South

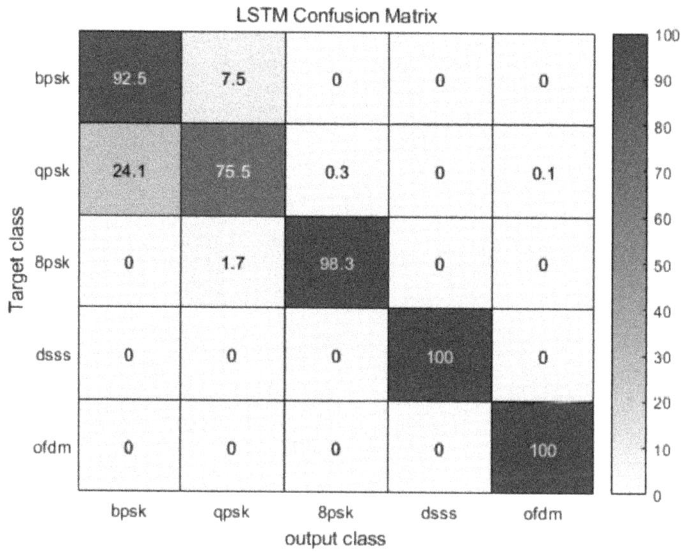

Figure 10.14. Confusion matrix of simulation testing results (SNR $= -5\,\text{dB}$).

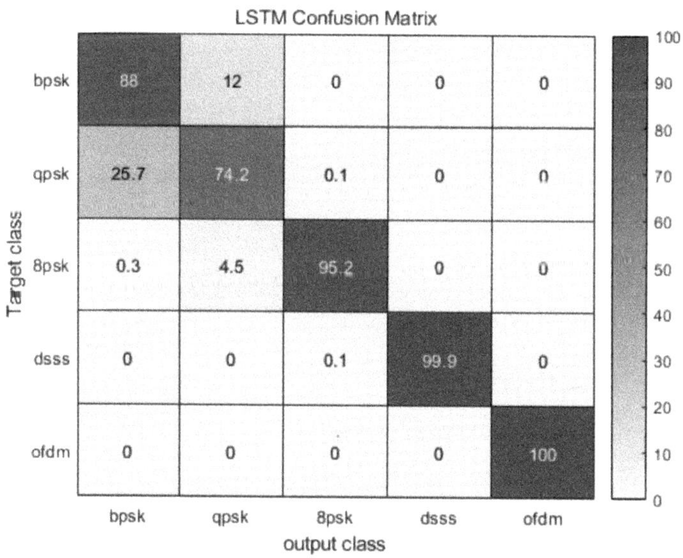

Figure 10.15. Confusion matrix of simulation testing results (SNR $= -10\,\text{dB}$).

Table 10.3. The positions of different modulation modes.

Position	Songhua River	Bayuquan	Russian Island	South China Sea	Arctic
BPSK	√	√	√	√	√
QPSK	√	√	√	√	√
8PSK	√	√	√	√	√
DSSS	√	√		√	√
OFDM	√		√	√	√

Table 10.4. Simulation test results.

Type	P_m	R_m	$F1_m$
BPSK	92.5%	75.3%	83.0%
QPSK	92.3%	76.4%	83.6%
8PSK	72.6%	98.5%	83.6%
DSSS	99.9%	80.2%	89.0%
OFDM	83.5%	100%	91.0%

China Sea, and the Arctic from 2016 to 2018. Five communication signals, BPSK, QPSK, 8PSK, DSSS, and OFDM, were identified. Table 10.3 shows the positions of different modulation modes.

In the experiments, the dataset was divided into a training set and a testing set. The samples in the training set and the samples in the testing set were mutually exclusive and completely non-overlapping. The signals in the dataset were obtained from multiple experiments and intercepted from over 700 different environmental reception signals. In the dataset, the computational complexity of the signals was high, with multi-frequency signals ranging from 2 kHz to 35 kHz. The SNRs of the dataset samples were set from 0 to 15 dB.

10.4.4.2 *Measurement results*

Table 10.4 shows that the classification performance of the DSSS signals and the OFDM signals is better than that of the MPSK signals. From the perspective of $F1_m$ value, the OFDM and the DSSS signals have better recognition performance, while the MPSK signals have average recognition performance. As for recall and precision, both the BPSK and the QPSK signals have high precision, reaching

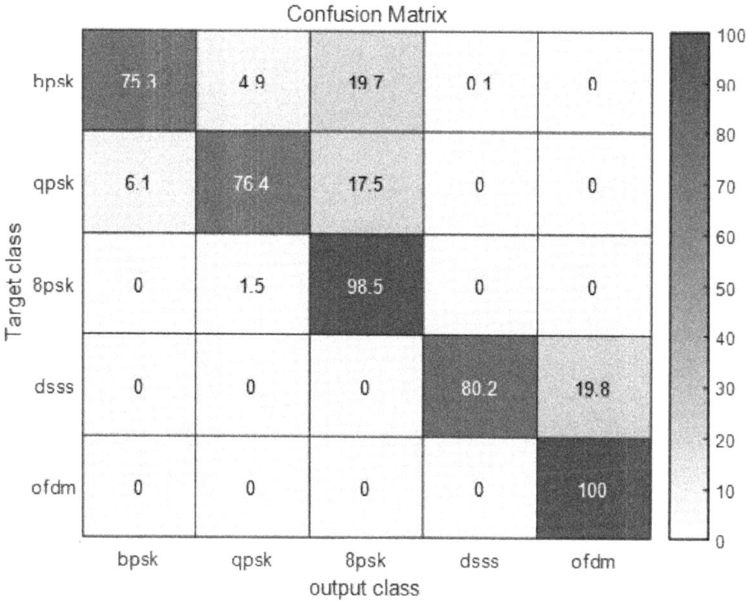

Figure 10.16. Confusion matrix of simulation test results.

up to 90%, but both signals have low recall. The recall rate of 8PSK signals can reach 98.5%, but their accuracy is relatively low.

Figure 10.16 shows the confusion matrix of the experimental results. It can be concluded that MPSK signals are prone to misclassification, while the DSSS signals may be misclassified as the OFDM signals.

Overall, for the DSSS and the OFDM signals, the LSTM classifiers based on instantaneous features have good classification and recognition performance, while the classification and recognition of the MPSK signals need further improvement. The experimental results indicate that the LSTM classifier based on instantaneous features has certain practical value for identifying underwater non-cooperative communication signals under single modulation mode. Due to the small amount of data in the field of underwater acoustic communication, continuous learning of testing data from different sea areas is required, and the data from different sea areas needs to be continuously supplemented to continue training the classifier.

10.5 Example 13: Fingerprint Outdoor Localization Method Using Hybrid Long Short-Term Memory (LSTM) Mechanism

The Internet of Things (IoT) technology integrates wireless fidelity (Wi-Fi), long-term evolution (LTE), radio frequency identification (RFID), frequency modulation (FM), inertial sensors, and other technologies, which can be applied to location tracking, traffic prediction, channel estimation, traffic management, and scheduling in wireless technology.

At present, there are various wireless measurement methods for outdoor positioning, such as Time of Arrival (ToA), Time Difference of Arrival (TDoA), Angle of Arrival (AoA), and Received Signal Strength (RSS). The ToA and TDoA require satellite clocks for synchronization, whereas the AoA, directional antenna arrays for Base Stations (BSs) and/or Access Points (APs) are also required. On the other hand, the RSS-based positioning technology does not require additional hardware, has low cost, low power consumption, simple deployment, and is not too complicated to obtain user location.

However, due to environmental factors such as temperature, pressure, sound, and humidity, as well as the number of available neighboring base stations and access points, especially the fluctuation of signals over time, the positioning method based on the RSS is inaccurate.

In order to improve positioning accuracy, Support Vector Regression (SVR), integrated k-Nearest Neighbor (kNN), and Genetic Algorithm (GA) positioning techniques have been proposed. However, due to the influence of signal fluctuations on pedestrian walking patterns and equipment-dependent positioning accuracy, key features need to be extracted from the RSS values of each area point's fluctuations. Nevertheless, deep learning (DL) methods can automatically learn more complex features in dynamic and complex environments. Even if multiple hidden layers in the DNN structure can extract more relevant features through hierarchical learning of wave signals, there are some issues with gradient vanishing and exploding. Whereas the Accelerated Robust Feature (SURF) algorithm can be used to detect image features and match image similarity to locate mobile users in indoor and outdoor environments, the Random Sample Consensus (RANSAC) algorithm can also be used to remove noisy feature points, however, this positioning is based on classes rather than specific locations, and does not consider the scalability of the system.

In this section, a robust and accurate Fingerprint Outdoor User Positioning Scheme (FOPS) is introduced, which includes detectable wireless access points and the RSS values of three unmanned aerial vehicles as data sources for estimating the position information of the IoT devices in dynamic environments [201]. Wireless networks and orthogonal frequency division multiplexing (OFDM) data are integrated to minimize fluctuations of signals and outliers in dynamic wireless environments. In addition, Linear Discriminant Analysis (LDA) is applied to select the optimal subset of access points to reduce computation time, save storage space, avoid overfitting, and improve system performance. Therefore, the integration of the LDA and the LSTM network-based location services has been introduced for the IoT devices in wireless networks and cellular working environments.

10.5.1 Dataset and model

The experimental data was collected from the real environment of National Taipei University of Science and Technology in Taiwan. The working environment included Light Of Sight (LOS) and Non-Light Of Sight (NLoS) locations, where the environment was divided into a $2\,\text{m} \times 2\,\text{m}$ square grid to collect the required dataset. Both wireless signals from reachable wireless access points and orthogonal frequency division multiplexing signals from three unmanned aerial vehicle base stations as cellular data were collected, as shown in Table 10.5. The drone base station was placed on the top of each building, with a distance of about 400 meters between each of those buildings. In each remote positioning system, from reachable access points to three drone base stations, sensor technology was used to scan the RSS values, which were then stored in a central server (i.e., PC). Due to collecting datasets from all accessible access points, it is difficult to distinguish relevant access points from open environments. At each grid point, 35 RSS values were collected regularly every 3 seconds to mitigate signal fluctuations. If there was a zero reading, its value was set to 100 dBm.

Here AP_N is the number of reachable access points, $BS_1, \ldots, BS_3$ represent the number of base stations, N denotes the number of records, RSS_n is the record of the received signal values, and

Table 10.5. Signal fingerprint data structure.

	Object	
Value of RSS vector	X-Y coordinate	
$(RSS_1^{AP_1}, \ldots, RSS_1^{AP_N}, RSS_1^{BS_1}, \ldots, RSS_1^{BS_3})$	x_i^1	y_j^1
$(RSS_2^{AP_1}, \ldots, RSS_2^{AP_N}, RSS_2^{BS_1}, \ldots, RSS_2^{BS_3})$	x_i^2	y_j^2
$\ldots$	$\ldots$	$\ldots$
$(RSS_{N-1}^{AP_1}, \ldots, RSS_{N-1}^{AP_N}, RSS_{N-1}^{BS_1}, \ldots, RSS_{N-1}^{BS_3})$	x_i^{m-1}	y_j^{m-1}
$(RSS_N^{AP_1}, \ldots, RSS_N^{AP_N}, RSS_N^{BS_1}, \ldots, RSS_N^{BS_3})$	x_i^m	y_j^m

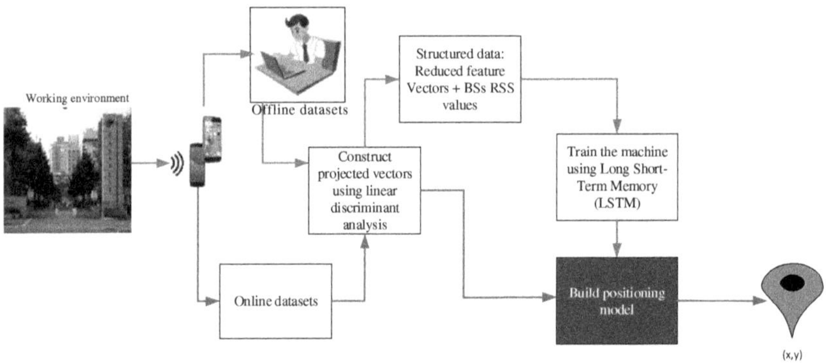

Figure 10.17. The proposed architecture of FOPS.

(x_i^m, y_j^m) is the x and y coordinates of specific Reference Points (RPs).

The overall architecture of the FOPS is shown in Figure 10.17. When Wi-Fi is combined with OFDM signal values, a robust and accurate fingerprint positioning scheme is constructed in dynamic and harsh outdoor spaces.

Due to the use of all reachable access points, a dimensionality reduction mechanism is employed for reducing computational complexity, minimizing memory requirements, simplifying highly overfitting problems, and achieving better accuracy of the system. The LDA method is used to reduce dimensionality and extract more distinctive features. Based on terrain subsidence and LoS considerations, the collected receiving signals are classified into four categories. Therefore, the LDA method uses Equations (10.5.1) and (10.5.2) to construct a low-dimensional space by minimizing intra-class variance and maximizing inter-class variance, respectively, to transform the original high-dimensional data matrix:

$$\sigma_w^2 = \sum_{C=1}^{4} \sum_{i=1}^{N_c} (\boldsymbol{D}_{i,c} - \boldsymbol{m}_c)(\boldsymbol{D}_{i,c} - \boldsymbol{m}_c)^{\mathrm{T}}, \tag{10.5.1}$$

$$\sigma_b^2 = \sum_{C=1}^{4} N_c(\boldsymbol{m}_c - \boldsymbol{m})(\boldsymbol{m}_c - \boldsymbol{m})^{\mathrm{T}}, \tag{10.5.2}$$

where σ_b^2 and σ_w^2 represent inter-class and intra-class variance, respectively, $\boldsymbol{m}_c$ is the mean of each class, $\boldsymbol{m}$ is the global mean of all data, $\boldsymbol{D}_{i,c}$ are the ith sample in the cth class, and N_c represents the total number of samples of the cth class. Second, Equation (10.5.2) is used to construct the transformation matrix, i.e.,

$$\boldsymbol{W} = (\boldsymbol{\sigma}_b^2)^{-1}\boldsymbol{\sigma}_b^2. \tag{10.5.3}$$

Then, the feature vectors are sorted in descending order based on their eigenvalues in order to select the eigenvalue of the transformation matrix $\boldsymbol{W}$, and k eigenvectors from all eigenvectors (where k is the dimensionality of the new feature space, $k \leq K$, where K is the number of original features). The selected features should contain more than 90% of the original data content, which means that the selected extremely low-dimensional space retains most of the valuable information for positioning, i.e.,

$$\frac{\lambda_1 + \lambda_2 + \cdots + \lambda_k}{\lambda_1 + \lambda_2 + \cdots + \lambda_k + \cdots + \lambda_K} = \frac{\sum_{i=1}^{k} \lambda_i}{\sum_{j=1}^{K} \lambda_j}. \tag{10.5.4}$$

Therefore, based on Equation (10.5.4), three features are selected from wireless access points, as shown in Figure 10.18. The first three principal components (features) in Figure 10.18 can retain most of the signal information without losing the original data content. Finally, Equation (10.5.5) is used to project the original data ($\boldsymbol{D}$) onto a new lower k-dimensional space, i.e.,

$$FinalData(\boldsymbol{y}) = \boldsymbol{DXW}, \tag{10.5.5}$$

where $\boldsymbol{D}$ is the $N \times K$-dimensional matrix representing N samples and $\boldsymbol{y}$ in the new space is the $N \times k$-dimensional sample.

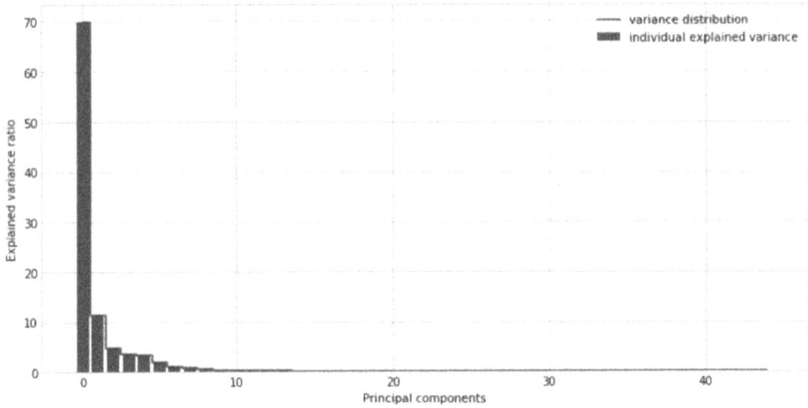

Figure 10.18. Feature selection of Wi-Fi data based on LDA.

The used dataset is collected over time in a specific RP (Reference Point) to adapt to fluctuations in the receiving signal values. In order to learn more abstract features from sequence data and establish accurate localization models, it is an effective method to use the LSTM network, which has the ability to learn from a large number of fluctuating signals and the noisy RSS samples in harsh dynamic environments. In addition, it prevents gradient vanishing and explosion problems by introducing gates. The LSTM network has multiple gates, each of which is used to control access to storage units. Besides, the proposed method can remember both long-term and short-term values of several time steps. To this end, a FOPS model based on the LSTM network is established by combining the simplified features with the OFDM signals as input using the LDA method. Therefore, in this section, a total of 6 features will be taken as inputs for the network architecture. The scale of each feature in the used data sources is different. Before inputting it into the LSTM network, the training data is standardized to fundamentally accelerate the training process and improve the performance of the network. So, Equation (10.5.6) is used to perform minimum-maximum normalization scaling on the data source; all feature values are changed from zero to one. The normalization formula for feature data $\boldsymbol{D}$ is given by

$$N_i = \frac{\boldsymbol{D}_i - \min(\boldsymbol{D})}{\max(\boldsymbol{D}) - \min(\boldsymbol{D})}, \tag{10.5.6}$$

where N_i is the ith normalization data of the feature data $\boldsymbol{D}$, $\min(\boldsymbol{D})$ and $\max(\boldsymbol{D})$ are the minimum value and maximum one of the feature data $\boldsymbol{D}$, and $\boldsymbol{D}$ is given by $\boldsymbol{D} = (\boldsymbol{D}_1, \ldots, \boldsymbol{D}_i, \ldots, \boldsymbol{D}_N)$.

The LSTM network here is a 1_3_1 network, which represents the number of input layers, LSTM layers, and output layers, respectively. To avoid overfitting issues during training the FOPS model, the trial and error method is used to select the optimal hyperparameters for the number of LSTM layers and the number of neurons in each LSTM layer. Different numbers of neurons at each layer are selected to learn the network and measure system performance, which are repeatedly selected until the best accuracy results are obtained during both training and testing phases. Let 512_256_256 represent the number of neurons in the first, second, and third layers of the LSTM network, respectively. In addition, a 20% loss rate is used to reduce overfitting in each LSTM layer, while the *Adam optimization algorithm* is utilized in training to find the optimal parameters (network weights) and make the network converge. To accelerate the training process and further reduce overfitting, the Batch Normalization (BN) method and Mean Square Error (MSE) loss function are utilized to obtain the optimal weights, in order to minimize the *Loss* between the expected output and the output of the LSTM network. Finally, the activation function ReLU of each layer in the LSTM network and the Softmax of the output layer are employed for estimating the location of the IoT devices. The selected training hyperparameters are shown in Table 10.6.

Table 10.6. Training hyperparameters.

Category	Value
Algorithm	LSTM
LSTM layer	3
LSTM nodes	512,256,256
Dropout	0.2
Batch size	300
Num. of Epoch	30
Optimizer	Adam
Loss	MSE
Activation function	ReLU

In order to evaluate the performance of the system, a series of experiments were conducted and the accuracy of the system was evaluated using different indicators. First, Root MSE (RMSE) index is used to evaluate the accuracy error of the system and is defined as

$$RMSE = \sqrt{\sum_{i=1}^{N} \frac{(act_i - pre_i)^2}{N}}, \tag{10.5.7}$$

where N is the number of testing points and act_i and pre_i are the ith ground truth and predicted distance at the ith testing point, respectively. Second, based on the system's prediction error evaluation, the error distribution in each testing point is defined as

$$PE_i = \sqrt{(X_{act_i} - X_{pre_i})^2 + (Y_{act_i} - Y_{pre_j})^2}, \tag{10.5.8}$$

where PE_i is the ith prediction error of each testing point, X_{act_i} and X_{pre_i} are the X coordinates of the ith actual and predicted user positioning on the ground, and Y_{act_i} and Y_{pre_j} are the Y coordinates of the jth actual and predicted user positioning on the ground, respectively.

To implement the proposed system, Python 3.6 and the Keras library with TensorFlow as the backend are used.

10.5.2 Simulation experiments and result analyses

The wireless access points were evaluated on an experimental dataset using received signal data from detectable wireless access points and three drone base stations. Wireless signals were collected from wireless access points used for communication purposes within buildings, while orthogonal frequency division multiplexing signals were collected from three fixed location drone base stations placed at the top of each building, at a height of 40 meters from ground users. Since most fixed base stations were deployed at a height of 30 to 50 meters, the average height was taken in this section. In the online phase, the current location of the IoT devices measured by online RSS was determined and compared with wireless maps. To evaluate the performance of the proposed system, 53 RPs were randomly selected as testing points and the FOPS was evaluated using two different scenarios. This included a positioning system using the original

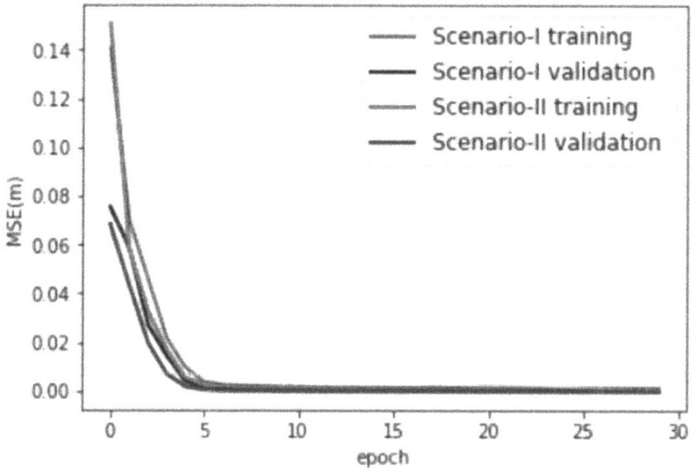

Figure 10.19. LSTM loss function.

dataset (scenario I) and a positioning system using the LDA method to extract the most representative features from the original dataset (scenario II) to reduce time and resource consumption. Figure 10.19 shows the performance evaluation results of training and validation accuracy for scenario I and scenario II. From Figure 10.19, it can be shown that there is no difference in training and validation accuracy so that the overfitting of the LSTM network is not a problem for the used training data, as the training and validation losses follow the same decreasing law. Furthermore, in both cases, the training loss decreases as the number of training iterations increases. However, scenario II has smaller positioning errors and fewer iterations in comparison with scenario I. Therefore, scenario II is significantly better than scenario I, which extracts more important features through the LDA method to separate the location of each user. In addition, in scenario II, due to the reduction of features, the training and localization time are also reduced.

Table 10.7 shows the FOPS performance evaluation indicators. From Table 10.7, there are some results as follows: For scenario I and scenario II, for estimation errors less than 0.75 meters, the positioning accuracies of FOPS are 76.19% and 81.83%, respectively. For positioning errors less than 1.5 meters, the positioning accuracies for scenario I and scenario II are 95.98% and 99.04%, respectively.

Table 10.7. Estimation errors under different scenarios.

Scenarios	Range estimation errors (m)			Average errors (m)	RMSE	Minimum errors (m)	Maximum errors (m)
	≤0.75 m	≤1 m	≤1.5 m				
Scenario I (%)	76.19%	88.31%	95.98%	0.53	0.036	0.016	1.82
Scenario II (%)	81.83%	91.90%	99.04%	0.46	0.031	0.0055	1.70

And the RMSE and average error for scenario I and scenario II are 0.036 m and 0.53 m, and 0.031 m and 0.46 m, respectively. In addition, the maximum errors for scenario I and scenario II are 1.82 m and 1.70 m, respectively. From these results, it can be seen that when unnecessary features are removed in scenario II, the system performance is significantly improved. Due to the variation of received signals over time, even at the same location, the signals still fluctuate. Before inputting into the LSTM network, the LDA method is used to remove outliers and redundant features to extract the most relevant features, to make it easy to identify the location of each user in a dynamic environment. Therefore, in scenario II, the accuracy and efficiency of localization estimation have been improved compared to scenario I.

Figure 10.20 shows the positioning performance of each testing point coordinate. The black dashed line represents the actual user coordinates (RPs), while the red and blue dashed lines represent the actual values corresponding to scenario I and scenario II, respectively. From Figure 10.20, we can know that the performance results at various testing points for scenario I are not consistent, for example, the predictions on the X-axis are not appropriate from 5 m to 7 m and from 9 m to 16 m. However, in scenario II, the estimation of user localization is consistent and more suitable than that in scenario I, as the LDA method extracts more important features to distinguish each location. Therefore, the FOPS provides accurate positioning solutions in dynamic wireless environments.

In summary, in this section, the Linear Discriminant Analysis (LDA) method is used to preprocess, the LSTM network is used as the main positioning algorithm, signals are collected from reachable access points, and unmanned aerial vehicle base stations are temporarily deployed to collect orthogonal frequency division multiplexing signals. In the preprocessing stage, the LDA is used to reduce the dimensionality of wireless sensor network data features and remove

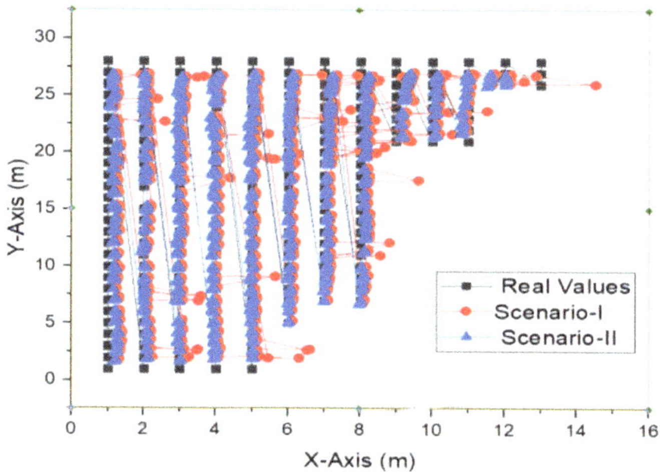

Figure 10.20. Error distribution of each test point.

outliers. Afterwards, the LSTM network is used to locate the position of the mobile user. To evaluate the performance of the proposed system, the performance of two methods is compared in different scenarios. The results indicate that the fusion of the LDA method and the LSTM network has better performance and can provide accurate positioning services.

References

[1] Hinton, G.E., Salakhutdinov, R.R. Reducing the dimensionality of data with neural networks. *Science*, 2006, 313(5786): 504–507.

[2] Chen, Y., Luo, T., Liu, S., *et al.* Da Dian Nao: a machine-learning supercomputer. *Proc. of the 47th Annual IEEE/ACM International Symposium on Microarchitecture*, Washington D.C.: IEEE Computer Society.

[3] Le, Q.V., Ngiam, J., Coates, A., *et al.* On optimization methods for deep learning. *Proc. of the 28th International Conference on Machine Learning*, Bellevue, WA, USA, IEEE, 2011: 265–272.

[4] Graves, A., Mohamed, A.R., Hinton, G.E. Speech recognition with deep recurrent neural networks. *Proc. of IEEE International Conference on Acoustics, Speech and Signal Processing*, Vancouver, BC, Canada, IEEE, 2013: 6645–6649.

[5] Yadan, O., Adams, K., Taigman, Y., *et al.* Multi-GPU training of ConvNets. https://arXiv.org/abs/1312.5853, 2015.

[6] Rosenblatt, F. The perceptron-a perceiving and recognizing automaton. Ithaca, NY: Cornell Aeronautical Laboratory, 1957.

[7] Bourlard, H., Kamp, Y. Auto-association by multilayer perceptrons and singular value decomposition. *Biological Cybernetics*, 1988, 59(4–5): 291–294.

[8] Lei, Y., Qin, L. Image classification algorithm based on deep learning. 2025 IEEE 5th International Conference on Electronic Technology, Communication and Information (ICETCI), Changchun, China, IEEE, 2025: 1225–1230.

[9] Alwani, M., Chen, H., Ferdman, M., *et al.* Fused-layer CNN accelerators. *Proc. of the 49th Annual IEEE/ACM International Symposium on Microarchitecture*, Washington D.C.: IEEE Computer Society, 2016: 1–12.

[10] Zeiler, M.D., Taylor, G.W., Fergus, R. Adaptive deconvolutional networks for mid and high-level feature learning. *Proc. of IEEE International Conference on Computer Vision*, Piscataway, NJ: IEEE, 2011: 2018–2025.

[11] Yu, K., Lin, Y., Lafferty, J. Learning image representations from the pixel level via hierarchical sparse coding. *Proc. of IEEE Conference on Computer Vision and Pattern Recognition*, Washington D.C., IEEE, 2011: 1713–1720.

[12] Hinton, G.E. A practical guide to training restricted Boltzmann machines. *Neural Networks: Tricks of the Trade.* Berlin: Springer, 2012: 599–619.

[13] Ackley, D.H., Hinton, G.E., Sejnowski, T.J. A learning algorithm for Boltzmann machines. *Cognitive Science*, 1985, 9(1): 147–169.

[14] Lee, H., Grosse, R., Ranganath, R., *et al.* Convolutional deep belief networks for scalable unsupervised learning of hierarchical representations. *Proc. of the 26th Annual International Conference on Machine Learning*, New York: ACM Press, 2009: 609–616.

[15] Krizhevsky, A., Hinton, G.E. Using very deep autoencoders for content based image retrieval. *Proc. of the 19th European Symposium on Artificial Neural Networks*, Bruges, Belgium, IEEE, 2011: 489–494.

[16] Zhang, J., Shan, S., Kan, M., *et al.* Coarse-to-fine auto-encoder networks (CFAN) for real-time face alignment. *Proc. of European Conference on Computer Vision-ECCV2014*, Springer International Publishing, 2014: 1–16.

[17] Mcculloch, W.S., Pitts, W. A logical calculus of the ideas immanent in nervous activity. *Bulletin of Mathematical Biology*, 1990, 52(1/2): 99–115.

[18] Glorot, X., Bengio, Y. Understanding the difficulty of training deep feedforward neural network. *Proc of the 13th International Conference on Artificial Intelligence and Statistics*, Chia Laguna Resort, Sardinia, Italy, IEEE, 2010, 9: 249–256.

[19] Li, Z., Liu, F., Yang, W., Peng, S., *et al.* A survey of convolutional neural networks: analysis, applications, and prospects. *IEEE Transactions on Neural Networks and Learning Systems*, 2022, 33(12): 6999–7019.

[20] Zhao, J. Regularization of neural network and its application in geological prediction. Doctoral Dissertation. Dalian University of Technology, 2020 (In Chinese).
[21] Fukushima, K. Neocognitron: a self-organizing neural network model for a mechanism of pattern recognition unaffected by shift in position. *Biological Cybernetics*, 1980, 36(4): 193–202.
[22] Chen, J. Research on deep learning regularization techniques and their applications in the field of driving safety risks. Doctoral Dissertation. University of Science and Technology of China, 2019.
[23] Li, H. *The Research of Random Effects Panel Data Models Based on Bayesian Method*. Doctoral Dissertation. Central China Normal University, 2020 (In Chinese).
[24] Liu, Z., Guo, Y., Gao, M., *et al.* Multi-wavelet fuzzy neural network blind equalization algorithm in underwater acoustic channel. *Acta Armamentarii*, 2010, 31(9): 1137–1144.
[25] Liu, N., Wang, Z., Wei, H. Trigonometric coordinate transformation blind equalization algorithm based on bi-direction long and short-term memory neural networks. *IEEE Access*, 2024, 12: 30653–30660.
[26] Takayoshi, Y., Hironori, F., Hironobu, F. Multiple skip connections of dilated convolution network for semantic segmentation. 2018 25th IEEE International Conference on Image Processing (ICIP), Athens, Greece, IEEE, 2018: 1593–1597.
[27] Yu, X., Zhao, J. Wide separate 3D convolution for video super resolution. 2019 IEEE Global Conference on Signal and Information Processing (GlobalSIP), Ottawa, ON, Canada, IEEE, 2019: 1–5.
[28] Lu, Y., Lu, G., Lin, R., *et al.* SRGC-Nets: Sparse repeated group convolutional neural networks. *IEEE Transactions on Neural Networks and Learning Systems*, 2020, 31(8): 2889–2903.
[29] Yu, Z., Yu, J., Fan, J., *et al.* Multi-modal factorized bilinear pooling with co-attention learning for visual question answering. 2017 IEEE International Conference on Computer Vision (ICCV), Venice, Italy, IEEE, 2017: 1839–1849.
[30] Zhang, F., Zhang, C., Zhang, L., *et al.* Jelly impurity detection based on scalable overlapping slide pooling SSD network. *Journal of Signal Processing*, 2020, 36(11): 1811–1819 (In Chinese).
[31] Zhou, T. Research on image enhancement algorithms based on deep learning. Master Thesis. Nanjing University of Information Science and Technology, 2020 (In Chinese).
[32] He, K., Sun, J., Tang, X. Guided image filtering. *IEEE Transactions on Pattern Analysis and Machine Intelligence*, 2012, 35(6): 1397–1409.

[33] Wu, H., Zheng, S., Zhang, J., *et al.* Fast end-to-end trainable guided filter. 2018 IEEE/CVF Conference on Computer Vision and Pattern Recognition, Salt Lake City, UT, USA, IEEE, 2018: 191.
[34] Guo, Y., Zhou, T. Image enhancement algorithm based on deep reinforcement adversarial learning. *Journal of Yanzhou University (Natural Science Edition)*, 2020, 23(2): 42–48 (In Chinese).
[35] Gong, H. Research on remote sensing image fusion and classification based on deep learning. Master Thesis. Nanjing University of Information Science and Technology, 2019 (In Chinese).
[36] Zhang, X., Zou, J., He, K. Accelerating very deep convolutional networks for classification and detection. *IEEE Transactions on Pattern Analysis and Machine Intelligence*, 2015, 38(10): 1943–1955.
[37] Srivastava, N., Hinton, G.E., Krizhevsky, A. Dropout: A simple way to prevent neural networks from overfitting. *Journal of Machine Learning Research*, 2014, 15(1): 1929–1958.
[38] Choi, J., Yu, K., Kim, Y. A new adaptive component-substitution-based satellite image fusion by using partial replacement. *IEEE Transactions on Geoscience and Remote Sensing*, 2011, 49(1): 295–309.
[39] Zhu, W. *Research on Blur Removal Based Method on Convolutional Neural Network*. Master Thesis. Nanjing University of Information Science and Technology, 2020 (In Chinese).
[40] Krishnan, D., Tay, T., Fergus, R. *Blind Deconvolution Using a Normalized Sparsity Measure*. CVPR 2011, Colorado Springs, CO, USA, IEEE, 2011: 233–240.
[41] Qi, Q., Guo, J., Jin, W. Attention network for non-uniform deblurring. *IEEE Access*, 2020, 8: 100044–100057.
[42] Liu, R., Cai, W., Li, G., Ning, N., Jiang, Y. Hybrid dilated convolution guided feature filtering and enhancement strategy for hyperspectral image classification. *IEEE Geoscience and Remote Sensing Letters*, 2022, 19: 5508105.
[43] Szegedy, C., Vanhoucke, V., Ioffe, S., *et al.* Rethinking the inception architecture for computer vision. Proceedings of the IEEE Conference on Computer Vision and Pattern Recognition, Las Vegas, NV, USA, IEEE, 2016: 2818–2826.
[44] Huang, G., Liu, Z., Van Der Maaten, L., *et al.* Densely connected convolutional networks. 2017 IEEE Conference on Computer Vision and Pattern Recognition (CVPR), Honolulu, HI, USA, IEEE, 2017, 1(2): 3.
[45] Wan, S., Tang, S., Xie, X., *et al.* Deep convolutional-neural-network-based channel attention for single image dynamic scene blind deblurring. *IEEE Transactions on Circuits and Systems for Video Technology*, 2021, 31(8): 2994–3009.

[46] Zhang, K., Zhai, S., Lu, H. Load prediction based on depthwise separable convolution model. 2021 4th International Conference on Mechatronics, Robotics and Automation (ICMRA), Zhanjiang, China, IEEE, 2021: 75–79.

[47] Xie, S., Girshick, R., Dollár, P., *et al.* Aggregated residual transformations for deep neural networks. 2017 IEEE Conference on Computer Vision and Pattern Recognition (CVPR), Honolulu, HI, USA, IEEE, 2017: 5987–5995.

[48] Wang, Y.V., Kim, S.H., Lyu, G., *et al.* Nowcasting heavy rainfall with convolutional long short-term memory networks: A pixelwise modeling approach. *IEEE Journal of Selected Topics in Applied Earth Observations and Remote Sensing*, 2024, 17(4): 8424–8433.

[49] Chen, C., Li Shen, Liu, W., Luo, Z.Q. Efficient-Adam: communication-efficient distributed Adam. *IEEE Transactions on Signal Processing*, 2023, 71(9): 3257–3266.

[50] Guo, Y., Li, C., Liu, Q. R^2N: a novel deep learning architecture for rain removal from single image. *CMC*, 2019, 58(3): 829–843.

[51] Pan, J., Sun, D., Pfister, H., *et al.* Blind image deblurring using dark channel prior. *Proceedings of the IEEE Conference on Computer Vision and Pattern Recognition*, Las Vegas, NV, USA, IEEE, 2016: 1628–1636.

[52] Chakrabarti, A. A neural approach to blind motion deblurring. European Conference on Computer Vision. Springer, Cham, 2016: 221–235.

[53] Gong, D., Yang, J., Liu, L., *et al.* From motion blur to motion flow: A deep learning solution for removing heterogeneous motion blur. 2017 IEEE Conference on Computer Vision and Pattern Recognition (CVPR), Honolulu, HI, USA, IEEE, 2017, 1(2): 3086–3096.

[54] Branikas, E., Murray, P., West, G. A novel data augmentation method for improved visual crack detection using generative adversarial networks. *IEEE Access*, 2023, 11: 22051–22059.

[55] Kuang, Y., Lan, T., Peng, X., *et al.* Unsupervised multi-discriminator generative adversarial network for lung nodule malignancy classification. *IEEE Access*, 2020, 8: 77725–77734.

[56] Kong, L., Zhao, T. A model-based policy search method based on conditional generative adversarial network. *Journal of Tianjin University of Science and Technology*, 2021, 36(1): 68–75 (In Chinese).

[57] Wang, Q., Guo, Y., Shen, Y., *et al.* Multi-layer graph attention network for sleep stage classification based on EEG. *Sensors*, 2022, 22(23): 9272.

[58] Sutton, R.S., Barto, A.G. *Reinforcement Learning: An Introduction.* Cambridge: MIT Press, 1998.

[59] Wan, L., Lan, X., Zhang, H., Zheng, N. A review of deep reinforcement learning theory and application. *Pattern Recognition and Artificial Intelligence*, 2019, 32(1): 67–82 (In Chinese).

[60] Lillicrap, T.P., Hunt, J.J., Pritzel, A., *et al.* Continuous control with deep reinforcement learning. *Computer Science*, 2015, 8(6): A187.

[61] Song, Y., Scaramuzza, D. Policy search for model predictive control with application to agile drone flight. *IEEE Transactions on Robotics*, 2022, 38(4): 2114–2130.

[62] Sehnke, F., Osendorfer, C., Ruckstiess, T., *et al.* Parameter exploring policy gradients. *Neural Networks*, 2010, 23(4): 551–559.

[63] Williams, R.J. Simple statistical gradient-following algorithms for connectionist reinforcement learning. *Machine Learning*, 1992, 8(3/4): 229–256.

[64] Deng, J., Guo, J., Yang, J., *et al.* ArcFace: additive angular margin loss for deep face recognition. *IEEE Transactions on Pattern Analysis and Machine Intelligence*, 2022, 44(10): 5962–5979.

[65] Civitelli, E., Sortino, A., Lapucci, M., *et al.* A robust initialization of residual blocks for effective resnet training without batch normalization. *IEEE Transactions on Neural Networks and Learning Systems*, 2025, 36(1): 1947–1952.

[66] Mao, X., Li, Q., Xie, H., *et al.* Least squares generative adversarial networks. 2017 IEEE International Conference on Computer Vision (ICCV), Venice, Italy, IEEE, 2017: 2813–2821.

[67] Demirel, H., Anbarjafari, G. Image resolution enhancement by using discrete and stationary wavelet decomposition. *IEEE Transactions on Image Processing*, 2011, 20(5): 1458–1460.

[68] Zhen, C., Yang, Y., Li, Y., Zhong, J. Atmospheric turbulence image restoration based on multi-scale generative adversarial network. *Computer Engineering*, 2021, 47(11): 227–233 (In Chinese).

[69] Kupyn, O., Budzan, V., Mykhailych, M., *et al.* Deblur GAN: blind motion deblurring using conditional adversarial networks. Proceedings of the IEEE Conference on Computer Vision and Pattern Recognition, Salt Lake City, UT, USA, IEEE, 2018: 8183–8192.

[70] Isola, P., Zhu, J.Y., Zhou, T., *et al.* Image-to-image translation with conditional adversarial networks. Proceedings of the IEEE Conference on Computer Vision and Pattern Recognition, Coimbatore, India, IEEE, 2017: 1125–1134.

[71] Chang, Y.L., Liu, Z.Y., Lee, K.Y., *et al.* Free-form video inpainting with 3D gated convolution and temporal PatchGAN. 2019 IEEE/CVF International Conference on Computer Vision (ICCV), Seoul, Korea, IEEE, 2019: 9065–9075.

[72] Xiong, Y., Chen, D., Wang, X. Super-resolution image reconstruction algorithm based on multi-scale back projection. *Computer Engineering*, 2020, 46(7): 251–259 (In Chinese).

[73] Hu, J., Shen, L., Sun, G. Squeeze-and-excitation networks. Proceedings of the IEEE Conference on Computer Vision and Pattern Recognition, Beijing, China, IEEE, 2018: 7132–7141.

[74] Milletari, F., Navab, N., Ahmadi, S.A. V-net: fully convolutional neural networks for volumetric medical image segmentation. 2016 Fourth International Conference on 3D Vision (3DV), Stanford, CA, USA: IEEE, 2016: 565–571.

[75] Levin, A., Weiss, Y., Durand, F., *et al.* Understanding and evaluating blind deconvolution algorithms. IEEE Conference on Computer Vision and Pattern Recognition, Miami, FL, USA, IEEE, 2009: 1964–1971.

[76] Zheng, S., Zhu, Z., Cheng, J., *et al.* Edge heuristic GAN for non-uniform blind deblurring. *IEEE Signal Processing Letters*, 2019, 26(10): 1546–1550.

[77] Gupta, A., Joshi, N., Zitnick, C.L., *et al.* Single image deblurring using motion density functions. ECCV, Springer, Berlin, Heidelberg, 2010: 171–184.

[78] Chen, L., Sun, Q., Wang, F. Attention-adaptive and deformable convolutional modules for dynamic scene deblurring. *Information Sciences*, 2021, 546(6): 368–377.

[79] Pan, J., Sun, D., Pfister, H., *et al.* Deblurring images via dark channel prior. *IEEE Transactions on Pattern Analysis and Machine Intelligence*, 2017, PP(99): 1–1.

[80] Xu, L., Ren, J.S., Liu, C., Jia, J. Deep convolutional neural network for image deconvolution. *Advances in Neural Information Processing Systems*, 2014: 1790–1798.

[81] Schuler, C.J., Hirsch, M., Harmeling, S., *et al.* Learning to deblur. *IEEE Transactions on Pattern Analysis and Machine Intelligence*, 2016, 38(7): 1439–1451.

[82] Hua, W., Zhang, C., Xie, W., *et al.* Polarimetric SAR image classification based on ensemble dual-branch cnn and superpixel algorithm. *IEEE Journal of Selected Topics in Applied Earth Observations and Remote Sensing*, 2022, 15: 2759–2772.

[83] Li, C., Guo, Y., Liu, Q., *et al.* DR-Net: a novel generative adversarial network for single image deraining. *Security and Communication Networks*, 2018, 2018: 1–14.

[84] Hinton, G.E. Training products of experts by minimizing contrastive divergence. *Neural Computer*, 2002, 14(8): 1771–1800.

[85] Lee, H., Ekanadham, C., Ng, A.Y. Sparse deep belief net model for visual area V2. Proceedings of Advances in Neural Information Processing Systems, NY, USA, IEEE, 2008: 873–880.

[86] Zhou, L., Liu, K., Lyu, H. Mechanism of sparse restricted Boltzmann machine based on competitive learning. *Journal of Computer Applications*, 2018, 38(7): 1872–1876 (In Chinese).

[87] Yin, J., Li, W.W., Yang, D.H., Yan, H. Improved model based on classification restricted Boltzmann machine. *Journal of Chinese Computer Systems*, 2018, 39(7): 1415–1420 (In Chinese).

[88] Chu, J., Wang, H., Meng, H., Jin, P., *et al.* Restricted Boltzmann machines with Gaussian visible units guided by pairwise constraints. *IEEE Transactions on Cybernetics*, 2019, 49(12): 4321–4335.

[89] Cho, K.H., Raiko, T., Ilin, A. Parallel tempering is efficient for learning restricted Boltzmann machines. International Joint Conference on Neural Networks (IJCNN), Barcelona, Spain, IEEE, 2010: 1–8.

[90] Ren, J., Li, X., Haupt, J. Robust PCA via tensor outlier pursuit. 2016 50th Asilomar Conference on Signals, Systems and Computers, Pacific Grove, CA, USA, IEEE, 2016: 1744–1749.

[91] Zhang, M. Research on blood cell detection algorithm based on semi-supervised learning. Master Thesis. Nanjing University of Information Science and Technology, 2024 (In Chinese).

[92] Lu, M., Huang, J.Z., Qian, X. Sparse exponential family principal component analysis. *Pattern Recognition*, 2016, 60: 681–691.

[93] Cheng, Z.D., Zhang, Y.J., Fan, X. Criteria for 2DPCA superior to PCA in image feature extraction. *Chinese Journal of Engineering Mathematics*, 2009, 26(6): 951–961 (In Chinese).

[94] Shi, Z.G. A human face recognition method of 2DPCA based on modular residual image. *Journal of Inner Mongolia Normal University*, 2015, 44(3): 380–384.

[95] Wang, H.X. 2DPCA with L1-norm for simultaneously robust and sparse modelling. *Neural Networks*, 2013, 46(10): 190–198.

[96] Li, W. The image feature extraction algorithm based on the DWT and the improved 2DPCA. *Applied Mechanics and Materials*, 2014, 556–562: 5042–5045.

[97] Liu, S., Feng, L., Qiao, H. Scatter balance: an angle-based supervised dimensionality reduction. *IEEE Transactions on Neural Networks and Learning Systems*, 2015, 26(2): 277–289.

[98] Mahanta, M.S., Plataniotis, K.N. Ranking 2DLDA features based on fisher discriminance. IEEE International Conference on Acoustics, Speech and Signal Processing (ICASSP), 2014: 8307–8311.

[99] Mashhoori, A., Zolghadri, M.J. Block-wise two-directional 2DPCA with ensemble learning for face recognition. *Neurocomputing*, 2013, 108(5): 111–117.

[100] Song, H.F., Chen, G.S., Jing, W.P., *et al.* Restricted Boltzmann machine algorithm for image classification and its parallel implementation based on $(2D)^2$PCA. *Journal of Applied Sciences*, 2018, 36(3): 495–504 (In Chinese).

[101] Shvachko, K., Kuang, H., Radia, S. The Hadoop distributed file system. 2010 IEEE 26th Symposium on Mass Storage Systems and Technologies (MSST), Incline Village, NV, USA, IEEE, 2010: 1–10.

[102] Vemula, S., Crick, C. Hadoop image processing framework. IEEE International Congress on Big Data, New York, NY, USA, IEEE, 2015: 506–513.

[103] Kachris, C., Sirakoulis, G.C., Soudris, D. A map reduce scratchpad memory for multi-core cloud computing applications. *Microprocessors and Microsystems*, 2015, 39(8): 599–608.

[104] Huang, X.X., Boulgouris, N.V. Gait recognition with shifted energy image and structural feature extraction. *IEEE Transactions on Image Processing*, 2012, 21(4): 2256–2268.

[105] Choudhury, S.D., Tjahjadi, T. Silhouette-based gait recognition using Procrustes shape analysis and elliptic Fourier descriptors. *Pattern Recognition*, 2012, 45(9): 3414–3426.

[106] Li, K., Cao, K. Gait feature extraction based on restricted Boltzmann machine and its recognition method. *Journal of Hebei University (Natural Science Edition)*, 2019, 39(6): 657–666.

[107] Liu, L. *Research of Gait Recognition Based on Outermost Contour.* Master Thesis. Shandong University, Shangdong Province,China, 2012.

[108] Qin, S., Zhang, Y.D., Amin, M.G. DOA estimation of mixed coherent and uncorrelated targets exploiting coprime MIMO radar. *Digital Signal Processing*, 2017, 61: 26–34.

[109] He, S., Chen, H. Closed-form DOA estimation using first-order differential microphone arrays via joint temporal-spectral-spatial processing. *IEEE Sensors Journal*, 2017, 17(4): 1046–1060.

[110] Capon, J. High-resolution frequency-wavenumber spectrum analysis. *Proceedings of the IEEE*, 1969, 57(8): 1408–1418.

[111] Schmidt, R. Multiple emitter location and signal parameter estimation. *IEEE Transactions on Antennas and Propagation*, 1986, 34(3): 276–280.

[112] Roy, R., Kailath, T. ESPRIT-estimation of signal parameters via rotational invariance techniques. *IEEE Transactions on Acoustics, Speech, and Signal Processing*, 1989, 37(7): 984–995.

[113] Malioutov, D., Cetin, M., Willsky, A.S. A sparse signal reconstruction perspective for source localization with sensor arrays. *IEEE Transactions on Signal Processing*, 2005, 53(8): 3010–3022.
[114] Liu, Z.M., Huang, Z.T., Zhou, Y.Y. An efficient maximum likelihood method for direction-of-arrival estimation via sparse Bayesian learning. *IEEE Transactions on Wireless Communications*, 2012, 11(10): 1–11.
[115] Xiao, X., Zhao, S., Zhong, X., *et al.* A learning-based approach to direction of arrival estimation in noisy and reverberant environments. IEEE International Conference on Acoustics, Speech and Signal Processing, South Brisbane, QLD, Australia, IEEE, 2015: 2814–2818.
[116] Faye, A., Youm, A.B., Ndaw, J.D. LVQ based DOA estimation. Computational Intelligence, Communication Systems and Networks (CICSyN), 2013 Fifth International Conference on, Madrid, Spain, IEEE, 2013: 245–250.
[117] Hennig, P., Martin, K. Quasi-Newton method: A new direction. *Journal of Machine Learning Research*, 2013, 14(Mar): 843–865.
[118] Zhou, M., Hannah, L., David Dunson, D., *et al.* Beta-negative binomial process and poisson factor analysis. Proceedings of the Fifteenth International Conference on Artificial Intelligence and Statistics, La Palma, Canary Islands, Spain, PMLR, IEEE, 2012, 22: 1462–1471.
[119] Luo, Y., Jia, B., Qiu, X., Deng, P., Wu, Q. Pilots' brain fatigue state inference based on Gamma deep belief network. *Acta Electronica Sinica*, 2020, 48(6): 1062–1071 (In Chinese).
[120] Dyk, D.A.V., Meng, X.L. The art of data augmentation. *Journal of Computational and Graphical Statistics*, 2001, 10(1): 1–50.
[121] Blackwell, Macqueen, J. Ferguson distributions via Polya urn schemes. *The Annals of Statistics*, 1973, 1(2): 353–355.
[122] Li, R., Wu, X., Tian, H., *et al.* Hybrid memetic pretrained factor analysis-based deep belief networks for transient electromagnetic inversion. *IEEE Transactions on Geoscience and Remote Sensing*, 2022, 60: 5920120.
[123] Huang, L., Lu, J., Zhang, Z., Zhang, X. Engine fault detection based on PKCA and deep belief network. *Journal of Wuhan University of Technology (Transportation Science and Engineering)*, 2019, 43(9): 920–925 (In Chinese).
[124] Mo, Y., Yao, L., Wang, J., Hu, Q., *et al.* Engine fault diagnosis based on principal component analysis and BP neural network. *Journal of Wuhan University of Technology (Transportation Science and Engineering)*, 2016, 40(2): 85–88 (In Chinese).

[125] Deng, L., Yu, D. Deep learning: methods and applications. *Foundations and Trends in Signal Processing*, 2014, 7(2): 197–387.

[126] Fischer, A., Igel, C. An introduction to restricted Boltzmann machines. *Progress in Pattern Recognition, Image Analysis, Computer Vision, and Applications*. Berlin: Springer, 2012.

[127] Tran, V.T., Althobiani, F., Ball, A. An approach to fault diagnosis of reciprocating compressor valves using Teager-Kaiser energy operator and deep belief networks. *Expert Systems with Applications*, 2014, 41(9): 4113–4122.

[128] Yang, Y., Zhang, N., Cheng, J. Global parameters dynamic learning deep belief networks and its application in rolling bearing life prediction. *Journal of Vibration and Shock*, 2019, 38(10): 199–208 (In Chinese).

[129] Li, Y., Wang, L., Jiang, L. Rolling bearing fault diagnosis based on DBN algorithm improved with PSO. *Journal of Vibration and Shock*, 2020, 39(5): 89–97 (In Chinese).

[130] Sun, M., Tian, Y. Parameter optimization of deep belief network based on chaos immune algorithm. *Computer Engineering and Design*, 2019, 49(9): 2643–2649 (In Chinese).

[131] Salakhutdinov, R., Murray, I. On the quantitative analysis of deep belief networks. Helsinki: ACM, 2008.

[132] Shi, Y.H., Eberhart, R.C. Parameter selection in particle swarm optimization. Berlin, Heidelberg: Springer Berlin Heidelberg, 1998.

[133] Guo, X., He, L. ADHD discrimination based on social network. 2014 International Conference on Cloud Computing and Big Data, Wuhan, China, IEEE, 2014: 55–61.

[134] Bledsoe, J.C., Xiao, D., Chaovalitwongse, A., *et al.* Diagnostic classification of ADHD versus control support vector machine classification using brief neuropsychological assessment. *Journal of Attention Disorders*, 2020, 24(11): 1547–1556.

[135] Duda, M., Ma, R., Haber, N., Wall, D. Use of machine learning for behavioral distinction of autism and ADHD. *Translational Psychiatry*, 2016, 6: 732.

[136] Sachnev, V. An efficient classification scheme for ADHD problem based on binary coded genetic algorithm and McFIS. 2015 International Conference on Cognitive Computing and Information Processing (CCIP), Noida, India, IEEE, 2015: 1–6.

[137] Kuang, D., He, L. Classification on ADHD with deep learning. 2014 International Conference on Cloud Computing and Big Data (CCBD), Wuhan, China, IEEE, 2014: 27–32.

[138] Igual, L., Soliva, J.C., Hernández-Vela, A., *et al.* Supervised brain segmentation and classification in diagnostic of attention-deficit/

hyperactivity disorder. 2012 International Conference on High Performance Computing and Simulation (HPCS), Madrid, Spain, IEEE, 2012: 182–187.

[139] Tenev, A., Markovska-Simoska, S., Kocarev, L., *et al.* Machine learning approach for classification of ADHD adults. *International Journal of Psychophysiology*, 2014, 93: 162–166.

[140] Abibullaev, B., An, J. Decision support algorithm for diagnosis of ADHD using electroencephalograms. *Journal of Medical Systems*, 2012, 36: 2675–2688.

[141] Liu, G., Lu, W., Qiu, J., Shi, L. Identifying individuals with attention-deficit/hyperactivity disorder based on multisite resting-state functional magnetic resonance imaging: a radiomics analysis. *Human Brain Mapping*, 2023, 44(8): 3433–3445.

[142] Saeed, F., Kianian, S. Diagnosis of attention deficit hyperactivity disorder using deep belief network based on greedy approach. 5th International Symposium on Computational and Business Intelligence, Dubai, United Arab Emirates, IEEE, 2017: 96–99.

[143] Sun, W., Shao, S., Zhao, R., *et al.* A sparse autoencoder-based deep neural network approach for induction motor faults classification. *Measurement*, 2016, 89: 171–178.

[144] Xu, J., Xiang, L., Liu, Q., *et al.* Stacked sparse autoencoder (SSAE) for nuclei detection on breast cancer histopathology images. IEEE International Symposium on Biomedical Imaging, Beijing, China, IEEE, 2014: 999–1002.

[145] Wang, S., Chen, H., Li, X., Shu, X. Conditional variational automatic encoder method for stochastic scenario generation of wind power and photovoltaic system. *Power System Technology*, 2018, 42(6): 1860–1869 (In Chinese).

[146] Hou, X., Shen, L., Sun, K., *et al.* Deep feature consistent variational autoencoder. 2017 IEEE Winter Conference on Applications of Computer Vision (WACV), Santa Rosa, CA, USA, IEEE, 2017: 1133–1141.

[147] Wang, H., Chen, Q. Transient stability assessment method of electric power systems based on stacked variational auto-encoder. *Electric Power Automation Equipment*, 2019, 39(12): 134–140 (In Chinese).

[148] She, B., Tian, F., Liang, W. Fault diagnosis based on a deep convolution variational auto-encoder network. *Chinese Journal of Scientific Instrument*, 2018, 39(10): 28–37 (In Chinese).

[149] Ghiassi, M., Saidane, H., Zimbra, D.K. A dynamic artificial neural network model for forecasting time series events. *International Journal of Forecasting*, 2005, 21(2): 341–362.

[150] Du, X., Vasudevan, R., Johnson-Roberson, M. Bio-LSTM: A biomechanically inspired recurrent neural network for 3-D pedestrian pose and gait prediction. *IEEE Robotics and Automation Letters*, 2018, 4(2): 1501–1508.

[151] Zen, H., Sak, H. Unidirectional long short-term memory recurrent neural network with recurrent output layer for low-latency speech synthesis. 2015 IEEE International Conference on Acoustics, Speech and Signal Processing (ICASSP), South Brisbane, QLD, Australia, IEEE, 2015: 4470–4474.

[152] Peng, D., He, Y., Xu, Y., Zhu, Q. Research and chemical application of data feature extraction based AANN ELM neural network. *CIESC Journal*, 2012, 63(9): 2920–2925 (In Chinese).

[153] Zhu, B., Qiao, J. Features extracted from auto-encoder based echo state network and its applications to process modeling. *CIESC Journal*, 2019, 70(12): 4470–4776 (In Chinese).

[154] Hotelling, H. Relations between two sets of variates. *Biometrika*, 1936, 28(3-4): 321–377.

[155] Cao, K.A., Martin, P.G., Robert-Granié, C., *et al.* Sparse canonical methods for biological data integration: application to a cross-platform study. *BMC Bioinformatics*, 2009, 10(1): 1–17.

[156] Witten, D.M., Tibshirani, R.J. Extensions of sparse canonical correlation analysis with applications to genomic data. *Statistical Applications in Genetics and Molecular Biology*, 2009, 8(1): 1–27.

[157] Witten, D.M., Robert, T., Trevor, H. A penalized matrix decomposition, with applications to sparse principal components and canonical correlation analysis. *Biostatistics*, 2009, 10(3): 515–534.

[158] Lin, D., Calhoun, V.D., Wang, Y.P. Correspondence between fMRI and SNP data by group sparse canonical correlation analysis. *Medical Image Analysis*, 2014, 18(6): 891–902.

[159] Du, L., Huang, H., Yan, J., *et al.* Structured sparse canonical correlation analysis for brain imaging genetics: an improved GraphNet method. *Bioinformatics*, 2016, 32(10): 1544–1551.

[160] Kim, M., Ji, H.W., Youn, J., *et al.* Joint-connectivity-based sparse canonical correlation analysis of imaging genetics for detecting biomarkers of parkinson's disease. *IEEE Transactions on Medical Imaging*, 2019, PP(99): 1–1.

[161] Hu, W., Lin, D., Cao, S., *et al.* Adaptive sparse multiple canonical correlation analysis with application to imaging (EPI) genomics study of schizophrenia. *IEEE Transactions on Biomedical Engineering*, 2018, 65(2): 390–399.
[162] Andrew, G., Arora, R., Bilmes, J., *et al.* Deep canonical correlation analysis. International Conference on Machine Learning (ICML), Atlanta, GA, USA, 2013: 1047–1255.
[163] Li, G., Han, D., Liu, Q., Wang, C., Li, Y. Classification of schizophrenia based on deep canonically correlated sparse autoencoder. *Chinese Journal of Medical Physics*, 2020, 37(3): 391–396 (In Chinese).
[164] Alghayadh, F.Y., Ramesh, J.V.N., Keshta, I. Quantum target recognition enhancement algorithm for UAV consumer applications. *IEEE Transactions on Consumer Electronics*, 2024, 70(3): 5553–5560.
[165] Miao, Z., Zhuang, Y., Li, W. Infrared target modeling method based on double adversarial autoencoding network. *Acta Optica Sinica*, 2020, 40(11): 111002-(1-8) (In Chinese).
[166] Wang, Q., Tian, X.M. Soft sensing based on improved GA-LSSVM. *Transactions of Beijing Institute of Technology*, 2012, 32(10): 1031–1035 (In Chinese).
[167] Li, H., Yu, Y., Shen, X., Huang, L. Research on the soft sensing technology based on KPCA and SVR for fermentation tank. *Process Automation Instrumentation*, 2018, 39(2): 12–16.
[168] Sun, W., Shao, S., Zhao, R., *et al.* A sparse auto-encoder-based deep neural network approach for induction motor faults classification. *Measurement*, 2016, 89: 171–178.
[169] Bengio, Y., Lamblin, P., Popovici, D., *et al.* Greedy layer wise training of deep networks. The 19th International Conference on Neural Information Processing Systems, Massachusetts, USA: MIT Press, 2007: 153–160.
[170] Luo, Y., Wan, Y. A novel efficient method for training sparse autoencoder. International Congress on Image and Signal Processing, Hangzhou, China, IEEE, 2014: 1019–1023.
[171] Simard, P., Steinkraus, D., Platt, J. Best practices for convolutional neural networks applied to visual document analysis. The 7th International Conference on Document Analysis and Recognition, Edinburgh, UK, IEEE, 2003: 958–963.
[172] Li, C., Sheng, V.S., Jiang, L., *et al.* Noise filtering to improve data and model quality for crowd sourcing. *Knowledge-Based Systems*, 2016, 107: 96–103.

[173] Huang, G.B., Zhu, Q.Y., Siew, C.K. Extreme learning machine: theory and applications. *Neurocomputing*, 2006, 70(1/2/3): 489–501.

[174] Charama, L.L., Zhou, H., Huang, G.B. Representational learning with ELMs for big data. *IEEE Intelligent Systems*, 2013, 28(6): 31–34.

[175] Sun, K., Zhang, J., Zhang, C. Generalized extreme learning machine autoencoder and a new deep neural network. *Neurocomputing*, 2017, 230: 374–381.

[176] Duan, M., Li, K., Li, K. An ensemble CNN2ELM for age estimation. *IEEE Transactions on Information Forensics and Security*, 2018, 13(3): 758–772.

[177] Vincent, P., Latochelle, H., Lajoie, I., *et al.* Stacked denoising autoencoders: learning useful representations in a deep network with a local denoising criterion. *The Journal of Machine Learning Research*, 2010, 11(12): 3371–3408.

[178] Chen, X., Li, M., Yang, X. Stacked denoise autoencoder based feature extraction and classification for hyperspectral images. *Journal of Sensors*, 2016, 2016: 1–10.

[179] Guo, Y., Zhang, H. Modulation recognition algorithm based on improved LDA and autoencoders. *Journal of System Simulation*, 2021, 33(2): 494–500 (In Chinese).

[180] Zhang, H. Research on modulation recognition based on autoencoders and convolutional neural networks. Master Thesis. Nanjing University of Information Science and Technology, 2020.

[181] Guo, Y., Wang, X. Modulation signal classification algorithm based on denoising residual convolutional neural network. *IEEE Access*, 2022, 10: 121733–121740.

[182] Aranganayagi, S., Thangavel, K. Clustering categorical data using silhouette coefficient as a relocating measure. 2007 International Conference on Computational Intelligence and Multimedia Applications, Sivakasi, Tamil Nadu, India, IEEE, 2007, 2: 13–17.

[183] Zhang, W., Zhang, P., He, X., *et al.* Convolutional neural network based two-layer transfer learning for bearing fault diagnosis. *IEEE Access*, 2022, 10: 109779–109794.

[184] Zhao, J., Zhang, C.Y., Liu, C., Zhou, H., *et al.* Recurrent neural networks with recursive least squares. *Acta Automatica Sinica*, 2022, 48(8): 2050–2061 (In Chinese).

[185] Connor, J.T., Martin, R.D., Atlas, L.E. Recurrent neural networks and robust time series prediction. *IEEE Transactions on Neural Networks*, 1994, 5(2): 240–254.

[186] Wang, Z., Yao, X., Huang, Z., Liu, L. Deep echo state network with multiple adaptive reservoirs for time series prediction. *IEEE Transactions on Cognitive and Developmental Systems*, 2021, 13(3): 693–704.

[187] Williams, R.J., Peng, J. An efficient gradient-based algorithm for on-line training of recurrent network trajectories. *Neurocomputering*, 1990, 2(4): 490–501.

[188] Liu, H., Shi, S., Xu, X., *et al.* A non-intrusive load identification method based on RNN model. *Power System Protection and Control*, 2019, 47(13): 162–171 (In Chinese).

[189] Sunil Kumar, K.H., Chetan, H., Indumathi, G. A novel image compression approach using DTCWT and RNN encoder. 2020 IEEE Bangalore Humanitarian Technology Conference (B-HTC), Vijiyapur, India, IEEE, 2020: 1–4.

[190] Cheng, H., Wang, S., Ke, Z., *et al.* A deep recursive cascaded convolutional network for parallel MRI. *Chinese Journal of Magnetic Resonance*, 2019, 36(4): 437–445 (In Chinese).

[191] Liu, X. Research on Tibetan speech recognition based on bidirectional recurrent neural networks. Master Thesis. Qinghai Normal University, 2020.

[192] Guan, S., Yang, G., Li, H., Fu, Y. Hyperspectral image classification based on 3-D convolutional recurrent neural network. *Laser Technology*, 2020, 44(4): 485–491.

[193] Xu, R., Liang, S. A method of traffic volume prediction based on bilinear recurrent neural networks. *China Transportation Review*, 2018, 40(11): 70–75 (In Chinese).

[194] Li, C. Research on image deraining and destriping based on deep neural network. Doctoral Dissertation. Nanjing University of Information Science and Technology, 2020.

[195] Li, H. Research on underwater image enhancement using a deep adversarial network. Master Thesis. Nanjing University of Information Science and Technology, 2020.

[196] Anastasiadis, A.D., Magoulas, G.D., Vrahatis, M.N. New globally convergent training scheme based on the resilient propagation algorithm. *Neurocomputering*, 2005, 64(Mar): 253–270.

[197] Sajjadi, M.S.M., Scholkopf, B., Hirsch, M. Enhancenet: single image super-resolution through automated texture synthesis. Proceedings of the IEEE International Conference on Computer Vision, Venice, Italy, IEEE, 2017: 4491–4500.

[198] Kim, J., Lee, J.K., Lee, K.M. Deeply-recursive convolutional network for image super-resolution. The IEEE Conference on Computer Vision and Pattern Recognition, Las Vegas, NV, USA, IEEE, 2016:

1637–1645; Tarekegn, G.B., Lin, H.P., Adege, A.B., *et al.* Applying long short-term memory (LSTM) mechanisms for fingerprinting outdoor positioning in hybrid networks. 2019 IEEE 90th Vehicular Technology Conference (VTC2019-Fall), Honolulu, HI, USA, IEEE, 2019: 1–5.

[199] Xiao, P. Method of improving the quality of infrared cloud image and its application in precipitation estimation. Doctoral Dissertation. Nanjing University of Information Science and Technology, 2020.

[200] Yu, X., Li, L., Yin, J., Shao, M., Han, X. Modulation pattern recognition of non-cooperative underwater acoustic communication signals based on LSTM network. 2019 IEEE International Conference on Signal, Information and Data Processing (ICSIDP), Chongqing, China, IEEE, 2019: 1–5.

[201] Tarekegn, G.B., Juang, R.T., Hsin-Piao Lin, H.P., *et al.* DFOPS: deep-learning-based fingerprinting outdoor positioning scheme in hybrid networks. *IEEE Internet of Things Journal*, 2021, 8(5): 3717–3729.

Index

B

C

www.ingramcontent.com/pod-product-compliance
Ingram Content Group UK Ltd.
Pitfield, Milton Keynes, MK11 3LW, UK
UKHW062100120926
8588IPUK00007B/68

9 789819 824175